A Flora
of
Glacier National Park,
Montana

Also available from OSU Press

Flora of Mount Rainier National Park by David Biek
Flora of Steens Mountain by Donald H. Mansfield

A Flora
of
Glacier National Park, Montana

by
Peter Lesica

with illustrations by
Debbie McNeil

Oregon State University Press

Corvallis

Cover photograph: *Xerophyllum tenax* by Peter Lesica.

The paper in this book meets the guidelines for permanence and durability of the Committee on Production Guidelines for Book Longevity of the Council on Library Resources and the minimum requirements of the American National Standard for Permanence of Paper for Printed Library Materials Z39.48-1984.

Library of Congress Cataloging-in-Publication
Lesica, Peter.
  A flora of Glacier National Park, Montana / Peter Lesica with illustrations by Debbie McNeil.
      p. cm.
Includes bibliographical references (p.   ).
  ISBN 0-87071-538-0 (alk. paper)
  1. Botany—Montana—Glacier National Park. 2. Plants—Identification.
3. Glacier National Park (Mont.) I. Title.
  QK171 .L46 2002
  581.9786'52—dc21
                                  2001006412

Oregon State University Press
101 Waldo Hall
Corvallis OR 97331-6407
541-737-3166 • fax 541-737-3170
OREGON STATE
UNIVERSITY   http://osu.orst.edu/dept/press

DEDICATION

This book is dedicated to all those friends who helped me explore the back country of Glacier National Park and patiently waited while I keyed out a plant, jotted down notes, or probed through the tundra. Jito, Ann, Kathy, Anne, Lynn, Rosalind, the Daves, Suzanne and Helen, thanks for the great times in Great Country.

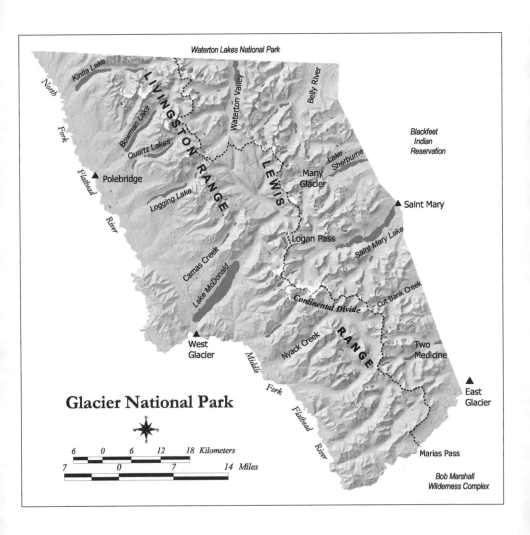

Waterton Lakes National Park

Kintla Lake

North Fork

Flathead River

LIVINGSTON RANGE

Bowman Lake

Quartz Lakes

▲ Polebridge

Logging Lake

Waterton Valley

Belly River

LEWIS

Blackfeet Indian Reservation

Lake Sherburne

Many Glacier

▲ Saint Mary

Logan Pass

Saint Mary Lake

Camas Creek

Lake McDonald

Continental Divide

Cut Bank Creek

RANGE

▲ West Glacier

Middle Fork

Nyack Creek

Two Medicine

▲ East Glacier

Flathead River

Marias Pass

Bob Marshall Wilderness Complex

Glacier National Park

6   0   6   12   18 Kilometers

7   0   7   14 Miles

CONTENTS

Acknowledgments 8

Introduction 9

    Climate 9

    Geology 11

    Vegetation 12

    History of Botanical Exploration 22

    Methods 24

    How to Use This Book 25

    Floristic Synopsis 26

    Floristic Plant Geography 27

References 36

Glossary 45

Key to the Families 54

The Flora

    Pteridophytes: Ferns and Allies 65

    Gymnosperms: Conifers 88

    Flowering Plants: Dicots 98

    Flowering Plants: Monocots 382

Index 495

ACKNOWLEDGMENTS

Many members of the Glacier National Park staff helped with this project over the years. Many thanks to Kathy Ahlenslager, Jen Asebrook, Ann DeBolt, Shannon Kimball, Laurie Kurth, Ellen Seely, and Dierdre Shaw. Tara Williams and Jack Potter put a great deal of effort into getting and keeping this project off the ground. This book would still be just an idea without their help. LeeAnn Simpson of the Glacier Natural History Association facilitated publication of the second checklist and the financial arrangements for this book. Richard Minicke prepared the map.

Funding and support were provided by grants from the National Park Foundation, National Fish and Wildlife Foundation, Montana Native Plant Society, Glacier Natural History Association, Glacier National Park Associates, Montana Natural Heritage Program, and Glacier National Park.

This book has been greatly improved by the careful editing of Anne Garde, Kenton Chambers, Jo Alexander, and Jane Jacobi. Kathy Ahlenslager, Jen Asebrook, Jim Habeck, Shannon Kimball, Rachel Potter, Chris Weiss, and Tara Williams reviewed and commented on portions of the manuscript.

Christian Damm, Jerry DeSanto, Tara Luna, Dave Shea, and Toby Spribille made valuable collections and shared their knowledge of the Park flora with me.

Mary Sloan, Rachel Potter, Dee Strickler, Ed Pedersen, and other members of the Montana Native Plant Society helped prepare specimens for the Glacier Park herbarium.

Lois Arnow, Ted Barkley, Rupert Barneby, Randy Bayer, Ken Chambers, William Cody, Art Cronquist, Robert Dorn, Barbara Ertter, Richard Halse, Matt Lavin, Tony Reznicek, Ernie Schuyler, John Semple, Rob Soreng, Herb and Florence Wagner, and Michael Windham were kind enough to give me much-needed guidance in large and difficult taxonomic groups. Paul Peterson and Rusty Russell helped me obtain critical Standley collections from the Smithsonian herbarium.

INTRODUCTION

Glacier National Park is located in northwest Montana along the main range of the Rocky Mountains south of the Canadian border. The Park adjoins the western edge of the Northern Great Plains on the east, while to the west, across the North Fork of the Flathead River, lies the inland maritime Whitefish Range. The Flathead Valley and Flathead Lake, the largest freshwater lake in western North America, are located just to the southwest. Glacier Park encompasses approximately 4000 km^2 (ca. 1,000,000 acres), nearly all of which is mountainous terrain. Mountains rise to elevations of 2450-3050 m (8,000-10,000 ft). There are two north-south mountain ranges in Glacier Park, the Lewis and Livingston ranges. The Livingston Range is found in the northwest part of the Park, while the Lewis Range traverses the entire east side. The Continental Divide (between the Pacific and Atlantic oceans) follows the crest of the Livingston Range in the north and then traverses the crest of the Lewis Range. The Northern Divide (between the Arctic and Atlantic oceans) begins in the center of the Lewis Range at Triple Divide Peak. Valleys are generally 920-1250 m (3,000-4,000 ft) west of the Continental Divide and 1250-1500 m (4,000-5,000 ft) east of the Divide.

Glacier was declared a national park by an act of Congress in 1910. Waterton Lakes National Park, Glacier's sister park, is adjacent to the north in Alberta, Canada. The Bob Marshall Wilderness Complex administered by the U. S. Forest Service is immediately south of the Park, separated by the Middle Fork of the Flathead River and U. S. Highway 2. The Blackfeet Indian Reservation adjoins the Park on the east side, and Flathead National Forest is to the west, separated by the North Fork of the Flathead River and an associated road. Glacier Park's pristine quality and vast biological wealth led the United Nations Education, Scientific and Cultural Organization (UNESCO) to designate Waterton-Glacier International Peace Park a Biosphere Reserve in 1976, and a World Heritage Site in 1995.

Glacier is primarily a wilderness park. With the exception of the Going-to-the-Sun Road crossing the Continental Divide from east to west, roads are found only on the periphery. On the other hand, there are more than 700 miles of trails in the Park. With the exception of reduced fire frequency and introduction of exotic plants, the vegetation of Glacier Park has been little changed by European settlement.

Climate

The climate of Glacier National Park is transitional between northern maritime and northern continental (Finklin 1986). It is most strongly influenced by Pacific storms from the west and arctic air from the north, mediated by the high peaks of the Continental Divide. A strong Pacific storm track follows the jet stream along the Canadian border during much of the year and brings a warm, moist influence to Glacier and the adjacent area. Most of this Pacific moisture falls along or west of the Continental Divide, and little remains by the time the storms reach the east edge of the Park. Cold arctic air masses moving south from continental Canada are usually confined east of the Continental Divide, with some cold air flushing west through mountain passes. These prevailing weather patterns cause a pronounced difference between the east and west edges of the Park, but these differences are lost as one approaches the Continental Divide from either side. Finklin's (1986) weather summary for the Park provides the following data.

Glacier Park temperatures are generally warmer in the western valleys than east of the Divide. January is normally the coldest month, while July is the warmest. Average minimum temperature for January was -11C (12F) at West Glacier and ca. -14C (7F) at St. Mary and East Glacier in 1951-1980. Average maximum for July was 26C (78F) at West Glacier and 24C (76F) and 23C (74F) at St. Mary and East Glacier respectively. Temperatures are generally cooler at higher elevations, averaging 8C cooler per 1000 m (4F per 1000 ft). However, temperature inversions, in which valleys are cooler than peaks, often occur at night during the summer and during the day in the winter.

Annual precipitation is generally similar on the two sides of Glacier Park, but decreases from south to north. West of the Divide, annual precipitation was 75 cm (30 in) and 58 cm (23 in) at West Glacier and Polebridge respectively in 1951-1980, while east of the Divide East Glacier, St. Mary and Lake Sherburne averaged 76 cm (31 in), 66 cm (23 in) and 59 cm (23 in) respectively. Precipitation is much higher closer to the Continental Divide. Many Glacier, just a short distance west of Lake Sherburne and at nearly the same elevation, receives nearly twice the precipitation, and Grinnell Glacier and high elevations along the Divide receive 250 cm (100 in) or more. At least half the precipitation falls during November through March. Half of the precipitation falls as snow near the edges of the Park, while 70% is snow at higher elevations. Snowfall may occur during any month of the year at higher elevations. West of the Continental Divide, average relative humidity remains near 80% over the entire year, while it is below 50% during summer months east of the Divide.

The eastern edge of the Park receives more sunshine and higher winds than the lower elevations west of the Divide. West Glacier averaged 132 clear days per year, while Browning, just east of East Glacier, averaged 164 during the period of 1921-48. Strong winds are common east of the Continental Divide, while winds west of the Divide are rather calm. Exceptionally strong, warm (chinook) winds occur along the east front of the mountains in winter; wind speeds of 160 km/h (100 mph) have been recorded. Valleys west of the Divide average winds of 10 km/h (6 mph) during winter months, while east of the Divide wind speeds average twice that. Prevailing wind direction is from the west to southwest.

Warmer, less windy winters coupled with cloudier, more humid summers provide a more temperate climate for vegetation west of the Continental Divide in Glacier. The large winter fluctuations in temperature associated with chinook winds east of the Divide are particularly detrimental to woody plants with living tissue above ground. These differences are less pronounced close to the Continental Divide, where the climate and vegetation of headwater basins are similar on both sides of the Divide.

Geology

Glacier National Park is a land of mountains carved by glaciers. Geologic formations in the Park are predominantly sedimentary with minor intrusions of igneous rock. Most of these formations are part of the Belt Series dating from the Precambrian age, 800 million to more than a billion years ago. They include limestones and dolomites of the Altyn and Helena (formerly Siyeh) formations and red and green mudstones (argillites) of the Grinnell and Appekunny formations. Shaly mudstones of Cretaceous age, 70-100 million years old, outcrop along the east and south edges of the Park. All of these formations were deposited in shallow seas. These sedimentary deposits were folded and uplifted 65-70 million years ago, and huge areas of old Belt rocks slid as a whole eastward on top of younger Cretaceous formations. Along this overthrust zone, older rocks lie on top of younger, contrary to what is normally expected. A more detailed account of Glacier's geology is given by Raupp et al. (1983).

Geology can influence vegetation where the derived soils are poorly developed and still reflect the chemical composition of the parent rock. Some species of plants favor calcium-rich soils, while others prefer more acidic substrates. The limestones and dolomites of the Altyn and Siyeh formations are more calcareous, with a higher pH than that of granitic rocks or most mudstones. Furthermore, the white limestones of the Altyn Formation, found around St. Mary, are more calcareous than the dirty yellow or grey Siyeh Formation in the center of the Park. A number of plants occur only on soils derived from the softer, younger, shaly Cretaceous mudstones outcropping in the Marias Pass area.

The uplifted rocks of Glacier Park have been dramatically carved by the action of ice and water, and this activity continues today. During the last ice age, Glacier Park was almost entirely buried beneath huge valley glaciers. These glaciers carved the sheer cliffs with amphitheater-like cirques at their bases and trough-shaped valleys that characterize the Park today. Glaciers deposited huge bands of rocky debris (moraine) along their edges and across their valleys as they retreated. These moraines sometimes dammed valleys, forming the long finger lakes common on the Park's west side. All of the Park's glaciers may have disappeared during a warm period about 8000 years ago (Carrara 1989). There are currently about 40 active glaciers in the Park, but they are small and have been shrinking for at least the past 150 years (Raupp et al. 1983). The glaciers continue to erode small areas, but water, snow and wind now play a more significant role. Streams and rivers erode their banks, especially during rapid spring snowmelt. Wind causes erosion of sparsely vegetated slopes and ridges. Avalanches are common in the Park and prevent many slopes from developing forest vegetation.

The physiography of Glacier affects the vegetation even more than the geology. The deep glacial valleys with high, steep headwalls near the Continental Divide provide warm, snowy, wet environments typical of more coastal ranges. Rock slides, talus slopes, and avalanche chutes form on steep slopes left by the glaciers. These physically harsh habitats are favorable to some plants and not others. Fresh moraine is home to many early successional species that thrive on moist, poorly developed, gravelly soil.

Vegetation

Vegetation of Glacier National Park is diverse, due primarily to the great variation in elevation that influences many environmental factors affecting plant growth. Temperature generally decreases while precipitation and wind increase with elevation. However, aspect greatly modifies these effects in dissected mountainous terrain. North- and east-facing slopes and valley bottoms are cooler because they receive less direct sunlight and may be wetter when wind redistributes snow to lee slopes. Merriam (1890) recognized that increases in elevation generally result in vegetation changes similar to those found when travelling north in the Northern Hemisphere at the same elevation, and this is the basis of his life zone classification. Four Merriam life zones occur in Glacier Park: Transition, Canadian, Hudsonian, and Arctic-Alpine (Standley 1921). I use a three-zone classification in this book to describe the distribution of plants. It is based on elevation and corresponds to the Merriam system. I further divide the subalpine zone into lower and upper zones that roughly correspond to Merriam's Canadian and Hudsonian zones respectively.

This Book	Merriam Zones	Elevation Range
Montane	Transition	915-1675 m (3,000-5,500 ft)
Subalpine		1525-2285 m (5,000-7,500 ft)
(lower)	Canadian	
(upper)	Hudsonian	
Alpine	Arctic-Alpine	>1980 m (>6,500 ft)

The Montane Zone

The climate of the montane zone is characterized by relatively warm temperatures. Grassland vegetation is common on both sides of the Continental Divide. However, montane forests are rare on the east side where lower subalpine forests extend to valley bottoms on cool slopes, and most warm slopes support grasslands.

Riparian Forests

Riparian forests occur along rivers and major streams and are well developed only in the Montane Zone. Both the North Fork and the Middle Fork of the Flathead River, on the west and south boundaries respectively, are unregulated by dams and display good examples of riparian succession. Black cottonwood (*Populus balsamifera* ssp. *trichocarpa*) is the dominant species in young and middle-age stands. It establishes on newly created gravel bars along with willow (*Salix exigua, S. drummondiana, S. eriocephala*), red osier dogwood (*Cornus stolonifera*), and a diverse assemblage of colonizing species such as *Agrostis stolonifera, Astragalus alpinus, A. vexilliflexus, A. robbinsii, Dryas drummondii, Chamerion latifolium,* and *Equisetum* spp. (Malanson and Butler 1991). Other deciduous shrubs or small trees colonize shortly afterward (Foote 1965). These include mountain maple (*Acer glabrum*), alder (*Alnus incana*), hawthorn (*Crataegus douglasii*), river birch (*Betula occidentalis*), silverberry (*Elaeagnus commutata*), and chokecherry (*Prunus virginiana*). Soil develops as stands age and species of more upland habitats become important. White spruce (*Picea glauca*), ponderosa pine (*Pinus ponderosa*), and western red cedar (*Thuja plicata*) share the canopy with cottonwood. Herbaceous species such as *Aster* spp., *Cerastium arvense, Galium triflorum, Heracleum spondylium, Osmorhiza chilensis, Poa pratensis,* and *Smilacina stellata* become more common. If flooding or channel migration fails to cause severe disturbance, cottonwood will eventually die out, and forests will resemble mesic upland coniferous forests.

Grasslands

Grasslands are dominated by cool season bunchgrasses. They occur in a broad band along the mountain front east of the Continental Divide, such as along St. Mary Lake, in the Blacktail Hills near Marias Pass, and north of Polebridge west of the Divide. Many of these communities have a significant shrub component and could be better thought of as shrublands. Polebridge grasslands have been considered most closely allied to Palouse prairie (Koterba and Habeck 1971), but they are floristically indistinguishable from foothills grasslands common throughout Montana's intermountain valleys east of the Continental Divide (Weaver 1980). Dominant native grasses include rough fescue (*Festuca scabrella*), Idaho fescue (*F. idahoensis*), bluebunch wheatgrass (*Elymus spicatus*), oatgrass (*Danthonia intermedia*), and needlegrass (*Stipa* spp.). *Lupinus sericeus, Achillea millefolium, Eriogonum* spp., *Geranium viscosissimum, Geum triflorum,* and *Arnica* spp. are common broad-leaved herbs. Mountain big sagebrush (*Artemisia tridentata* ssp. *vaseyana*) is common in the Polebridge grasslands. It is killed by fire, so its abundance is probably determined primarily by fire frequency. Grasslands east of the Divide have varying amounts of shrubs including shrubby cinquefoil (*Pentaphylloides fruticosa*), serviceberry (*Amelanchier alnifolia*), Canada buffaloberry (*Shepherdia canadensis*), and bearberry (*Arctostaphylos uva-ursi*). These become more common in the subalpine zone. Kentucky bluegrass (*Poa pratensis*) and timothy (*Phleum pratense*) are common exotic grasses, especially on the east side of the Park where timothy dominates large areas to the exclusion of most other species.

The mesic grasslands of Glacier Park, especially on the west side, are being invaded by trees, primarily lodgepole pine (*Pinus contorta*) and aspen (*Populus tremuloides*). Historically, fires were frequent around Polebridge (Barrett et al. 1991) and on the east side of the Park (Barrett 1993, 1997) and were probably the most important factor in maintaining the prairies, although relatively low precipitation and well-drained soils may also have played a role (Koterba and Habeck 1971). Fires around Polebridge in 1988 killed sagebrush and returned advancing lodgepole forest to grassland.

Aspen Forest
Aspen forest is common in the montane zone, especially east of the Divide where aspen groves intermingle with grassland to form extensive parklands (Lynch 1955). Many of these groves are probably in the lower subalpine zone, but they are all treated here for convenience. West of the Continental Divide, aspen stands are small and usually associated with stream corridors or depressions on slopes otherwise dominated by conifers. Examples are common along the North Fork Flathead Inside Road. Similar sites support aspen east of the Divide, but there stands may be extensive, occupying depressions and cool slopes in the hills along the east edge of the Park. Aspen forest is leafless until late spring and canopies are not dense, so a good deal of light reaches the ground. As a result, understory vegetation is usually luxuriant. Driest stands have snowberry (*Symphoricarpos* spp.), serviceberry, Oregon grape (*Berberis repens*), and rose (*Rosa acicularis*) dominant in the understory. Wetter stands east of the Divide often have black cottonwood in the canopy and an understory of tall herbs and grasses such as *Elymus glaucus, Bromus* spp., *Geranium richardsonii, Osmorhiza occidentalis, Heracleum spondylium,* and *Angelica arguta.* A dense flowering of glacier lilies (*Erythronium grandiflorum*) can be seen in some stands before the trees leaf out.

Aspen stands, especially west of the Divide, often have conifers, such as lodgepole pine, spruce (*Picea* spp.) and subalpine fir (*Abies lasiocarpa*), mixed in. This indicates that many aspen groves are a successional stage eventually leading to spruce/fir forest (Pfister et al. 1977). Periodic fires may be maintaining aspen in many areas (Barrett 1993). Around Polebridge numerous aspen sprouts occur around older trees killed by the Red Bench fire of 1988. Fire is considered essential for maintaining aspen farther south in the Rocky Mountains (DeByle et al. 1987), but it may not be so important in the aspen parklands of Canada and Montana (Lynch 1955). Aspen stands may be at least partially defoliated during outbreaks of tent caterpillars (*Malacosoma* spp.).

Wetlands

Wetlands, such as wet meadows, swamps, marshes, and fens, are usually small but common in the montane zone. These wetlands are associated with glacial ponds and lakes and with riparian areas and beaver impoundments. Examples can be seen above Lake McDonald and along the Camas Road. Small wet meadows dominated by *Calamagrostis canadensis* often occur in depressions in coniferous forest and along lake margins. Swamps dominated by alder and willow (*Salix drummondiana, S. geyeriana,* or *S. bebbiana*) occur most commonly along streams or around beaver impoundments. Marsh vegetation develops on saturated to flooded mineral soil, often around the shallow margins of lakes or ponds. Marshes are typically dominated by *Carex utriculata* and/or *Equisetum fluviatile.* Fen vegetation develops on wet organic soils of glacial depressions or gentle slopes associated with groundwater seepage. Fens are most common west of the Divide, often on nearly level drainage divides. They are commonly dominated by sedges, especially *Carex lasiocarpa, C. buxbaumii,* and *C. utriculata.* Although wetlands occupy only a small amount of surface area in Glacier Park, they provide a large diversity of habitats and support a large number of species not found elsewhere.

Coniferous Forest

Coniferous forest of the montane zone occurs primarily on warm slopes and river terraces along the west and, to a lesser extent, east edges of the Park. The most common montane forests are dominated by Douglas fir (*Pseudotsuga menziesii*), lodgepole pine, paper birch (*Betula papyrifera*), and spruce. Western larch (*Larix occidentalis*) is common west of the Divide. These forests occur on warm slopes along forks of the Flathead River and their larger tributaries and in the lower St. Mary and Two Medicine valleys. The shrub understory includes snowberry (*Symphoricarpos albus*), Oregon grape, mountain maple, spiraea (*Spiraea betulifolia*), thimbleberry (*Rubus parviflorus*), and huckleberry (*Vaccinium membranaceum*). *Calamagrostis rubescens, Aster conspicuous, Thalictrum occidentale,* and *Arnica cordifolia* are common in the ground layer; *Linnaea borealis* and *Clintonia uniflora* occur in moister sites. Spruce (*Picea engelmannii*) and subalpine fir become more common at the upper limits of the montane zone. Forests dominated by Ponderosa pine occur on broad terraces above the North Fork Flathead River south of Logging Creek. Most stands have Douglas fir and western larch younger than the pine and understories similar to typical Douglas-fir forests described above (Lunan and Habeck 1973).

River terraces often support moist to wet forest dominated by spruce (*Picea glauca*), often with scattered cottonwood, western larch (*Larix occidentalis*), white pine (*Pinus monticola*), or Douglas fir. Moist stands support shrubs such as serviceberry, thimbleberry, huckleberry and a rich ground layer often including *Cornus canadensis, Linnaea borealis, Clintonia uniflora,* and *Aralia nudicaulis.* Wet stands have standing water between trunks for much of the year. Common understory plants include *Cornus stolonifera, Rubus pubescens, Equisetum arvense, E. scirpoides,* and *Mitella nuda.*

Exceptionally warm, moist forests, rich in species, occur along Lake McDonald. Western red cedar and western hemlock (*Tsuga heterophylla*) dominate these stands, but large white pine, larch, and Douglas fir are also common. Lodgepole pine, paper birch, grand fir, and even cottonwood can be found in these stands (Habeck 1968). Pacific yew (*Taxus brevifolia*), huckleberry, rose (*Rosa gymnocarpa*) and spiraea are common shrubs. The ground layer is luxuriant with many species of mosses and vascular plants, including *Clintonia uniflora, Cornus canadensis, Linnaea borealis, Aralia nudicaulis, Viola orbiculata, Goodyera oblongifolia, Chimaphila umbellata,* and *Tiarella trifoliata.*

The fire history of montane forests west of the Continental Divide is complex (Barrett et al. 1991). A mixture of crown fires and ground fires predominated in the North Fork Flathead Valley resulting in some stands of old ponderosa pine, larch, and Douglas fir with open understories as well as stands of dense lodgepole pine. However, fire suppression during the past century has allowed young spruce, subalpine fir, and Douglas fir to invade formerly open stands (Lunan and Habeck 1973). The moister forests near West Glacier experienced fewer ground fires and less frequent crown fires. The old western larch, white pine, and Douglas fir in these forests have persisted since just after the last catastrophic fire as the more shade-tolerant spruce, western hemlock, and western red cedar regenerate beneath them.

The Subalpine Zone

Grasslands
Grasslands occur from the montane zone to above treeline east of the Continental Divide in Glacier Park. In fact, along the east front of the mountains north of Two Medicine Lake, grasslands extend unbroken from the plains to the summits of Spot and Mad Wolf Mountains, gradually becoming alpine turf dominated by sedges and forbs above treeline. At subalpine elevations, grasslands are usually restricted to south- or west-facing slopes or areas of great wind exposure. Subalpine grasslands are similar in composition to those described for the montane zone. At higher elevations *Festuca idahoensis* and *Elymus trachycaulus* become more common, while *Festuca scabrella* and *Elymus spicatus* decrease. Shrubs, such as shrubby cinquefoil and Canada buffaloberry, develop a short stature, and dwarf shrubs, such as bearberry and alpine dryad (*Dryas octopetala*), become more common.

Meadows
Meadows are usually dominated by broad-leaved herbaceous plants and sedges. Dry meadows occur on steep, usually warm slopes with stony, poorly-developed soil subject to frequent downhill movement. Plant cover is sparse. *Phacelia hastata, Penstemon ellipticus, Hedysarum sulphurescens, Cirsium hookerianum, Potentilla glandulosa, Aquilegia flavescens, Epilobium alpinum,* and *Eriogonum flavum* are some of the common forbs. *Calamagrostis purpurascens, Trisetum spicatum,* and *Festuca idahoensis* are common grasses, and shrubby cinquefoil is a frequent shrub.

Wet meadows are found on gentle to level terrain with adequate snow cover. Soils are relatively deep with a well-developed organic horizon and remain wet or moist during much of the short growing season. Shortly after snowmelt, the early ephemerals, such as *Erythronium grandiflorum, Claytonia lanceolata,* and *Ranunculus eschscholtzii,* begin flowering. Shortly thereafter large, showy wildflowers, such as *Arnica* spp., *Senecio triangularis, Erigeron peregrinus, Mimulus lewisii, Polygonum bistortoides, Valeriana sitchensis,* and *Castilleja* spp., begin flowering and continue through the rest of the summer (DeBolt and Lesica 1986). *Carex paysonii, Juncus drummondii,* and *Phleum alpinum* are common grass-like plants in these habitats. High subalpine meadows are common around the visitor's center at Logan Pass.

Warm, moderately steep, subalpine slopes may support extensive meadows dominated by beargrass (*Xerophyllum tenax*). Extensive meadows of this type occur along the Garden Wall north of Logan Pass. Upon closer inspection, most beargrass meadows are seen to be burned-off forests that are regenerating slowly or not at all due to the harsh environment. Most of the species common in these meadows also occur in open, subalpine forests:

huckleberry, thimbleberry, mountain ash (*Sorbus scopulina*), *Arnica* spp., *Chamerion angustifolium, Veratrum viride,* and *Thalictrum occidentale.* In some years, beargrass meadows provide spectacular wildflower displays.

Avalanche Chutes

Avalanche chutes are common on steep, usually warm slopes in the subalpine zone, especially west of the Continental Divide (Butler 1979) and can easily be seen from Going-to-the-Sun Road. They are formed by large amounts of snow sliding rapidly downhill, often in the spring. Trees with rigid trunks are broken off by the moving snow, so the centers of avalanche paths are dominated by herbaceous plants and shrubs with flexible stems. Avalanche chutes are usually associated with long, steep ravines (Butler 1979) that are moister than the adjacent slopes. Common shrubs include green alder (*Alnus viridis*), serviceberry, spiraea, thimbleberry, Scouler willow (*Salix scouleriana*), and elderberry (*Sambucus racemosa*). Associated with the shrubs are many tall, luxuriant, moisture-loving, herbaceous plants, including *Heracleum spondylium, Chamerion angustifolium, Urtica dioica, Veratrum viride, Angelica arguta,* and *Elymus glaucus.* Avalanche chutes gradually merge into subalpine forest at the margins (Malanson and Butler 1984).

Forests

Forests of the subalpine zone are the most common habitat in Glacier National Park. Subalpine forests are dominated by subalpine fir, Engelmann spruce, and/or lodgepole pine. Lower subalpine forests may also have Douglas fir, western larch, and white pine, while higher forests often have whitebark pine (*Pinus albicaulis*). These mesic forests usually have abundant shrubs in the understory, including fool's huckleberry (*Menziesia ferruginea*), thimbleberry, mountain ash, spiraea, and huckleberry. Common ground layer species include *Arnica cordifolia* or *A. latifolia, Chimaphila umbellata, Linnaea borealis,* and *Thalictrum occidentale.* Relatively warm, moist stands, usually at lower elevations, are characterized by the shrubs Utah honeysuckle (*Lonicera utahensis*), fool's huckleberry, and mountain lover (*Paxistima myrsinites*), as well as the herbaceous plants *Clintonia uniflora, Tiarella trifoliata,* and *Trisetum cernuum.* As stands become colder and drier, often at higher elevations, these species decrease while the whortleberries (*Vaccinium myrtilloides, V. scoparium*), *Hieracium albiflorum,* and *Xerophyllum tenax* increase. Forests near timberline that receive heavy snowfall have open canopies with dense ground layers dominated by *Phyllodoce* spp., *Xerophyllum tenax, Erythronium grandiflorum,* and *Luzula hitchcockii.*

Subalpine larch (*Larix lyallii*) dominates scattered, small stands on cool, snowy, high-elevation sites with stony soil (Arno and Habeck 1972). These trees do not become stunted, even at the highest elevations. Subalpine fir and Engelmann spruce often occur in these stands as well. The understory usually has *Xerophyllum tenax, Luzula hitchcockii, Juncus drummondii,* and *Carex nigricans.* Most stands are near or west of the Continental Divide in the north half of the Park, although one stand occurs as far east as the east side of Waterton Lake. The discontinuous distribution of subalpine larch in the Park is anomalous.

Subalpine fir, spruce, and whitebark pine become stunted and dwarfed at upper treeline due to ice-scouring wind or heavy snow accumulations (Arno and Hammerly 1984). These "krummholz" forests are usually sparse and discontinuous, interspersed with alpine tundra or heath. Tree cover breaks the strong, high-elevation wind, allowing snow accumulation and providing plentiful moisture where there is adequate soil. Typical subalpine forest plants, such as *Thalictrum occidentale, Senecio triangularis, Xerophyllum tenax, Erigeron peregrinus, Valeriana sitchensis, Gentiana calycosa, Erythronium grandiflorum,* and *Luzula piperi* are common in the ground layer of these protected sites (Habeck 1969). Heath, dominated by the dwarf shrubs *Phyllodoce* spp. and *Salix arctica,* develops under sparse krummholz where snow accumulations are deeper. Good examples of moist krummholz occur around Logan Pass. Krummholz developed on skeletal soils of very wind-exposed sites often has an understory of dwarf shrubs such as alpine dryad and bearberry and dwarfed plants of common juniper (*Juniperus communis*), shrubby cinquefoil, and Canada buffaloberry. These drier elfin forests are most abundant east of the Continental Divide and can be seen along the trail to Scenic Point above Two Medicine.

Fire has influenced the composition and structure of subalpine forests in Glacier Park. This influence has generally been greater in the lower subalpine zone. Testimony is given by the large areas dominated by lodgepole pine along the North Fork of the Flathead River and the extensive brush fields on steep, warm slopes such as those west of Logan Pass or along the Middle Fork of the Flathead. Large areas of lower subalpine forest are strongly dominated by lodgepole pine, a short-lived tree that usually requires fire to open cones and release seeds to germinate in recently burned soil (Muir and Lotan 1985). Engelmann spruce and subalpine fir can invade these forests and would eventually come to dominate, but short fire intervals maintain the lodgepole dominance. Unlike the coniferous trees that are killed by fire, nearly all the understory shrubs and herbaceous plants can survive and resprout from stems or roots protected underground. So while fire drastically changes the overstory, the understory recovers quickly and stays much the same. Subalpine forests demonstrate a pattern of mixed-severity fires with stand-replacing fires predominating in most areas (Barrett 1993, 1997).

Large areas of the Park have burned during the past century. Upper subalpine forests have longer fire-free intervals, but the infrequent fires that do occur promote whitebark pine by providing sites for recruitment and reducing competition from subalpine fir (Arno 1986). Whitebark pine, in turn, provides a more benign microclimate for the establishment of subalpine fir.

Disease and insects have also played an important role in shaping Glacier's forests. Mountain pine beetles (*Dendroctonus ponderosae*) generally attack stands of lodgepole pine that are 80 or more years old, killing most trees. Fallen and standing dead lodgepole increase the likelihood of fire. Mountain pine beetles will also attack whitebark pine, especially trees weakened by disease or competition from fir trees (Keane and Arno 1993). A more serious threat to whitebark pine is the introduced fungal disease white pine blister rust (*Cronartium ribicola*). Blister rust has spread rapidly in the Northern Rockies in the past 20 years (Keane and Arno 1993). The majority of the whitebark trees in the Park are infected, and many have already died (Kendall 1998). As a result, upper subalpine forests show increasing dominance by subalpine fir.

The Alpine Zone

Although nearly one third of Glacier Park is above treeline, there is relatively little alpine vegetation because most of this zone is too steep or too snowy for soil development. Sites are too cold in the summer, too windy, or too snowy to support forests. Scattered trees can be found in the shelter of boulders or steep slopes. The few recognizable plant communities that repeat across the landscape are highly variable . Sparsely vegetated habitats are better thought of as loose assemblages rather than communities.

Fellfields

Fellfields dominated by alpine dryad are the most extensive alpine vegetation in the Park. These low plant communities develop on stony, usually calcareous soils of exposed slopes and ridges. Plant cover is low. Alpine dryad often occurs on slopes subject to downward movement due to frost heaving. It tends to form stripes and a stair-step ground pattern perpendicular to the slope (Bamberg and Major 1968, Zwinger and Willard 1972). These patterns are seen clearly at Siyeh and Piegan passes. The willows *Salix arctica* and *S. reticulata* are other common dwarf shrubs in this vegetation. *Carex nardina, C. rupestris,* and *Poa alpina* are common sedges and grasses. *Minuartia obtusiloba* and *Silene acaulis* are common cushion plants. Other common broad-leaved species include *Hedysarum sulphurescens, Polygonum viviparum, Potentilla diversifolia,* and *Smelowskia calycina.*

Turf

On more protected slopes, soils become deeper and fellfields grade into turf. Vegetation is relatively dense, similar to grassland vegetation, but plants are rarely greater than 12 cm (5 in) high. Dry turf communities are dominated by grasses or sedges including *Festuca brachyphylla, Poa alpina, Carex phaeocephala, Luzula spicata,* and *L. piperi. Arnica mollis, Cerastium arvense, Potentilla diversifolia, Ranunculus eschscholtzii, Silene acaulis,* and *Solidago multiradiata* are common broad-leaved species. The dwarf willow *Salix arctica* may also be common. Moist slopes with subirrigated, highly organic soil may develop below permanent snowfields. These relatively rare sites support moist turf dominated by the dwarf shrubs, alpine dryad, and the willow *Salix reticulata* among the sedges *Carex scirpoidea, C. rupestris, C. capillaris, C. paysonis,* and *Kobresia simpliciuscula. Polygonum viviparum, Solidago multiradiata,* and *Anemone parviflora* are common broad-leaved species.

Snow Fields

Two minor types of alpine vegetation are associated with late-persisting snow fields: heath and sedge snowbed. Alpine heath often occurs on shallow soil of rock ledges (Choate and Habeck 1967). It is dominated by ericaceous, dwarf shrubs: *Phyllodoce empetriformis, P. glanduliflora, Cassiope tetragona,* and *Kalmia microphylla.* The dwarf willow, *Salix arctica,* is also common. *Carex paysonis* and *C. podocarpa* are common sedges. The forbs *Erythronium grandiflorum, Arnica latifolia, Senecio cymbalarioides,* and *Hypericum formosum* may be scattered throughout the heath. Sedge snowbeds occur in relatively level areas where snow lies the latest. This vegetation is species-poor and strongly dominated by *Carex nigricans. Juncus drummondii* and *Deschampsia atropurpurea* are other common grass-like plants. *Sibbaldia procumbens* and *Hieracium triste* may occur in openings in the sod.

Talus

Talus and scree slopes are common above treeline in Glacier Park. Because these slopes are near the angle of repose, the surface is always shifting, making it difficult for most plants to take root. However, plants that can grow downslope with the surface rock experience a surprisingly benign environment, especially on warm slopes. Most talus slopes have a surface layer of loose rock over deep, mineral soil. The surface rock allows rain and snowmelt into the soil below but acts as a mulch by preventing evaporation and keeping the soil moist and warm. Plant cover is usually very sparse and the vegetation is better considered an assemblage rather than a community. Nonetheless, several distinctive species, such as *Crepis nana, Erigeron lanatus, Eriogonum androsaceum, Stellaria americana, Polemonium viscosum, Claytonia megarhiza,* and *Chamerion latifolium* occur only on talus slopes (or river gravels). Alpine dryad, *Saxifraga bronchialis, Potentilla glandulosa,* and *Eriogonum ovalifolium* are also common on shifting talus but are found in other habitats as well.

History of Botanical Exploration

Botanical exploration of Glacier Park's rugged terrain was minimal before the beginning of the twentieth century. R. S. Williams lived in Columbia Falls and Great Falls and helped survey the east boundary of Glacier National Park in 1897. He collected vascular plants as well as mosses in what was to become the Park and the surrounding area in 1887 and from 1892 to 1897. He later moved east to work for the New York Botanical Garden. Levi M. Umbach of Northwestern University in Illinois collected plants in Glacier Park from 1901 to 1903. Many of the trails were not maintained at that time, so Umbach collected mainly around the towns of East Glacier and West Glacier. He also visited the high country near Sperry Glacier and the Two Medicine Valley. Also in 1901 Frederick K. Vreeland, an electrical engineer from New York City and P. A. Rydberg's companion during his Colorado explorations, collected plants around Lake McDonald and donated them to the New York Botanical Garden. Marcus Jones, one of the West's most famous botanists of the early twentieth century, spent the summers of 1908 and 1909 in western Montana and collected plants near Sperry Glacier and north of Logan Pass (Jones 1910). Morton Elrod, a biology professor at the University of Montana in Missoula, established a biological research station on Flathead Lake, just south of Glacier, in 1899. From then through 1928 he collected plants in Glacier Park in addition to working there as chief naturalist for eight summers. Joseph E. Kirkwood, another professor at the University of Montana, collected plants in the Park in 1910. Joseph Blankenship, the botany professor at Montana State University in Bozeman, collected plants for a short while in the Park in 1903. A. S. Hitchcock, one of the country's foremost agrostologists, spent three weeks in 1914 collecting grasses in Glacier. A summary of early botanical exploration in Glacier Park is given by Standley (1921).

The most significant Glacier Park plant collections were made by Paul Standley during the summer of 1919 while working for the Smithsonian Institution. He spent most of that summer on the east slope, working out of the Many Glacier Valley. Standley spent only three weeks west of the Divide in the vicinity of West Glacier and Lake McDonald (Standley 1921) because access by road was poor or nonexistent at the time. In spite of having to do field work during a drought year, Standley collected ca. 1000 specimens and catalogued the vast majority of species currently known to occur in the Park. His book, Flora of Glacier National Park, Montana, was published in 1921 by the Smithsonian Institution (Standley 1921) and is still a useful reference.

Publication of Standley's book sparked interest in the Park, and many academic botanists began to visit. Bassett Maguire, curator of the herbarium at Utah State University, spent time in Glacier National Park in 1932 and again in 1934. He collected around many of the low-elevation lakes on both sides of the Divide and found many aquatic and wetland plants previously unrecorded for the Park (Maguire 1934, 1939). William T. McLaughlin of Northwestern University collected plants in the Park from 1930 to 1934 and found a few plants previously unreported (McLaughlin 1935). C. Leo Hitchcock collected plants in Glacier Park during the summer of 1933 while he was a professor at the University of Montana in Missoula. LeRoy H. Harvey was professor of Botany and curator of the herbarium at the University of Montana in Missoula from 1946 through 1977. He collected hundreds of specimens throughout Glacier Park, mainly in the 1950s, in preparation for writing a flora that was never completed (Harvey 1954). Frederick J. Hermann, curator of the U. S. Forest Service Herbarium and author of a manual on Rocky Mountain sedges, collaborated with Harvey, collecting vascular plants (mainly sedges) in the Park in 1955 (Hermann 1956). Alfred Schuyler, a curator at the Academy of Natural Sciences of Philadelphia, collected wetland and aquatic plants on the west side of the Park during the summers when he taught summer classes at the University of Montana Biological Station in 1978-1995 (Schuyler 1980, 1982). Several students, including Sam Bamberg, Richard Pemble, Irene Sammons, and Christian Damm, made plant collections as part of their graduate work (Bamberg and Pemble 1968).

Many of Glacier Park's biologists, rangers, and naturalists have also made significant plant collections, most of which are housed in the herbarium at West Glacier. Mona Myatt made numerous plant collections while working on a project to map Glacier's vegetation and fire history during the mid-1970s. J. L. McMullen was an interpretive naturalist for five summers during 1947-54; three of those were spent in the Two Medicine Valley where he took some of the first collections from that area. Dave Shea, a ranger in the northeast portion of Glacier Park for many years, found a number of Park records. Jerry DeSanto was a ranger in the Park for over 20 years. He found several species previously unreported and deposited many collections in the Park herbarium (Lesica et al. 1993). He also wrote and privately published a book on the Park's alpine flora (DeSanto 1989).

Fred Hermann returned to Glacier Park in 1962, 1966 and 1968 to collect mosses and publish a checklist (Hermann 1969). Bruce McCune, Roger Rosentretter, and Ann DeBolt collected lichens in the Park during the 1970s and 80s and published an annotated checklist (DeBolt and McCune 1993).

Many Glacier Park plant collections made by Elrod, Harvey, C. L. Hitchcock, Jones, Kirkwood, McMullen, Pemble, Sammons, Schuyler, Umbach, and Williams are deposited at the University of Montana herbarium (MONTU). Collections of A. S. Hitchcock and Standley are at the Smithsonian.

Methods

I began work on this book in 1984 when I spent three months in West Glacier working in the herbarium, compiling an inventory of specimens and preparing a checklist of vascular plants (Lesica 1985). I have continued to collect published information on the Park's flora. Since then I have hiked nearly every trail in Glacier Park with a good deal of bushwhacking as well, collecting hundreds of specimens and taking notes on the flora and vegetation. Other botanists and I collected many new records for the Park during this time (Lackschewitz et al. 1988; Lesica 1991; Lesica et al. 1986, 1993, 1998; Lesica and Stickney 1994), leading to a revised checklist (Lesica 1996). I have observed nearly all of the plants treated in this book in the field, in or near the Park.

Only taxa for which there is a verified specimen are included in this book as part of the Glacier Park flora. Most taxa are verified by specimens at the Glacier Park herbarium or the herbarium at the University of Montana. Literature reports by knowledgeable botanists that are supported by vouchers were accepted at face value. I attempted to verify dubious reports, and in some cases these turned out to be erroneous and were not included.

Morphological descriptions were taken from monographic treatments and then revised using herbarium material collected in and around Glacier Park. I examined and took measurements from approximately 5,000 specimens. Habitat information and distribution within the Park were taken from these specimens and personal experience. "East" and "West" refer to east and west of the Continental Divide respectively. Global distribution information was taken from a number of floristic monographs and is the range of the species as a whole. Ethnobotanical information was obtained from Johnston (1987). Infraspecific taxa (i.e., subspecies, varieties) are mentioned in the second paragraph of the species description. Morphological, ecological, or distributional differences are described when they are thought to be significant in the area.

Original illustrations were prepared from living and pressed specimens collected in or around Glacier National Park. Nearly 350 species are illustrated, approximately one-third of the flora. Most genera in the Park are represented by at least one illustration. In addition, we have tried to illustrate the most common species, those most likely to be seen by short-term visitors to the Park. Illustrations can sometimes be used to help determine the family or genus of an unknown plant, making identification a much easier task.

Taxonomic Philosophy

Application of molecular genetics and methods of modern cladistics to plant systematics has resulted in a good deal of recent change in taxonomic nomenclature. In most cases I have not had the time or competence to evaluate the primary literature justifying these changes. Rather, I have chosen to follow treatments provided in recently published floristic monographs (e.g., Cronquist et al. 1986-97, Hickman 1993, Flora of North America Editorial Committee 1993-2000) or by experts participating in the Flora of North America project. These treatments are cited following the family or genus descriptions. In a few cases I have chosen to follow a treatment at odds with that currently in favor based on my knowledge of the literature and the plants in the field.

I have attempted to include taxonomic synonyms used by floristic manuals pertinent to Glacier Park, including Standley (1921), Hitchcock and Cronquist (1973), Dorn (1984), Moss and Packer (1983), and Kuijt (1982). I hope that including these synonyms will help relieve the distress felt by many of us who have trouble keeping up with the rapid taxonomic flux. When both a varietal and subspecific epithet were available for a particular taxon, I usually chose to use the subspecies. I have included one or more common names only when I believe they are truly in vernacular use.

How to Use This Book

All vascular plants currently known to occur in Glacier National Park are included in this manual's keys and descriptions. I attempted to broaden the applicability of the book by including brief mention of species not currently known from Glacier but known to occur in adjacent mountainous areas. To this end I consulted Kuijt's (1982) Flora of Waterton Lakes National Park and a checklist of vascular plants for Flathead National Forest prepared by Maria Mantas (1999). These additional species are usually listed following genus descriptions. The vast majority of plants occurring in the mountains of southeast British Columbia, southwest Alberta, and much of Flathead National Forest in northwest Montana are included here. Many plants occurring at low elevations in the Flathead Valley or on the Great Plains of the Blackfeet Indian Reservation will not be found in this book. As a general rule, this manual will be less useful the farther one travels from Glacier Park.

The book is organized phylogenetically by four major groups of vascular plants: Pteridophytes (ferns and allies), Gymnosperms (conifers), Dicot Flowering Plants, and Monocot Flowering Plants. Families within these four major groups are arranged alphabetically. Genera and species are arranged alphabetically within families.

Plants can be identified by means of "dichotomous keys." The first "couplet" (a pair of identical numbers) consists of two alternate, mutually exclusive statements. The reader must choose the statement that best describes the plant in hand. Following the correct statement is either the name of the family, genus or species to which the plant belongs, or the number of another couplet. For example:

1. Plants slender, trailing subshrubs; flowers nodding ....*Linnaea*
1. Plants shrubs with erect stems 2

If the first (1) is correct, then the plant is in the genus *Linnaea*. If the second (1) is correct, then the reader must go to the second couplet (beginning with 2) and again choose the correct statement and continue until a name rather than a number is obtained. The key to the families is used to determine the correct family. Under the description for that family will be a key to the genera used to determine the correct genus. Under the genus description will be a key to the species that finally gives the name (scientific binomial) of the plant. The illustration may aid in this process. A small number of abbreviations are used in the keys and descriptions: ca. (circa), approximately; >, greater than; ≥, greater than or equal to; ≤, less than; <, less than or equal to.

Once the name of the plant is obtained, the description for that species in the text should be checked against the plant in hand to make sure it is correctly identified. Characters common to all members of a family or genus are given under the descriptions for that family and genus and are usually not repeated in the species descriptions, so it is a good idea to read descriptions for the family and genus as well as species. Unfamiliar botanical terminology is defined in the glossary. Geographic ranges are given for the species as a whole, not just the variety or subspecies occurring in the Park.

Floristic Synopsis

Families	86
Genera	345
Species	1132
Taxa (subspecies, varieties)	1182
Introduced species	127
Native species	1005

Largest families:	*Native*	*Introduced*
Asteraceae	124	26
Cyperaceae	96	0
Poaceae	88	18
Rosaceae	46	0
Brassicaceae	40	12
Scrophulariaceae	39	10
Ranunculaceae	32	0
Saxifragaceae	31	0
Fabaceae	29	0
Onagraceae	26	0

Annuals and biennials	88	61
Herbaceous perennials	804	64
Shrubs and vines	93	1
Trees	20	1

Floristic Plant Geography

A primary goal of biogeography as well as ecology is understanding the distribution of organisms across landscapes and continents. Methods range from molecular genetics to analysis of fossils (Myers and Giller 1988, Axelrod and Raven 1985). A common method in plant geography has been floristic analysis: the classification of the flora into groups sharing distinct geographic patterns (Hultén 1937, McLaughlin 1989). These patterns generally correspond to "floristic regions," areas with a distinct flora thought to have evolved under relatively static climatic and soil conditions over long periods of time (Gleason and Cronquist 1964). A species whose geographic range largely corresponds to one of these patterns is considered to have affinities with that floristic region and is assumed to have evolved primarily with other species of the same affinity. It is assumed that plants came to be in a given area by either migrating there or evolving in place. Knowing the geographic patterns underlying a local flora provides clues to evolution and important historic migrations (Hultén 1937, Cain 1944). Compiling a floristic monograph of Glacier Park provides an opportunity to analyze the flora quantitatively and gain new understanding of the evolution of Park's vegetation.

Geographic Patterns in Glacier National Park

Much of the flora of Glacier National Park can be classified into one of four broad geographic patterns: Arctic-alpine, Boreal, Cordilleran, and Great Plains. Delineation of floristic provinces follows Gleason and Cronquist (1964). Within three of these provinces, I recognize subprovinces that roughly correspond to floristic areas described by McLaughlin (1989). There is usually little overlap between provinces; however, there is often a good deal of overlap among subprovinces within a province. Nonetheless, the majority of species can be placed in a subprovince with little uncertainty. There are, however, many species too widespread to discern a geographic affinity; i.e., they are common throughout two or more broad phytogeographic provinces. There are 224 such widespread species in Glacier's flora. In addition, there are 127 species thought to be introduced in the Park; these are addressed later.

Cordilleran Floristic Province	385
Cascade Mountains Subprovince	77
Rocky Mountain Subprovince	274
Southern Rocky Mountain Subprovince	4

Northern Rocky Mountain Subprovince   30
Boreal Floristic Province   305
   Circumboreal Subprovince   148
   American Boreal Subprovince   157
Arctic-alpine Floristic Province   82
   Circumpolar Arctic-alpine Subprovince   49
   American Arctic-alpine Subprovince   33
Great Plains Floristic Province   9

Cordilleran Flora

Cordilleran species (385) occur in the mountainous regions of western North America from Alaska south to California, New Mexico, Arizona, and sometimes northern Mexico. This region is characterized by bunchgrass grasslands at the lowest elevations, coniferous forest at mid-elevations, and alpine tundra above treeline. Cascade Mountains species (77) occur primarily west of the Rocky Mountains from Alaska or British Columbia south to California and east occasionally to northern Idaho and northwest Montana. Rocky Mountain species (274) are distributed along the main chain of the Rocky Mountains from Alaska or British Columbia and Alberta south to Colorado and Utah. The geographic center of Southern Rocky Mountain species (4) is in Utah and southern Wyoming south to Arizona and New Mexico. Northern Rocky Mountain species (30) are endemic to southeast British Columbia, eastern Washington, northeast Oregon, southwest Alberta, western Montana, and northern Idaho. It is likely that some of these species evolved in the area of Glacier Park.

The Cordilleran Flora is the most well represented in Glacier Park. This is not surprising since the Park lies in the center of the western North American Cordillera. Over two-thirds of these species have a Rocky Mountain Cordilleran distribution. More than three-quarters of the species with Cordilleran affinity occur in the montane zone; nearly half can be

Table 1. Percentage of native species with Cordilleran, Boreal, and Arctic-alpine floristic affinities found in three vegetation zones in Glacier National Park (percentages add to > 100 because some species occur in more than one zone). Species with widespread and Great Plains distributions are not included except in the overall "Park" statistics.

Zone	Cordilleran	Boreal	Arctic-alpine	Park
	(385)	(305)	(82)	(1005)
Montane	81%	94%	8%	83%
Subalpine	48%	41%	17%	38%
Alpine	30%	17%	99%	27%

Table 2. Percentage of native species with Cordilleran, Boreal and Arctic-alpine floristic affinities found in five habitat types in Glacier National Park (percentages add to > 100 because some species occur in more than one zone). Species with widespread and Great Plains distributions are not included except in the overall "Park" statistics. "Rocky" refers to cliffs, talus, and fellfields.

Zone	Cordilleran	Boreal	Arctic-alpine	Park
	(385)	(305)	(82)	(1005)
Forest	40%	40%	1%	34%
Grass/Turf	50%	24%	80%	48%
Wetland	16%	49%	7%	28%
Riparian	14%	22%	7%	17%
Rocky	20%	9%	51%	17%

found in the subalpine; and nearly one-third occur near or above treeline (Table 1). Half of the Park's Cordilleran species occur in grassland or turf. Dominant montane grassland species include *Elymus spicatus, Festuca idahoensis, Bromus carinatus, Stipa nelsonii,* sagebrush (*Artemisia tridentata*), *Geranium viscosissimum, Arnica sororia,* and *Balsamorhiza sagittata.* Forty percent of Cordilleran species can be found in forests, including many of the Park's most common such as subalpine fir (*Abies lasiocarpa*), Engelmann spruce (*Picea engelmannii*), whitebark pine (*Pinus albicaulis*), ponderosa pine (*P. ponderosa*), lodgepole pine (*P. contorta*), and Douglas fir (*Pseudotsuga menziesii*). Many Cordilleran shrubs are common in the forest understories. Included are Rocky Mountain maple (*Acer glabrum*), Utah honeysuckle (*Lonicera utahensis*), fool's huckleberry (*Menziesia ferruginea*), huckleberry (*Vaccinium membranaceum*), whortleberries (*V. myrtillus, V. scoparium*), mountain ash (*Sorbus scopulina*), and spiraea (*Spiraea betulifolia*). Many Cordilleran plants are common in the forest ground layer as well, such as *Osmorhiza occidentalis, Aster conspicuus, A. engelmannii, Senecio triangularis, Aquilegia flavescens, Thalictrum occidentale,* and *Prosartes trachycarpa.* Relatively few Cordilleran species occur in wetland habitats compared to the Park flora as a whole (Table 2).

Like the Cordilleran Flora as a whole, the Cascade Mountain Flora is best represented in low-elevation forests, with 83% of the species occurring in the montane zone and 66% being found in forests. On the other hand, only 10% of Cascadian species occur above treeline, and only 23% are associated with grassland or turf vegetation. Many of these plants find suitable habitat in the relatively warm, moist valleys west of the Divide, where a humid maritime climate prevails. These forests are often dominated by Cascadian species such as western red cedar (*Thuja plicata*), western hemlock (*Tsuga heterophylla*), western white pine (*Pinus monticola*), and western larch (*Larix occidentalis*). Other Cascadian species, such as baldhip rose (*Rosa gymnocarpa*), and oceanspray (*Holodiscus discolor*), are common shrubs, and *Viola orbiculata, Tiarella trifoliata*, and *Clintonia uniflora* are abundant in the understory. Cascadian species found in subalpine forests include subalpine larch (*Larix lyallii*), *Veratrum viride*, and *Xerophyllum tenax.*

Relatively few of Glacier's plants are endemic to the Northern Rocky Mountains even though the Park is in this floristic subprovince (McLaughlin 1989). In contrast to Cascadian species, members with this geographic pattern are most likely found in open, high-elevation habitats. Over half of the species occur near or above treeline and one-third are restricted to the alpine. Only 16% of the Northern Rocky Mountain endemics are found in forests, while two-thirds occur in the shallow soil of cliffs, fellfields, talus or stony slopes, and nearly as many occur in grasslands and turf. *Aquilegia jonesii, Antennaria aromatica, Arnica alpina, Eriogonum androsaceum,* and *Papaver pygmaeum* are found in alpine fellfields, while *Erigeron lanatus* and *Stellaria americana* specialize in talus.

Very few plants have a clearly discernable affinity with the southern Rocky Mountains. Most are rare in the Park and occur near Marias Pass or along the east front of the Mountains. These are *Carex elynoides, Saxifraga subapetala,* and *Valeriana edulis.*

Boreal Flora

Boreal species (305) are found in the northern coniferous forests of Canada and adjacent U. S., between the treeless arctic to the north and the Deciduous Forest and Great Plains floristic provinces to the south. These species commonly occur well south of Canada in the coniferous forests of the continent's mountain ranges. Circumboreal species (148) are found in the northern coniferous forests of Europe and Asia as well as North America, while American Boreal species (157) occur only in North America, across Canada or nearly so.

Plants with a Boreal distribution are very common in Glacier Park. Boreal species occur most frequently in forests and wetlands of the montane zone. Over 90% of these plants are found in the montane zone, while only 17% occur above treeline (Table 1). Nearly half of the Park's Boreal species occur in wetlands, and 40% are found in forests (Table 2). Only a few of the Park's Boreal species are trees; white spruce (*Picea glauca*) and black cottonwood (*Populus balsamifera*) are found on riparian terraces, while aspen (*P. tremuloides*) and paper birch (*Betula papyrifera*) are early successional forest species. On the other hand, many of the Park's forest understory shrubs have Boreal affinities. These include common juniper

(*Juniperus communis*), green alder (*Alnus viridis*), serviceberry (*Amelanchier alnifolia*), buckthorn (*Rhamnus alnifolia*), snowberry (*Symphoricarpos albus*), swamp currant (*Ribes lacustre*), Canada buffaloberry (*Shepherdia canadensis*), and elderberry (*Sambucus racemosa*). The ground layer of many forests is also dominated by Boreal plants, especially ferns and ericaceous species such as *Lycopodium* spp., *Athyrium filix-femina*, *Dryopteris* spp., *Gymnocarpium disjunctum*, *Polystichum lonchitis*, *Chimaphila umbellata*, *Orthilia secunda*, and *Arctostaphylos uva-ursi*. The majority of Circumboreal plants occur in wetlands such as marshes, lake shores, and fens. *Equisetum* spp., *Potamogeton* spp., *Sparganium* spp., and 20 species of wetland sedges (*Carex*) are all widespread in the Boreal zone around the world.

Fens are wetlands with perennially saturated organic soil. They are uncommon in Glacier Park and occur primarily west of the Divide. Ninety-seven species of plants commonly occur in Glacier's fens, and over 80% of these have a Boreal distribution. Thirty species are restricted to fens in Glacier Park, and only 3 of these do not have Boreal affinities. Obligate fen species are among Glacier's rarest plants; they include *Carex chordorhiza*, *C. limosa*, *C. livida*, *C. paupercula*, *C. rostrata*, *C. tenuiflora*, *Dulichium arundinaceum*, *Eriophorum spp.*, *Scheuchzeria palustris*, and the carnivorous plants *Drosera* spp. and *Utricularia* spp.

Arctic-Alpine Flora

Arctic-alpine species (82) are widespread near or north of the Arctic Circle (67north latitude), north of the general distribution of forest in northern Canada, Greenland and Alaska. They also occur near or above treeline in the mountains of eastern and/or western North America. Dominant vegetation is tundra. Circumpolar Arctic-alpine species (49) have an Arctic-alpine distribution in Europe and Asia as well as North America, while American Arctic-alpine species (33) occur only in North America or sometimes adjacent Asia.

Nearly all species with an Arctic-alpine distribution occur near or above treeline in Glacier Park. Plantago canescens occurs in the arctic and along the east front of the Canadian Rockies in exposed montane grasslands rather than alpine turf. Nearly half of Arctic-alpine species are found only along or east of the Continental Divide, while only one (*Gentiana glauca*) is found exclusively west of the Divide. Alpine dryad (*Dryas octopetala*) and dwarf willows (*Salix arctica, S. reticulata*), the dominant subshrubs in Glacier's alpine vegetation, have a circumpolar distribution. *Minuartia obtusiloba, Ranunculus eschscholtzii, Silene acaulis, Oxyria digyna, Polygonum viviparum, Carex nigricans, Festuca brachyphylla*, and *Poa alpina* are common herbaceous Arctic-alpine species.

Great Plains Flora
Great Plains species (9) are found primarily in the semi-arid grasslands at the center of the continent, east of the Rocky Mountains and west of the Mississippi River. Although Glacier is directly adjacent to the Great Plains Floristic Province, few Great Plains species occur in the Park because the mountains rise abruptly, and the Park receives much more precipitation than the Great Plains grasslands that dominate just a few miles to the east. The few plants with a Great Plains distribution in Glacier occur along the east edge of the Park. They are rare or uncommon, and all but one (*Carex sartwellii*) are found in montane grasslands. Examples include *Rosa arkansana, Liatris punctata, Astragalus flexuosus,* and *Carex heliophila.*

Summary
Glacier Park's flora is dominated by species with Cordilleran and Boreal distributions. Cordilleran plants are well distributed at all elevations and in all habitats; however, Cascadian species are prevalent in moist, low-elevation forests west of the Divide, while Northern Rocky Mountain endemics are most abundant in open habitats at higher elevations east of the Divide. Boreal plants are best represented in forests and especially wetlands. Not surprisingly, Arctic-alpine plants are found mainly above treeline, with more species east than west of the Divide.

Local Plant Geography

In the Northern Rockies, the Continental Divide sharply demarcates the boundary between the semi-arid continental climate of the Great Plains Floristic Province and the temperate maritime climate of the Northern Rocky Mountains to the west. The west side is generally more humid and less windy than east of the Divide. Paul Standley (1921) was the first to describe how these environmental differences affected the floras between the two sides of the Park. There are 154 plant species known to occur only on the west side of the Continental Divide in Glacier Park and 227 on the east side or along the Divide. The distribution of these species suggests that Boreal and Cascadian montane forest and wetland plants are more prevalent west of the Divide, while Arctic-alpine and Northern Rocky Mountain endemics of open habitats are more common on the east side.

Over one-third of the westside-restricted species have a Boreal distribution, but only 1% are Arctic-alpine, and only 9% occur above the montane zone. Nearly half of them occur in forests, and over one-third are found in wetlands; however, only 21% occur in grasslands or turf, and only 4% are found in rocky cliff, talus, and fellfield habitats. Plants restricted to the east side of the Park generally have a different ecology and phytogeographic affinity. Only 9% of these are found in forest, but two-thirds occur in grassland and turf, and nearly one-quarter in rocky habitats.

Plants with an arctic-alpine or boreal distribution each account for 15% of these eastside-restricted species. Circumboreal plants are well represented on both sides of the Divide; however, the distribution between floristic subprovinces is very different. There are 26 species of Cascadian plants restricted to the west side, but only 7 east of the Divide. On the other hand, there are 15 Northern Rocky Mountain endemics restricted to the east side but only 3 on the west side.

Historical Plant Geography

Boreal, Cordilleran, and Arctic-alpine floras had achieved much of their modern character by the end of the Tertiary, before the start of the Pleistocene (Axelrod and Raven 1985, Löve and Löve 1974, Davis 1981, Hultén 1937). Glacier Park lies along the main chain of the Rocky Mountains in the middle of the Cordilleran Floristic Province and just southwest of the Boreal Floristic Province, so it is not surprising that Glacier's flora is dominated by species with Cordilleran, Boreal, and Arctic-alpine distributions. It is the Northern Rocky Mountain endemics that raise the most interesting questions. Why are there so few endemics? Why do they have such limited distributions? Why do so many occur in open, high-elevation habitats?

Endemism, the occurrence of species with an exclusively local geographic distribution, generally declines from southwest to northeast in Montana (Lesica et al. 1984). Much of northern Montana and the Northern Rocky Mountain Floristic Province, including Glacier Park, was almost completely covered with ice during the Pleistocene glaciations (Perry 1962). Presumably all the vegetation including narrowly endemic species was destroyed by the glaciers, and during warmer interglacial periods the barren landscapes were recolonized by more widespread plants that survived south or perhaps north of the glaciers (Davis 1981, Hultén 1937). The low number of Northern Rocky Mountain endemics in Glacier Park is likely attributable to glaciation.

Some of the Northern Rocky Mountain endemics may be relics of a much larger distribution, while others have likely evolved in the Northern Rockies and never had a larger distribution (Cain 1944). *Papaver pygmaeum* is part of an ancient circumpolar arctic-alpine complex of very similar species (Welsh et al. 1987). Presumably Glacier's pygmy poppy is an isolate from this complex that survived only in and around the Park. On the other hand, several Northern Rocky Mountain endemic genera have actively spawned many local endemics throughout the western Cordillera. These include *Allium, Antennaria, Astragalus, Calochortus, Erigeron, Eriogonum, Lomatium,* and *Penstemon.* Northern Rocky Mountain species in these genera likely evolved relatively recently and have had little chance to attain a larger geographic distribution.

The few extant Northern Rocky Mountain endemics must have survived glaciation near the glacial front, perhaps in an ice-free area just east of the Park (Perry 1962, Carrara 1989). Paleoecologists believe that the environment near the ice front was cold with stony, poorly developed soil, similar to that of modern-day arctic tundra (Davis 1981). Plants adapted to this environment would have a better chance of persisting, so it is probably no accident that all but one of the 30 endemics in Glacier Park (Table 3) are found in cold grassland, turf or talus, cliffs, fellfields, or gravelly streambanks. *Trisetum orthochaetum,* a forest plant, is a rare hybrid likely of recent origin (Hitchcock et al. 1969).

Introduced Plants

More than half of the Park's introduced plants are in five families, and these are among the ten largest families in the native flora. Although similar to natives at the family level, 60% of Glacier's introduced plants belong to alien genera. Whereas annuals or biennials are poorly represented in the native flora (9%), they account for nearly half (47%) of the introduced species. On the other hand, woody plants are poorly represented in the exotic (2%) compared to the native (11%) flora. Only four introduced species occur in the subalpine, and none have been collected above treeline,

Table 3. Plant species endemic to the Northern Rocky Mountains.

Allium fibrillum	*Hedysarum sulphurescens*
Antennaria aromatica	*Lathyrus bijugatus*
Angelica dawsonii	*Lomatium sandbergii*
Aquilegia jonesii	*Papaver pygmaeum*
Arnica alpina	*Penstemon albertinus*
Astragalus bourgovii	*Penstemon ellipticus*
Astragalus vexilliflexus	*Penstemon lyallii*
Calochortus apiculatus	*Phacelia lyallii*
Carex platylepis	*Physaria didymocarpa*
Conimitella williamsii	*Physaria saximontana*
Douglasia montana	*Senecio megacephalus*
Epilobium suffruticosum	*Stellaria americana*
Erigeron lackschewitzii	*Suksdorfia violacea*
Erigeron lanatus	*Townsendia parryi*
Eriogonum androsaceum	*Trisetum orthochaetum*

a typical pattern for the Northern Rockies (Forcella and Harvey 1983). Half of the introduced species known in the Park have been found only west of the Continental Divide, while only 10% have been collected only on the east side.

The introduction of plants is considered one of the most serious problems faced by Glacier Park managers (Lange 1991). Although 126 species of exotics have been recorded, only a handful have caused significant adverse effects. Most exotics (60%) have been found only in disturbed habitats such as along roads or around buildings. Roads are particularly important weed staging habitats because of their disturbed periphery and the number of visitors that use them each year (Tyser and Worley 1992, Lesica et al. 1993). Of the remaining 50 species that do occur in native vegetation, only 10 are thought to be able to invade undisturbed communities and replace native species. Grasslands are the most vulnerable vegetation type. Large infestations of spotted knapweed (*Centaurea maculosa*), leafy spurge (*Euphorbia esula*), and St. Johnswort (*Hypericum perforatum*) occur in montane grasslands, while meadow hawkweed (*Hieracium caespitosum*), dalmatian toadflax (*Linaria dalmatica*), and sulphur cinquefoil (*Potentilla recta*) are locally common. Spotted knapweed is currently the most widespread and troublesome weed in the Park (Tyser and Key 1988).

Three other species invade wetlands and riparian areas: Canada thistle (*Cirsium arvense*), tall buttercup (*Ranunculus acris*), and common tansy (*Tanacetum vulgare*). Only Canada thistle is widespread. I have observed this plant, which has light, wind-born seeds, on fresh beaver dams far from roads or even a trail. Reed canarygrass (*Phalaris arundinacea*) is native in the Northern Rockies; however, invasive domestic cultivars capable of replacing native vegetation may also have been introduced (Merigliano and Lesica 1998).

There are currently no serious infestations of exotics in the closed-canopy coniferous forests that dominate Glacier's landscape and much of the Northern Rockies (Forcella and Harvey 1983, Weaver and Woods 1986). However, *Veronica officinalis*, a lawn weed with creeping stems, thrives in dense shade and occurs in undisturbed moist, montane forest in many places west of the Divide in northwest Montana and Glacier Park. It is similar in appearance to our common native, *Linnaea borealis*, and may be able to compete with it in these habitats.

Timothy (*Phleum pratense*), Kentucky bluegrass (*Poa pratensis*), and smooth brome (*Bromus inermis*) are also common in the Park's grasslands, especially on the east side. These exotics have been seeded along roads (Tyser and Worley 1992) and were also probably intentionally introduced by outfitters in the early half of the twentieth century to increase forage for horses. Infestations of these plants may be due more to overgrazing and seeding than to their inherent invasiveness.

REFERENCES

Adams, R. P. 1993. *Juniperus*. Pages 412-20 in Flora of North America Editorial Committee, Flora of North America, Vol. 2. Oxford University Press, New York, NY.

Aiken, S. G. 1981. A conspectus of *Myriophyllum* (Haloragaceae) in North America. Brittonia 37: 57-69.

Aiken, S. G. and S. J. Darbyshire. 1990. Fescue Grasses of Canada. Publication 1844/E, Agriculture Canada, Ottawa, Ontario, Canada.

Arno, S. F. 1986. Whitebark pine cone crops: a diminishing source of wildlife food? Western Journal of Applied Forestry 1: 92-94.

Arno, S. F. and J. R. Habeck. 1972. Ecology of alpine larch (*Larix lyallii* Parl.) in the Pacific Northwest. Ecological Monographs 42: 417-50.

Arno, S. F. and R. P. Hammerly. 1984. Timberline. Mountain and arctic forest frontiers. The Mountaineers, Seattle, WA.

Arnow, L A. 1981. *Poa secunda* Presl versus *Poa sanddbergii* Vasey (Poaceae). Systematic Botany 6: 412-21.

Axelrod, D. I. and P. H. Raven. 1985. Origins of the cordilleran flora. Journal of Biogeography 12: 21-47.

Bamberg, S. A. and J. Major. 1968. Ecology of the vegetation and soils associated with calcareous parent materials in three alpine regions of Montana. Ecological Monographs 42: 417-50.

Bamberg, S. A. and R. H. Pemble. 1968. New records of disjunct arctic-alpine plants in Montana. Rhodora 70: 103-72.

Barkworth, M. E. and D. R. Dewey. 1985. Genomically based genera in the perennial Triticeae of North America: identification and membership. American Journal of Botany 72: 767-76.

Barkworth, M. E. and J. Everett. 1987. Evolution in the Stipeae: identification and relationships of its monophyletic taxa. Pages 251-64 in T. R. Soderstrom, K. W. Hilu, C. S. Campbell and M. E. Barkworth (eds.), Grass systematics and evolution. Smithsonian Institution Press, Washington, D.C.

Barkworth, M. E., J. McNeil and J. Maze. 1979. A taxonomic study of *Stipa nelsonii* (Poaceae) with a key distinguishing it from related taxa in western North America. Canadian Journal of Botany 57: 2539-53.

Barneby, R. C. 1989. Intermountain flora, Volume 3, part B, Fabales. New York Botanical Garden, Bronx, NY.

Barrett, S. W. 1993. Fire history of southeastern Glacier National Park: Missouri River drainage. Glacier National Park, West Glacier, MT.

Barrett, S. W. 1997. Fire history of Glacier National Park: Hudson Bay drainage. Glacier National Park, West Glacier, MT.

Barrett, S. W., S. F. Arno and C. H. Key. 1991. Fire regimes of western larch-lodgepole pine forests in Glacier National Park, Montana. Canadian Journal of Forestry Research 21: 1711-20.

Bayer, R. J. 1990. Patterns of clonal diversity in the *Antennaria rosea* (Asteraceae) polyploid agamic complex. American Journal of Botany 77: 1313-19.

Bayer, R. J. and G. L. Stebbins. 1993. A synopsis with keys for the genus *Antennaria* (Asteraceae: Inuleae: Gnaphaliinae) of North America. Canadian Journal of Botany 71: 1589-1604.

Boivin, B. and D. Löve. 1960. *Poa agassizensis,* a new prairie bluegrass. Canadian Naturalist 88: 173-80.

Boraiah, G. and M. Heimburger. 1964. Cytotaxonomic studies on New World *Anemone* section *Eriocephalus* with woody rootstocks. Canadian Journal of Botany 42: 891-922.

Brooks, R. E. and S. E. Clemants. 2000. *Juncus.* Pages 211-254 in Flora of North America Editorial Committee, Flora of North America, Vol. 22. Oxford University Press, New York, NY.

Brown, G. K. 1993. *Pyrrocoma.* Pages 330-331 in J. C. Hickman (ed.), The Jepson Manual. University of California Press, Berkeley, CA.

Butler, D. R. 1979. Snow avalanche path terrain and vegetation, Glacier National Park, Montana. Arctic and Alpine Research 11: 17-32.

Cain, S. A. 1944. Foundations of plant geography. Harper Brothers, New York, NY.

Carrara, P. E. 1989. Late Quaternary glacial and vegetative history of the Glacier National Park region, Montana. U. S. Geological Survey Bulletin 1902, Denver, CO.

Chambers, K. L. 1993. *Claytonia.* Pages 898-900 in J. C. Hickman (ed.), The Jepson Manual. University of California Press, Berkeley, CA.

Chambers, K. L. and S. Sundberg. 1998. Oregon vascular plant checklist: Asteraceae. Oregon Flora Project, Oregon State University, Corvallis, OR.

Chinnappa, C. C. and J. K. Morton. 1991. Studies on the *Stellaria longipes* complex (Caryophyllaceae) — taxonomy. Rhodora 93: 129-35.

Choate, C. M. and J. R. Habeck. 1967. Alpine plant communities at Logan Pass, Glacier National Park. Proceedings of the Montana Academy of Sciences 27: 36-54.

Cronquist, A. 1955. Vascular plants of the Pacific Northwest. Part 5: Compositae. University of Washington Press, Seattle, WA.

Cronquist, A., A. H. Holmgren, N. H. Holmgren, P. K. Holmgren, and R. Barneby. 1986-97. Intermountain Flora, Volumes 1-6. New York Botanical Garden, Bronx, NY.

Cronquist, A., N. H. Holmgren and P. K. Holmgren. 1997. Onagraceae. Pages 172-244 in A. Cronquist, N. H. Holmgren and P. K. Holmgren, Intermountain Flora Volume 3, Part A, Subclass Rosidae. New York Botanical Garden, Bronx, NY.

Davis, M. B. 1981. Quaternary history and the stability of forest communities. Pages 134-153 in D. C. West et al. (eds.), Forest Succession: concepts and applications. Springer-Verlag, New York, NY.

Dean, M. L. 1966. A biosystematic study in the genus *Aster*, section Aster, in western North America. Ph.D. dissertation, Oregon State University, Corvallis, OR.

DeBolt, A. and P. Lesica. 1986. Alpine plants of Glacier National Park. Pages 58-62 in J. Williams (ed.), Rocky Mountain alpines. Denver Botanic Gardens, Denver, CO.

DeBolt, A. and B. McCune. 1993. Lichens of Glacier National Park, Montana. Bryologist 96: 192-204.

DeByle, N. V., C. D. Bevins and W. C. Fischer. 1987. Wildfire occurrence in aspen in the interior western United States. Western Journal of Applied Forestry 2: 73-76.

DeSanto, J. 1989. Alpine wildflowers of Glacier National Park, Montana and Waterton Lakes National Park, Alberta. Privately published.

DeSanto, J. 1993. Bitterroot. Lere Press, Babb, MT.

Dorn, R. D. 1984. Vascular plants of Montana. Mountain West Publishing, Cheyenne, WY.

Dorn, R. D. 1995. A taxonomic study of *Salix* section Cordatae subsection Luteae (Salicaceae) Brittonia 47: 160-74.

Dorn, R. D. and J. L. Dorn. 1997. Rocky Mountain region willow identification field guide. USDA Forest Service Region 2 Publication RR-97-01, Denver, CO.

Dunlop, D. A. and G. E. Crow. 1999. The taxonomy of *Carex* section Scirpinae (Cyperaceae). Rhodora 101: 163-99.

Elisens, W. J. and J. G. Packer. 1980. A contribution to the taxonomy of the *Oxytropis campestris* complex in northwestern North America. Canadian Journal of Botany 58: 1820-31.

Elvander, P. E. 1984. The taxonomy of *Saxifraga* (Saxifragaceae) Section Boraphila Subsection Integrifoliae in western North America. Systematic Botany Monographs 3: 1-44.

Finklin, A. I. 1986. A climatic handbook for Glacier National Park with data for Waterton Lakes National Park. USDA Forest Service Intermountain Research Station General Technical Report INT-204. Ogden, UT.

Flora of North America Editorial Committee. 1993-2000. Flora of North America, Volumes 1-3, 22. Oxford University Press, New York, NY.

Foote, G. G. 1965. Phytosociology of the bottomland hardwood forests in western Montana. M. A. thesis, University of Montana, Missoula.

Forcella, F. and S. J. Harvey. 1983. Eurasian weed infestation in western Montana in relation to vegetation and disturbance. Madrono 30: 102-9.

Fritz-Sheridan, J. K. 1988. Reproductive biology of *Erythronium grandiflorum* varieties *grandiflorum* and *candidum* (Liliaceae). American Journal of Botany 75: 1-14.

Furlow, J. J. 1997. Betulaceae. Pages 507-538 in Flora of North America Editorial Committee, Flora of North America, Vol. 3. Oxford University Press, New York, NY.

Gillett, J. M. 1957. A revision of the North American species of *Gentianella* Moench. Annals of the Missouri Botanical Garden 44: 195-269.

Gleason, H. A. and A. Cronquist. 1964. The natural geography of plants. Columbia University Press, New York, NY.

Gould, F. W. 1947. Nomenclatural changes in *Elymus* with a key to the California species. Madrono 9: 120-27.

Graff, P. W. 1922. Unreported plants for Glacier National Park. Bulletin of the Torrey Botanical Club 49: 175-81.

Greene, C. W. 1993. *Calamagrostis*. Pages 1243-46 in J. C. Hickman (ed.), The Jepson Manual. University of California Press, Berkeley, CA.

Habeck, J. R. 1968. Forest succession in the Glacier Park cedar-hemlock forest. Ecology 49: 872-80.

Habeck, J. R. 1969. A gradient analysis of a timberline zone at Logan Pass, Glacier Park, Montana. Northwest Science 43: 65-73.

Haber, E. and J. E. Cruise. 1974. Generic limits in the Pyroloideae (Ericaceae). Canadian Journal of Botany 52: 877-83.

Hartman, R. L. 1993. Caryophyllaceae. Pages 475-497 in J. C. Hickman (ed.), The Jepson Manual. University of California Press, Berkeley, CA.

Harvey, L. H. 1954. Additions to the flora of Glacier National Park, Montana. Proceedings of the Montana Academy of Sciences 14: 23-25.

Hauke, R. L. 1993. Equisitaceae. Pages 76-84 in Flora of North America Editorial Committee, Flora of North America, Vol. 2. Oxford University Press, New York, NY.

Hawksworth, F. G. and D. Wiens. 1972. Biology and classification of dwarf mistletoes (Arceuthobium). USDA Agriculture Handbook No. 401. Washington, D.C.

Hermann, F. J. 1956. Range extensions in northwestern plants. Rhodora 58: 278-79.

Hermann, F. J. 1969. The bryophytes of Glacier National Park, Montana. Bryologist 72: 358-76.

Hermann, F. J. 1970. Manual of the carices of the Rocky Mountains and Colorado Basin. USDA Forest Service Agriculture Handbook No. 374, Washington, D.C.

Hermann, F. J. 1975. Manual of the rushes (*Juncus* spp.) of the Rocky Mountains and Colorado Basin. USDA Forest Service General Technical Report RM-18, Fort Collins, CO.

Hickman, J. C. (ed). 1993. The Jepson Manual. Higher Plants of California. University of California Press, Berkeley, CA.

Hickman, J. C. 1993. *Polygonum*. Pages 886-891 in J. C. Hickman (ed.), The Jepson Manual. University of California Press, Berkeley, CA.

Hitchcock, A. S. and A. Chase. 1950. Manual of grasses of the United States, second edition. USDA Misc. Pub. 200, Washington, D.C.

Hitchcock, C. L. and A. Cronquist. 1961. Vascular plants of the Pacific Northwest. Part 3: Saxifragaceae to Ericaceae. University of Washington Press, Seattle, WA.

Hitchcock, C. L. and A. Cronquist. 1964. Vascular plants of the Pacific Northwest. Part 2: Salicaceae to Saxifragaceae. University of Washington Press, Seattle, WA.

Hitchcock, C. L., A. Cronquist and M. Ownbey. 1969. Vascular plants of the Pacific Northwest. Part 1: Vascular cryptogams, gymnosperms and monocotyledons. University of Washington Press, Seattle, WA.

Hitchcock, C. L. and A. Cronquist. 1973. Flora of the Pacific Northwest. University of Washington Press, Seattle, WA.

Holmgren, N. H. 1997. Rosaceae. Pages 64-158 in A. Cronquist, N. H. Holmgren and P. K. Holmgren, Intermountain Flora Volume 3, Part A, Subclass Rosidae. New York Botanical Garden, Bronx, NY.

Hultén, E. 1937. Outline of the history of arctic and boreal biota during the Quaternary Period. Verlag Von Cramer, New York, NY.

Hultén, E. 1968. Flora of Alaska and neighboring territories. Stanford University Press, Stanford, CA.

Iltis, H. H. 1965. The genus *Gentianopsis* (Gentianaceae): transfers and phytogeographic comments. Sida 2: 129-54.

Isely, D. 1998. Native and naturalized Leguminosae (Fabaceae) of the United States. M. L. Bean Museum, Brigham Young University, Provo, UT.

Johnston, A. 1987. Plants and the Blackfoot. Occasional Paper No. 15, Lethbridge Historical Society, Lethbridge, Alberta, Canada.

Jones, M. E. 1910. Montana botany notes. Biological Series No. 15, University of Montana, Missoula, MT.

Keane, R. E., S. F. Arno. 1993. Rapid decline of whitebark pine in western Montana: evidence from 20-year remeasurements. Western Journal of Forestry 8: 44-47.

Keil, D. J. et al. 1993. Asteraceae [Compositae]. Pages 174-360 in J. C. Hickman (ed.), The Jepson Manual. University of California Press, Berkeley, CA.

Kellogg, E. A. 1985. A biosystematic study of the *Poa secunda* complex. Journal of the Arnold Arboretum 66: 201-42.

Kendall, K. C. 1998. Whitebark pine. Pages 483-85 in M. J. Mac et al. (eds.), Status and trends of the nation's biological resources. USDI U. S. Geological Survey, Reston, VA.

Koterba, W. D. and J. R. Habeck. 1971. Grasslands of the North Fork Valley, Glacier National Park, Montana. Canadian Journal of Botany 49: 1627-36.

Kuijt, J. 1982. A flora of Waterton Lakes National Park. University of Alberta Press, Edmonton, Alberta, Canada.

Lackschewitz, K., P. Lesica and J. S. Shelly. 1988. Noteworthy collections: Montana. Madrono 35: 355-58.

Lange, D. E. 1991. Exotic vegetation management plan. Glacier National Park, West Glacier, MT.

Lesica, P. 1985. Annotated checklist of vascular plants of Glacier National Park, Montana. Monograph No. 4, Montana Academy of Sciences, Supplement to the Proceedings Vol. 44.

Lesica, P. 1991. Noteworthy Collections: Montana. Madrono 38: 297-98.

Lesica, P. 1996. Checklist of the vascular plants of Glacier National Park, Montana. Second edition. Glacier National History Association, West Glacier, Montana.

Lesica, P., G. Moore, K. M. Peterson and J. H. Rumely. 1984. Vascular plants of limited distribution in Montana. Monograph No. 2, Montana Academy of Sciences, Supplement to the Proceedings Vol. 43.

Lesica, P. and P. F. Stickney. 1994. Noteworthy collections: Montana. Madrono 41: 229-31.

Lesica, P., K. Lackschewitz, J. Pierce, S. Gregory and M. O'Brien. 1986. Noteworthy collections: Montana. Madrono 33: 310-12.

Lesica, P., K. Ahlenslager and J. DeSanto. 1993. New vascular plant records and the increase of exotic plants in Glacier National Park, Montana. Madrono 40: 126-31.

Lesica, P., P. Husby and S. V. Cooper. 1998. Noteworthy collections: Montana. Madrono 45: 328-30.

Lewis, H. and J. Szweykowski. 1964. The genus *Gayophytum* (Onagraceae). Brittonia 16: 343-91.

Löve, A. and D. Löve. 1974. Origin and evolution of the arctic and alpine floras. Pages 571-604 in J. D. Ives and R. G. Barry (eds.), Arctic and alpine environments. Methuen, London, Great Britain.

Luer, C. A. 1975. The native orchids of the United States and Canada. New York Botanical Garden, Bronx, NY.

Lunan, J. S. and J. R. Habeck. 1973. The effects of fire exclusion on ponderosa pine communities in Glacier National Park, Montana. Canadian Journal of Forestry Research 3: 574-79.

Lynch, D. 1955. Ecology of the aspen groveland in Glacier County, Montana. Ecological Monographs 25: 321-44.

Maguire, B. 1934. Distributional notes concerning plants of Glacier National Park, Montana. Rhodora 36: 305-8.

Maguire, B. 1939. Distributional notes concerning plants of Glacier National Park, Montana. II. Rhodora 41: 504-8.

Malanson, G. P. and D. R. Butler. 1991. Floristic variation among gravel bars in a subalpine river, Montana, U. S. A. Arctic and Alpine Research 23: 273-78.

Mantas, M. 1999. Vascular plant checklist for the Flathead National Forest, Montana. Flathead National Forest, Kalispell, MT.

McGraw, J. B. 1985. Experimental ecology of *Dryas octopetala* ecotypes. III. Environmental factors and plant growth. Arctic and Alpine Research 17: 229-39.

McLaughlin, S. P. 1989. Natural floristic areas of the western United States. Journal of Biogeography 16: 239-48.

McLaughlin, W. T. 1935. Notes on the flora of Glacier National Park, Montana. Rhodora 37: 362-65.

Merigliano, M. F. and P. Lesica. 1998. The native status of reed canary grass in the Inland Northwest, U. S. A. Natural Areas Journal 18: 223-30.

Morton, J. K. and R. K. Rabeler. 1989. Biosystematic studies on the *Stellaria calycantha* (Caryophyllaceae) complex. I. cytology and cytogeography. Canadian Journal of Botany 67: 121-27.

Moss, E. H. and J. G. Packer. 1983. Flora of Alberta, second edition. University of Toronto Press, Toronto, Ontario, Canada.

Muir, P. S. and J. E. Lotan. 1985. Disturbance history and serotiny of *Pinus contorta* in western Montana. Ecology 66: 1658-68.

Mulligan, G. A. 1995. Synopsis of the genus *Arabis* (Brassicaceae) in Canada, Alaska and Greenland. Rhodora 97: 109-63.

Murray, D. F. 1969. Taxonomy of *Carex* sect. Atratae (Cyperaceae) in the Southern Rocky Mountains. Brittonia 21: 55-76.

Myers, A. A. and P. S. Giller (eds.). 1988. Analytical biogeography. Chapman and Hall, London, Great Britain.

Ownbey, M. 1959. *Castilleja.* Pages 295-326 in Hitchcock, C. L., A. Cronquist, M. Ownbey and J. W. Thompson, Vascular plants of the Pacific Northwest. Part 4: Ericaceae to Campanulaceae. University of Washington Press, Seattle, WA.

Perry, E. S. 1962. Montana in the geologic past. Bulletin 26, Montana Bureau of Mines and Geology, Butte, MT.

Pfister, R. D., B. L. Kovalchik, S. F. Arno and R. C. Presby. 1977. Forest habitat types of Montana. USDA Forest Service General Technical Report INT-34, Ogden, UT.

Porsild, A. E. and W. J. Cody. 1980. Vascular plants of continental Northwest Territories, Canada. National Museums of Canada, Ottawa, Ontario, Canada.

Price, R. A. 1993. *Draba.* Pages 416-20 in J. C. Hickman (ed.), The Jepson Manual. University of California Press, Berkeley, CA.

Pringle, J. S. 1997. *Clematis.* Pages 158-76 in Flora of North America Editorial Committee, Flora of North America, Vol. 3. Oxford University Press, New York, NY.

Raupp, O. B., R. L. Earhart, J. W. Whipple and P. E. Carrara. 1983. Geology along Going-to-the-Sun Road, Glacier National Park, Montana. Glacier Natural History Association, West Glacier, MT.

Rollins, R. C. 1993. The Cruciferae of continental North America. Stanford University Press, Stanford, CA.

Rydberg, P. A. 1910. Studies of the Rocky Mountain flora XXI. Bulletin of the Torrey Botanical Club 37: 144.

Schuyler, A. E. 1980. *Carex chordorrhiza* in Glacier National Park, Montana. Rhodora 82: 519.

Schuyler, A. E. 1983. Distributional notes on northwestern Montana aquatic vascular plants - 1982. Bartonia 49: 52-54.

Semple, J. C. 1996. A revision of *Heterotheca* sect. *Phyllotheca* (Nutt.) Harms (Compositae: Astereae): the prairie and montane goldenasters of North America. University of Waterloo Biological Series 37: 1-164.

Shinwari, Z. K., R. Terauchi, F. H. Utech and S. Kawano. 1994. Recognition of the New World *Disporum* section *Prosartes* as *Prosartes* (Liliaceae) based on the sequence data of the rbcL gene. Taxon 43: 353-66.

Sinnott, Q. P. 1985. A revision of *Ribes* subg. *Grossularia* (Mill.) pers. sect. *Grossularia* (Mill.) Nutt. (Grossulariaceae) in North America. Rhodora 87: 189-286.

Smith, A. R. 1993a. Dryopteridaceae. Pages 246-307 in Flora of North America Editorial Committee, Flora of North America, Vol. 2. Oxford University Press, New York, NY.

Smith, A. R. 1993b. Polypodiaceae. Pages 312-30 in Flora of North America Editorial Committee, Flora of North America, Vol. 2. Oxford University Press, New York, NY.

Smith, A. R. 1993. Thelypteridaceae. Pages 206-22 in Flora of North America Editorial Committee, Flora of North America, Vol. 2. Oxford University Press, New York, NY.

Soltis, D. E., R. K. Kuzoff, E. Conti, R. Gornall and K. Ferguson. 1996. MatK and rbcL gene sequence data indicate that *Saxifraga* (Saxifragaceae) is polyphyletic. American Journal of Botany 83: 371-82.

Soreng, R. J. 1985. *Poa* L. in New Mexico, with a key to middle and southern Rocky Mountain species (Poaceae). Great Basin Naturalist 45: 395-422.

Soreng, R. J. 1991. Notes on infraspecific taxa and hybrids in North American *Poa* (Poaceae). Phytologia 71: 390-413.

Standley, L. A. 1985. Systematics of the Acutae group of *Carex* (Cyperaceae) in the Pacific Northwest. Systematic Botany Monographs 7: 1-106.

Standley, P. C. 1921. Flora of Glacier National Park, Montana. Contributions to the U. S. National Herbarium 22: 235-438.

Strong, M. T. 1994. Taxonomy of *Scirpus, Trichophorum* and *Schoenoplectus* (Cyperaceae) in Virginia. Bartonia 58: 29-68.

Swab, J. C. 2000. *Luzula.* Pages 255-267 in Flora of North America Editorial Committee, Flora of North America, Vol. 22. Oxford University Press, New York, NY.

Taylor, R. L. 1965. The genus *Lithophragma* (Saxifragaceae). University of California Publications in Botany 37: 1-122.

Tutin, T. G., V. H. Heywood, N. A. Burges, D. M. Moore, D. H. Valentine, S. M. Walters and D. A. Webb. 1976. Flora Europaea, Vol. 4. Cambridge University Press, Cambridge, Great Britain.

Tyser, R. W. and C. H. Key. 1988. Spotted knapweed in natural area fescue grasslands: an ecological assessment. Northwest Science 62: 151-59.

Tyser, R. W. and C. A. Worley. 1992. Alien flora in grasslands adjacent to road and trail corridors in Glacier National Park, Montana (U. S. A.). Conservation Biology 6: 253-62.

Valdespino, I. A. 1993. Selaginellaceae. Pages 38-63 in Flora of North America Editorial Committee, Flora of North America, Vol. 2. Oxford University Press, New York, NY.

Vander Kloet, S. P. 1988. The genus *Vaccinium* in North America. Canadian Government Publishing Centre, Ottawa, Ontario, Canada.

Wagner, W. H. and J. M. Beitel. 1993. Lycopodiaceae. Pages 18-37 in Flora of North America Editorial Committee, Flora of North America, Vol. 2, Oxford University Press, New York, NY.

Wagner, W. H., R. C. Moran and C. R. Werth. 1993. Aspleniaceae. Pages 228-45 in Flora of North America Editorial Committee, Flora of North America, Vol. 2, Oxford University Press, New York, NY.

Wagner, W. H. and F. S. Wagner. 1993. Ophioglossaceae. Pages 85-106 in Flora of North America Editorial Committee, Flora of North America, Vol. 2, Oxford University Press, New York, NY.

Warnock, M. J. 1997. *Delphinium*. Pages 196-240 in Flora of North America Editorial Committee, Flora of North America, Vol. 3. Oxford University Press, New York, NY.

Weaver, T. 1980. Climates of vegetation types of the Northern Rocky Mountains and adjacent plains. American Midland Naturalist 103: 392-98.

Weaver, T. and B. Woods. 1988. Exotic plant invasion of *Thuja-Tsuga* habitats, N.W. Montana. Pp. 111-19 in L. K. Thomas (ed.), Management of exotic species in natural communities, Vol. 5, Proceedings of the Conference on Science in the National Parks. U. S. National Park Service, Fort Collins, CO.

Weber, W. A. 1987. Colorado flora: western slope. Colorado University Associated Press, Boulder, CO.

Welsh, S. L., N. D. Atwood, L. C. Higgins and S. Goodrich. 1987. A Utah flora. Great Basin Naturalist Memoir 9: 1-894.

Whittemore, A. T. 1997. *Ranunculus*. Pages 88-135 in Flora of North America Editorial Committee, Flora of North America, Vol. 3. Oxford University Press, New York, NY.

Windham, M. D. 1993. Pteridaceae. 1993. Pages 122-86 in Flora of North America Editorial Committee, Flora of North America, Vol. 2. Oxford University Press, New York, NY.

Zwinger, A. H. and B. E. Willard. 1972. Land above the trees. University of Arizona Press, Tucson, AZ.

GLOSSARY

Abortive. Partly or completely undeveloped or unformed.

Achene. Dry, 1-seeded fruit with a hard coat; achenes often resemble seeds.

Adventive. Introduced, not native but self-perpetuating without cultivation.

Alkaline. Salty with the dominant salt being calcium carbonate; alkaline soils are often derived from limestone or dolomite.

Alpine. Above the altitudinal limit of trees.

Alternate. Arranged 1-per-node; see opposite or whorled.

Ament. A small, tightly clustered spike of unisexual flowers with bracts but lacking petals and sepals; catkin.

Annual. A plant that completes its lifecycle in 1 year or less.

Anther. The sac at the end of the stamen that contains the pollen.

Aphyllous. Without true leaves but sometimes with rudimentary bracts.

Apical. Of the apex or tip.

Appressed. Lying pressed up against a surface or structure.

Arcuate. Curved like a bow.

Asexual. Reproducing vegetatively without the sexual unification of egg and sperm.

Awn. A stiff bristle-like appendage.

Axil. Juncture between a leaf and stem or any two organs.

Axillary. In the axil.

Axis. A central straight line or linear structure about which there is radial symmetry.

Banner. The upper petal of a legume (Fabaceae) flower, often large and partly folded up.

Barb. A swollen point like a fish hook.

Beak. A firm, elongate structure tip or projection.

Biennial. A plant that is vegetative its first year, then flowers and dies the second; sometimes such plants may live more than 2 years, but flowering is always followed by death.

Bilateral symmetry. Capable of only a single division into two halves that are mirror images of each other (see radial symmetry).

Bisexual. Both male and female flower parts in the same flower (bisexual flower) or on the same plant (bisexual plant).

Blade. The usually broad and sometimes flat portion of the leaf; does not include the stem-like petiole, if one exists.

Boreal. Referring to the northern part of the northern hemisphere.

Bract. Reduced or modified leaf subtending a flower or group of flowers.

Bulb. Underground bud covered by fleshy scales, like an onion.

Bulblet. A small bulb-like organ produced with the roots or among the flowers.

Calcareous. With high concentrations of calcium; usually referring to soils derived from limestone or dolomite or water that has been in contact with these rocks or soils.

Callus. The hardened base of a grass flower.

Calyx. The collective sepals; the outermost, usually green series of flower parts.

Canopy. Referring to the branches and leaves above; e.g., grasses form a canopy above mosses but shrubs form a canopy above the grasses.

Capsule. A dry, many-seeded fruit with more than 1 chamber, at least at the base.

Carr. Shrub-dominated vegetation with wet, peaty soil.

Catkin. A small, tightly clustered spike of unisexual flowers with bracts but lacking petals and sepals; ament.

Chlorophyll. The green pigment in most plants that captures light energy to produce sugars from water and CO_2.

Circumboreal. Occurring around the globe in the northern part of the northern hemisphere; i.e., across Canada, Alaska, and northern Europe and Asia.

Circumpolar. Around the North Pole.

Clasping. Partly surrounding; the base of a clasping leaf may go halfway around the stem rather than being attached at a single point.

Cleistogamous. Referring to a flower that never opens and is self-fertilized, often below ground.

Clonal. Referring to a clone.

Clone. A colony of plants that arose vegetatively (without sexual reproduction) from a single individual, often through underground stems (rhizomes).

Collar. Top of a grass leaf sheath where it joins the blade.

Compound. Composed of two or more similar elements; a compound leaf is divided into two or more leaflets; a twice compound leaf is divided into leaflets that are themselves divided into leaflets.

Concave. Shallowly hollowed out like a saucer or bowl.

Cone. A usually cylindrical reproductive structure consisting of a central axis with spirally arranged scales, each subtending seed- or pollen-bearing structures; occurring in non-flowering plants.

Congeneric. In the same genus.

Congested. Crowded.

Corm. Bulb-like enlargement of stem base.

Corky. Spongy but dry.

Corolla. The bowl- or tube-like union of the petals; the petals collectively.

Crown. The collective branches and leaves of a tree. The top of a structure; e.g., the upper portion of the root where it joins the stem is the root crown.

Deciduous. Falling off, usually at the end of the growing season.

Decumbent. Basal portion on the ground but becoming vertical toward the tip.

Dicot. A plant whose germinated seedlings have 2 leaf-like cotyledons before the first true leaves are formed; true leaves have net-like venation; flower parts usually in 2s or 4s or 5s.

Dimorphic. With two forms.

Dioecious. Plants with either male or female flowers only.

Discoid. In the Asteraceae; flower heads with disk flowers only or apparently so.

Disk. In the Asteraceae; the part of the flower head composed of disk flowers.

Disk flower. In the Asteraceae; a flower with a tubular corolla.

Dissected. Divided into many segments.

Drupe. A fruit with a fleshy exterior and a stony interior surrounding a single seed; e.g., a peach.

Drupelet. Small drupe.

Early successional. Occurring in plant assemblages occurring soon after a disturbance such as fire or a landslide or on newly created substrate such as glacial moraine.

Ecotype. A genetically distinct population or group of populations adapted to a particular environment; e.g., different ecotypes of Douglas fir occur in the relatively wet Cascade Mtns. and the drier Rocky Mtns.

Ellipsoid. Elliptical in outline, like a football.

Elliptic. A rounded figure that is about 1.5 times as long as wide, like the outline of a football.

Elongate. Lengthened, long.

Emergent. At least the upper portion of the plant emerging above the water's surface.

Endemic. Restricted in occurrence to a particular area; in this book usually the Northern Rocky Mtns.

Entire. With a smooth edge, without teeth, lobes or divisions.

Equitant. Each leaf folded in half lengthwise and partly enfolding the base of the one above; like an iris.

Evergreen. Remaining green through one or more winters.

Exotic. Introduced; indigenous or native outside of a particular area; for this book, outside the Northern Rocky Mtns., but usually outside of North America.

Exserted. Projecting beyond the surrounding organ.

Fascicles. Clusters or groups.

Fellfield. Wind-exposed habitat with well-drained, stony soil and sparse vegetation dominated by cushion plants, low grass-like plants and lichens.

Fen. Vegetation dominated by grass-like plants with wet, peaty soil.

Fertile. Bearing flowers or reproductive structures as opposed to vegetative shoots that have no flowers.

Filament. The stalk of a stamen which usually has an anther on its tip.

Filiform. Thread-like.

Floret. A grass flower composed of the lemma, palea and enclosed male and female parts.

Follicle. A dry, usually many-seeded fruit similar to a capsule but with only 1 chamber.

Foliolate. Referring to the number of leaves or leaflets.

Frond. A fern leaf.

Fruit. A mature ovary including seeds and any enclosing tissue.

Funnelform. With the shape of a funnel or cone, broad at the top and narrowed to the base.

Galea. A hood-shaped flower part, formed by the union of the upper 2 petals in the Scrophulariaceae.

Gametophyte. In the ferns and related families; within the same species, the plant that produces spores (sporophyte) is separate from that which produces eggs and sperm (gametophyte). These two "generations" alternate; gametophytes produce eggs and sperm that unite to form a sporophyte that produces spores that germinate to become gametophytes.

Gemmae. Tiny, wing-like, rudimentary plants produced vegetatively.

Genus. A group of related species, or sometimes a genus may contain only 1 species.

Glabrate. At first hairy but becoming hairless.

Glabrous. Hairless and smooth.

Gland. An organ usually recognized by its secretion of tiny droplets of oil, resin, etc., often at the ends of hairs.

Glandular. Bearing glands.

Glaucous. Covered with a white or bluish, waxy powder that rubs off; e.g., a plum.

Globose. Spherical, like a globe.

Glume. One of two scale-like bracts at the base of a grass spikelet composed of 1-many flowers (florets).

Grassland. Vegetation dominated by grasses.

Head. A cluster of unstalked flowers or fruits at the tip of a stalk.

Heath. Vegetation dominated by low, often evergreen shrubs in the Ericaceae.

Herbage. Vegetative, usually green plant parts; leaves and stems.

Herbs. Non-woody plants with stems that die back to the ground each winter.

Herbaceous. Dominated by non-woody plants.

Hip. Fleshy, spherical or ellipsoid fruit of a rose plant (Rosa).

Humic. Highly organic; humic soil is composed mainly of decomposing vegetation.

Hummock. A small hill about the size of a basketball or automobile tire.

Hybrid. Progeny of two dissimilar parents, usually two different species.

Hypanthium. Tube- or cup-like structure formed by fusion of the basal portions of the calyx, corolla and stamens.

Incised. Deeply and sharply cut into angular divisions.

Indusium. Membranous outgrowth of a fern leaf margin that at least partly covers the spores.

Inferior ovary. Petals, sepals and stamens are attached at the top of an inferior ovary (see superior ovary).

Inflorescence. Flowering portion of the plant.

Inodorous. Without smell.

Insectivorous. Trapping and digesting insects and other small animals.

Intergrade. Characters of one species merge toward those of another species in some areas, sometimes due to hybridization.

Involucel. Series of bracts subtending the umbels of the umbel (secondary umbels of a compound umbel).

Involucral. Pertaining to an involucre.

Involucre. A series of bracts surrounding a flower or flower cluster.

Keel. A ridge along the outside axis of a fold, midrib or union of two sides. In the Fabaceae; the boat-shaped union of the two lowest petals of a legume flower.

Lanceolate. Lance-shaped; longer than wide, widest near the base with rounded edges, narrowed to a point at the tip.

Lateral. At the side, along the margin.

Leaflet. A separate, leaf-like segment of a compound leaf. Leaves are attached to stems that have a bud at the tip; leaflets are attached to the axis (rachis) of a compound leaf, which has a leaflet or coiled tendril at the tip instead of a bud.

Lemma. The outermost of the pair of scale-like grass flower parts surrounding the ovary and/or stamens.

Ligulate. In the Asteraceae; a head with all strap-shaped ray flowers; disk flowers are absent.

Ligule. Commonly in grasses and sedges; a short-hairy or membranous projection on the inside of the leaf sheath at the point where it joins the blade.

Linear. Long and narrow and often straight.

Lobe. Rounded or pointed segment of a leaf, petal or fruit that projects or stands out from the margin.

Local. Occurring in only a limited number of places; the opposite of widespread.

Locally common. Found in only a few places but common in those places.

Loment. In the Fabaceae; an elongate fruit (legume) with a single series of seeds and a constriction between each seed.

Longitudinal. Pertaining to the long axis of a 2-dimensional figure; lengthwise.

Marsh. Wetland with predominantly mineral (non-organic) soil dominated by non-woody plants.

Meadow. Moist, treeless vegetation with an abundance of broad-leaved, herbaceous plants and/or sedges.

Megaspore. The larger of two spore types produced by some non-flowering plants (see microspore).

Membranous. Papery, thin and partly transparent.

Mericarp. 1-seeded fruit that is a part of what appears to be the whole fruit (schizocarp).

Mesic. Pertaining to optimal moisture for plant growth; not too dry and not too wet.

Microspore. The smaller of two spore types produced by some non-flowering plants (see megaspore).

Midrib. The central vein of a leaf, leaflet, scale or bract.

Monocot. A plant whose germinated seedlings have a single leaf-like cotyledon before the first true leaves are formed; leaves undivided with parallel veins; flower parts usually in 3's or 6's.

Monoculture. Vegetation composed of a single species or nearly so.

Montane. Referring to the lowest vegetation zone defined by elevation in Glacier Park; foothills and lower mountain (see Introduction).

Moraine. Stony detritus left behind by glacial advances and retreats.

Morphology. Form and structure.

Mycorrhizae. Symbiotic relationship in which a fungus penetrates plant roots obtaining carbon compounds from the plant while donating water and nutrients.

Naked. Lacking some structures such as hairs, scales, leaves; e.g, a naked stem lacks leaves.

Node. A stem "joint," the point of attachment for leaves or side branches.

Nutlet. Hard-coated, 1-seeded fruit that does not open upon maturity.

Oblong. Longer than wide with nearly parallel sides and blunt ends.

Obovate. Egg-shaped with the narrow end at the base.

Opposite. Paired at the nodes; in opposite leaves there are 2 leaves arising opposite each other at a single node on the stem.

Orbicular. Circular or nearly so.

Ovate. Egg-shaped in outline; elliptic but with one wide end and one narrow end.

Ovoid. Egg-shaped.

Palea. The innermost of the pair of scale-like grass flower parts surrounding the ovary and/or stamens.

Palmate. Leaflets arranged with their stalks all attached at a single point on the tip of the petiole, like fingers to the palm of a hand.

Pappus. In the Asteraceae; the outer series of flower parts (calyx), usually composed of several to many bristles or scales attached to the top of the fruit.

Parasitic. Gaining water and/or nutrients through their roots from another plant.

Parasitize. To be a parasite on another plant.

Pemmican. Meat and berries combined and dried. Native Americans made pemmican to preserve meat.

Pendant. Hanging downward.

Perennial. A plant that persists through at least one winter and flowers in at least two growing seasons.

Perigynium. In the genus Carex; the sac-like flower part surrounding the seed-like achene.

Persistent. Remaining attached for a long time, not readily falling away.

Petal. One member of the series of flower parts just inside the sepals, often colored and showy; collectively the corolla.

Petiolate. Having a petiole.

Petiole. The stalk of a leaf, attaching the leaf blade to the stem.

Pinnae. In fern families; the primary divisions of the frond.

Pinnate. With leaflets, lobes or veins arising along opposite sides of a long axis.

Pinnule. In the fern families; the ultimate divisions of the frond.

Pistil. Collectively the female organs of a flower, including one or more ovaries, styles and stigmas.

Pistillate. Referring to a flower with female but without male parts.

Plaits. Flattened folds as in an accordion.

Pod. A dry fruit that opens at maturity.

Pollen. Dust-like particles that contain sperm.

Polyploidy. Having two or more times the standard or primitive number of chromosomes, often associated with hybridization.

Pome. A fleshy fruit with several seed chambers; e.g., an apple.

Prostrate. Lying on the ground.

Pubescence. Hairiness.

Pubescent. Hairy.

Radial symmetry. Symmetrical about a central point so that any division through that point produces two halves that are mirror images of each other (see bilateral symmetry).

Radiate. In the Asteraceae; pertaining to flower heads that contain both strap-shaped ray flowers on the outside and disk flowers in the center of the head.

Ray. In the Asteraceae; a flower with a strap-shaped corolla.

Receptacle. Expanded end of the flower stalk to which are attached 1-many flowers; in the Asteraceae; the base of the flower head from which the flowers arise.

Recurved. Bent downward.

Reduced. Smaller than usual.

Reflexed. Bent abruptly downward.

Rhizomatous. Having rhizomes.

Rhizome. An underground, often horizontal stem that may give rise to many individual-appearing plants.

Root. The underground portion of a plant, parts of which may store food absorb water and nutrients from the soil.

Root crown. An underground stem base that surmounts the root(s).

Rosette. A cluster of leaves all radiating out from a central point of attachment.

Scale. A small, thin structure, like a fish scale.

Schizocarp. A compound fruit that splits into 2 to many, 1-seeded fruits (mericarps).

Scree. A substrate of sandy soil and small loose stones.

Seed. Mature egg containing the embryo.

Sepal. A member of the outer series of flower parts, collectively the calyx, often green.

Serotinous. Late; usually flowering or opening late; serotinous cones open many years after maturity.

Sessile. Without a stalk.

Sheath. A tubular covering; in grasses the leaf sheath surrounds and encloses the stem.

Simple. Not divided or lobed.

Sinus. Notch between two lobes or teeth.

Slough. Body of water or wetland adjacent to a river or stream.

Sorus (plural, sori). In the fern families; cluster of spores.

Spade-shaped. Broadly heart-shaped; shaped like the spade symbol in a deck of playing cards.

Spatulate. Spatula shaped, strap-shaped with a rounded end and tapered gradually to the base.

Species. A group of closely related populations that share distinct characters that separate them from other such groups.

Sphagnum. A distinct group (genus) of mosses found primarily in wetlands with organic soils; they acidify the soil and are capable of existing with very low levels of nutrients.

Spike. An elongate inflorescence bearing unstalked flowers.

Spikelet. A small spike in the grass and sedge families.

Sporangiophore. Stem branches bearing sporangia.

Sporangium. A spore case.

Spore. A simple 1-celled reproductive body, capable of giving rise to a new individual.

Sporophore. In the Ophioglossaceae; the distinct portion of the frond that bears the spores.

Sporophyll. Spore-bearing leaf in non-flowering plants.

Spur. A long, narrow appendage, often from a petal or sepal.

Stamen. The male, pollen-producing organ of a flower, composed of the filament and anther.

Staminate. With stamens but without pistils; staminate flowers function only as males.

Staminode. A sterile stamen, one that does not produce pollen.

Stem. Main axis of the plant; it has nodes from which stems leaves and roots may arise.

Sterile. Lacking reproductive potential.

Stigma. The terminal portion of the pistil (female organ) that is receptive to pollen.

Stipules. A pair of scale- or leaf-like appendages at the base of the leaf where it joins the stem.

Stolon. A stem that grows horizontally along the surface of the ground, giving rise to roots and shoots.

Stomates. Tiny openings, mainly on the surface of leaves, that allow air to pass to interior tissue.

Strobilus. A cone-like reproductive structure with spore-bearing leaves radiating from a central axis.

Style. The usually tubular portion of the pistil between the ovary and the stigma. Pollen is received on the stigma and releases a tube that carries sperm down the style into the ovary.

Subalpine. Referring to the middle vegetation zone defined by elevation in Glacier Park; higher mountains but still within the growing limits of trees (see Introduction).

Subequal. Nearly equal.

Subtend. To be immediately beneath.

Succession. The more-or-less predictable sequence of vegetation types following disturbance or on newly created substrate; e.g., glacial moraine.

Succulent. Thick, fleshy and often juicy.

Superior ovary. Petals, sepals and stamens are attached at the base of a superior ovary (see inferior ovary).

Symbiotic. Pertaining to a partnership that is mutually beneficial to both parties.

Talus. Loose stones prone to shifting, usually on a steep slope near the angle of repose.

Taproot. Primary root along the main axis of the plant when all other roots are much smaller.

Tendril. Slender outgrowth of a leaf or stem that usually twists and curls.

Terminal. Referring to the last or end segment or structure.

Tepal. Both sepals and petals when they are not easily distinguishable from each other.

Trophophore. In the Ophioglossaceae; the portion of the frond that is more leaf-like and usually does not bear spores.

Tubercle. A bulge or crown on a fruit or other structure.

Tuberous. Thickened like a potato.

Tundra. Low vegetation above the limit of trees, usually dominated by sedges and/or broad-leaved plants.

Turf. Dense, low vegetation usually dominated by grasses or grass-like plants developed on moist sites.

Turion. A small, bulb-like plantlet produced vegetatively on the stem or, more commonly, among the roots.

Tussock. A distinct tuft or hummock formed by a grass or sedge in which all the stems arise from a well-defined central area.

Ultimate. Terminal; referring to the last or end segment or structure.

Umbel. An inflorescence with the flower stalks ascending from the same point at the tip of a stem or branch.

Umblet. Secondary umbels; the umbels at the end of each stalk of a primary umbel.

Unisexual. Having only 1 sex, either male or female.

Ventral. The side of a leaf or fruit closest to or facing toward the main stem.

Whorl. A group of 3 or more leaves or branches all attached at the same level of attachment (node) on the stem.

Wing. Thin, flat expansion or appendage of a structure such as a fruit or stem. In the Fabaceae, the two side petals of the flower.

KEY TO THE FAMILIES

1. Plants reproducing by spores (ferns, horsetails, clubmosses and allies) 2
1. Plants reproducing by seeds (conifers, flowering plants) 13

2. Stems grooved lengthwise, jointed and easily pulled apart at the nodes; leaves reduced to papery scales; plants apparently of only simple or branched stems .... **Equisetaceae**
2. Stems not as above; leaves not papery 3

3. Plants a stemless cluster of linear leaves with white spores at the base where they join together; aquatic .... **Isoetaceae**
3. Plants not as above, usually with stems 4

4. Leaves small, needle- or scale-like; stems often resembling small conifer branches 5
4. Leaves larger, fern-like 6

5. Plants < 2 cm tall, usually on rocks .... **Selaginellaceae**
5. Plants mostly > 2 cm tall, in soil of forests and meadows ...... **Lycopodiaceae**

6. Spores clustered on a specialized branch-like portion of the leaf .... **Ophioglossaceae**
6. Spores borne on leaves arising directly from the roots; fertile leaves mostly similar to the vegetative leaves (true ferns) 7

7. Spore clusters (sori) borne along the leaf or leaflet margins, elongate and usually partly covered by the inrolled leaflet edge 8
7. Sori borne on veins between midrib and margin of leaf or leaflet, often round in outline 9

8. Rhizomes hairy, not scaly; leaves > 30 cm long-hairy beneath .... **Dennstaedtiaceae**
8. Rhizomes scaly; leaves often < 30 cm long, often not hairy beneath .... **Pteridaceae**

9. Spore clusters (sori) elongate in outline 10
9. Sori round or crescent-shaped in outline 11

10. Leaves once pinnately divided, < 15 cm long .... **Aspleniaceae**
10. Leaves > 1 time pinnately divided, > 15 cm long .... **Dryopteridaceae**

11. Leaves deeply lobed but not fully divided into leaflets .... **Polypodiaceae**
11. Leaves divided into leaflets 1-4 times 12

12. Margins and veins of leaflet undersides with spreading, needle-like hairs .... **Thelypteridaceae**
12. Undersides of leaflets lacking spreading needle-like hairs **Dryopteridaceae**

13. Plants lacking flowers; trees or shrubs with needle- or scale-like, mostly evergreen leaves (conifers) 14
13. Plants with flowers; leaves usually broad or grass-like, mostly not evergreen (Angiosperms—flowering plants) 17

14. Leaves scale-like, pressed flat to the stem .... **Cupressaceae**
14. Leaves needle-like 15

15. Fruit a cone of dry seed-bearing scales .... **Pinaceae**
15. Fruit berry-like, dry or juicy 16

16. Fruit with a red, fleshy outer covering .... **Taxaceae**
16. Fruit dry, purplish .... **Cupressaceae**

17. Plants truly aquatic with submerged or floating leaves that become limp when withdrawn from the water (emergent plants with self-supporting stems are not included) .... Group 1
17. Plants emergent not obligately aquatic 18

18. Plants herbaceous; leaves unlobed and undivided with parallel veins, often grass-like; petals and sepals mostly in 3's or 6's or lacking (Monocots) ....
.... Group 2
18. Plants herbaceous or woody; leaves undivided, divided or lobed, usually with net-veination; petals and sepals in 2's, 4's, or 5's (Dicots) 19

19. Plants trees, shrubs or woody vines, stems woody well above ground level .
.... Group 3
19. Plants herbaceous or woody only at the base 20

20. Flowers without distinct whorls of petals and sepals that are differentiated from each other (petals or sepals or both lacking or petals and sepals identical) .... Group 4
20. Flowers with petals and sepals different from each other 21

21. Petals separate from each other all the way to the base .... Group 5
21. Petals united, at least toward the base .... Group 6

Group 1, Aquatic Plants

1. Plants floating, without stems, of few fronds < 2 cm across .... **Lemnaceae**
1. Plants larger; stems often present or plant rooted in the bottom 2

2. Leaves > 10 cm long, floating, heart-shaped; flowers > 3 cm across ....
.... **Nymphaeaceae**
2. Leaves and flowers smaller 3

3. Stems lacking; leaves grass-like and basal 4
3. Plants with leafy stems 7

4. Leaves with a sac of white spores at the base .... **Isoetaceae**
4. Leaves without a spore sac at the base 5

5. Leaves narrow but with a blade distinctly wider than the petiole ....
.... **Scrophulariaceae**
5. Leaves grass-like, without a distinct blade and petiole 6

6. Leaves needle-like, round in cross section .... **Eleocharis**
6. Leaves flattened, vegetative .... **Sagittaria** or **Sparganium**

7. Submerged leaves highly dissected into hair-like segments........................... 8
7. Submerged leaves linear to elliptic but not dissected................................. 10

8. Some or all leaves with tiny, egg-shaped bladders .... **Lentibulariaceae**
8. Leaves without bladders 9

9. Leaves 1 per node; flowers conspicuous with colored petals **Ranunculaceae**
9. Leaves whorled (> 2 per node); flowers inconspicuous, without petals.........
.... **Haloragaceae**

10. Leaves whorled (> 2 per node).. 11
10. Leaves alternate or opposite (1 or 2 per node) 12

11. Leaves 3-4 per node .... **Hydrocharitaceae**
11. Leaves mostly 6 per node .... **Hippuridaceae**

12. Submerged leaves opposite (2 per node) .... **Callitrichaceae**
12. Submerged leaves alternate (1 per node) 13

13. Main leaf veins branched off the midvein .... **Polygonaceae**
13. Main leaf veins parallel to each other... 14

14. Leaves with a pale membranous appendage surrounding the stem (stipule)
.... **Potamogetonaceae**
14. Stipule lacking 15

15. Floating leaves with elliptic or arrow-shaped blades .... **Alismataceae**
15. Floating leaves grass-like.. 16

16. Leaves flat where they join the stem .... **Poaceae**
16. Leaves V-shaped in cross section where they meet the stem ...... **Typhaceae**

Group 2, Monocots

1. Flowers unisexual, borne in dense clusters 2
1. Flowers bisexual, often not densely clustered 3

2. Stems mostly 3-sided; each flower subtended by 1 (rarely 2), apparent, scale-like bract (sedges) .... **Cyperaceae**
2. Stems round in cross section; flower bracts minute and inconspicuous ....
.... **Typhaceae**

3. Stamens and ovaries enclosed by 1-2 bracts, scales or sacs; petals and sepals lacking (grasses and sedges)... 4
3. Flowers with 3 or 6 petals or sepals 5

4. Stems mostly solid and 3-sided; each flower subtended by 1 (rarely 2), scale-like bract (sedges) .... **Cyperaceae**
4. Stems mostly round in cross section and swollen at the leaf nodes; each flower enclosed by 2 bracts (grasses)... **Poaceae**

5. Leaves whorled(> 2 per node)... **Rubiaceae** (a dicot with 3 petals per flower)
5. Leaves 1-2 per node (except Trillium with a single whorl of 3 leaves) 6

6. Base of petals and sepals attached at the top of the (inferior) ovary 7
6. Base of petals and sepals attached at the base of the (superior) ovary 8

7. Flowers bilaterally symmetrical; one petal different than the other 3 ....
.... **Orchidaceae**
7. Flowers radially symmetrical; all petals identical .... **Iridaceae**

8. Sessile flowers borne on upper portion of a leafless stem ..... **Juncaginaceae**
8. Inflorescence not as above... 9

9. Fruit a berry... **Liliaceae**
9. Fruit a capsule or a cluster of coated seeds (achenes) 10

10. Fruit a wheel-like or globose cluster of achenes .... **Alismataceae**
10. Fruit a capsule 11

11. Fruit a group of 3 1- or 2-seeded capsules .... **Scheuchzeriaceae**
11. Fruit a single capsule (often 3-chambered) 12

12. Petals and sepals brown or green; leaves grass-like .... **Juncaceae**
12. Petals and sepals white, yellow or blue (except Asparagus with scale-like leaves) .... **Liliaceae**

Group 3, Woody Plants

1. Trailing or climbing vines 2
1. Trees or shrubs with rigid stems 4

2. Leaves alternate (1 per node) .... **Solanaceae**
2. Leaves opposite (2 per node) 3

3. Leaves elliptic with entire margins .... **Caprifoliaceae**
3. Leaves divided into leaflets .... **Ranunculaceae**

4. Leaves opposite or whorled (> 1 per node) 5
4. Leaves alternate (1 per node) 20

5. Leaves divided into leaflets .... **Caprifoliaceae**
5. Leaves with entire, toothed or lobed margins but not divided into leaflets .. 6

6. Some leaves with 3-5 maple-like lobes 7
6. Leaf margins entire or toothed but not lobed 8

7. Fruit berry-like .... **Caprifoliaceae**
7. Fruit dry, winged .... **Aceraceae**

8. Leaves scale-like, < 5 mm long, 4 per node .... **Ericaceae**
8. Leaves not as above 9

9. Leaves with pointed teeth on the margins 10
9. Leaves with smooth or wavy margins 15

10. Flowers tubular, 2-lipped, > 2 mm long .... **Scrophulariaceae**
10. Flowers not as above 11

11. Leaves < 3 cm long 12
11. Some leaves > 3 cm long 13

12. Flowers and fruits in leaf axils .... **Celastraceae**
12. Flowers and fruits paired on stem tips .... **Caprifoliaceae**

13. Inflorescence with > 12 flowers or fruits .... **Rhamnaceae**
13. Inflorescence with < 12 flowers or fruits 14

14. Fruit berry-like; petals < 4 mm long .... **Rhamnaceae**
14. Fruit a capsule; petals > 10 mm long .... **Hydrangeaceae**

15. Flowers tubular, 2-lipped, > 2 mm long .... **Scrophulariaceae**
15. Flowers not as above 16

16. Twigs reddish .... **Cornaceae**
16. Twigs gray or brown 17

17. Leaves covered with mealy, brown or silver scales at least below ....
.... **Elaeagnaceae**
17. Leaves not mealy or scaly 18

18. Leaves leathery, deep green; flowers pink, saucer-shaped .... **Ericaceae**
18. Leaves not leathery; flowers not as above 19

19. Petals united into an urn-shaped or tubular corolla .... **Caprifoliaceae**
19. Petals separate; corolla saucer-shaped .... **Hydrangeaceae**

20. Leaves divided into leaflets 21
20. Leaves with entire, toothed or lobed margins but not divided into leaflets ..
.... 24

21. Leaflets with spine-tipped teeth on the margins .... **Berberidaceae**
21. Leaflets without spine-tipped teeth 22

22. Stems with spines or prickles .... **Rosaceae**
22. Stems unarmed 23

23. Leaflets 3 .... **Anacardiaceae**
23. Leaflets > 3 .... **Rosaceae**

24. Stems with thorns, spines or prickles 25
24. Stems unarmed 27

25. Stems with thorns (sharp branch tips); leaves not lobed .... **Rosaceae**
25. Stems with spines or prickles; leaves lobed 26

26. Some leaf blades > 15 cm wide .... **Araliaceae**
26. Leaf blades < 15 cm wide .... **Grossulariaceae**

27. Leaves < 3 mm wide, evergreen, leathery .... **Ericaceae**
27. Leaves > 3 mm wide or not leathery and evergreen 28

28. Leaf margins entire or nearly so 29
28. Leaf margins toothed or lobed, sometimes obscurely so 34

29. Plants mat-forming; woody stems close to the ground .... **Polygonaceae**
29. Woody stems mostly erect, ascending 30

30. Leaves with resinous dots or covered with silvery scales beneath 31
30. Leaves without scales or resin dots beneath 32

31. Leaves covered with silvery scales beneath .... **Elaeagnaceae**
31. Leaves with resinous dots beneath .... **Ericaceae**

32. Flowers unisexual in cylindrical catkins; winter buds covered by 1 scale ....
   .... **Salicaceae**
32. Flowers bisexual; winter buds covered by > 1 scale 33

33. Inflorescence with > 20 flowers or fruits .... **Rhamnaceae**
33. Inflorescence of < 20 flowers or fruits .... **Ericaceae**

34. Leaves with 3 lobes and widest at the tip; foliage with a sage odor ....
   .... **Asteraceae**
34. Leaves not as above 35

35. Plants alpine, mostly < 10 cm tall 36
35. Plants mostly > 10 cm tall 37

36. Flowers unisexual, borne in cylindrical catkins .... **Salicaceae**
36. Flowers bisexual, solitary at stem tips .... **Rosaceae**

37. Leaves palmately lobed (like a maple) 38
37. Leaf margins toothed but not lobed 39

38. Flowers with 5 stamens .... **Grossulariaceae**
38 Flowers with > 5 stamens .... **Rosaceae**

39. Flowers unisexual, borne in cylindrical catkins 40
39. Flowers not clustered in catkins 41

40. Female catkins brittle and cone-like; leaves < 2 times as long as wide;
   winter buds covered by > 1 scale .... **Betulaceae**
40. Female catkins soft, without brittle scales; leaves often > 2 times as long as
   wide; winter buds covered by 1 scale .... **Salicaceae**

41. Petals united, at least toward the base .... **Ericaceae**
41. Petals separate 42

42. Stamens 5 .... **Rhamnaceae**
42. Stamens 10 .... **Rosaceae**

Group 4, Dicots with Petals and/or Sepals Lacking or Undifferentiated

1. Few to many flowers clustered in heads; each head subtended by a cup- or vase-shaped whorl(s) of bracts and appearing like a single flower ....
.... **Asteraceae**
1. Flowers not as above 2

2. Plants parasitic on conifers; stems jointed; leaves reduced .... **Viscaceae**
2. Plants not as above 3

3. Plants with milky sap and glabrous foliage .... **Euphorbiaceae**
3. Plants with clear sap or hairy foliage 4

4. Middle and lower leaves opposite or whorled (> 1 per node) 5
4. Middle and lower leaves alternate (1 per node) or all in a basal rosette 11

5. Plants with sharp, stinging hairs on stems and leaves .... **Urticaceae**
5. Plants without stinging hairs 6

6. Leaves in whorls of 6; flowers in leaf axils; wet areas .... **Hippuridaceae**
6. Plants not as above 7

7. Middle stem leaves divided or lobed 8
7. Middle stem leaves with entire, toothed or wavy margins 9

8. Pistils and/or stamens numerous .... **Ranunculaceae**
8. Flowers with 1 pistil and 3 stamens .... **Valerianaceae**

9. Pistils and/or stamens > 10 each .... **Ranunculaceae**
9. Pistils and stamens 10 or fewer 10

10. Leaves mostly whorled (> 2 per node); ovary inferior .... **Rubiaceae**
10. Leaves opposite (2 per node); ovary superior .... **Caryophyllaceae**

11. Leaves divided or deeply lobed (> halfway to midvein) 12
11. Leaves with entire, toothed or wavy margins 17

12. Stamens > 10 per flower 13
12. Stamens 1-10 14

13. Plants with milky sap .... **Papaveraceae**
13. Plants with clear sap .... **Ranunculaceae**

14. Flowers green; leaves thick and often covered with white scales ....
.... **Chenopodiaceae**
14. Flowers with colored sepals or petals; leaves not covered with white scales
.... 15

15. Petals not all the same, united at the base .... **Fumariaceae**
15. Petals all similar, separate to base or nearly so 16

16. Flowers and fruits in a dense cylindrical spike .... **Rosaceae**
16. Inflorescence not a dense spike .... **Apiaceae**

17. Stamens > 10 per flower .... **Ranunculaceae**
17. Flowers with 4-10 stamens 18

18. Flowers with 2 sepals and 2 purple stamens in a dense spike ....
.... **Scrophulariaceae**
18. Flowers and inflorescence not as above 19

19. Fruit nearly orbicular, flattened with an apical notch .... **Cruciferae**
19. Fruit not as above 20

20. Flower bracts or sepals with a spine tip .... **Amaranthaceae**
20. Flower bracts and sepals without spine tips 21

21. Membranous appendages (stipules) sheathing the stem above leaf
attachment .... **Polygonaceae**
21. Sheathing stipules lacking 22

22. Rhizomatous plants with fleshy leaves and berry-like fruit .... **Santalaceae**
22. Plants not as above 23

23. Flowers green; leaves often thick or covered with white scales but not hairy
.... **Chenopodiaceae**
23. Flowers white, yellow or pink, or if green, then foliage hairy 24

24. Flowers stalked in an umbrella-like inflorescence (umbel) ..... **Polygonaceae**
24. Flowers not in umbels... **Saxifragaceae**

Group 5, Dicots with Sepals Differentiated from the Separate Petals

1. Flowers resembling a daisy or dandelion (these are actually clusters of
flowers each with united petals).. **Asteraceae**
1. Flowers not as above 2

2. Flowers bilaterally symmetrical; petals not all identical 3
2. Flowers radially symmetrical petals all identical 9

3. Flowers with > 10 stamens .... **Ranunculaceae**
3. Flowers with 10 or fewer stamens 4

4. Petals 4 5
4. Petals 5 6

5. Base of petals attached at base of (superior) ovary .... **Fumariaceae**
5. Petals attached to top of the (inferior) ovary .... **Onagraceae**

6. Flowers with 5 stamens .... **Violaceae**
6. Flowers with 10 stamens 7

7. Flowers pea-like; petals very different, lowest 2 united to form a canoe-
shaped keel containing the stamens.. **Fabaceae**
7. Petals saucer- to cup-shaped; petals not very different 8

8. Petals attached at the base of the (superior) ovary .... **Ericaceae**
8. Petals attached along the upper surface of the ovary .... **Saxifragaceae**

9. Flowers with > 10 stamens 10
9. Flowers with 10 or fewer stamens 14

10. Stem leaves opposite (2 per node) .... **Hypericaceae**
10. Stem leaves alternate (1 per node) or in a basal rosette.......................... 11

11. Anther filaments united into a tube that surrounds the style ..... **Malvaceae**
11. Stamens not fused into a tube 12

12. Flowers having 1 pistil with 3-8 stigmas at the tip .... **Portulacaceae**
12. Flowers having > 1 pistil 13

13. Basal portion of sepals united to form a cup fused with the basal part of
the ovary (hypanthium) .... **Rosaceae**
13. Sepals usually separate to the base; hypanthium lacking .... **Ranunculaceae**

14. Petals and sepals arising from the top of the (inferior) ovary 15
14. Petals and sepals arising from the base of the (superior) ovary 21

15. Flowers with 2 or 4 each of petals, sepals and stamens.......................... 16
15. Flowers with 5 petals and sepals and 5 or 10 stamens 17

16. Fruit berry-like; flowers in a hemispheric cluster subtended by 4 petal-like
bracts .... **Cornaceae**
16. Fruit a capsule; inflorescence not as above .... **Onagraceae**

17. Leaves divided or lobed at least 1/3 the way to the midvein 18
17. Leaves with entire, toothed or shallowly lobed margins 20

18. Inflorescence umbrella-like, flower stalks all joined at a common point ... 19
18. Inflorescence not as above.. **Saxifragaceae**

19. Fruit fleshy, berry-like .... **Araliaceae**
19. Fruit of 2 dry seeds .... **Apiaceae**

20. Styles 2; stamens 10 .... **Saxifragaceae**
20. Styles > 2 or stamens > 10 .... **Rosaceae**

21. Basal portion of sepals united to form a cup fused with the basal part of
the ovary (hypanthium) 22
21. Sepals usually separate to the base; hypanthium lacking 23

22. Styles 2; stamens 10 .... **Saxifragaceae**
22. Styles > 2 or stamens >10 .... **Rosaceae**

23. Styles > 5 per flower .... **Ranunculaceae**
23. Styles 0-5 24

24. Sepals 2-3 .... **Portulacaceae**
24. Sepals 4-5 25

25. Leaves divided or lobed at least halfway to the midvein 26
25. Leaves with entire, toothed or shallowly lobed margins 30

26. Flowers with 4 petals 27
26. Flowers with 5 petals 28

27. Leaves with 3 leaflets... **Capparaceae**
27. Leaves not as above... **Brassicaceae**

28. Style 1 .... **Geraniaceae**
28. Styles 2-many 29

29. Styles 2; stamens 10 .... **Saxifragaceae**
29. Styles > 2 or stamens > 10 .... **Rosaceae**

30. Flowers with 4 petals 31
30. Flowers with 5 petals 32

31. Stems and leaves white to red but not green .... **Ericaceae**
31. Stems and leaves green .... **Brassicaceae**

32. Leaves fleshy; each flower with 4-5 pistils .... **Crasssulaceae**
32. Plants not as above 33

33. Flowers unisexual, with stamens or pistils only 34
33. Flowers bisexual 35

34. Leaves alternate; flowers red .... **Crassulaceae**
34. Leaves opposite; flowers not red .... **Caryophyllaceae**

35. Leaves basal, with long, purplish, sticky hairs for catching insects ....
.... **Drosseraceae**
35. Leaves not as above 36

36. Leaves opposite or whorled (> 1 per node) 37
36. Leaves alternate (1 per node) 38

37. Leaves finely translucent-dotted (hold leaf up to light and look with a 10X
lens) .... **Hypericaceae**
37. Leaves not translucent-dotted .... **Caryophyllaceae**

38. Leaves linear; flowers with 5 stamens .... **Linaceae**
38. Plants not as above 39

39. Flowers with 1 style 40
39. Flowers with 2-many styles 41

40. Leaves deeply lobed or divided .... **Geraniaceae**
40. Leaves not deeply divided .... **Ericaceae**

41. Styles 2; stamens 10 .... **Saxifragaceae**
41. Styles > 2 or stamens > 10 .... **Rosaceae**

Group 6, Dicots with Sepals Differentiated from the United Petals

1. Few to many flowers clustered in heads surrounded by bracts forming a cup-
or vase-shaped involucre, appearing like a single daisy- or dandelion-like
flower .... **Asteraceae**
1. Flowers not as above 2

2. Stems and leaves white to purple but not green 3
2. Stems and leaves green 4

3. Plants < 12 cm high; corolla > 8 mm long .... **Orobanchaceae**
3. Plants > 15 cm high; corolla 5-8 mm long .... **Ericaceae**

4. Plants with milky sap 5
4. Plants with clear, watery sap 7

5. Corolla attached on top of the (inferior) ovary .... **Campanulaceae**
5. Corolla attached at the base of the ovary 6

6. Plants erect with opposite leaves .... **Apocynaceae**
6. Plants twining with alternate leaves .... **Convolvulaceae**

7. Flowers unisexual, with stamens or pistils only .... **Caryophyllaceae**
7. Flowers bisexual 8

8. Anthers more numerous than lobes of the corolla (petals) 9
8. Anthers as numerous or fewer than corolla lobes 17

9. Flowers radially symmetrical; petals all identical 10
9. Flowers bilaterally symmetrical; 1 or more petals different than the others 14

10. Stamens > 10, united to form a tube surrounding the style ...... **Malvaceae**
10. Stamens 2-10, not united into a tube 11

11. Leaves fleshy and succulent 12
11. Leaves flat, not fleshy and succulent 13

12. Sepals 2 .... **Portulacaceae**
12. Sepals 4-5 .... **Crassulaceae**

13. Leaves opposite (2 per node) .... **Caryophyllaceae**
13. Leaves alternate (1 per node) or all basal .... **Ericaceae**

14. Flowers with > 10 stamens .... **Ranunculaceae**
14. Stamens 10 or fewer 15

15. Flowers with 9-10 anthers .... **Fabaceae**
15. Flowers with 4-8 anthers 16

16. Flowers with 6 anthers .... **Fumariaceae**
16. Flowers with 4 (or apparently 8) anthers .... **Scrophulariaceae**

17. Corolla attached on top of the (inferior) ovary 18
17. Corolla attached at the base of the (superior) ovary 23

18. Leaves opposite or whorled (> 1 per node) 19
18. Leaves alternate (1 per node) or all basal 22

19. Leaves whorled (> 2 per node), at least in part .... **Rubiaceae**
19. Leaves opposite (2 per node) 20

20. Ovary and fruit with hooked hairs .... **Rubiaceae**
20. Ovary and fruit without hooked hairs 21

21. Flowers or fruits 2 per stem .... **Caprifoliaceae**
21. Flowers or fruits many .... **Valerianaceae**

22. Leaves divided into 3 leaflets; petals covered with hair-like projections ....
.... **Menyanthaceae**
22. Leaves not divided into 3 leaflets .... **Campanulaceae**

23. Flowers bilaterally symmetrical; 1 or more petals different than the others ..
.... 24
23. Flowers radially symmetrical; petals all the same 27

24. Flowers with 5 anther-bearing stamens .... **Scrophulariaceae**
24. Flowers with 2-4 anther-bearing stamens 25

25. Ovary and fruit 4-lobed; stem 4-angled 26
25. Ovary and fruit unlobed or 2-lobed; stem usually not 4-angled ....
.... **Scrophulariaceae**

26. Upper 2 petals slightly larger than lower 3 .... **Verbenaceae**
26. Corolla with an upper hood and 3-lobed lower lip .... **Lamiaceae**

27. Anther-bearing stamens 2-4, fewer than lobes of corolla 28
27. Anther bearing stamens as many as corolla lobes or > 4 34

28. Flowers with 4 anther-bearing stamens 29
28. Flowers with 2-3 anther-bearing stamens 31

29. Ovary and fruit 4-lobed; stem 4-angled 30
29. Ovary and fruit unlobed or 2-lobed; stem usually not 4-angled ....
.... **Scrophulariaceae**

30. Anthers as long as or longer than their stalks........................ **Verbenaceae**
30. Anthers shorter than their stalks .... **Lamiaceae**

31. Sepals 2; leaves linear, fleshy .... **Portulacaceae**
31. Plants not as above.. 32

32. All leaves basal; petals thin and papery .... **Plantaginaceae**
32. Plants with some stem leaves; petals not papery 33

33. Ovary and fruit 4-lobed; stem 4-angled .... **Lamiaceae**
33. Ovary and fruit unlobed or 2-lobed; stem mostly not 4-angled ....
.... **Scrophulariaceae**

34. Ovary and fruit 4-lobed or -grooved, splitting into 4 nutlets at maturity
    (sometimes 1 or 2 are missing) 35
34. Ovary and fruit not 4-lobed 36

35. Stamens 4; leaves opposite.. **Lamiaceae**
35. Stamens 5; at least some leaves alternate .... **Boraginaceae**

36. Stems twining; leaf blades arrow-shaped .... **Convolvulaceae**
36. Plants not as above.. 37

37. Flowers with 2 sepals.. **Portulacaceae**
37. Sepals > 2 38

38. Leaves opposite or whorled (> 1 per node) 39
38. Leaves alternate or basal 40

39. Style branched into 3 stigmas .... **Polemoniaceae**
39. Style unbranched.. **Gentianaceae**

40. Stamens 4; petals thin and papery; flowers in a dense spike ....
.... **Plantaginaceae**
40. Flowers not as above 41

41. Style branched into 3 stigmas .... **Polemoniaceae**
41. Style unbranched or 2-branched 42

42. Leaves all basal; flowers borne on a naked stem .... **Primulaceae**
42. Leafy stems present 43

43. Sepals united for ≥ 1/2 their length.. **Solanaceae**
43. Sepals united for < 1/3 their length 44

44. Leaves opposite .... **Primulaceae**
44. Leaves alternate 45

45. Style branched into 2 stigmas .... **Hydrophyllaceae**
45. Style unbranched... **Scrophulariaceae**

PTERIDOPHYTES
FERNS AND ALLIES

ASPLENIACEAE: SPLEENWORT FAMILY

Asplenium L., Spleenwort

Asplenium trichomanes-ramosum L. [*A. viride* Hudson]. Evergreen perennial with short, scaly rhizomes; leaves all alike, clustered, 5-10 cm long, petiole shorter than the blade; leaf blade linear, once pinnate, glabrous or sparsely hairy; leaf segments (pinnae) ovate, 6-21 pairs, with toothed margins; spore clusters (sori) 4-8, linear, borne near center of pinnae, thinly covered by indusium that opens along the side; gametophyte green, terrestrial. Reference: Wagner et al. (1993).

Rare in moist to wet, subalpine and alpine limestone crevices near or east of the Divide. Circumboreal south at scattered localities to CA, CO, WI, NY.

DENNSTAEDTIACEAE: BRACKEN FAMILY

Pteridium Gled. ex Scopoli., Bracken

Pteridium aquilinum (L.) Kuhn., Bracken. Perennial with vigorously spreading rhizomes that lack scales; leaves all alike, scattered, 30-150 cm tall, petiole ca. same length as blade; leaf blade twice pinnately divided; ultimate segments (pinnules) narrowly lance-shaped, pinnately lobed, upper surface glabrous but hairy below, margin of fertile pinnules inrolled and partly covering continuous line of spores; gametophyte green and terrestrial. Fig. 1.

Abundant in dry to moist montane forest as well as meadows and slopes within forested landscapes; East, West. Cosmopolitan.

DRYOPTERIDACEAE: WOOD FERN FAMILY

Perennials with creeping or short, erect rhizomes; leaves all alike (except *Drypoteris cristata*), forming "fiddleheads" as they expand, petioles scaly; leaf blades 1-5 times pinnately divided; spores borne in clusters (sori) along veins on underside of ultimate leaf divisions (pinnules), usually covered by a papery indusium; indusium disintegrating with age; gametophyte green, terrestrial. Reference: Smith (1993a).

Members of this family were formerly placed in a single family (Polypodiaceae sensu lato) that included all of our ferns except grapeferns, moonworts and adder's-tongue (Ophioglossaceae).

1. Frond first divided into 3 parts .... **Gymnocarpium**
1. Primary divisions of frond > 3 2

2. Fronds once pinnate; leaflets toothed or lobed .... **Polystichum**
2. Fronds 2-3 times pinnately divided at least below 3
3. Fronds mostly < 25 cm long 4
3. At least some fronds > 25 cm long 5
4. Fronds scattered on a rhizome without conspicuous old leaf bases ....
.... **Cystopteris**
4. Fronds clustered on a scaly rhizome clothed in old leaf bases .... **Woodsia**
5. Sori mostly longer than wide, crescent-shaped; indusium elongate and
   attached on 1 side when present .... **Athyrium**
5. Sori round in outline; indusium round or heart-shaped and appearing to arise
   like an umbrella from the center of the sorus .... **Dryopteris**

Athyrium Roth., Lady Fern

Perennials with short, nearly vertical, scaly rhizome, covered with old leaf bases; leaves clustered and 2-4 times pinnately divided into ultimate segments (pinnules) with toothed margins; spore clusters (sori) round to narrowly elliptic; thin flap covering sori (indusium) persistent and attached at the side or absent.

1. Thin, pale flap of tissue (indusium) present (many may have fallen); plants
   usually > 60 cm tall, forests and open habitats .... *A. filix-femina*
1. Indusium lacking; plants usually < 60 cm, open, rocky habitats ... *A. alpestre*

Athyrium alpestre (Hoppe) Clairv. [*A. distentifolium* Tausch ex Opiz]. Leaves ascending or erect, densely clustered, 25-60 cm long, petiole shorter than the blade; leaf blade narrowly lance-shaped, twice pinnately divided; pinnules deeply lobed; sori round or elliptic, borne at sinuses of pinnule lobes; indusium small or absent.

Common in rockslides and stony soil of subalpine and alpine meadows and along streams, less common in cliff crevices; East, West. Our plants are var. *americanum* Butters. Circumboreal south to CA, CO. This is our only large fern occurring above treeline; it usually has a distinctive yellowish-green color, and Standley (1921) reports that it has a slight odor of balsam.

Athyrium filix-femina (L.) Roth ex Mertens., Lady Fern. Leaves ascending or erect, 30-150 cm long, petiole much shorter than the blade; leaf blade narrowly elliptic, twice pinnately divided; pinnules deeply lobed; sori round to elliptic and borne between midvein and margin of pinnule; indusium often curved with marginal hairs. Fig. 2. Color plate 13.

Abundant in moist to wet forest, margins of meadows, along streams and avalanche slopes; montane and subalpine; East, West. Our plants are var. *cyclosorum* Rupr. Circumboreal south to CA and CO. This species can be confused with *Dryopteris filix-mas*.

Cystopteris Bernh., Bladder Fern

Perennials with scaly rhizomes; leaves all alike, 2-4 times pinnately divided; spore clusters (sori) circular and borne on veins on underside of ultimate leaf segments (pinnules), partly covered by thin, pale, cup-like indusium attached at the base.

1. Leaf blade nearly as wide as long; rare .... **C. montana**
1. Leaf blade 2-3 times as long as wide; common .... **C. fragilis**

Cystopteris fragilis (L.) Bernh., Fragile Fern [*Filix fragilis* (L.) Griseb.]. Leaves loosely clustered, glabrous or slightly glandular, up to 30 cm long, petiole shorter or longer than the blade; leaf blade narrowly elliptic, once pinnately divided; pinnae deeply lobed with toothed margins; indusium with glands. Fig. 3.

Common in dry to wet, shallow or stony soil and crevices of at least partially shaded slopes and outcrops at all elevations; East, West. Cosmopolitan south to CA, NM, NE. At lower elevations this species often grows with the similar *Woodsia* spp., but the latter have more clustered leaves with numerous old stems.

Cystopteris montana (Lam.) Bernh. Leaves scattered on a long rhizome, 10-30 cm long, petiole longer than the blade; leaf blade triangular, twice pinnately divided below but once pinnate above; lowest pinnae ca. as large as rest of blade; pinnules deeply lobed with toothed margins; indusium with gland-tipped hairs.

Rare, reported from wet cliffs near Gunsight Pass (McLaughlin 1935). Circumboreal to WA, MT, Que.; disjunct in CO. The species superficially resembles *Gymnocarpium* spp.

Dryopteris Adanson., Shield Fern
Perennials with short, scaly rhizomes, commonly covered with old leaf bases; leaves with scaly petioles; leaf blades 2-3 times pinnately divided and lobed; spore clusters (sori) circular, borne on veins midway between midvein and margins of fertile ultimate segments (pinnules), covered by a thin, pale, heart-shaped indusium attached at the sinus.

Dryopteris carthusiana and the more common *D. expansa* have been considered one species by previous authors and called by various other names.

1. Leaves broadly ovate to triangular in outline 2
1. Leaves narrowly elliptic or lance-shaped 3
2. First leaflet (pinnule) pointing down on lowest branch of leaf (pinnae) at least twice as long and wide as the opposing, upward-pointing one ... **D. expansa**
2. First downward-pointing pinnule on lowest pinnae less than twice as long and wide as opposing one .... **D. carthusiana**
3. Leaves all alike, 1-2 times pinnately divided and lobed .... **D. filix-mas**
3. Leaves dimorphic, sterile ones arching, fertile ones erect, once pinnately divided, rare .... **D cristata**

Dryopteris carthusiana (Vill.) Fuchs [*D. austriaca* (Jacq.) Schinz & Thell., D. spinulosa (Muell.) Watt]. Leaves all alike, 15-60 cm long, petiole shorter than blade; leaf blade light green, ovate, non-glandular, 2-3 times pinnately divided; basal pinnules usually longest; indusium lacking glands. Fig. 4.

Uncommon in shady, mesic, montane forests; West. Circumboreal; B.C. to Newf. south to WA, MT, NE, MO, SC.

Dryopteris cristata (L.) Gray, Buckler Fern. Leaves dimorphic, 35-70 cm long, petiole ca. half as long as blade; sterile leaves evergreen, spreading; fertile leaves, erect, longer than sterile; leaf blades narrowly lance-shaped, once pinnately divided, not glandular; pinnae pinnately lobed, fertile ones twisted upward; indusium lacking glands.

Rare along forested margins of montane fens; West. Circumboreal south to ID, IL, GA. This fern is uncommon even at the few sites where it occurs.

Dryopteris expansa (Presl) Fraser-Jenkins & Jermy [*D. dilatata* (Hoffm.) Gray, *D. austriaca* (Jacq.) Schinz & Thell. misapplied]. Leaves all alike, up to 90 cm long, petiole ca. half the blade length; leaf blade green, triangular, 3 times pinnately divided, glandular or not; indusium with or without glands.

Common in mesic, shady, montane forests; West, less common East. AK to CA, MT and Ont. to Newf., Greenl.; Europe. Often with *D. filix-mas, Athyrium filix-femina* and *Gymnocarpium disjunctum*.

Dryopteris filix-mas (L.) Schott, Male Fern. Leaves all alike, 30-100 cm long, petiole ca. 1/3 length of blade; leaf blade narrowly elliptic, green, 1-2 times pinnately divided, not glandular; indusium lacking glands.

Common in mesic forests and in moist to wet rock crevices, montane and subalpine; East, West. Circumboreal to AZ, NM, WI, Ont., N. S. Superficially resembles *Athyrium filix-femina*, but the latter has thinner leaves and more divided pinnae. Cliff-dwelling plants of exposed sites have more leathery leaves.

Gymnocarpium Newman, Oak Fern

Plants with long, slender, scaly rhizomes; leaves scattered, erect, petiole longer than blade; leaf blade triangular, twice pinnately divided below but once pinnate above; lowest pair of pinnae similar to the rest of blade; ultimate segments (pinnules) pinnately lobed or toothed, glabrous or slightly glandular; spore clusters (sori) circular, borne on veins on underside of fertile pinnules; indusium lacking.

1. Pinnules opposite each other at the base of the pinnae, deeply lobed and
   unequal in length .... **G. disjunctum**
1. Pinnules opposite each other at the base of the pinnae, merely toothed and
   nearly equal in length .... **G. dryopteris**

Gymnocarpium disjunctum (Rupr.) Ching. Leaves to 40 cm high; pinnules mostly pinnately lobed; pinnules opposite each other at the base of the pinnae unequal, the upper shorter. Fig. 5.

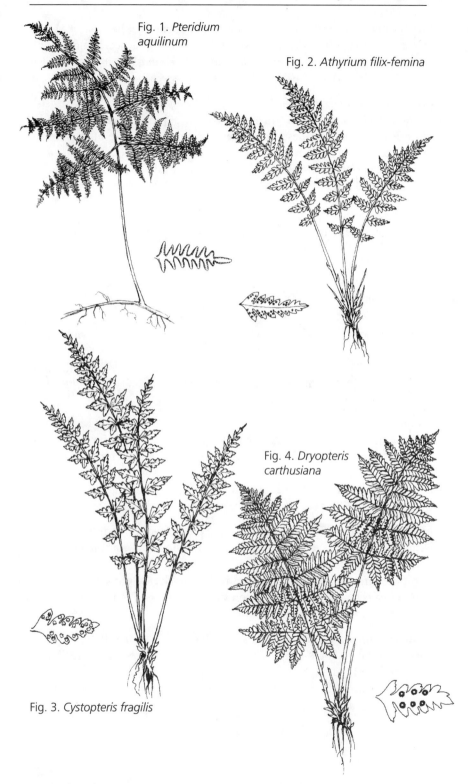

Fig. 1. *Pteridium aquilinum*

Fig. 2. *Athyrium filix-femina*

Fig. 3. *Cystopteris fragilis*

Fig. 4. *Dryopteris carthusiana*

Abundant in moist forest, montane and subalpine; East, West. AK to Greenl. south to OR, AZ, NM, IL, WV. Often forming carpets in old-growth forest with *Dryopteris* spp. and *Athyrium filix-femina*. Previously included under *G. dryopteris*.

Gymnocarpium dryopteris (L.) Newman [*Dryopteris linnaeana* C. Christ.]. Similar to G. disjunctum, smaller; leaves to 30 cm high; most pinnules merely toothed; pinnules opposite each other at the base of the pinnae nearly equal.

Uncommon in moist forest in the McDonald Valley. AK to Greenl. south to WA, ID, MT, IL, WV.

Polystichum Roth, Sword Fern, Holly Fern

Evergreen perennials with short, stout, scaly, nearly vertical rhizomes; leaves clustered, all alike, petioles scaly, much smaller than blade; leaf blades once pinnately divided, the divisions (pinnae) lobed or toothed; spore clusters (sori) in 1-2 rows on each side of midvein on underside of fertile pinnae, partly covered by pale, thin umbrella of tissue (indusium) arising from the center.

1. Leaf divisions (pinnae) deeply lobed and spiny .... *P. andersonii*
1. Pinnae with toothed or spiny margins but not lobed 2

2. Pinnae at least 3 times as long as wide, rare ....*P. munitum*
2. Pinnae less than 3 times as long as wide, common ....*P. lonchitis*

Polystichum andersonii Hopkins. Leaves arching, 30-100 cm long, commonly with 1-several scaly, vegetative buds (bulblets) on central axis among upper pinnae; leaf blade narrowly elliptic or lance-shaped; pinnae narrowly lance-shaped, deeply divided with sharp-pointed lobes; indusium with sparse marginal hairs.

Rare in moist forest and thickets of the McDonald and Many Glacier valleys. AK to OR, MT. The presence of bulblets is diagnostic.

Polystichum lonchitis (L.) Roth, Holly Fern. Leaves erect, 5-50 cm long; leaf blade nearly linear; pinnae broadly lance-shaped, with spiny margins; indusium with jagged margins. Fig. 6.

Abundant on wet, rocky slopes, avalanche chutes, cliffs and in moist forest at all elevations. This is our only common sword fern. AK to Greenl. south to CA, AZ, CO, WI, Que.

Polystichum munitum (Kaulf.) Presl. Leaves arching, 50-120 cm long; leaf blade narrowly lance-shaped; pinnae narrowly lance-shaped to linear, minutely toothed, slightly spiny; indusium with fringed margins.

Rare in moist montane forest in southwest portion of the Park. B.C. to CA, MT.

Woodsia R. Br., Cliff Fern

Perennials with short, densely scaly rhizomes clothed in old leaf bases; leaves clustered, 1-2 times pinnately divided; spore clusters (sori) circular, borne between midvein and margins of pinnules; pale, membranous indusium attached basally and dissected into lobes spreading over the sorus.

The clustered, persistent old leaf bases distinguish these species from *Cystopteris fragilis.*

1. Leaf axis (rachis) and underside glabrous or with glandular hairs only ....
.... ***W. oregana***
1. Rachis and leaves with glandular and non-glandular hairs .... ***W. scopulina***

Woodsia oregana D.C. Eat. Leaves glabrous or with glandular hairs; petioles ca. as long as blades; leaf blades linear to narrowly elliptic, 7-20 cm long, lower pairs of pinnae widely separated; pinnules toothed and lobed.

Uncommon on dry, rocky, montane and subalpine slopes; East. Our plants are subsp. oregana. B.C. to Ont. South to CA, OK, MI.

Woodsia scopulina D.C. Eat. Similar to *W. oregana*; lower leaf surfaces and main axis (rachis) with glandular and non-glandular hairs. Fig. 7.

Common on dry, rocky, montane slopes and cliff crevices; East, West. AK to Que. south to CA, NM. Often conspicuous among lichens and mosses in stabilized talus.

EQUISETACEAE: HORSETAIL FAMILY

Equisetum L., Horsetail, Scouring Rush

Rhizomatous perennials with hollow stems with series of longitudinal ridges terminating at the nodes in papery sheaths with tooth-like projections; branches absent or in whorls at the nodes; leaves small, whorled and sometimes with tips breaking off; spores all alike, borne on umbrella-like sporangiophores in successive whorls organized into terminal, cone-like strobili; fertile stems green and similar to sterile ones or non-green and lacking branches; gametophyte tiny and green. Reference: Hauke (1993).

E. ferrisii Clute, a hybrid between *E. hyemale* and *E. laevigatum*, and *E. trachyodon* A. Br., a hybrid between *E. hyemale* and *E. variegatum*, could also occur in the Park.

1. Stems deep green, leathery, evergreen, all alike, unbranched or branched at the base; strobili sharp-pointed 2
1. Stems medium or light green, somewhat succulent, annual, branched or unbranched; strobilus tip rounded 4

2. Stems robust, > 5 mm diameter, ridges 14-40 .... ***E. hyemale***
2. Stems < 4 mm diameter, ridges 3-12 3

3. Teeth of stem sheaths 3; stems mostly prostrate and twisted .... ***E. scirpoides***
3. Teeth of sheaths 4-10; stems erect or ascending, not twisted . ***E. variegatum***

4. Stems with whorls of branches at some nodes 5
4. Stems unbranched or branched at base 10
5. Branches themselves branched .... *E. sylvaticum*
5. Branches simple 6
6. Stem central cavity ca. 4/5 diameter of stem .... *E. fluviatile*
6. Stem cavity < 2/3 diameter of stem 7
7. Spores usually deformed, abortive .... *E. litorale*
7. Spores symmetrical, not deformed 8
8. Central stem cavity < 1/3 stem diameter; branches with small central cavity .
.... *E. palustre*
8. Central cavity > 1/3 stem diameter; branches solid 9
9. First branch segment longer than the nearest stem sheath .... *E. arvense*
9. First branch segment the same length or shorter than the nearest stem
sheath .... *E. pratense*
10. Teeth of sheaths quickly deciduous, sheaths apparently without teeth; moist
to dry habitats .... *E. laevigatum*
10. Teeth of sheaths apparent; habitat inundated most of growing season ....
.... *E. fluviatile*

Equisetum arvense L., Common Horsetail. Stems annual, dimorphic; sterile stems, 2-70 cm tall, with 10-12 ridges, green and branched; sheaths green with dark teeth; fertile stems tan, unbranched, shorter than sterile stems; strobili 5-35 mm long, blunt. Fig. 8.

Abundant in moist to wet soil of meadows, forests, stream banks and lake shores throughout all vegetation zones; East, West. Fertile stems are often apparent only early in the growing season. Cosmopolitan; throughout most of temperate N. America. This extremely variable species can be aggressive in moist, disturbed habitats, including gardens.

Equisetum fluviatile L. Stems annual, all alike, branched or not, with 9-25 ridges, up to 1 m tall; sheaths green with black teeth; strobili blunt.

Common in shallow water of montane and lower subalpine ponds, lakes, sloughs, marshes and fens; East, West. Circumboreal to WA, IL, PA. This species often grows in monocultures or with cattail and larger sedges. Strobili are borne in summer.

Equisetum hyemale L., Scouring Rush [*E. praealtum* Raf.]. Stems all alike, leathery, dark green, evergreen, unbranched, 18-100 cm long with 14-40 ridges; sheaths green becoming gray with black lower band and black, deciduous teeth; strobili 10-25 mm with sharp-pointed tips.

Common in moist, montane grasslands, meadows, aspen groves, and streambanks; East, West. Circumboreal to most of temperate N. America. The large, deep green, leathery stems with ≥ 14 ridges are diagnostic.

Equisetum laevigatum A. Braun [*E. kansanum* Schafn.]. Stems all alike, usually unbranched, annual, 20-100 cm long with 10-32 ridges; sheaths

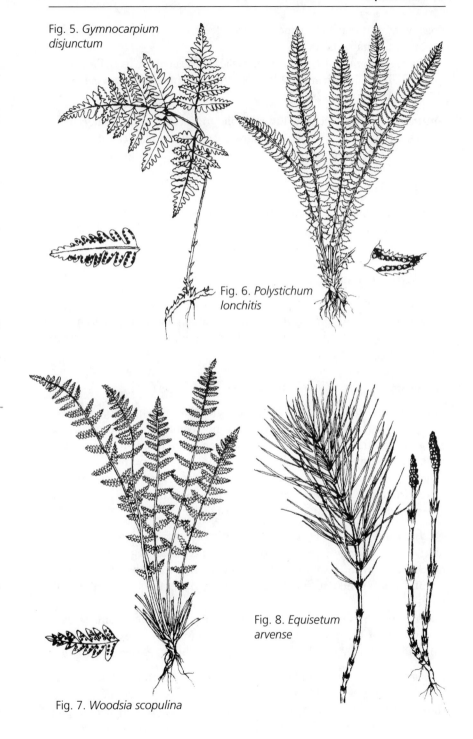

Fig. 5. *Gymnocarpium disjunctum*

Fig. 6. *Polystichum lonchitis*

Fig. 7. *Woodsia scopulina*

Fig. 8. *Equisetum arvense*

green with terminal black band, the black teeth quickly deciduous; strobili 10-25 mm, usually blunt.

Uncommon in moist, montane grasslands and aspen groves; East, West. Stems are lighter green and less leathery than the more common *E. hymale.* B.C. to Que. south to CA, TX, IL.

Equisetum litorale Kuehl. ex Rupr. Similar to *E. arvense* but usually larger; strobilus-bearing and sterile stems similar; spores misshapen.

Occasional in wet ground and swampy thickets (Standley 1921). AK to Newf. south to OR, IL, MD. A sterile hybrid between *E. arvense* and *E. fluviatile* (Hauke 1993).

Equisetum palustre L. Stems all alike, annual, 5-20 cm tall, with 4-10 ridges, branches absent or whorled on upper half; sheaths green with black teeth outlined in white; strobili with blunt tips, maturing in summer.

Uncommon in wet soil of montane to alpine stream banks and seeps; East, West. Circumboreal to OR, MT, WI, PA.

Equisetum pratense Ehrh. Stems annual, delicate, dimorphic; sterile stems 5-30 cm tall, ascending or erect, with 10-18 ridges, pale green and branched; sheaths green, the teeth with a dark center and light margins; fertile stems tan, unbranched; strobili blunt.

Uncommon in shallow water of montane seeps, swamps, and stream margins; East, West. Circumboreal to B.C., MT, IL, NY.

Equisetum scirpoides Michx. Stems all alike, dark green, evergreen, 5-15 cm long, prostrate or ascending, twisted, mostly unbranched or branched at the base, with 6 ridges; sheaths with a green base, central black band and pale, bristle-tipped teeth; strobili 5-10 mm long, sharp-pointed.

Locally common in damp soil of montane spruce forests; West, expected East. Circumboreal to WA, UT, IL, NY. The short and thin, twisted, dark green, leathery, prostrate stems are unmistakable.

Equisetum sylvaticum L. Stems annual, dimorphic; sterile stems, 3-70 cm tall, with 10-18 ridges, green and twice branched; sheaths green below and brown above with teeth often united into a few broad lobes; fertile stems tan and unbranched at first, later becoming green and branched; stobili blunt.

Wet meadows and shallow water of marshes, ponds and stream backwater in the montane zone; locally common West, rare East. Circumboreal to WA, MT, IA, VA. The feathery branches are distinctive.

Equisetum variegatum Schleich. ex Weber & Mohr. Stems all alike, with 5-12 ridges, evergreen, 8-25 cm tall, branched at the base, erect or ascending; sheaths green with a terminal black band, teeth black with white margins and a hair-like tip; strobili 5-10 mm with a sharp-pointed tip. Fig. 9.

Locally common in wet, often calcareous, gravelly soil along montane to alpine seeps, streams and lakes; East, West. Our plants are subsp. *variegatum.* Circumboreal south to OR, UT, IL, NY. Strobili are rare at high elevations.

ISOETACEAE: QUILLWORT FAMILY

Isoetes L., Quillwort

Stemless, aquatic perennials with short, bulbous, 2-lobed rootstocks and fibrous roots; leaves grass-like with dilated bases that clasp the stock; spore cases borne on the inner face of each leaf base and covered with a transparent membrane; small, light tan microspores (male) and larger, white megaspores (female) are borne on alternating whorls of leaves; gametophyte tiny and non-green.

Plants collected early in the growing season have smooth, immature spores that cannot be used for identification. Plants are often found floating after being unearthed by ducks.

1. Megaspores densely covered with spines .... *I. echinospora*
1. Megaspores with bumps or low ridges .... *I. bolanderi*

Isoetes bolanderi Engelm. Leaves 5-10 cm long, tapering to a long point; spore clusters 3-6 mm long; megaspores 0.3-0.4 mm wide.

Locally common in unconsolidated mud in shallow water of montane or subalpine lakes; known only from Dutch Lakes but probably in other lower subalpine lakes; West. B.C., Alta. south to CA, AZ, NM.

Isoetes echinospora Durieu. Leaves 7-15 cm long with a blunt tip; spore clusters 4-7 mm long; megaspores 0.3-0.6 mm wide.

Locally common in unconsolidated mud in shallow water of montane and subalpine lakes. Maguire's (1939) report of this species for Trout Lake is our only record. Circumboreal to CA, ID, CO, OH, NY.

LYCOPODIACEAE: CLUB-MOSS FAMILY

Terrestrial evergreen herbs with ascending or (more often) prostrate stems on or just below the ground; spores numerous, all alike; gametophyte subterranean and non-green. Reference: Wagner & Beitel (1993).

In addition to the following species, *Lycopodiella innundata* (L.) Holub is found south of the Park in montane fens and could occur in the Park west of the Divide.

1. Distinct horizontal stems on or just below ground absent; wing-like gemmae present among leaves .... **Huperzia**
1. Horizontal stems present; gemmae lacking 2
2. Ultimate shoots 2-3 mm thick, 4-angled, round or flattened; leaves in 4-5 ranks and closely appressed to the stem .... **Diphasiastrum**

2. Ultimate shoots 3-8 mm thick, rounded; leaves in ≥ 6 ranks, spreading or ascending .... **Lycopodium**

Diphasiastrum Holub., Club-moss, Running Pine

Low, perennial, evergreen herbs with scattered, upright, branched or unbranched shoots arising from spreading, prostrate stems; upright shoots mostly flattened or 4-angled; leaves scale-like, appressed to the branches, 4-ranked and overlapping; powder-like spores borne in capsules on reduced, broadly to narrowly fan-shaped leaves (sporophylls) grouped together to form cone-like strobili.

All three species were formerly placed in *Lycopodium.*

1. Cone-like strobili 2-3 on a stalk clearly differentiated from shoots; shoots very flattened.. ***D. complanatum***
1. Strobili sessile on shoot tips; shoots little flattened 2

2. Ultimate branches of upright shoots appearing winged and 4-angled in cross section; leaves 4-ranked and closely appressed........................... ***D. alpinum***
2. Ultimate branches round in cross section; leaves 5-ranked, spreading ....
.... ***D. sitchense***

Diphasiastrum alpinum (L.) Holub [*Lycopodium alpinum* L.]. Prostrate stems just below ground; upright shoots clustered, 4-10 cm high with branches 4-angled and appearing winged; fertile shoots above the sterile; leaves 2-3 mm long, lance-shaped, 4-ranked, overlapping, the margins turned under; strobili 5-15 mm long, solitary and sessile; sporophylls nearly triangular, 2-4 mm long with wavy margins.

Common in moist, humic turf near or above treeline, often with *Phyllodoce, Vaccinium* or dwarf *Salix* in areas where snow lies late; East, West. AK to WA, MT; Greenl to Que.; Europe, Japan.

Diphasiastrum complanatum (L.) Holub [*Lycopodium complanatum* L.]. Prostrate stems on or just below ground surface; upright shoots 10-25 cm high with strongly flattened branchlets; leaves 4-ranked, appressed, clasping the stem, overlapping; lateral leaves linear, underside leaves much shorter; stalk, 1-8 cm high, with 1-3 cone-like strobili borne at the top; sporophylls 2-3 mm long, fan-shaped with an abrupt point. Fig.10. Color plate 10.

Common in moist montane and lower subalpine forests; East, West. AK to Greenl. south to OR, MT, WI, NH. Upright shoots resemble twigs of western red cedar. This and *Lycopodium annotinum* are our 2 most common club-mosses.

Diphasiastrum sitchense (Rup.) Holub [*Lycopodium sitchense*]. Vegetatively similar to *D. alpinum;* prostrate stems often on soil surface; branches round in cross section; leaves 5-ranked and somewhat spreading.

Rare in moist, humic turf or open subalpine forest; East, West. AK to Newf. south to OR, MT, NY; Japan. Can easily be confused with the more common *D. alpinum* without close inspection.

Huperzia Bernhardi, Fir-moss

Low, evergreen perennials with clusters of erect or ascending stems; rhizomes lacking; ascending or spreading leaves taper from base, densely set, giving the appearance of a large moss or small spruce branch, stomates on both sides; small wing-like gemmae borne among leaves; spore cases borne on undifferentiated leaves in zones alternating with sterile leaves.

Our 3 species were previously called *Lycopodium selago*; they have distinct habitats.

1. Leaves broader in upper half; forests .... *H. occidentalis*
1. Leaves widest below middle; meadows, wetlands 2

2. Mature shoots 12-25 cm long; gemmae produced in 1 terminal, annual whorl .... *H. miyoshiana*
2. Mature shoots 8-11 cm long; gemmae throughout mature leaves ....
.... *H. haleakalae*

Huperzia haleakalae (Bracken.) Holub [*Lycopodium selago* misapplied]. Leaves 3-6 mm, entire, ascending or appressed to the stem, yellowish below, lustrous.

Local in moist, humic soil of subalpine and alpine meadows, turf and fens near the Divide, often near tree islands where snow lies late; frequently growing with *Phyllodoce, Cassiope* or *Kalmia.* AK to WA, CO; Siberia; disjunct in HI.

Huperzia miyoshiana (Makino) Ching [*Lycopodium selago* misapplied]. Leaves 3-7 mm long, entire, spreading to erect, light green to yellowish, lustrous.

Rare around montane fens and marshes; known only from the McDonald Valley. AK to Newf., south to WA, ID, MT; Asia.

Huperzia occidentalis (Clute) Kartesz & Gandhi [*Lycopodium selago* misapplied]. Shoots prominently curved at the base; leaves reflexed to spreading, 4-10 mm, light green, lustrous.

Rare in moist montane forest in the McDonald Valley. AK to OR, MT.

Lycopodium L., Club-moss

Low, perennial, evergreen herbs with scattered, upright, branched or unbranched shoots arising from spreading prostrate stems or rhizomes; leaves ascending or spreading, narrow, densely set, resembling large mosses or small spruce branches; yellow, powder-like spores borne in capsules on reduced leaves (sporophylls) spirally arranged to form terminal cone-like strobili.

1. Cone-like strobili on stalks clearly differentiated from shoots; shoot leaves minutely hair-tipped 2
1. Strobili sessile on shoot tips; leaves not hair-tipped 3

2. Strobili mostly 1 per stalk; plants subalpine or alpine .... *L. lagopus*
2. Strobili 2-4 per stalk; plants montane or low subalpine .... *L. clavatum*

3. Upright shoots branched and tree-like; uncommon .... **L. dendroideum**
3. Upright shoots sparingly branched; common .... **L. annotinum**

Lycopodium annotinum L. Leafy stems creeping on soil surface; upright shoots 5-25 cm high, solitary but often branched near base; leaves dark green, spreading, narrowly lanceolate with shallowly toothed margins, 4-10 mm long; strobili solitary and sessile on stem tips; sporophylls with a narrow point. Fig. 11.

Common in mesic to wet, montane and lower subalpine forest, occasional above treeline; East, West. AK to Greenl. south to AZ, NM, OH, TN. This is our most common club-moss.

Lycopodium clavatum L. Leafy stems creeping on soil surface; upright shoots to 20 cm high, solitary but often with short, arm-like branches; leaves narrow with entire margins and tiny, hair-like tips, 4-6 mm long, spreading or ascending; 2-4 strobili, 2-7 mm long, on branched, terminal stalk sparsely set with yellowish, leaf-like bracts; sporophylls with toothed margins and a hair-like tip.

Rare around boggy margins of fens in the McDonald Valley. B.C. to CA, MT; Man. to Newf. south to OH, TN. The more common *L. annotinum* has sessile strobili.

Lycopodium dendroideum Michx. [*L. obscurum* L. var. *dendroideum* (Michx.) D.C. Eaton]. Prostrate stems below soil surface; upright shoots 10-20 cm, solitary, branched and tree-like; leaves 3-5 mm long, needle-like, spreading to ascending, pale green; strobili 1-5 cm long, solitary and sessile on upright branches; sporophylls yellowish, ca. 3 mm long and ovate with a short point and ragged margins.

Rare in moist montane forest in the Middle Fork Flathead and Waterton river valleys. AK to Newf. south to WA, WY, WV; Asia. The tree-like growth form is diagnostic.

Lycopodium lagopus (Laest. ex Hartm.) Zins. ex Kuzen. [*L. clavatum* var. *monostachyon* Hook. & Grev.]. Similar to L. clavatum; sparingly branched in lower half of upright shoots; leaves strongly ascending; strobili, solitary and stalked or, 2 and unstalked.

Rare, collected only in moist turf near Logan Pass. Circumboreal to B.C., Man., MI, NY; disjunct in MT.

OPHIOGLOSSACEAE: ADDER'S-TONGUE FAMILY

Perennials with unbranched stems and a single leaf; leaf divided into a sterile, photosynthetic trophophore and a fertile spore-bearing sporophore (except in *B. paradoxum*); trophophore leaf-like, simple, or pinnately lobed or divided; sporophore simple or pinnately branched; spores all alike; gametophyte fleshy, not green, subterranean.

1. Sterile leaf segment (trophophore) elliptic with entire margins ....
.... **Ophioglossum**
1. Trophophore lobed or divided (or absent) .... **Botrychium**

Botrychium Swartz., Grapefern, Moonwort

Fleshy herbs with erect stems and fleshy roots; trophophore (sterile blade) spreading or ascending; sporophore 1-5 times pinnately divided or lobed, attached to stem at or well above ground level; sporophore 1-3 times pinnate, bearing rows of globose sporangia containing numerous tiny, yellow spores.

Botrychium virginianum and B. multifidum (grapeferns) are large and easily seen; all other species (moonworts) are diminutive and inconspicuous. It is common for some species (e.g., B. lanceolatum) to have sporangia on the trophophore. B. watertonense Wagner, a hybrid between B. paradoxum and B. hesperium, is similar to the latter but with spore sacs along the margins of the lobes of the trophophore; it has been found near Lake Sherburne.

1. Trophophore (sterile leaf blade) branching from main stem at ground level or nearly so 2
1. Trophophore branching above ground level 3
2. Plant > 10 cm tall, trophophore mostly 3-4 times pinnately divided..............
.... **B. multifidum**
2. Plant < 10 cm tall, trophophore 1-2 times pinnate .... **B. simplex**
3. Trophophore 3-4 times pinnately divided, fern-like; plants usually > 10 cm tall .... **B. virginianum**
3. Trophophore 1-2 times pinnate; plants usually < 10 cm tall 4
4. Trophophore lacking; sporophore (spore-bearing portion of leaf) double.......
.... **B. paradoxum**
4. Trophophore present; sporophore solitary 5
5. Trophophore sessile, its outline wider than long .... **B. lanceolatum**
5. Trophophore usually stalked, longer than wide 6
6. Trophophore usually twice pinnately divided or lobed 7
6. Trophophore usually once pinnate 8
7. Basal pair of primary lobes noticeably longer than next pair above; lobes overlapping or nearly so .... **B. hesperium**
7. Basal lobes not much longer than adjacent pair, lobes not overlapping ....
.... **B. pinnatum**
8. Plants of deep forest (e.g., red cedar); trophophore deeply lobed but not fully divided to central axis ....**B. montanum**
8. Plants usually in less shady sites; trophophore divided to central axis 9
9. Lower trophophore lobes fan-shaped and overlapping .... **B. lunaria**
9. Trophophore lobes well separated .... **B. minganense**

Botrychium hesperium (Maxon & Clauson) Wagner & Lellinger [B. matricariifolium (Doll) Kock misapplied]. Plants 10-20 cm tall; trophophore gray-green, lance-shaped, up to 4 cm long, 1-2 times pinnately divided and lobed, primary lobes overlapping or nearly so; trophophore stalk 0-4 mm long; sporophore 1-2 times pinnate; 1-2 times length of trophophore.

Locally common in grasslands or low vegetation in gravelly soil of river terraces and slopes in montane zone; East, West. B.C., Alta. south to AZ, CO.

Botrychium lanceolatum (Gmel.) Angstrom. Plants 5-20 cm tall; trophophore dark green to yellow-green, triangular, up to 5 cm long, 1-2 times divided and lobed with 3-5 primary, almost overlapping lobes; trophophore stalk 0-1 mm long; sporophore 1-3 times pinnate, 1-2.5 times as long as trophophore.

Common beneath vegetation in montane and subalpine meadows and open forests; East, West. Ssp. *lanceolatum* has a broad, light green trophophore with middle segments > 2 mm wide exclusive of lobes; subsp. *angustisegmentum* (Pease & Moore) Clausen has slender, dark green trophophore with middle segments < 2 mm wide exclusive of lobes. AK to CA, NM; Newf. to TN. This is our only moonwort with the trophophore outline as wide as long.

Botrychium lunaria (L.) Swartz. Plants 3-15 cm tall; trophophore dark green, fleshy, oblong, up to 7 cm long, once pinnately divided into fan-shaped, overlapping pinnae; trophophore stalk 0-1 mm long; sporophore 1-2 times pinnate, 1-2 times as long as trophophore. Fig. 12.

Common in meadows, forest openings and on open slopes at all elevations; East, West. AK to Greenl. south to CA, NM, WI, PA.

Botrychium minganense Vict. Plants 4-15 cm tall; trophophore dull green, linear, up to 5 cm long, once pinnately divided, lobes not overlapping; trophophore stalk 0-2 cm long; sporophore usually once pinnate, 1.5-2.5 times as long as trophophore.

Rare in montane and subalpine meadows, thickets and forests; East, West. AK to Newf. south to CA, CO, WI, NY.

Botrychium montanum Wagner. Plants 4-12 cm tall; trophophore dull gray-green, fleshy, linear, up to 6 cm long, once pinnate into widely separate lobes; trophophore stalk 0.3-2 cm long; sporophore once pinnate, 1.5-4.5 times length of trophophore.

Uncommon in sparsely vegetated soil of moist (usually western red cedar) forest in the McDonald Valley. B.C. to CA, MT.

Botrychium multifidum (Gmel.) Rupr. [*B. silaifolium* Presl.]. Plants 10-40 cm tall; trophophore shiny green and leathery, evergreen, 4-25 cm long, wider than long, 2-4 times pinnately divided with overlapping primary lobes; trophophore stalk attached near ground level, 2-15 cm long; sporophore 2-3 times pinnate, ca. equal to length of trophophore.

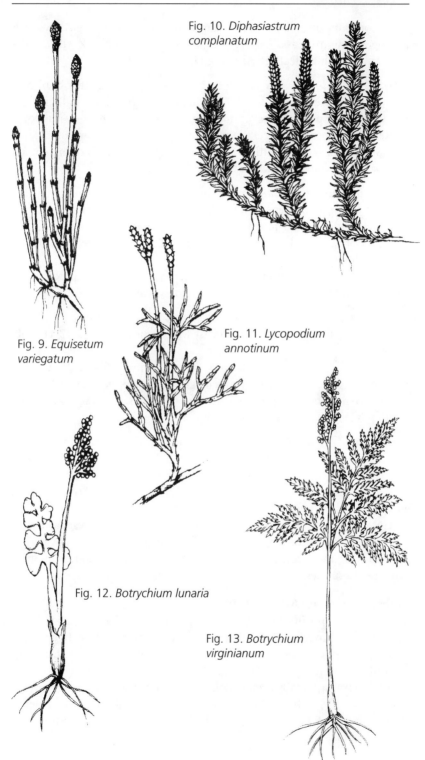

Fig. 10. *Diphasiastrum complanatum*

Fig. 9. *Equisetum variegatum*

Fig. 11. *Lycopodium annotinum*

Fig. 12. *Botrychium lunaria*

Fig. 13. *Botrychium virginianum*

Uncommon in moist or wet, organic soils of montane meadows, fens and open swamps; East, West. AK to Greenl. south to CA, AZ, IL, VA. The large size and shiny, leathery, dissected leaf is distinctive.

Botrychium paradoxum Wagner. Plants 7-15 cm tall; trophophores completely converted to a second sporophore; trophophore stalk half the length of fertile portion; sporophore double, once pinnate, 5-40 mm long.

Rare, often beneath vegetation, in montane grasslands and meadows; East, West. B.C. to Sask. south to OR, UT. The apparent rarity of this species may be partly due to its diminutive size.

Botrychium pinnatum St. John [*B. boreale* Milde misapplied]. Plants 5-15 cm tall; trophophore green, lance-shaped, up to 8 cm long, 1-2 times pinnately divided and lobed; primary lobes overlapping or nearly so; trophophore stalk 0-2 mm long; sporophore 2 times pinnate; 1-2 times as long as trophophore.

Locally common, often beneath vegetation in moist microsites in montane and subalpine forests and meadows; West, East. AK to CA, NV, CO.

Botrychium simplex E. Hitchc., Least Moonwort. Plants 3-13 cm tall; trophophore light green, lance-shaped, to 5 cm long, 1-2 times pinnately divided and lobed, not overlapping; trophophore stalk 0-3 cm long; sporophore once pinnate, 1-8 times length of trophophore.

Uncommon in montane grasslands and meadows; East, West. B.C., Alta. south to CA, CO; Ont., Newf. to IA, VA.

Botrychium virginianum (L.). Swartz, Virginia Grapefern. Plants 10-45 cm tall; trophophore without a stalk, pale green, thin, up to 25 cm long, broader than long, 3-4 times divided and lobed, primary lobes overlapping or nearly so; sporophore twice pinnate, 0.5-1.5 times length of trophophore. Fig. 13.

Common in dry to moist, montane forests and aspen groves; East, more common West. AK to Newf. to most of N. America and into S. America; Eurasia. Our most common and conspicuous *Botrychium* and the most widespread species in N. America.

Ophioglossum L., Adder's-tongue

Ophioglossum pusillum Raf. [*O. vulgatum* L. misapplied]. Plants 8-25 cm tall; trophophore elliptic, pale green, thin with a short stalk, to 10 cm long; sporophore undivided, rising above tropophore; sporophore tip long-conical, 1-4 cm long. July-Aug. Fig. 14.

Rare in moist, usually organic soil of montane meadows and fens; West. B.C. to Newf. south to CA, MT, IA, MD. Plants often grow in dense herbaceous vegetation, are inconspicuous, and may be more common than records indicate.

POLYPODIACEAE: POLYPODY FAMILY

Polypodium L., Polypody

Polypodium hesperium Maxon [*P. vulgare* L.]. Small perennial with long, creeping, scaly rhizomes; leaves scattered, all alike, 5-25 cm long, petiole shorter than the blade; leaf blade narrowly lance-shaped to ovate, glabrous, once pinnately divided almost to the central axis, divisions with shallowly wavy margins; spore clusters (sori) circular, borne midway between midvein and margins of leaf divisions; indusium lacking. Reference: Smith (1993b).

Uncommon in moist to wet, usually moss-covered boulders and cliffs in the montane and subalpine zones; West, more common East (Standley 1921). B.C., Alta. to Mex. Haufler et al. (1993) state that *P. hesperium* is rarely found on limestone, but this does not seem to apply in Glacier Park. High elevation plants are dwarfed (Jones 1910); the degree of licorice flavor in the roots is variable.

PTERIDACEAE: MAIDENHAIR FERN FAMILY

Perennials with leaves arising from a short rhizome or rootstock covered with scales or hairs; leaves usually scaly, pinnately or palmately divided; spores all alike; spore-bearing organs (sporangia) clustered into circular or elongate sori on the undersides of leaves, each usually with a membranous covering (indusium) at least when young; gametophytes small, green and terrestrial. Reference: Windham (1993).

Members of this family were formerly placed the Polypodiaceae *sensu lato* that included all of our ferns except grapeferns, moonworts and adder's-tongue (Ophioglossaceae).

1. Leaves fan-shaped in outline, appearing to be palmately branched ....
....**Adiantum**
1. Leaves linear to triangular, pinnately divided 2
2. Leaves all alike ....**Cheilanthes**
2. Leaves dimorphic, fertile leaves with longer and narrower ultimate segments than sterile leaves 3
3. Fertile leaves with sharp-pointed ultimate segments, petiole dark brown ......
....**Aspidotis**
3. Fertile leaves lacking sharp-pointed tips, petioles green or yellowish, at least above .... **Cryptogramma**

Adiantum L., Maidenhair Fern

Adiantum aleuticum (Rupr.) Paris [*A. pedatum* var. *aleuticum* Rupr.]. Plants with a short, stout, horizontal or ascending rhizome; leaves arching to erect, 10-60 cm tall with a glabrous, shiny, black petiole; leaf blades fan-shaped, appearing palmately divided into several pinnately divided branches (pinnae); pinnae with 15-35 leaflets (pinnules); pinnules light green, 10-20 mm long, fan-shaped to oblong with jagged front margins; sori crescent-shaped, borne on edges of pinnules, covered by inrolled pinnule margins. Fig. 15.

Common in wet rock crevices or shallow soil of cliffs, occasional in damp forest, usually in partial shade; montane and subalpine; East, West. AK to CA, AZ, CO; disjunct in Mex.; Newf. to PA. In Glacier Park maidenhair fern is most often associated with dripping cliff faces.

Aspidotis (Nutt. ex Hooker & Baker) Cope., Lace Fern

Aspidotis densa (Brack.) Lellin., Podfern [*Cheilanthes siliquosa* Maxon, *Cryptogramma densa* (Brack.) Diels., *Pellaea densa* (Brack.) Hook.]. Herbaceous perennial with short, branched rhizome covered with glossy, brown scales; leaves clustered, dimorphic, 8-25 cm long with long petioles and lance-shaped to triangular blades; sterile leaves 3-4 times pinnately divided into oblong ultimate lobes (pinnules), parsley-like; fertile leaves longer; sterile leaves leathery, 2-3 times pinnate with linear pinnules; pinnules with a prominent midrib and pale, papery inrolled margins partly covering the spore clusters (sori).

Rare in dry rock slides, or cliff crevices; montane and alpine; East, West. B.C. to CA, UT, WY; disjunct in Que. The fertile fronds resemble clustered pods of tumble mustard or tansy mustard.

Cheilanthes Swartz., Lip Fern

Cheilanthes gracillima D.C. Eaton, Lace Fern. Herbaceous perennial with short, branched rhizomes covered with brown scales; leaves all alike, clustered, linear, 5-25 cm long, petiole ca. as long as the blade; leaf blade twice pinnate, basal ultimate divisions (pinnules) often lobed; pinnules oblong, glabrous above but densely covered with hair-like brown scales below, margins rolled under, partially covering the spore clusters (sori).

Common in dry to moist rock crevices and steep talus slopes, montane to near treelines; East, West. B.C., Alta. south to CA, NV, UT. *Cheilanthes feei* Moore occurs on Madison limestone south of the Park. It apparently does not favor the less calcareous limestones of the Belt Series, but it could occur in the Divide Mountain area.

Cryptogramma R. Br., Rock Brake, Parsley Fern

Small plants with horizontal rhizomes; leaves glabrous, 2-3 times pinnately divided and lobed, dimorphic; fertile fronds with linear ultimate segments (pinnules); sterile leaves with broader pinnules, shorter than fertile fronds; sori concealed under inrolled pinnule margins.

1. Leaves clustered among old leaf petioles; inrolled margins of fertile pinnules meeting on the underside .... *C. acrostichoides*
1. Leaves scattered; inrolled fertile pinnule margins not meeting below ....
   .... *C. stelleri*

Cryptogramma acrostichoides R. Br. [*C. crispa* var. *acrostichoides* (R. Br.) C. B. Clarke]. Rhizome short, covered with old leaf bases; sterile leaves 5-25 cm long, petiole ca. half the length, blades 2-3 times pinnately divided, leathery, parsley-like, ultimate lobes elliptic; fertile leaves up to 30 cm long, mostly twice pinnate, ultimate segments linear with yellowish margins rolled under and often meeting in the center. Fig. 16.

Common in shallow soil of rock outcrops and rock slides, less common on wet cliffs at all elevations; East, West. AK to Ont. south to CA, NM, MN, MI; Asia. The clustered dimorphic leaves with numerous old, broken petioles at the base are diagnostic.

Cryptogramma stelleri (Gmel.) Prantl. Rhizome long and slender, easily broken; leaves scattered; sterile leaves 5-15 cm long, petiole ca. half the length, mostly twice pinnate with lobed, ovate ultimate segments; fertile leaves at least half again as long as sterile ones, mostly twice pinnate, ultimate segments linear with pale margins rolled under but not meeting in the center.

Uncommon in wet crevices of cliffs and outcrops among moss and other ferns in the montane and subalpine zones; East, West. AK to UT, CO; Ont. to Newf. south to IL, WV. The sparse growth form makes this species inconspicuous.

SELAGINELLACEAE: SPIKE-MOSS FAMILY

Selaginella Beauv., Spike-moss

Evergreen perennials with leafy prostrate stems and ascending or erect shoots; fibrous roots arise from points of stem branching; leaves stiff and spirally arranged on stem (ours); spores borne on somewhat differentiated leaves (sporophylls) spirally arranged into cone-like strobili, solitary and sessile at the top of fertile shoots; megaspores borne in lower portion of strobili, microspores in upper portion. Reference: Valdespino (1993).

1. Leaf bases abruptly joined to the stem; leaves and stem different color ....
.... *S. wallacei*
1. Leave bases gradually merging with stem; leaves and stem same color 2
2. Leaf tips white .... *S. scopulorum*
2. Leaf tips yellow.. *S. standleyi*

Selaginella scopulorum Maxon [*S. densa* var. *scopulorum* (Maxon) Tryon]. Plants forming cushion-like to dense mats; leaves stiff, strap-shaped with a groove on the back and distinct white, pointed tip, margins entire or with short, stiff hairs, 2-4 mm long including the tip, arranged in 4-6 ranks; strobili 1-2 cm long, sporophylls lance-shaped with entire margins on upper half and a short tip. Fig. 17.

Abundant in shallow, stony, sparsely vegetated soil of grasslands, meadows, fellfields and rock outcrops at all elevations; East, West. B.C. to

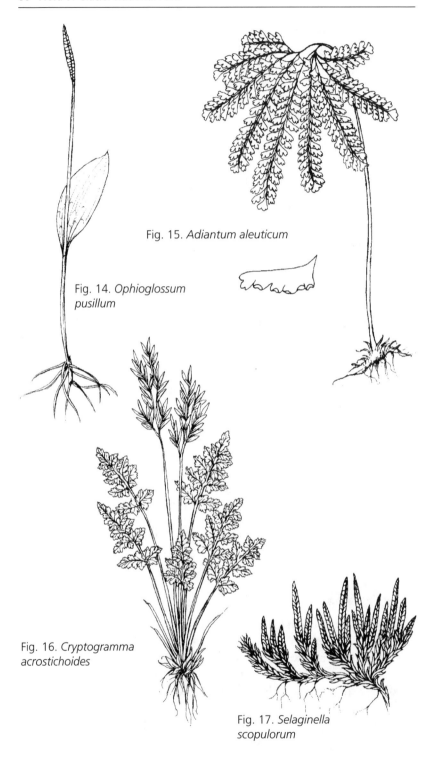

Fig. 15. *Adiantum aleuticum*

Fig. 14. *Ophioglossum pusillum*

Fig. 16. *Cryptogramma acrostichoides*

Fig. 17. *Selaginella scopulorum*

CA, AZ, NM. Paul Standley collected the type specimen at Cracker Lake. In less exposed habitats, the growth form of *S. scopulorum* may approach that of *S. wallacei*. Plants were used by the Blackfeet for inducing labor.

Selaginella standleyi Maxon [*S. densa* var. *standleyi* (Maxon) Tryon]. Similar to *S. scopulorum*; plants forming small mats; leaves with a yellow, translucent tip; sporophylls with small teeth towards the tip.

Uncommon in exposed, stony soil of fellfields and outcrops near or above treeline; East, West. AK to CO along the main range of the Rocky Mtns. Named for Paul Standley who wrote the first Flora of Glacier National Park and collected the type specimen at Sexton Glacier. Less common than *S. scopulorum*, even at high elevations.

Selaginella wallacei Hieron [*L. montanensis* Hieron.]. Plants usually loosely branched; leaves stiff, strap-shaped with a groove on the back and a distinct tip, 2-4 mm long including the tip, margins with spreading teeth; strobili 1-3 cm long, sporophylls lance-shaped with a short tip.

Common in sparsely vegetated soil of rock outcrops, and rocky slopes, especially talus, at all elevations; East, West. B.C., Alta. south to CA, MT. This species usually occurs in more protected sites than the other species and generally has a more loosely branched growth form, but growth form depends on habitat and is not always a reliable diagnostic character.

THELYPTERIDACEAE: MARSH FERN FAMILY

Phegopteris (Presl) Fee., Beech Fern
Phegopteris connectilis (Michx.) Watt [*Phegopteris polypodioides* Fee, *Thelypteris phegopteris* (L.) Slosson]. Perennial with long rhizomes; leaves all alike, scattered, 15-40 cm long, the petiole longer than blade; leaf blade lance-shaped, pinnately divided, lowest divisions (pinnae) reflexed, upper divisions united at base; pinnae pinnately lobed, margins and veins beneath with spreading hairs; spore clusters (sori) circular, borne on fertile pinnule margins; gametophyte green, terrestrial. Reference: Smith (1993c).

Rare in crevices of wet, shaded, cliffs of subalpine basins near Sperry Glacier (Jones 1910, McLaughlin 1935). AK to Greenl. south to OR, MT, Sask., MO, VA. Often growing with *Adiantum aleuticum*. Throughout most of its range, including other sites in Montana, this fern occurs in moist soil of mesic forests.

GYMNOSPERMS
CONIFERS

CUPRESSACEAE: CEDAR FAMILY

Evergreen shrubs or trees with small, simple, thick, scale- or needle-like leaves; pollen and seeds borne in separate cones on branch tips; pollen cones small, solitary, with umbrella-shaped stamens; seed cones with 4-6 dry scales (*Thuja*) or 3-8 fleshy, united scales, resembling a berry (*Juniperus*); usually 1-2 seeds borne at the base of each scale.

1. Seed cones fleshy, nearly round, berry-like; shrubs or small trees; some leaves usually ascending and needle-like .... **Juniperus**
1. Seed cones woody, dry, with spreading scales; trees; all leaves scale-like and appressed .... **Thuja**

Juniperus L., Juniper

Evergreen shrubs or trees with small, stiff leaves, opposite or in whorls of 3; pollen cones and seed cones on separate plants (ours); pollen cones with 12-16 stems in 2's or 3's; seed cones globose, fleshy, berry-like, maturing the second summer, with mostly 1-5 wingless seeds.

Although the juniper fruit appears to be a berry, close examination will reveal signs of the joined cone scales. Immature plants of *J. scopulorum* may superficially resemble *J. communis* by having needle-like leaves, but these are not jointed at the base as in *J. communis*. The presence of a nearby adult tree is often the best clue.

1. Shrub or small tree, usually with a single trunk and ascending branches ....
.... ***J. scopulorum***
1. Shrubs, with spreading branches near the base 2
2. Some leaves, especially on slow-growing branches, scale-like and appressed, covering the stem .... ***J. horizontalis***
2. All leaves spreading and needle-like .... ***J. communis***

Juniperus communis L., Common Juniper. [*J. sibirica* Burgst.]. Prostrate to erect and rounded shrub 100 cm high with brown, fibrous bark; leaves stiff, needle-like, 7-12 mm long, whitish on upper surface but green below, borne in whorls of 3, jointed where they meet the stem; seed cones bluish-black, 6-9 mm long; seeds 2-3. Fig. 18.

Abundant in drier forests and on open slopes or outcrops in the montane and subalpine zones; East, West. Our plants have been considered var. *montana* Aiton; however, Adams (1993) asserts that they are var. *depressa* Pursh. AK to Greenl. south to CA, AZ, NM, IL, SC. Plants may be prostrate in exposed habitats. "Berries" of this species are used to flavor gin.

Juniperus horizontalis Moench, Creeping Juniper. Prostrate shrub with trailing branches and brown, stringy bark; leaves opposite, 2 kinds: scale-like, entire-margined, glandular ca. 2 mm long, and appressed or needle-like, 4-8 mm long and ascending, the latter usually on vigorous stems; foliage bright to bluish green but often turning purplish in winter; seed cones blue-black, 5-7 mm long; seeds 1-6.

Locally common, especially in shaly soil, on exposed slopes and flats along the east front of the mountains to above treeline, East. AK to Lab., south to CO, IL, NY. Plants in sheltered sites may become semi-erect.

Juniperus scopulorum Sargent, Rocky Mountain Juniper. Erect shrub or small tree up to ca. 6 m tall with stringy, reddish-brown bark, ascending branches and a conical crown; leaves green but often with a whitish, waxy coat (glaucus) otherwise similar to *J. horizontalis*; seed cones blue with glaucus bloom, 5-6 mm long; seeds 1-2.

Uncommon, mainly on older terraces along the Middle and North forks of Flathead River and on rocky, montane slopes of the McDonald Valley. B.C., Alta., south to AZ, NM.

Thuja L., Arborvitae

Thuja plicata Donn ex D. Don., Western Red Cedar. Large evergreen tree to 45 m tall with reddish-brown, fibrous bark and a conical crown with a drooping tip; lower trunk of older trees broad and often buttressed; branchlets pendant, arrayed in flat sprays; leaves glossy green, scale-like, opposite, overlapping, appressed against the flattened twigs; pollen cones 1-3 mm long, reddish; seed cones ellipsoid, 8-12 mm long, with 4-6 pairs of woody, overlapping scales, the top and bottom pairs sterile, maturing the first summer; fertile scales with 2 seeds each.

Common in moist to wet, montane forest, especially along streams; West, especially in the McDonald Valley and adjacent drainages. AK to CA, ID, and MT. Western red cedar is long-lived and shade-tolerant and often occurs with cottonwood and spruce along streams and with western hemlock in the wet forests of the McDonald Valley. The tree contains anti-fungal compounds that make the wood resistant to rot. In eastern N. America the name "red cedar" is applied to *Juniperus virginiana*.

PINACEAE: PINE FAMILY

Evergreen or deciduous trees with resinous and aromatic sap; leaves needle-like, spirally arranged, solitary or borne in clusters; pollen and seed cones borne on the same plant; pollen cones small with spirally arranged stamens; seed cones large and woody, formed of spirally arranged scales, maturing in 1 to many seasons; fertile scales bearing 2 seeds on the inner surface and subtended by a papery bract; seeds winged in most species.

These are the dominant plants in most of Glacier Park, defining its boreal nature. The aromatic sap makes them highly flammable, especially during late summer of dry years, and fire has played a major role in the evolution of many species and their associated communities.

1. Leaves borne in clusters (fascicles) 2
1. Leaves borne singly, not fascicled 3
2. Leaves >5 per cluster, autumn deciduous .... **Larix**
2. Leaves 2-5 per cluster, evergreen .... **Pinus**
3. Old needles breaking off above the base, causing young branches to be rough where needles have fallen 4
3. Leaves falling from the base, young branches smooth where needles have fallen 5
4. Needles sharp-pointed, not flattened .... **Picea**
4. Needles flattened with a rounded tip .... **Tsuga**
5. Cones erect and disintegrating on the tree; terminal buds rounded ..... **Abies**
5. Cones pendant, falling whole after seeds are shed; terminal buds sharp-pointed .... **Pseudotsuga**

Abies Miller. Fir

Evergreen trees with conical or spire-like crown and thin, smooth bark; branches whorled; primary limbs branched in one plane to form large, flat sprays; leaves flat, single, spirally arranged, persisting 5 or more years, leaving a slight, nearly circular depression on twigs after falling; buds globose and resinous; cones borne on year-old twigs; pollen cones clustered; seed cones cylindrical, mature in 1 season, disintegrating on the tree; scales fan-shaped, pubescent, falling individually; seeds winged.

1. Leaves with blunt or retuse tip, lines of white stomates on lower surface only, borne in a single plane so twigs appear flat; uncommon .... **A. grandis**
1. Leaves with pointed tip, curving upward and not in 1 plane; white stomate lines on both surfaces; abundant .... **A. lasiocarpa**

Abies grandis (Douglas ex D. Don) Lindley, Grand Fir. Medium-size tree up to ca. 50 m tall with a conical crown and branches that bend down and then out; bark gray with reddish furrows, becoming brown in older trees; leaves 2-6 cm long, borne in a single plane opposite each other on the twigs, stomates on underside only; seed cones usually green, 6-11 cm long.

Uncommon in moist, montane forest in the McDonald Valley and Middle Fork Flathead drainage. B.C. to CA, ID, MT. Grand fir is shade tolerant and grows rapidly, but it is susceptible to rotting fungi and thus relatively short-lived. Grand fir cones are borne at the tops of the trees and disintegrate in place, so they are rarely seen unless cut by squirrels.

Abies lasiocarpa (Hooker) Nutt., Subalpine Fir [*A. bifolia* A. Murray]. Generally a small tree up to ca. 30 m tall with a narrow crown; bark gray but splitting to reveal brownish layer beneath; leaves 2-4 cm long, turned upward, stomates on both surfaces; seed cones deep blue, 6-10 cm long. Fig. 20.

Abundant forest tree from upper montane zone to treeline and commonly forming hedge-like krumholz stands in the alpine zone, the dominant plant of the subalpine zone both sides of the Divide. Yuk. to CO, AZ and NM. Subalpine fir is shade-tolerant but relatively short-lived and readily killed by fire. Near treeline lower limbs pressed against the ground by snow pack can root and send up new shoots, thus forming clumps of stems. Lower limbs under snow are susceptible to snow mold, a fungus that gives the leaves a sooty, matted appearance. Blackfeet Indians used the resin for incense and to treat fevers and colds. *Abies bifolia* is the Rocky Mountain segregate of *A. lasiocarpa*, separable on chemical and minor morphological characters.

Larix Miller, Larch

Deciduous trees with open crowns, whorled branches, and scaly bark that thickens with age; short, nubbin-like spur shoots prominent on twigs and bearing clusters of pale green leaves and/or cones; pollen cones solitary on spur shoots; seed cones solitary, maturing first season; scales thin, ovoid, pubescent; bracts projecting beyond scale in a long awn; seeds winged.

1. Young branches densely covered with long, tangled hairs; seed cones > 35 mm long; trees of the subalpine and alpine zones .... *L. lyallii*
1. Young branches glabrous or somewhat covered with short hairs; seed cones < 35 mm long; trees of the montane zone .... *L. occidentalis*

Larix lyallii Parl., Subalpine Larch. Small trees to 20 m tall with open, often asymmetrical crowns; bark covered with red- to purple-brown flakes; leaves 30-40 per spur, 4-angled, 2-4 cm long; seed cones 35-45 mm long, yellow to purplish-green.

Locally common near treeline where snow lies late; East, West. Alta., B.C., WA, ID, MT. Usually the dominant tree where it occurs. Stands occur as far west of the Divide as Boulder Pass and east to near Goat Haunt Lake.

Larix occidentalis Nutt., Western Larch. Large trees to 50 m; older trees without branches for most of their height; bark of mature trees thick, deeply furrowed, covered with cinnamon-colored plates; leaves 15-30 per spur, 2-4 cm long, 3-angled; seed cones brown to reddish, 25-30 mm long. Fig. 21.

Common in montane forests west of the Divide, especially on cool slopes. B.C. to OR, ID, MT. Barrett (1997) reports a few trees from the St. Mary Valley. Recruitment of western larch occurs only in open ground, usually after fire. It is long-lived, and the thick bark of older trees resists ground fires, so although it is an early seral tree, it dominates in many forests. Needles turn yellow before falling and provide much of the autumn color in northwest Montana. Old larch trees often have rectangular cavities near the base, excavated by pileated woodpeckers in search of wood ants.

Picea A. Dietrich, Spruce

Evergreen trees with thin, scaly bark, whorled branches and a conical crown; leaves 4-angled with stomates on all surfaces, rigid, sharp-pointed, single, spirally arranged and spreading out in all directions from the twigs; pollen cones clustered; seed cones pendant, ovoid to cylindric, maturing in 1 season; scales thin, subtended by smaller bract; seeds 2 per scale and prominently winged.

Our two species of spruce hybridize extensively. Pure *P. glauca* is most likely encountered in the lower valleys west of the Divide, while pure *P. engelmannii* is the common spruce of higher elevations. Hybrids with intermediate seed cone characters are most likely to be encountered in the montane zones, especially west of the Divide.

1. Twigs finely pubescent; cone scales with wavy-margined tip . **P. engelmannii**
1. Twigs glabrous; cones scales with entire margins .... **P. glauca**

Picea engelmannii Parry ex Engelmann, Engelmann Spruce. Trees up to ca. 50 m tall with gray to reddish brown bark and narrow, spire-like crowns; leaves blue-green, 2-3 cm long; seed cones 3-7 cm long, yellow- to purplish-brown; scales widest above middle with wavy margins toward the tip. Fig. 22.

Common in subalpine forests, less common along streams and other moist areas in the montane zone, occasionally forming krumholz hedges at or above treeline. Our plants are var. *engelmannii* occurring from B.C., Alta. to AZ, NM, Mex.

Picea glauca (Moench) Voss, White Spruce. [*P. canadensis* (Miller) B. S. & P.] Small tree to 30 m tall with gray-brown bark and a conical crown; leaves 1-2 cm long; seed cones 25-60 mm long; scales fan-shaped, widest near the entire-margined tip.

Common along streams and in swamp forests where water is near the surface throughout the year; West. AK to Lab. south to MT, CO, SD, WI, NY. Most trees referable to this species have some characters intermediate to *P. engelmannii*.

Pinus L., Pine

Evergreen trees with conical to flat-topped crowns; groups of 2-5 needle leaves borne on nubbin-like, spur shoots (fascicles), these subtended by sheathing scale leaves at the base; pollen cones densely clustered at the base of current year's growth; seed cones ovoid to cylindric, maturing the second season, opening at maturity or much later (serotiny); scales numerous, spirally arranged, woody, often with a thickened tip; seeds 2 per scale.

Whitebark and limber pine have ascending branches and flat-topped crowns that more resemble broad-leaved trees. These two species as well as western white pine (soft pines) are succumbing to an epidemic of the white pine blister rust, a European disease that has gooseberries and currants

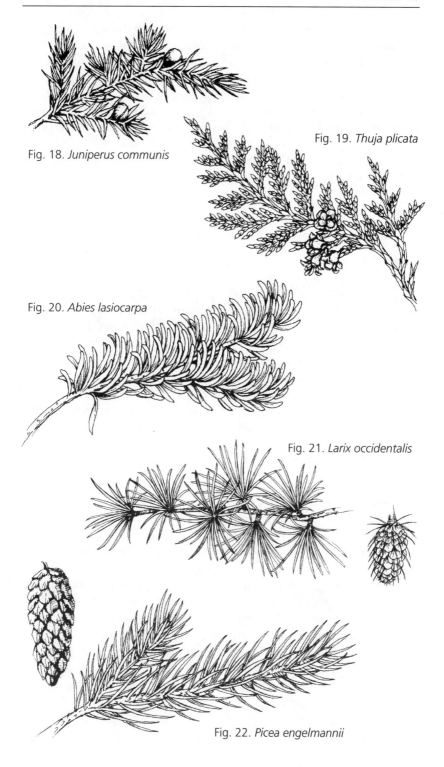

Fig. 18. *Juniperus communis*

Fig. 19. *Thuja plicata*

Fig. 20. *Abies lasiocarpa*

Fig. 21. *Larix occidentalis*

Fig. 22. *Picea engelmannii*

(*Ribes* spp.) as an intermediate host. Distinctive whitebark pine snags are commonly seen throughout the Park's high country.

1. Needles in clusters of 5; cone scales lacking a terminal prickle (soft pines)... 2
1. Needles in clusters of 2-3; cones scales with a terminal prickle (hard pines) . 4

2. At least some needles > 7 cm long; cones ca. 3 times as long as wide; low elevations west of the Divide.. *P. monticola*
2. Needles < 7 cm long; cones ca. 2 times as long as wide; high elevation or east of the Divide 3

3. Cones 8-25 cm long, falling intact after shedding seed; scales thinner toward the tip than base.. *P. flexilis*
3. Cones 4-8 cm long, seldom falling intact; scales thicker toward the tip than the base.. *P. albicaulis*

4. Needles in clusters of 2, 3-6 cm long ....*P. contorta*
4. Needles mostly in clusters of 3, > 10 cm long .... *P. ponderosa*

Pinus albicaulis Engelmann, Whitebark Pine. Small trees to 20 m tall with smooth, light gray bark, ascending branches, and a rounded or flat-topped crown; leaves yellow green, 3-7 cm long, 5 per fascicle; seed cones ovoid, 4-8 cm long, remaining on the tree and closed until opened and/or dislodged by squirrels or birds; scales thin at the base but thickened toward the tip; seeds wingless, 7-11 mm long.

Common in subalpine forest and in tree islands near treeline; East, West. B.C., Alta. south to CA, NV, and WY. Twisted and stunted krumholz forms can often be found in sheltered sites above treeline. Clark's nutcracker routinely opens the cones, collects the "pine nuts" and buries them for later use. In this way whitebark pine seed is dispersed. Pine squirrels collect and cache the cones, and bears often raid these caches; however, it is unlikely that these mammals are as effective at dispersing seed as the nutcracker.

Pinus contorta Douglas ex Loudon, Lodgepole Pine. Small, slender tree to 20 m with thin, scaly brown or gray bark and whorled horizontal branches forming a conical crown; leaves yellow-green, 5-8 cm long, 2 per fascicle; seed cones ovoid but asymmetrical, 3-6 cm long, maturing the second season; scales tongue-shaped with a spine tip; seeds with a conspicuous wing. Fig. 23.

Abundant in the montane and subalpine zones in mixed or nearly pure stands; East, West. Our plants are var. *latifolia* Engelm. AK to CA, UT, CO, SD. Lodgepole pine is short-lived and shade-intolerant. It is well adapted to fire. Some seed cones open upon maturity and fall, while others remain on the tree until opened by heat (serotinous). Trees are easily killed by fire, but afterward millions of winged seeds are shed, and even-age stands of lodgepole often develop. Serotinous cones may remain on the trees for so long that they become embedded in wood. Native Americans used the poles for tipis.

Pinus flexilis James, Limber Pine. Small tree to 15 m tall, very similar to *P. albicaulis*; leaves green, 5 per fascicle, 4-7 cm long; seed cones ovoid, 5-12 cm long, maturing the second season and falling shortly after opening; scales rhombic, thinner towards the tip; seeds 10-15 mm long, wingless.

Uncommon mostly on exposed slopes and ridges, montane and subalpine; primarily East, but Standley (1921) reports it for Granite Park. B.C., Alta. south to CA, AZ, NM. Unlike whitebark pine, limber pine cones open, drop their seed, then fall intact. Many mature trees have a carpet of old cones beneath.

Pinus monticola Douglas ex D. Don., Western White Pine. Large tree to 50 m tall with thin bark, smooth at first becoming scaly with age; crown conical to rounded; leaves light blue-green, 5-10 cm long, 5 per fascicle; seed cones narrowly elliptic, 15-25 cm long, maturing the second season; scales thin, tongue-shaped; seeds with conspicuous wings.

Common in mixed coniferous, montane forest; West. B.C. to CA, NV, MT. Western white pine is intolerant of shade and long-lived although prone to blister rust and sensitive to fire.

Pinus ponderosa Douglas ex Loudon, Ponderosa Pine. Large trees to 50 m tall; bark of old trees thick, furrowed, covered with scales that resemble pieces of a jigsaw puzzle; crown open and rounded with spreading branches; leaves yellow-green, 12-20 cm long, mainly 3 but often 2 per fascicle, clustered on branch ends; seed cones broadly ovoid, 8-15 cm long, maturing the second season; scales thick with a terminal prickle; seeds with a conspicuous wing.

Locally common in the valley zone along the North Fork Flathead River and in the lower McDonald drainage, often associated with grasslands. It is long-lived, fire resistant, shade intolerant and appears to thrive in drier habitats. Our plants are var. *ponderosa*. B.C. to NE; Mex. In late spring and early summer when the sap is flowing and the sun warms the trunk, the bark of ponderosa pine may have an odor of vanilla.

Pseudotsuga Carriere, Douglas Fir.
Pseudotsuga menziesii (Mirbel) Franco, Douglas Fir [*P. mucronata* (Raf.) Sudw.]. Large evergreen tree up to 50 m tall with spreading branches and a narrow to broadly conical crown; bark of older trees thick, furrowed and gray with reddish brown between furrows; leaves yellow- to blue-green, single, spirally arranged and spreading, 2-3 cm long, stomates lacking on upper surface; buds conical; seed cones narrowly ellipsoidal, 4-7 cm long, maturing the first season and shed entire; scales broadly rhombic, stiff, pubescent, subtended by a longer 3-lobed bract; seeds winged.

Abundant tree of dry to mesic, montane forests, less common in the subalpine zone; East, West. Our plants are var. *glauca* (Mayr) Franco. B.C.

to Alta. south to CA, TX, Mex. Douglas fir is somewhat tolerant of shade, fast-growing, long-lived, and fire-resistant when old. The 3-lobed bract with the long, narrow, central lobe is diagnostic. This is the most important lumber tree in the Pacific Northwest. It is an important early seral species throughout the Park, but is rarely the indicated climax.

Tsuga (Endlicher) Carriere, Hemlock

Tsuga heterophylla (Raf.) Sargent, Western Hemlock. Medium-size evergreen trees to 35 m tall with drooping branches and a narrow crown; bark brown, thin and scaly; leaves shiny yellow-green above with broad, white stomatal bands below, flattened, 10-20 mm long, single and borne in a single plane or nearly so; buds ovoid; pollen cones solitary; seed cones green to brown with age, narrowly ellipsoid, 15-25 mm long, maturing the first season; scales oblong-ovoid; seeds winged. Fig. 25.

Common in mesic to wet, montane forests in the McDonald Valley; locally common in moister sites in other valleys west of the Divide. AK to CA, ID, MT. Western hemlock is very shade-tolerant but not resistant to fire and is intolerant of drought. Hemlock seedlings may survive in the deep shade of canopy trees for decades. Growing tips of old and young trees droop.

TAXACEAE: YEW FAMILY

Taxus L., Yew

Taxus brevifolia Nutt., Pacific Yew. Evergreen shrub or small tree to 5 m tall with spreading or drooping branches forming an open, conical crown; bark thin and brown or purplish; leaves yellow-green above, paler beneath, 14-18 mm long, single but borne opposite each other in 1 plane; pollen and seed cones borne on separate plants; pollen cones small, globose; seed cones reduced to a single seed surrounded by a red, fleshy, pea-size, berry-like cup (aril); seed 5-6 mm long. Fig. 26.

Locally common, often forming dense thickets, in wet, montane forest; West, rare East. AK to CA, ID, and MT. Yew is very shade-tolerant, grows slowly, and is easily killed by fire; thus, it is usually found only in old forests or sites protected from fire. Native Americans used the wood to make bows.

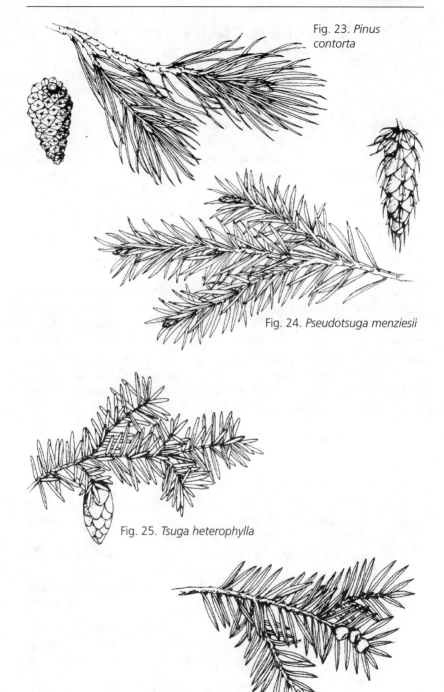

Fig. 23. *Pinus contorta*

Fig. 24. *Pseudotsuga menziesii*

Fig. 25. *Tsuga heterophylla*

Fig. 26. *Taxus brevifolia*

FLOWERING PLANTS: DICOTS

ACERACEAE: MAPLE FAMILY

Acer L., Maple

Acer glabrum Torrey, Rocky Mountain Maple [*A. douglasii* Hook.]. Large shrub or small tree up to 6 m high, with gray bark and reddish, glabrous stems; leaves and branches arising opposite each other; leaves petiolate, heart-shaped, 3-12 cm across, 3- to 5-lobed, with twice-toothed margins; flowers, unisexual, 5-10 mm across, borne in an open, dichotomously branched inflorescence at the base of petioles; sepals 5, separate; petals 5 or lacking; male flowers with 8-10 stamens and reduced ovary; female flowers with 2 styles and non-functional stamens; fruit 2-seeded, separating at maturity, each with a divergent, crescent-shaped wing 15-35 mm long. May-June. Fig. 27.

Var. *douglasii* (Hook.) Dippel, with shallowly lobed leaves and sometimes an arborescent growth form, occurs primarily at low elevations, West. Var. *glabrum*, with more deeply lobed leaves, is abundant in avalanche chutes and moist to wet forests, especially along streams and on rocky slopes in the montane and lower subalpine zones; East, West. AK to Alta., south to CA and NM; NE. Norway maple (*A. platanoides*) is planted as an ornamental in West Glacier. Although it escapes cultivation in the Missoula area, I have seen no evidence that it does so in Glacier Park.

AMARANTHACEAE: PIGWEED FAMILY

Amaranthus L., Pigweed

Annual plants with alternate, petiolate leaves and erect or prostrate stems; flowers unisexual, inconspicuous, subtended by persistent bracts, borne in tight clusters or spikes; petals lacking; sepals 3-5; stamens 2-5; ovary superior with 2-3 stigmas; fruit an ovoid, 1-seeded capsule.

These species are weeds in warmer regions, but they do not appear to be invasive in our area.

1. Stems erect, coarsely hairy above .... ***A. retroflexus***
1. Stems prostrate, glabrous or nearly so .... ***A. graecizans***

Amaranthus graecizans L. [*A. blitoides* Watson]. Stems prostrate, branched, often purplish; leaves spoon-shaped; male and female flowers on same plant, densely clustered in leaf axils; sepals 5, 1-2 mm long; stamens 3; bracts awn-tipped.

Rare in disturbed ground near West Glacier (Standley 1921). Native to arid regions of western N. America but probably adventive in Glacier Park.

Amaranthus retroflexus L. Stems erect, branched, 15-50 cm tall; herbage rough and hairy; leaf blades ovate, 4-10 cm long; male and female flowers on separate plants, borne in stout spikes in leaf axils; sepals papery, 2-4 mm long; stamens 5; bracts awn-tipped, twice as long as sepals.

Rare in severely disturbed soil, usually in developed areas in the valleys; East, West. Throughout much of N. America but probably adventive in Glacier Park.

ANACARDIACEAE: SUMAC FAMILY

Toxicodendron P. Mill., Poison Ivy
Toxicodendron rydbergii (Small ex Rydb.) Greene, Poison Ivy [*Rhus radicans* L.]. Stems solitary, woody at the base, arising from underground stems; leaves alternate, divided into 3; leaflets ovate with wavy margins, glabrous, 3-12 cm long, tapered to a point; flowers in tightly branched inflorescences in leaf axils; petals 5, separate, ca. 3 mm long, yellowish; sepals 5, ca. 1 mm long; ovary 1, superior; stigmas 3; fruit a yellowish berry (drupe), 5-6 mm, with 1 seed. June. Fig. 28.

Uncommon in stony soil in the flood zone along the forks of the Flathead River and rare on warm, montane slopes in open forest; West. B.C. to N. S. south to OR, AZ, TX, TN, VA. Poison ivy appears to respond positively to mild disturbance such as flooding or gradual slumping. Oils of this plant often cause dermatitis.

APIACEAE: PARSLEY FAMILY

Herbaceous perennials or biennials with alternate or basal, usually divided or deeply lobed leaves; flowers borne on stalks originating from the same point and forming simple or compound umbrella-like inflorescences (umbels); each secondary umbel (umblet) usually subtended by bracts (involucel); flowers small, radially symmetrical with 5 calyx lobes, 5 separate petals and 5 stamens; ovary inferior, 2-celled; fruit of 2 halves joined face-to-face (schizocarp); each half (mericarp) 1-seeded, flattened or hemispheric in cross section with 5 ribs or wings on the outer face.

Flower morphology throughout the family is relatively undifferentiated; mature fruit and often also flowers are required for genus and species identification. *Ligusticum canbyi* Coult. & Rose, similar to *Angelica* but with spherical, narrowly winged fruits, occurs just south of the Park.

1. Basal leaves undivided 2
1. Lowest leaves divided 3

2. Basal leaves heart-shaped, toothed; some stem leaves divided .... **Zizia**
2. Leaves entire-margined .... **Bupleurum**

3. Fruit with hooked prickles... **Sanicula**
3. Fruit lacking prickles 4

4. At least some ultimate leaflets > 15 mm wide 5
4. Ultimate leaflets all < 15 mm wide; leaves often fern-like 8

5. Leaves with 3 leaflets.. **Heracleum**
5. Some leaves with > 3 leaflets... 6

6. Fruit linear, ca. 3 times as long as wide .... **Osmorhiza**
6. Fruit ovate or elliptic, < 2 times as long as wide 7

7. Fruit heart-shaped with corky marginal wings .... **Pastinaca**
7. Fruit ovate or elliptic with papery wings .... **Angelica**

8. Fruit club-shaped, > 3 times as long as wide .... **Osmorhiza**
8. Fruit narrowly elliptic or wider 9

9. Leaflets linear, > 5 times as long as wide 10
9. Leaflets < 5 times as long as wide.. 13

10. Plants with small bulbs in the upper leaf axils................................... **Cicuta**
10. Plants lacking bulbils.. 11

11. Flowers yellow, fruit flattened on front and back .... **Lomatium**
11. Flowers white, fruit rounded on back 12

12. Leaflets with toothed margins, wet habitats .... **Sium**
12. Leaflets with entire margins, moist to dry sites .... **Perideridia**

13. Fruit flattened on front and back with marginal wings .... **Lomatium**
13. Fruit rounded on the back 14

14. Flowers yellow; plants usually < 20 cm tall .... **Musineon**
14. Flowers white; plants usually > 30 cm tall 15

15. Stem with a swollen base; lateral veins of leaflets terminating in the sinuses
   between teeth .... **Cicuta**
15. Stem without a swollen base, lateral leaf veins otherwise 16

16. Leaves finely divided, fern-like; plants of dry to moist habitats .... **Carum**
16. Leaves once divided into long leaflets; wet sites .... **Sium**

Angelica L., Angelica

Taprooted perennials with glabrous, erect stems and twice pinnately divided
leaves with petioles that sheath the stem; umbels compound; bracts of
involucel linear or lacking; sepals minute, petals white or yellow; fruit
winged, flattened in cross section, and elliptic to ovate in outline.

   A. roseana Hend., with smaller leaves than those of *A. arguta*, occurs on
high-elevation rocky slopes west and south of the Park.

1. Flowers yellow; usually 1 umbel per plant .... ***A. dawsonii***
1. Flowers white; usually >1 umbel per plant .... ***A. arguta***

Angelica arguta Nutt., White Angelica [*A. lyallii* Wats.]. Stems robust, 50-
150 cm high; leaflets ovate to narrowly elliptic, sharply toothed to shallowly
lobed, 3-10 cm long; involucel inconspicuous or lacking; petals white; fruit
ovate, 4-7 mm long, glabrous, marginal wings much wider than central
ones. July-Aug.

   Common in avalanche areas and moist forests and meadows, especially
aspen groves and along streams; montane and subalpine; East, West. B.C.

and Alta. to CA, UT, WY. Stems and roots have an anise-like smell. The succulent plants are important spring forage for bears.

Angelica dawsonii Watson, Yellow Angelica. Stems 20-70 cm tall from a simple or branched root crown; leaflets lance-shaped, finely toothed, 2-5 cm long; bracts of umbel large and deeply toothed; bracts of involucel smaller and less divided; petals yellow; fruit glabrous, broadly elliptic, 4-7 mm long, marginal wings slightly wider than central ones. July-Aug. Fig. 29.

Common in montane and subalpine meadows and open forest, often with beargrass; East, West. B.C., Alta., ID, MT. This distinctive plant is endemic to northwest MT and adjacent Canada.

Bupleurum L.

Bupleurum americanum Coulter & Rose, American Thoroughwax. Perennial with 1-many stems from a taproot and branched crown; herbage glabrous and glaucus; stems 5-30 cm tall; leaves undivided, narrowly lance-shaped, entire-margined, 2-8 cm long; umbels subtended by lance-shaped, leaf-like bracts; umblets compact, subtended by involucel of united bracts 3-5 mm long; petals yellow; fruit oblong with raised ribs, glabrous, 3-4 mm long. June-Aug. Fig. 30.

Common in stony soil of grasslands and open, exposed slopes; montane to near or above treeline; near or east of the Divide. AK to ID, WY; Asia. Our only member of the Parsley Family without any divided leaves.

Carum L., Caraway

Carum carvi L., Caraway. Taprooted biennial with glabrous herbage and stems up to 80 cm tall; leaves fern-like, 3-4 times pinnately divided into small segments; petioles sheathing the stem; umbels open; bracts of umbel and umblets narrow or lacking; petals white; fruit glabrous, elliptic with evident ribs, 3-4 mm long. July-Aug.

Moist disturbed meadows, often along streams. Reported by Standley (1921) for the foot of Sherburne Lake, an area that is probably now under water. Native to Europe; introduced in Glacier Park. The aromatic seeds are used in cooking.

Cicuta L., Water Hemlock

Perennials with inflated stem bases and thickened roots; leaves 2-3 times pinnately divided; bracts of umbels few or lacking; involucel bracts narrow; petals white; fruit ovoid with thick ribs.

The roots of both species are highly poisonous.

1. Small bulbs present in axils of upper leaves .... **C. bulbifera**
1. Bulbils lacking .... **C. douglasii**

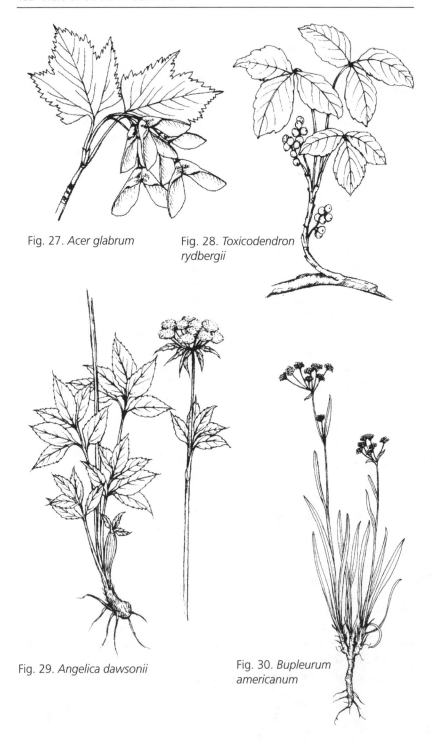

Fig. 27. *Acer glabrum*

Fig. 28. *Toxicodendron rydbergii*

Fig. 29. *Angelica dawsonii*

Fig. 30. *Bupleurum americanum*

Cicuta bulbifera L. Slender, glabrous plants with stems 30-70 cm tall; ultimate leaf segments linear and nearly entire, 1-4 cm long; uppermost leaves undivided, bearing small bulbs in the axils; fruit 1-2 mm long. July-Aug.

Locally common in a few fens in North Fork Flathead and McDonald valleys. AK to Newf. south to OR, MT, NE, LA, FL. Most plants lack flowers and fruit; the linear leaflets, bulbils and fen habitat are distinctive.

Cicuta douglasii (D. C.) Coulter & Rose [*C. occidentalis* Greene, *C. maculata* L. var. *angustifolia* Hooker]. Stout plants with stems 50-100 cm tall; ultimate leaf segments lance-shaped and sharply toothed, 2-6 cm long; fruit 2-4 mm long. July-Aug. Fig. 31.

Uncommon in saturated soil along montane streams, marshes and swamps; East, West. AK to Alta., CA, NM, Mex. This extremely poisonous plant can be distinguished from the more common *Sium suave* by having leaves with lateral veins that terminate in the sinuses between the teeth of the leaf margins.

Heracleum L.

Heracleum sphondylium L., Cow Parsnip [*H. lanatum* Michaux]. Robust perennial with hairy herbage and hollow stems 1-2 m tall from a taproot; leaves divided into 3 maple-leaf-like segments 10-30 cm long; petioles clasping the stem; umbels up to 20 cm across; umbels and umblets subtended by many linear bracts; petals white; fruit heart-shaped, 7-12 mm long, with narrow ribs and dark lines extending half way to the base. June-Aug. Fig. 32. Color plate 22.

Common in moist soil of avalanche slopes, thickets and open forest or aspen groves, often along streams in the montane and subalpine zones; East, West. The large size and 3-parted leaves are diagnostic. AK to Newf. south to CA, AZ, OH, GA; Asia. The succulent plants are important spring forage for bears. Blackfeet Indians used this plant in religious ceremonies and ate roasted young shoots.

Lomatium Rafinesque, Biscuitroot, Desert Parsley

Perennials with tuberous or woody roots; stems usually short or lacking; leaves dissected into small segments; flowers yellow or white (purple); bracts of umbel mostly absent; involucel usually present; fruit strongly flattened with marginal wings.

1. Leaves divided into linear segments > 1 cm long 2
1. Ultimate leaf segments < 1 cm long 3

2. Herbage glabrous; involucel lacking .... **L. ambiguum**
2. Herbage finely hairy; involucel of narrow bracts .... **L. triternatum**

3. Fruit densely short-hairy .... **L. foeniculaceum**
3. Fruit glabrous or roughened 4

4. Flowers white; fruit linear-elliptic .... *L. macrocarpum*
4. Flowers yellow; fruit elliptic to oval 5
5. Involucel bracts and ultimate leaf segments elliptic .... **L. cous**
5. Involucel bracts and ultimate leaf segments linear 6
6. Plants > 30 cm tall .... **L. dissectum**
6. Plants < 30 cm tall .... **L. sandbergii**

Lomatium ambiguum (Nutt.) Coulter & Rose. Plants to 20 cm high from a tuberous root; herbage glabrous; leaves twice divided into unequal, linear segments 1-8 cm long; rays of umbel to 10 cm long; involucel lacking; petals yellow; fruit narrowly elliptic, 5-12 mm long, glabrous with narrow wings. May.

Rare on rocky outcrops on montane, open slopes of the McDonald valley. B.C. to MT south to OR, UT, WY.

Lomatium cous (Wats.) Coulter & Rose Cous. Stemless, glabrous plants from tuberous-thickened roots; leaves 3-4 times pinnately divided into small, narrowly elliptic segments; petioles with sheathing bases; umbels on stalks to 15 cm tall, with 5-20 umblets,; involucel of persistent, elliptic bracts 2-5 mm long; petals yellow; fruit elliptic, glabrous, 5-8 mm long. May-June.

Locally common in shallow, stony soil of lower subalpine open slopes and ridges of the Middle Fork Flathead drainage. Alta. and Sask. to OR, ID, WY. Farther south, where the plant is more common, the root is an important food for bears and was used extensively by Native Americans.

Lomatium dissectum (Nutt.) Mathias & Constance, Fern-leaved Desert Parsley [*Leptotaenia multifida* Nutt.]. Plants 30-100 cm tall with sparsely leafy stems from a taproot and branched crown; leaves 2-4 times dissected into numerous segments ca. 1 mm wide, short hairy beneath; umbels with 10-30 umblets; involucel of linear bracts; petals yellow; fruit 8-12 mm long, oval with thick, narrow marginal wings. May-June.

Common in very rocky soil of montane and lower subalpine open slopes, thickets and forest; East, West. Our plants are var. *multifidum* (Nutt.) Mathias & Constance. B.C. to Sask. south to CA, AZ, CO. The combination of large size, highly dissected, fern-like leaves, and rocky habitat is diagnostic. Roots of the plant were used by Blackfeet in tonics and in tanning.

Lomatium foeniculaceum (Nutt.) Coulter & Rose. Plants 10-30 cm tall, stemless, covered with short, whitish hairs, from a narrow taproot; leaves 3-4 times pinnately dissected into short, linear segments; petiole sheaths purplish; involucel of small linear bracts; petals yellow; fruit elliptic with narrow marginal wings, 7-10 mm long, hairy. June.

Rare in open, sometimes shallow soil of montane eroding slopes or rock outcrops in the McDonald and Many Glacier valleys. Var *foeniculaceum* has united involucel bracts, while var. *macdougalii* (Coult. & Rose) Cronq. has separate, narrow involucel bracts. B.C. to Man. south to OR, AZ, TX, MO.

Lomatium macrocarpum (Nutt. ex Torrey & Gray) Coulter & Rose [*Cogswellia macrocarpa* (Nutt. ex T. & G.) Jones]. Plants nearly stemless from a thickened taproot, up to 50 cm tall in fruit; herbage short-hairy; leaves 2-3 times pinnately divided into lobed, lance-shaped segments; involucel of linear bracts, asymmetric; petals white; fruit 8-12 mm long, linear-elliptic with narrow marginal ribs, glabrous to sparsely hairy.

Common in montane grasslands or shallow soil of outcrops and rocky, open slopes, sometimes higher; East, West. B.C. to Man. south to CA, CO, SD. Our only white-flowered *Lomatium* blooms near ground level but flower stalks elongate to disperse seed.

Lomatium sandbergii Coulter & Rose [*Cogswellia sandbergii* (C. & R.) Jones]. Plants 7-25 cm high from a slender taproot, stems elongate in fruit; herbage glabrous or roughened; leaves 3-4 times divided into linear segments; petiole sheaths white-margined; involucel of filiform bracts; petals yellow; fruit rough-surfaced, narrowly elliptic with narrow wings, 5-10 mm long. June-July. Fig. 33.

Common in shallow soil of outcrops and open slopes and ridges in the subalpine and alpine zones; East, West. Endemic to northwest MT, adjacent ID, and Canada. Roots are eaten by grizzly bears.

Lomatium triternatum (Pursh) Coulter & Rose [*Cogswellia triternata* (Pursh) Jones]. Plants 20-40 cm tall, stemless from a narrow taproot; herbage glabrous or sparsely hairy; leaves 2-3 times divided into linear segments 1-10 cm long; involucel of filiform bracts; petals yellow; fruit narrowly elliptic with marginal wings, glabrous, 7-15 mm long. Fig. 34.

Common in montane grassland, meadows and open forest: East, West. Our plants are ssp. *triternatum*. B.C., Alta. south to CA, CO. Roots were eaten by the Blackfeet.

Musineon Rafinesque

Musineon divaricatum (Pursh) Nutt. ex Torr. & Gray. Plant 5-15 cm tall with short stems from a narrow taproot; leaves clustered at the base, up to 12 cm long, 1-2 times pinnately divided into toothed segments; umbel without bracts; involucel bracts narrow; flowers yellow; fruit narrowly ovate, 3-5 mm long, prominently ribbed. May-June.

Grasslands, exposed slopes and ridges near East Glacier and to be expected along the east edge of the Park. Alta. to Man. south to NV, CO, NE. This plant is sometimes confused with *Lomatium cous*, which has elliptic involucel bracts and more rounded ultimate leaf segments.

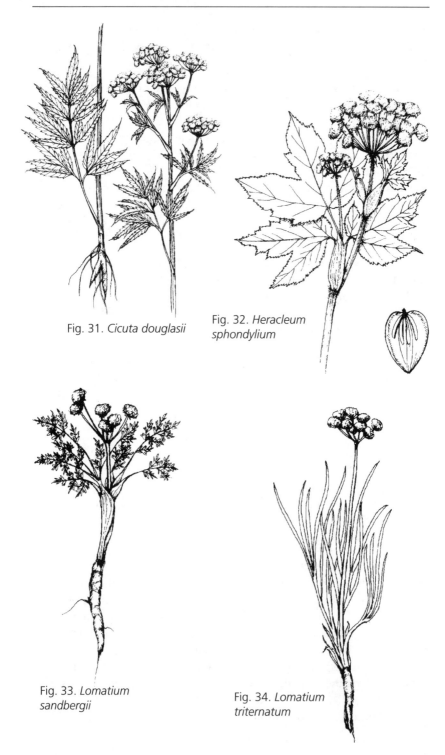

Fig. 31. *Cicuta douglasii*

Fig. 32. *Heracleum sphondylium*

Fig. 33. *Lomatium sandbergii*

Fig. 34. *Lomatium triternatum*

Osmorhiza Rafinesque, Sweet Cicely

Perennial, more-or-less aromatic herbs with a taproot and leafy stems; leaves petiolate, twice divided into 3's; flowers white, yellow or purplish, borne in few-flowered umbels, inconspicuous in flower, becoming open in fruit; bracts of umbels and umblets inconspicuous or lacking; fruit linear or club-shaped with narrow ribs, surmounted by the old style.

Our 3 white-flowered species are similar and difficult to distinguish although their habitats are somewhat different; O. *chilensis* is the most common; mature fruit is required for determination.

1. Flowers yellow; fruit glabrous .... ***O. occidentalis***
1. Flowers white; fruit hairy, at least below 2
2. Rays of umbel widely spreading; fruit abruptly rounded at the tip .... ***O. depauperata***
2. Rays of umbel ascending; fruit concavely tapering to the tip 3
3. Mature fruit > 12 mm long; styles forming a pointed beak .... ***O. chilensis***
3. Mature fruit < 12 mm long; styles forming a blunt beak .... ***O. purpurea***

Osmorhiza chilensis Hook. & Arn. [*O. divaricata* Nutt. ex T. & G., *O. brevipes* (Coult. & Rose) Suksdorf, O. berteroi D. C.]. Stems 30-80 cm tall, herbage glabrous to sparsely hairy; leaflets coarsely toothed to incised, 1-7 cm long; flowers white; fruit 12-18 mm long, hairy, gradually, concavely narrowed to the tip; styles cylindric, beak-like. June-July.

Common in dry to moist, montane forest, less common on avalanche slopes and along streams; East, West. AK to Newf. south to CA, AZ, MI, NH; S. America.

Osmorhiza depauperata Philippi. Stems 20-60 cm tall; herbage glabrous; leaves similar to O. chilensis; flowers white; fruit 10-15 mm long, hairy, abruptly rounded at the tip with inconspicuous styles.

Uncommon in mesic to wet forests and along streams; montane and lower subalpine; East, West. AK to Newf. south to CA, NM, MI, VT; S. America. June-Aug.

Osmorhiza occidentalis (Nutt. ex Torrey & Gray) Torrey [*Glycosoma occidentalis* Nutt. ex T. & G.]. Stems 30-100 cm tall, hairy at the nodes; herbage glabrous with an anise-like odor when crushed; leaflets toothed to lobed, 2-8 cm long; umbels more congested than other species; flowers yellow; fruit linear, glabrous, 12-20 mm long. July-Aug. Fig. 35.

Abundant in aspen groves and avalanche slopes and common in open forest and along streams in the montane and subalpine zones; East, less common West. B.C., Alta. to CA, CO. An important spring bear food.

Osmorhiza purpurea (Coult. & Rose) Suksdorf. Stems 20-60 cm tall; leaves similar to O. chilensis; flowers purplish to greenish-white; fruit 8-13 mm long, hairy, gradually, concavely narrowed to the flattened tip; styles inconspicuous. July.

Rare in moist forest in the McDonald Valley. AK to CA south to ID, MT.

Pastinaca L. Parsnip

Pastinaca sativa L., Wild Parsnip. Biennial herb, 30-100 cm tall from a thick taproot; leaves pinnately divided into toothed and lobed, ovate segments, those at the base up to 50 cm long; umbels and umblets without bracts; flowers yellow; fruits oval, 5-7 mm long, wing-margined and notched at both ends. June-July.

Moist, open or partially shaded, disturbed ground; reported for the east edge of the Park by Standley (1921) but not collected since. This is the wild form of cultivated parsnip introduced from Europe.

Perideridia Reichenb.

Perideridia gairdneri (Hook. & Arn.) Mathias, Yampah [*Carum gairdneri* (H. & A.) Gray]. Slender perennial herb with stems 30-80 cm tall from tuberous roots; leaves pinnately divided into long, linear segments 4-12 cm long; bracts of umbels and umblets narrow; flowers white; fruits glabrous, 2-3 mm long, orbicular with prominent ribs. June-July Fig. 36.

Montane and lower subalpine meadows and aspen groves; uncommon East, rare West. Our plants are ssp. *borealis* Chuang and Constance. B.C. to Sask., south to CA, SD. Blackfeet Indians ate the roots and used the plant in treating sore throats.

Sanicula L., Sanicle, Snakeroot

Glabrous perennials; leaves divided or lobed; umbels subtended by leaf-like bracts; umblets congested, subtended by an asymmetrical involucel of partly united bracts, composed of sessile, bisexual flowers and stalked, staminate flowers; fruits ovoid and covered with hooked prickles.

1. Lowest leaves with 3 deeply divided leaflets; flowers yellowish; plants usually < 25 cm tall .... *S. graveolens*
1. Lower leaves palmately divided into 5-7 toothed leaflets; flowers white; plants usually > 25 cm tall .... *S. marilandica*

Sanicula graveolens Poepp. ex D. C. Solitary stems 5-25 cm high from a taproot, taller in fruit; leaves divided into 3 deeply lobed and toothed leaflets; involucel of 6-10 narrow, pointed bracts ca. 1 mm long; flowers light yellow; fruits 3-5 mm long. May-June.

Locally common in vernally moist, rocky soil of warm, open, montane slopes in the Middle Fork Flathead drainage. B.C., Alta. to CA, WY.

Sanicula marilandica L., Black Snake-root. Stems 30-90 cm tall from fibrous roots; leaves palmately divided into 5-7 coarsely toothed segments; involucel of narrow bracts; flowers greenish white; fruits 4-6 mm long. June-July. Fig. 37.

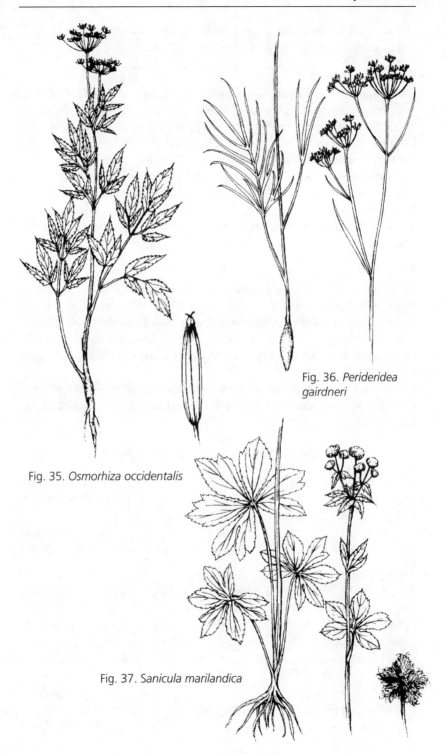

Fig. 35. *Osmorhiza occidentalis*

Fig. 36. *Perideridea gairdneri*

Fig. 37. *Sanicula marilandica*

Common in montane aspen groves, willow thickets and moist, open forest; East, West. B.C. to Newf. south to WA, NM, NE, FL.

Sium L.
Sium suave Walt., Water-parsnip [*S. cicutaefolium* Schrank]. Glabrous perennial with stems 50-100 cm tall from fibrous roots; leaves once pinnately divided into toothed, lance-linear leaflets up to 9 cm long; umbels subtended by lance-shaped, reflexed bracts; involucel bracts similar; flowers white; fruits elliptic with corky ribs, 2-3 mm long. July-Aug. Fig. 38.

Uncommon in shallow water and banks of montane ponds, swamps and slow streams; West, expected East. B.C. to Newf. south to CA, TX, SC. Submersed leaves, when present, are more finely dissected. Lateral leaf veins do not terminate between marginal teeth as in the poisonous *Cicuta douglasii.*

Zizia Koch
Zizia aptera (Gray) Fernald., Heart-leaved Alexanders [*Z. cordata* (Walt.) Koch]. Glabrous perennial with stems 20-40 cm tall from fibrous roots; leaf margins coarsely toothed; basal leaves petiolate, simple and heart-shaped; stem leaves 3-parted; umbels densely-flowered without bracts; involucel inconspicuous; flowers yellow; fruit glabrous, 2-4 mm long, elliptic with narrow ribs.

Uncommon in meadows and moist grassland along the east edge of the Park. B.C. to Que., south to OR, CO, GA. The difference between basal and stem leaves is diagnostic.

APOCYNACEAE: DOGBANE FAMILY

Perennial herbs with milky sap; leaves opposite with entire margins; flowers bisexual; corolla bell- to funnel-shaped, 5-lobed; calyx 5-lobed; stamens 5; ovary superior; style 1.

1. Flowers white or pink, to 15 mm long .... **Apocynum**
1. Flowers blue, ca. 2 cm long .... **Vinca**

Apocynum L., Dogbane, Indian Hemp
Perennial rhizomatous herbs; leaves petiolate, glabrous above but sometimes hairy below; flowers in open inflorescences in leaf axils; corolla 5-lobed, bell-shaped or tubular; ovaries 2 with 1 stigma; fruit 2 long, slender pods united at the base; seeds numerous, each with a tuft of long hair.

Hybrids between the following two species are called *A. medium* Greene and may occur in the Park.

1. Sepals > half length of pink corolla tube; leaves ovate **A. androsaemifolium**
1. Sepals < half length of whitish corolla tube; leaves narrowly lance-shaped ....
.... **A. cannabinum**

Apocynum androsaemifolium L. [*A. ambigens* Greene, *A. pumilum* (Gray) Greene]. Stems erect, 20-100 cm tall; leaves ovate; corolla pink, bell-shaped, 5-10 mm long with recurved lobes; pods 8-12 cm long, pendulous. July-Aug. Fig. 39.

Common in rocky soil of montane meadows and open forests, usually on slopes; East, West. Our plants are var. *androsaemifolium*. AK to Newf. south to CA, TX, GA.

Apocynum cannabinum L. Stems erect, 30-80 cm high; leaves narrowly lance-shaped; corolla tubular with erect lobes, greenish-white; pods 6-12 cm long. July-Aug.

Open or partially shaded, often disturbed areas in the montane zone. Standley (1921) reports large patches along Middle Fork Flathead River near West Glacier. Temperate N. America. Used by the Blackfeet to make cord.

Vinca L., Periwinkle

Vinca major L. Stems prostrate 10-50 cm long, rooting at the nodes; leaves ovate, 3-6 cm long; calyx lobes linear; flowers long-stalked in leaf axils; corolla blue, tubular, ca. 2 cm long, with flared lobes ca. 2 cm long; fruit mostly not developing. June.

Collected in open forest near Apgar Village. Introduced garden ornamental.

ARALIACEAE: GINSENG FAMILY

Perennial herbs or shrubs; leaves alternate, petiolate, deeply lobed or divided; flowers small, radially symmetrical, borne in dense, spherical clusters; petals 5, sepals 5 or lacking; stamens 5; ovary inferior; fruit a berry (drupe).

1. Spiny shrubs ≥ 1 m tall; leaves lobed .... **Oplopanax**
1. Low herbs; leaves twice divided .... **Aralia**

Aralia L.

Aralia nudicaulis L., Wild Sarsaparilla. Rhizomatous herb with short erect stems barely above ground level; leaves 1 per stem, long petiolate, twice divided, first into 3 then 3-5 parts; leaflets ovate to lance-shaped, 3-12 cm long with toothed margins; inflorescence of usually 3 flower clusters on a stalk shorter than the leaf; flowers greenish-white, 5-7 mm; berries dark purple, 4-8 mm wide. May-June. Fig. 40.

Common in moist, montane forest, West. B.C. to Newf. south to WA, CO, MO, GA. The roots have medicinal properties similar to true sarsaparilla.

Oplopanax (Torr. & Gray) Miq.

Oplopanax horridus (Smith) Miq., Devil's Club [*Echinopanax horridum* (Smith) Decne. & Planch.]. Deciduous shrub with densely spiny stems 1-2 m tall; leaves spiny beneath, 10-30 cm wide, with 5-9 pointed, maple-like lobes and toothed margins; inflorescence a raceme of dense flower clusters 8-20 cm long; flowers greenish, 5-6 mm; berries red, 4-6 mm long. June-July. Fig. 41.

Low areas of moist to wet forest, uncommon on avalanche slopes; montane and lower subalpine; common West, uncommon in deep valleys near the Divide East. AK to Alta. south to OR, ID, MT; MI, Ont. Unmistakable when encountered at close quarters.

ASTERACEAE: SUNFLOWER FAMILY

Herbs (1 shrub in *Artemisia*); flowers clustered in heads on a common receptacle, tightly surrounded by a cup- or tube-like involucre of papery to leaf-like bracts in 1-many series of different heights; receptacle often with small bracts or scales among flowers; flowers with an inferior, 1-seeded ovary; corolla tubular and 5-lobed (disk flower) or strap-like (ray flower); calyx modified into pappus of bristles, scales or lacking; stamens 5 with anthers united into a tube around style; fruit a dry, 1-seeded achene.

This is the largest family in the Park, containing some of the showiest wildflowers as well as some of the worst weeds. The flower heads are often mistaken for flowers. Flower heads are of 3 types: (1) ligulate, where all flowers have strap-shaped ray corollas, (2) discoid, where all flowers have tubular corollas, and (3) radiate, where central flowers are tubular and outer flowers are strap-shaped rays. The involucral bracts and pappus are important characters for delineating genera.

Saussurea americana D. C. Eat., tall with purple discoid heads and alternate, toothed leaves, is reported for Waterton Lakes Park (Kuijt 1982), and the dwarf alpine *S. densa* (Hook.) Rydb. occurs on limestone just south and north of the Park. *Tonestus lyalli* (Gray) Nels. [*Haplopappus lyallii* Gray] occurs at high elevations on limestone talus to the north and south but is conspicuously absent from Glacier Park. References: Chambers & Sundberg (1998), Cronquist (1955), Keil (1993).

Key to Artificial Groups of Genera:

1. Flower heads with ray flowers only; disk flowers absent; sap milky .. Group A
1. Disk flowers present; sap not milky 2

2. White-wooly herbs with discoid heads; at least upper portion of involucral bracts papery; pussytoes, everlasting, etc. .... Group B
2. Plants without the above combination of characters 3

3. Ray flowers lacking or apparently lacking .... Group C
3. Rays present 4

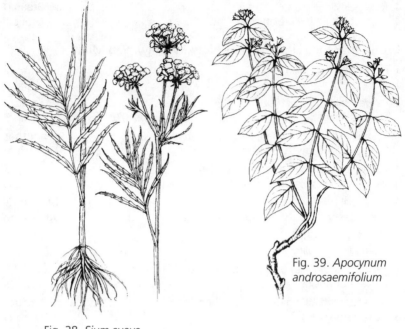

Fig. 38. *Sium suave*

Fig. 39. *Apocynum androsaemifolium*

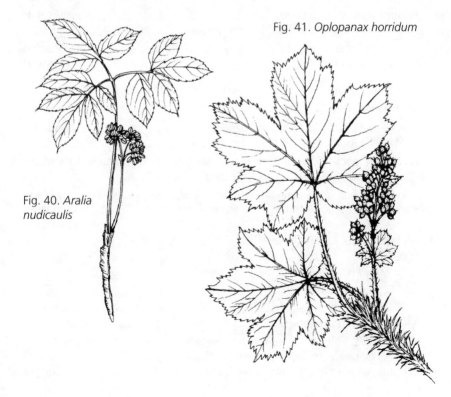

Fig. 41. *Oplopanax horridum*

Fig. 40. *Aralia nudicaulis*

4. Rays yellow .... Group D
4. Rays white, pink, blue, purple .... Group E

Group A: Heads Ligulate (Ray Flowers Only), Sap Milky

1. Flowers white, pink or blue 2
1. Flowers yellow or orange 4

2. Flowers blue .... **Lactuca**
2. Flowers white 3

3. Leaf blades arrow-shaped; involucre glabrous .... **Prenanthes**
3. Leaves oblong; involucre hairy .... **Hieracium**

4. Pappus of branched, feather-like bristles 5
4. Pappus of simple, unbranched bristles 6

5. Involucral bracts up to 15 mm long .... **Microseris**
5. Involucral bracts > 20 mm long .... **Tragopogon**

6. Stems with > 1 head 7
6. Flower heads solitary 10

7. Leaves with prickles on the margins or underside of midrib 8
7. Leaves not prickly 9

8. Achenes with a long beak below the pappus .... **Lactuca**
8. Achenes beakless .... **Sonchus**

9. Pappus white .... **Crepis**
9. Pappus brownish .... **Hieracium**

10. Involucral bracts subequal .... **Nothocalais**
10. Involucral bracts in at least 2 unequal series 11

11. Seeds (achenes) ribbed with bumps toward the top .... **Taraxacum**
11. Seeds smooth-ribbed .... **Agoseris**

Group B: White-Wooly Herbs with Discoid Heads; Upper Portion of Involucral Bracts Papery

1. Involucral bracts almost entirely papery, shiny white, glabrous ...... **Anaphalis**
1. At least part of involucral bracts colored or hairy 2

2. Plants perennial with basal leaves, some larger than stem leaves **Antennaria**
2. Larger basal leaves lacking; plants annual or perennial 3

3. Involucral bracts in 1 series; inflorescence long, narrow .... **Filago**
3. Involucral bracts > 1 series; inflorescence head-like .... **Gnaphalium**

Group C: Ray Flowers Lacking or Apparently Lacking

1. Disk flowers yellow or orange 2
1. Disk flowers white, pink, blue, purple, green 9

2. All leaves undivided and unlobed 3
2. At least some leaves divided or lobed 6

3. Plants taprooted annuals 4
3. Plants perennial 5

4. Plants glandular, sticky; heads few to several .... **Madia**
4. Plants hairy but not glandular; heads many .... **Conyza**

5. Heads 10-14 mm high .... **Arnica**
5. Heads < 5 mm high .... **Artemisia**

6. Heads numerous, 2-4 mm high .... **Artemisia**
6. Heads larger, > 5 mm high 7

7. Receptacle hemispheric, ultimate leaf segments filiform; herbage pineapple-scented .... **Matricaria**
7. Receptacle flat; ultimate leaf segments wider 8

8. Involucral bracts with papery tips, in > 1 series .... **Tanacetum**
8. Involucral bracts green, in 1 series .... **Senecio**

9. Leaf blades arrow-shaped or triangular, often with toothed margins 10
9. Leaves linear, lance-shaped, divided or lobed but not arrow-shaped 14

10. Leaves densely white-hairy beneath 11
10. Leaves green beneath 12

11. Plants of wet soil; basal leaves arising separately from stems .... **Petasites**
11. Forest plants; flower stems with basal leaves .... **Adenocaulon**

12. Heads < 5 mm high .... **Iva**
12. Heads > 5 mm high 13

13. Basal leaves present; involucre spiny .... **Arctium**
13. Basal leaves absent at flowering; involucre not spiny .... **Brickellia**

14. Leaves and stems spiny 15
14. Leaves and stems below inflorescence without spines 16

15. Pappus of minutely toothed bristles .... **Carduus**
15. Pappus of feathery bristles .... **Cirsium**

16. Heads < 5 mm high 17
16. Heads > 5 mm high 18

17. Lower leaves opposite .... **Ambrosia**
17. Leaves alternate .... **Artemisia**

18. Leaves lance-shaped or wider, often divided; some disk flowers expanded ..
.... **Centaurea**
18. Leaves linear; disk flowers tubular .... **Liatris**

Group D: Heads Radiate, Rays Yellow

1. Pappus of numerous fine bristles 2
1. Pappus of scales, spines, a crown, or lacking 6

2. Leaves opposite .... **Arnica**
2. Leaves alternate 3

3. Involucral bracts more-or-less in 1 equal series .... **Senecio**
3. Involucral bracts in > 1 overlapping series 4

4. Heads to 6 mm high .... **Solidago**
4. Heads > 6 mm high 5

5. Herbage glabrous or sparsely hairy; basal rosette present .... **Pyrrocoma**
5. Herbage long-hairy; clustered basal leaves lacking .... **Heterotheca**

6. Leafy stems lacking; basal leaves large, arrow-shaped .... **Balsamorhiza**
6. Leafy stems present 7

7. Leaves opposite .... **Bidens**
7. Leaves alternate 8

8. Disk flowers yellow 9
8. Disk flowers brown or purple 12

9. Basal leaves divided into linear segments .... **Hymenoxys**
9. Leaves not lobed or divided 10

10. Herbage hairy but not sticky or glandular .... **Helianthus**
10. Herbage sticky, glandular 11

11. Outer involucral bracts curled back .... **Grindellia**
11. Involucral bracts not recurved .... **Madia**

12. Basal leaves divided; disk conical.................................. **Ratibida**
12. Leaves entire or lobed; disk flat or hemispheric 13

13. Rays with 3-lobed tips .... **Gaillardia**
13. Rays with rounded or pointed tips .... **Helianthus**

Group E: Heads Radiate, Rays White, Pink, Blue or Purple

1. Pappus of numerous, thin bristles 2
1. Pappus of scales, spines, a crown, or lacking 5

2. Basal leaves large, arrow-shaped, white-wooly beneath .... **Petasites**
2. Leaves, linear to lance-shaped or divided 3

3. Heads solitary, ≥ 10 mm high, leaves to 5 mm wide .... **Townsendia**
3. Heads smaller or more numerous or some leaves larger 4

4. Involucral bracts in series of unequal length, often with a whitish base and
   green tip .... **Aster**
4. Involucral bracts mostly the same length, lacking a whitish base ..... **Erigeron**

5. Heads solitary 6
5. Stems with > 1 head 7

6. Flower stem leafless; rays ca. 1 cm long .... **Bellis**
6. Stems leafy; rays 1-2 cm long .... **Chrysanthemum**

7. True ray flowers lacking; outer disk flowers enlarged and appearing like rays;
   leaves toothed to once pinnately divided .... **Centaurea**
7. True rays present; leaves > 1 times pinnately divided 8

8. Rays 2-3 mm long; heads many in a flat-topped inflorescence .... **Achillea**
8. Rays 6-16 mm long; heads few..................................... **Matricaria**

Achillea L. Yarrow

Rhizomatous perennials; herbage wooly, aromatic; leaves alternate, 2-3 times pinnately dissected,into minute segments, appearing feathery; heads radiate, in an flat-topped inflorescence; involucral bracts brownish, papery, overlapping in 3-4 series; ray flowers, white, 2-3 mm long; disk flowers white; receptacle with scales; pappus lacking.

1. Flower heads 2-3 mm high ....*A. nobilis*
1. Flower heads 4-6 mm high ....*A. millefolium*

Achillea millefolium L. [*A. lanulosa* Nutt.]. Stems 10-60 cm tall; leaves narrowly lance-shaped, 2-15 cm long; heads 4-6 mm high; ray flowers 5-12, 2-3 mm long, longer than wide; disk flowers 10-30. June-Aug. Fig. 42.

Common in grasslands, meadows, open forest, talus, moraine, and exposed ridges at all elevations; East, West. Circumboreal and throughout temperate N. America. Our plants have been called ssp. *lanulosa* (Nutt.) Piper. High elevation plants have darker involucral bracts. Native Americans used this plant to treat consumption and headache.

Achillea nobilis L. Similar to *A. millefolium*; heads 2-3 high; rays ca. as wide as long.

Rare in gravelly soil along roads or streambanks near Polebridge. Introduced from Europe, probably during construction following the 1988 fire.

Adenocaulon Hooker, Trail Plant
Adenocaulon bicolor Hooker, Pathfinder. Fibrous-rooted perennial with erect stems 30-80 cm tall; leaves alternate, winged-petiolate, triangular with wavy margins, 3-15 cm wide, green and glabrous above, white-wooly below; heads small, few, in an open inflorescence; involucre of ca. 5 bracts, ca. 3 mm long; disk flowers whitish, ray flowers and pappus lacking; achenes club-shaped with glandular hairs on top, 5-8 mm. June-July. Fig. 43.

Dry to moist, montane forest, especially in lightly disturbed areas such as along trails; abundant West, locally common East. B.C. to Ont. south to CA, MT, SD. The glandular achenes are diagnostic and readily stick to clothing.

Agoseris Rafinesque, Mountain Dandelion
Stemless, taprooted, perennial herbs with milky sap, resembling dandelions; leaves basal, longer than wide; heads solitary, long-stalked, ligulate; involucral bracts narrowly lance-shaped, in several series, sometimes purple-spotted; receptacle naked; achenes club-shaped, often with a long beak on top; pappus of thin, white bristles, on the beak or top of achene.

Achene beaks elongate as fruits develop; mature achenes may be needed for positive identification. The annual *A. heterophylla* (Nutt.) Greene has been found just south of the Park.

1. Flowers orange, turning purplish with age; beak of mature achene ≥ half as long as the body .... ***A. aurantiaca***
1. Flowers yellow; achene beak < half as long as body .... ***A. glauca***

Agoseris aurantiaca (Hooker) Greene, Orange Mountain Dandelion [*A. graciliens* (Gray) Kuntz, *A. elata* (Nutt.) Greene misapplied, *A. graminifolia* Greene]. Plants 10-60 cm tall; leaves linear-oblong with entire to sparsely lobed margins; herbage mostly glabrous; flowers orange, drying purplish; achene 5-9 mm, shorter or as long as the beak. July-Aug.

Meadows and on open, rocky slopes at all elevations; common East, less so West. Our plants are var. *aurantiaca*. B.C., Alta. to CA and NM.

Agoseris glauca (Pursh) Raf. [*A. villosa* Rydb., *A. aspera* Rydb., *A. pumila* (Nutt.) Rydb., *A. scorzoneraefolia* (Schrad.) Greene]. Plants 10-70 cm tall; leaves linear to oblong, entire to toothed, 3-35 cm long, usually somewhat hairy; involucral bracts hairy; flowers yellow; achene 5-12 mm long including the beak if present. July-Aug. Fig. 44.

Var. *agrestis* (Osterh.) Q. Jones ex Cronq. is > 25 cm tall and is common in montane grasslands and along streams. Var. *dasycephala* (T. & G.) Jeps. is < 25 cm tall and occurs in meadows and on open, rocky, alpine and subalpine slopes. Standley (1921) reports var. *glauca*, with glabrous leaves, to be common in grasslands and meadows in the valley and montane zones. AK to Man. south to CA, AZ, MN.

Ambrosia L., Ragweed

Herbaceous with mostly opposite, deeply lobed leaves; heads unisexual; staminate heads in terminal spikes; pistillate heads below in axils of bracts, 1-flowered; corolla and pappus lacking; achene enclosed in involucre.

Standley (1921) reports both species occurring sparingly along railroad tracks at West Glacier; they have not been collected since.

1. Taprooted annual; leaves twice lobed; fruit with small spines ....
.... **A. artemisiifolia**
1. Rhizomatous perennial; leaves once lobed; fruit spineless .... **A. psilostachya**

Ambrosia artemisiifolia L. [*A. elatior* L.]. Taprooted annual 30-100 cm tall; leaves twice lobed; fruit with short spines on top. July-Sept.

Disturbed areas. Native to much of N. America; introduced in Glacier Park. Our plants are var. *elatior* (L.) Desc.

Ambrosia psilostachya D. C. Rhizomatous perennial 30-80 cm tall; leaves once lobed with wavy margins; fruit without spines. July-Sept.

Disturbed areas. Native to central N. America; introduced in Glacier Park.

Anaphalis D. C., Everlasting

Anaphalis margaritacea (L.) Bentham & Hooker, Pearly Everlasting. Rhizomatous perennial with erect stems 20-70 cm tall; herbage white-wooly; leaves alternate, without petioles, entire-margined, narrowly lance-shaped, 3-10 cm long; heads discoid, crowded in a hemispheric inflorescence; involucre 4-7 mm high, of several series of white, papery bracts, nearly obscuring the flowers; receptacle naked; flowers yellow, unisexual, male and female on separate plants; pappus of bristles. June-Aug. Fig. 45.

Common in montane meadows, open slopes and forests; East, West; AK to Lab. south to CA, NM, KS, NC. Especially abundant with moderate

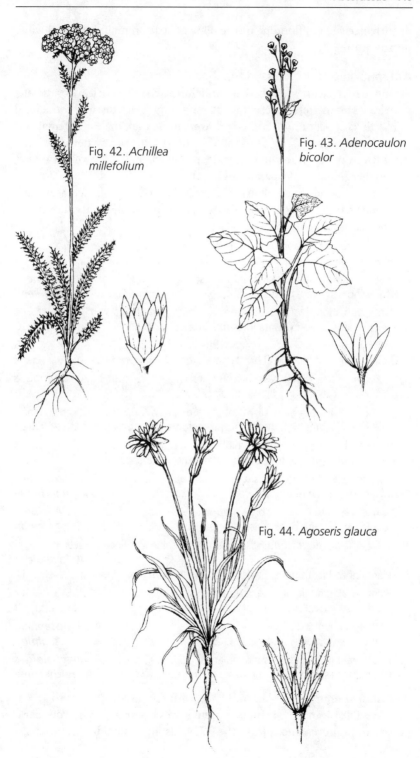

Fig. 42. *Achillea millefolium*

Fig. 43. *Adenocaulon bicolor*

Fig. 44. *Agoseris glauca*

disturbance such as flooding or fire. Flower heads remain fresh and white into autumn.

Antennaria Gaertn., Pussy-toes

Perennial herbs with wooly herbage and fibrous roots; basal leaves petiolate; stem leaves alternate, sessile, with entire margins; heads discoid, 1 to several in mostly compact inflorescences; flowers unisexual, male and female on separate plants; involucre of several series of papery bracts, often obscuring the flowers, male shorter than female; receptacle naked; corollas tubular, male wider than female; pappus of bristles.

In some species male plants are rare, and seed is produced asexually; this and the occurrence of polyploidy makes species delineation difficult. Reference: Bayer & Stebbins (1993).

1. Leaves green and glabrous or sparsely hairy on upper surface 2
1. Leaves wooly on both surfaces 3

2. Heads on long stalks in an open inflorescence .... *A. racemosa*
2. Heads short-stalked, clustered .... *A. howellii*

3. Plants with short, leafy stolons, forming mats (sometimes small) 4
3. Plants lacking stolons, sometimes rhizomatous or branched at the base.... 10

4. Upper portion of involucral bracts white or pink 5
4. Upper portion of involucral bracts dark green, brown, or black 7

5. Involucral bracts with a basal black or brown spot .... *A. corymbosa*
5. Involucral bracts lacking brown or black spot 6

6. Involucres 7-11 mm high; plants of dry grasslands .... *A. parvifolia*
6. Involucres 4-7 mm high; grasslands, meadows, and open forest ...... *A. rosea*

7. Involucral bracts brown; low to high elevation 8
7. Involucral bracts black to green-black; high elevation 9

8. Flowering stems glandular as well as hairy .... *A. aromatica*
8. Flowering stems without glands .... *A. umbrinella*

9. Stem leaves with a papery brown tip, rare .... *A. alpina*
9. Stem leaves lacking papery tips .... *A. media*

10. Plants rhizomatous, forming loose mats, basal leaves linear, erect ....
.... *A. luzuloides*
10. Plants non-rhizomatous, basal leaves lance-shaped or oblong 11

11. Plants < 20 cm tall, high elevation 12
11. Plants >20 cm tall, low to high elevation 13

12. Flower heads 1-2 .... *A. monocephala*
12. Flower heads mostly ≥ 3 .... *A. alpina*

13. Involucral bracts all whitish; grasslands .... *A. anaphaloides*
13. Involucral bracts black at the base, wetter sites .... *A. pulcherrima*

Antennaria alpina (L.) Gaertn. Plants with few stolons; stems to 15 cm high; basal leaves narrowly oblong; stem leaves linear with a narrow, dark, papery tip; inflorescence of mostly 3-7 heads; involucral bracts greenish-black.

Uncommon in moist turf of cool alpine slopes, often where snow lies late; East, West. Our plants are var. *canescens* Lange. Circumboreal south to WY. July-Aug. *A. monocephala*, which lacks stolons and has only 1-2 heads per stem, sometimes occurs with *A. alpina*.

Antennaria anaphaloides Rydb., Tall Pussy-toes. Stems 20-50 cm tall; basal leaves tufted, narrowly oblong, 5-15 cm long; stem leaves narrowly lance-shaped; heads numerous in a compact inflorescence; involucre pubescent, 5-8 mm high; bracts white, often with a basal black spot. June-Aug.

Our common tall pussy-toes of montane grasslands, occasionally higher; East, West. B. C to Sask. south to OR, UT, CO.

Antennaria aromatica Evert. Low, mat-forming plants; herbage glandular; fresh material smells like citronella; basal leaves spoon-shaped, 5-10 mm long; stem leaves narrower, often with a brown, papery tip; inflorescence of 2-5 clustered heads; involucre 5-7 mm high, greenish-brown. July-Aug.

Uncommon in calcareous gravel or talus near or above treeline; East. Alta. to OR and WY. This recently described species would key to *A. umbrinella* in older treatments.

Antennaria corymbosa E. Nels. Plants mat-forming, similar to *A. microphylla*; basal leaves narrowly oblong; involucres 4-5 mm high, bracts white with a brown or black spot just above the base.

Rare in moist to wet, montane meadows; collected by Umbach near St. Mary (Standley 1921). Alta. to Sask. south to CA, UT, CO.

Antennaria howellii Greene [*A. neglecta* Greene in part]. Mat-forming plants with stems 10-25 cm tall; basal leaves spoon-shaped, densely wooly beneath, green and sparsely hairy to glabrous above, to 2 cm wide; stem leaves linear with sharp-pointed tips; inflorescence of several clustered heads; female involucre 6-10 mm high, bracts narrow with white tips; male involucres shorter with broader bracts. June-July.

Common in montane open forest (especially lodgepole) and mesic grasslands; East, West. Our plants are ssp. *petaloidea* (Fern.) Bayer. Yuk. to Newf. south to CA, WY, SD.

Antennaria luzuloides Torrey & Gray. Loosely mat-forming with stems 10-40 cm tall; basal leaves erect, broadly linear, 3-8 cm long; stem leaves similar but smaller; inflorescence rather open; involucre 4-5 mm high; bracts glabrous with age, pale greenish-brown, paler above. June-July.

Shallow soil of montane grasslands and outcrops; East, West. The erect leaves and open inflorescence are distinctive. B.C., Alta. to CA and CO. Non-flowering plants often bear small plantlets among the upper leaves.

Antennaria media Greene [*A. alpina* (L.) Gaertn. var. *media* (Greene) Jeps., *A. pulvinata* Greene, *A. chlorantha* Greene]. Mat-forming plants to 10 cm tall; basal leaves oblong; stem leaves linear, without brown, papery tips;

inflorescence of 3-7 clustered heads; involucres 4-7 mm high, bracts greenish black, usually rounded at the tip; male plants rare. July-Aug.

Common in meadows and on moist slopes near or above treeline; East, West; uncommon in exposed sites lower down, East. Yuk. to CA, NM. Many authors have placed these plants in *A. alpina*; however, *A. media* as applied here is a polyploid, hybrid complex derived in part from the more northern *A. alpina* which is uncommon in the Park.

Antennaria monocephala D. C. Stems 5-10 cm tall from a simple or few-branched rootcrown; basal leaves 5-10 mm long, narrowly oblong; stem leaves linear with a narrow, dark, papery tip; inflorescence of 1 or sometimes 2 heads; involucral bracts black.

Uncommon in moist turf of cool alpine slopes; East, West. Our plants are ssp. *augustata* (Greene) Hult. AK to Lab. south to B.C., WY; Asia, Greenl. See *A. alpina*.

Antennaria parvifolia Nutt. [*A. aprica* Greene]. Low mat-forming plants with stems 5-20 cm tall; basal leaves 1-2 cm long, spoon-shaped with broad tips; stem leaves linear; inflorescence of several clustered heads; female involucres 7-11 mm high, bracts white or pink. June.

Rare in montane grassland and open forest; West, expected East. Similar to A. rosea but with broader basal leaves. B.C. to Man. south to WA, AZ, NE.

Antennaria pulcherrima (Hooker) Greene. Plants 20-50 cm tall; habit and foliage similar to *A. anaphaloides*; involucral bracts brown to black below with paler, pointed tips. July-Aug.

Wet soil of meadows, fens and turf; montane to near treeline; our one record from the Logan Pass area. AK to Newf. south to WA, CO, Man. The wet habitat most easily distinguishes this plant from *A. anaphaloides*. Standley's (1921) report of *A. lanata* (Hook.) Greene from near the Continental Divide has not been verified and is probably referable here.

Antennaria racemosa Hooker. Plants with long, leafy stolons; stems glandular above, 10-60 cm tall; basal leaves elliptic, 1-6 cm long, glabrous and green above, white-wooly below; stem leaves narrowly lance-shaped; heads on slender stalks in an open raceme; female involucres 5-8 mm high, bracts narrow and transparent. June-July. Fig. 46.

Abundant in montane to subalpine forests and meadows; East, West. B.C., Alta. to OR, WY. The open inflorescence is distinctive.

Antennaria rosea Greene, Rosy Pussy-toes [*A. microphylla* Rydb. misapplied]. Mat-forming plants with stems 10-30 cm tall; basal leaves narrowly oblong to ovate, 8-20 mm long; stem leaves linear; inflorescence of several clustered heads; involucre 4-7 mm high, bracts pink or sometimes white at the tip. May-July. Fig. 47.

Common in montane grasslands, meadows and open forest; East, West. AK to Lab. south to CA, NM and Man. *A. rosea sensu stricto* is a complex of asexual, polyploid hybrids without male plants derived from as many as eight other species (Bayer 1990). *A. microphylla*, a similar sexually reproducing species with white involucral bracts and glandular stems, may also occur in the Park, but I have seen no specimens.

Antennaria umbrinella Rydb. [*A. flavescens* Rydb.]. Mat-forming with stems 6-15 cm tall; basal leaves narrowly lance-shaped; stem leaves linear; inflorescence of 3-8 clustered heads; involucre 5-6 mm high, bracts rounded, brown, sometimes dirty white at the tip. July.

Common in grasslands, open slopes, ridges and fellfields at all elevations; East, expected West. AK to Que., south to CA, AZ.

Arctium L., Burdock
Arctium minus (Hill) Bernh. Large biennial with erect, branched stems to 120 cm tall; leaves alternate, petiolate, the blades arrow-shaped, to 30 cm long, glabrous above, thinly hairy below; several clusters of heads borne in leaf axils; heads discoid, globose, 15-25 mm wide; involucre of several series of spreading, spine-tipped bracts; receptacle bristly; corolla purple; pappus of short bristles. July-Aug.

Uncommon in disturbed, rich soil in the valleys. Introduced from Eurasia. The large, clinging burs are distinctive.

Arnica L., Arnica
Perennial herbs with rhizomes or elongate rootstocks; leaves undivided and opposite; heads radiate or discoid, solitary to several in open inflorescences; rays female, yellow or orange, conspicuous (or lacking); disk flowers yellow, bisexual; involucral bracts greenish, in 1-2 series, equal in length; receptacle naked; pappus of white-to-brown, simple-to-feathery bristles.

Glacier Park has a rich assemblage of *Arnica* species, many of which appear to intergrade. *A. diversifolia* is a hybrid complex and includes many different-appearing plants that are, to some extent, intermediate between *A. mollis, A. latifolia,* and *A. amplexicaulis*. Confident identification of all specimens may not be possible. Reference: Cronquist (1955).

1. Rays lacking, lowest flowers nodding .... ***A. parryi***
1. Rays present on most plants 2

2. Stem leaves mostly 1-4 pairs 3
2. Stem leaves > 4 pairs on well-developed plants 10

3. Pappus light brown with short side branches on each bristle 4
3. Pappus white, the bristles barbed but not truly branched 5

4. Heads hemispheric, often with > 90 disk flowers .... ***A. mollis***
4. Heads more conical with < 90 disk flowers .... ***A. diversifolia***

5. Basal leaf blades broadly arrow- or heart-shaped 6
5. Basal leaf blades lance-shaped to ovate, not indented at the base 7

6. Achene hairy to the base; involucre white-hairy at the base ...... **A. cordifolia**
6. Achene glabrous at the base; involucre only sparsely hairy at the base ....
.... **A. latifolia**

7. Plants densely long-wooly; high elevations.................................... **A. alpina**
7. Plants short-hairy to glabrous 8

8. Most heads with 7-10 rays, usually high elevations .... **A. rydbergii**
8. Heads with > 10 rays, usually lower 9

9. Tufts of brown hair among old leaf bases at ground level .... **A. fulgens**
9. Hair among leaf bases white or lacking... **A. sororia**

10. Involucral bracts with a small tuft of white hair at the tip .... **A. chamissonis**
10. Tip of involucral bracts no hairier than body 11

11. Leaf margins toothed, stems mostly solitary .... **A. amplexicaulis**
11. Leaves entire, stems usually clustered .... **A. longifolia**

Arnica alpina (L.) Olin [*A. angustifolia* Vahl, *A. tomentosa* Macoun]. Stems solitary, 5-25 cm tall; herbage long-hairy and glandular; leaves entire, narrowly lance-shaped, to 10 cm long, the basal clustered; stem leaves 1-3 pairs and much shorter; heads solitary, bell-shaped, 10-20 mm high; rays 9-12, 12-20 mm long with lobed tips; pappus barbed, white; achenes hairy. July.

Locally common in stony, often calcareous soil of exposed ridges and talus slopes near or above treeline near or east of the Divide. Our plants are var. *tomentosa* (Macoun) Cronq. Alta., B.C., MT.

Arnica amplexicaulis Nutt. Stems to 80 cm tall; herbage hairy and glandular to nearly glabrous; leaves 4-6 pairs, elliptic to broadly lance-shaped with toothed margins, 5-15 cm long, mostly without petioles; heads several, broadly bell-shaped, 9-15 mm high; rays 8-14, 10-15 mm long; pappus bristles brownish with numerous minute branches. July-Aug.

Common in moist montane forest and aspen groves; East, West. AK south to CA, MT. Often confused with its hybrid derivative, *A. diversifolia*. The former has leaves > 3 times as long as wide without petioles and occurs at relatively low elevations, while the latter has petiolate leaves < 3 times as long as wide and is mainly subalpine.

Arnica chamissonis Less. [*A. foliosa* Nutt.]. Stems 2-80 cm tall; leaves 5-10 pairs, lance-shaped, 5-20 cm long, only the lowest with petioles; heads several, broadly bell-shaped, 8-15 mm high; rays ca. 13, 10-15 mm long. Aug.

Moist soil of aspen groves, meadowy depressions and along streams; montane. Ssp. *foliosa* (Nutt.) Maguire with whitish, barbed pappus and entire leaf margins is common; East, West. The larger ssp. *chamissonis* with brownish, more feathery pappus and toothed leaves is uncommon, East. AK to Ont., south to CA, NM.

Arnica cordifolia Hook, Heart-leaf Arnica. Stems 15-50 cm tall, borne separately from the tufts of basal leaves; herbage hairy and often glandular; leaf blades heart-shaped, 4-12 cm long, 2-4 pairs on the stem, the lower with long petioles and toothed margins; heads 1-3, conical, 13-20 mm high; rays 10-15, 2-3 cm long; pappus white, barbed. June-Aug. Fig. 48. Color plate 38.

Dry to moist montane and subalpine forest; abundant West, less common East. The dwarf, high-elevation form has been called var. *pumila* (Rydb.) Maguire. AK south to CA, NM, MI. *Arnica cordifolia* is often confused with *A. latifolia*; both are common in the Park. In the former the lowest stem leaves are the largest, while the latter has largest leaves in the middle, and is less common in the montane zone west of the Divide.

Arnica diversifolia Greene. Stems 15-50 cm tall; herbage glandular-hairy to nearly glabrous; leaf blades heart-shaped to lance-shaped with toothed margins, mostly petiolate, 4-8 cm long, 2-4 pairs per stem, largest in the middle; heads usually several, conical, 8-15 mm high; rays 8-13, 1-2 cm long; pappus yellowish, feathery. July-Aug.

Common in moist, stony soil, of meadows and thickets, often along streams; montane, subalpine: East, West. AK to CA, UT. *A. diversifolia* is thought to be a hybrid complex involving *A. cordifolia, A. latifolia, A. mollis,* and *A. amplexicaulis*; some plants resemble *A. cordifolia,* while others are more like *A. mollis.*

Arnica fulgens Pursh. Stems 20-40 cm tall; herbage glandular-hairy; leaves narrowly oblong, 3-10 cm long, petiolate, entire, 2-4 pairs with tufted brown hair among leaf bases at ground level; heads solitary, hemispheric, 10-20 mm high; rays 10-23, to 20 mm long; pappus white or yellowish, barbed. June-July.

Common in montane grasslands; East, West. B.C. to Sask. south to CA, CO.

Arnica latifolia Bong. [*A. gracilis* Rydb.]. Stems 20-50 cm tall, often separate from basal leaves; herbage glandular-hairy to glabrous; leaf blades lance- to arrow-shaped with toothed margins, 2-18 cm long, 2-4 pairs, the lower petiolate; heads 1-3, broadly conical, 7-18 mm high; rays to 15 mm long; pappus white, barbed. June-July.

Moist forest, meadows and rocky slopes at all elevations; abundant East, less common west. AK south to CA, CO. *Arnica latifolia* tends to replace *A. cordifolia* east of the Divide and at higher elevations (see *A. cordifolia*). Smaller plants at high elevations tend to have only a single head.

Arnica longifolia D. C. Eat. Stems 30-70 cm tall, often clustered; herbage short-hairy, glandular; leaves narrowly lance-shaped with entire margins, 5-15 cm long, 5-7 pairs without petioles; heads several, narrowly bell-

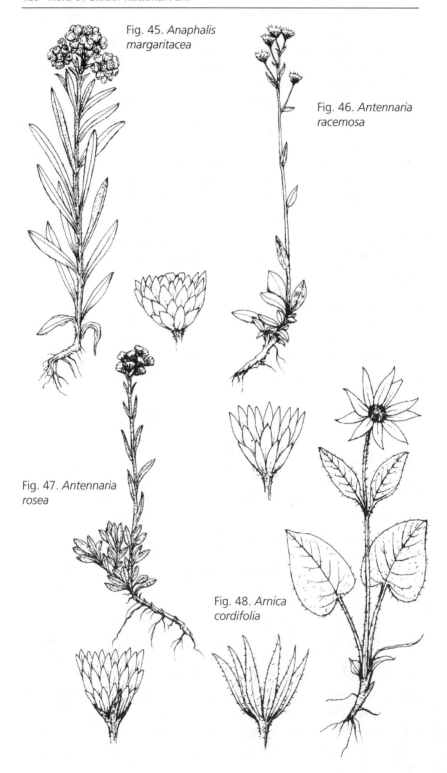

Fig. 45. *Anaphalis margaritacea*

Fig. 46. *Antennaria racemosa*

Fig. 47. *Antennaria rosea*

Fig. 48. *Arnica cordifolia*

shaped, 7-10 mm high; rays 8-13, to 15 mm long; pappus yellow to brownish, barbed. July-Aug.

Uncommon in very stony soil on slopes or along streams, usually near or above treeline, near or east of the Divide. AK to CA, CO.

Arnica mollis Hook. Stems 20-40 cm high, often separate from basal leaves; herbage hairy and glandular; leaves oblong-elliptic with minutely toothed margins, 3-10 cm long, 3-4 pairs, the lower petiolate; heads 1-3, 10-15 mm high, hemispheric; rays 12-18 up to 2 cm long; pappus brownish, feathery. June-Aug.

Moist soil of meadows, open forest, and along streams; East, West; common in the subalpine zone, less common lower and higher. AK to CA, CO. Reported to be more variable in other portions of its range (Cronquist 1955).

Arnica parryi Gray. Stems 20-60 cm tall; herbage glandular above; leaf blades lance-shaped with shallowly toothed margins, 3-10 cm long, 2-4 pairs, the lowest petiolate; heads several, bell-shaped, 10-16 mm high, lateral ones often nodding; rays lacking; pappus brownish, barbed. July-Aug.

Uncommon in meadows and aspen or lodgepole forest openings; montane and subalpine; near or east of the Divide. B.C., Alta. south to CA, CO.

Arnica rydbergii Greene. Stems 10-25 cm tall; often separate from basal leaves; herbage glandular and short-hairy; leaves narrowly elliptic with mostly entire margins, 2-5 cm long, 3-4 pairs, only the basal petiolate; heads 1-3, broadly conical, 9-16 mm high; rays ca. 8, to 2 cm long; pappus white, barbed. July-Aug.

Common in rocky soil of meadows and slopes; upper montane to alpine, near or east of the Divide. B.C., Alta. south to CA, CO.

Arnica sororia Greene. Similar to *A. fulgens*; leaves narrowly lance-shaped, sometimes with white hair among leaf bases; heads solitary or few; pappus white, barbed. June-July. Fig. 49.

Montane grassland; common in North Fork prairies, uncommon East. B.C., Alta. south to CA, WY.

Artemisia L., Sagebrush, Sagewort, Mugwort, Wormwood

Biennial or perennial, herbs or shrubs; herbage usually aromatic; leaves alternate; heads small, discoid, many in narrow to spreading inflorescences; flowers bisexual or female; heads mostly 2-4 mm high; involucral bracts papery, not green, in overlapping series; receptacle naked or hairy; pappus lacking; achenes usually glabrous.

A. tridentata, A. ludoviciana, and *A. frigida* were used ceremonially by native Americans.

1. Shrubs at least 40 cm high .... **A. tridentata**
1. Plants herbaceous or woody only at the base 2

2. Leaves glabrous 3
2. Leaves densely hairy at least below 5

3. Middle and upper leaves undivided .... **A. dracunculus**
3. Middle and upper leaves divided or dissected 4

4. Ultimate leaf segments 1 mm wide or less .... **A. campestris**
4. Ultimate leaf segments > 1 mm wide .... **A. biennis**

5. Leaves divided, ultimate segments 2 mm wide or less 6
5. Leaves entire or lobed with segments > 2 mm wide 7

6. Leaves 5-20 mm long, ultimate segments < 1 mm wide .... **A. frigida**
6. Some leaves > 25 mm long, segments ≥ 1 mm wide .... **A. campestris**

7. Leaves entire or once lobed .... **A. ludoviciana**
7. Leaves twice pinnate 8

8. Leaves green above, white-wooly below; montane or above ....
.... **A. michauxiana**
8. Leaves equally short-hairy above and below; disturbed montane areas ....
.... **A. absinthium**

Artemisia absinthium L., Wormwood. Perennial, 40-80 cm tall; herbage silvery-hairy; leaves 2-3 times pinnately divided into narrow segments, 3-8 cm long, reduced above; inflorescence branched, spreading. July-Aug.

Uncommon in disturbed areas along roads and near residential areas in the valleys, West. Introduced from Europe.

Artemisia biennis Willd. Taprooted annual or biennial with stems 30-100 cm tall; herbage glabrous, inodorous; leaves 5-15 cm long, pinnately divided into narrow lobes with toothed margins; inflorescence dense, spike-like. June-July.

Rare, reported for margins of shallow, temporary ponds near East Glacier (Standley 1921). Native to westen N. America but adventive elsewhere.

Artemisia campestris L. [*A. forwoodii* Wats., *A. spithamaea* Pursh]. Taprooted perennial or biennial with stems 10-50 cm tall; herbage glabrous or white-hairy, nearly inodorous; leaves to 7 cm long, twice pinnately divided into narrow segments, reduced above; inflorescence spike-like. June-July. Fig. 50.

The biennial ssp. *caudata* (Michx.) Hall & Clem. is uncommon in montane meadows and grasslands; East (Standley 1921). The perennial ssp. *borealis* (Pall.) Hall & Clem. with rosettes of basal leaves is locally common in stony soil of exposed ridges and slopes near or above treeline in the Two Medicine area. Circumboreal south to NM, TX, MI, VT. A collection of ssp. borealis (Umbach 612) has been mistaken for *A. norvegica*.

Artemisia dracunculus L., Tarragon [*A. dracunculoides* Pursh]. Rhizomatous perennial with stems 40-80 cm tall; herbage mostly glabrous;

leaves narrow, 2-6 cm long with entire margins or lowest with 1-2 long lobes; inflorescence branched, open, leafy. Aug.

Uncommon in montane grasslands and meadows; East, West. B.C. to Man. south to CA, TX, MO.

Artemisia frigida Willd., Fringed Sagewort. Mat-forming perennial; stems 10-30 cm tall, woody at the base; herbage silvery-hairy; leaves numerous, 5-12 mm long, 2-3 times divided into linear segments; inflorescence leafy, branched but narrow; heads yellow. July-Aug.

Rare in grasslands along the east edge of the Park. AK to Man. south to B.C., NM, TX, and IA; Eurasia.

Artemisia ludoviciana Nutt. [*A. flocosa* Rydb., *A. diversifolia, A. gnaphaloides* Nutt.]. Rhizomatous perennial with stems 30-80 cm tall; herbage white-hairy; leaves simple to deeply lobed, 3-8 cm long; inflorescence branched, narrowly conical; heads numerous. June-Aug.

Var. *ludoviciana*, with narrowly lance-shaped leaves and entire margins, is uncommon in montane grasslands, East. Var. *latiloba*, with lobed lower leaves, is common in grasslands, along streams and open rocky slopes; montane and subalpine; East, West. B.C. to Ont. south to CA, Mex. *A. tilesii*, a more northern species, has leaves similar in shape to *A. ludoviciana* var. *latiloba*, but they are green and glabrous above like those of *A. michauxiana*.

Artemisia michauxiana Besser [*A. discolor* Dougl.]. Perennial herb with stems 20-70 cm tall; leaves 2-5 cm long, 1-2 times pinnately divided with toothed margins, upper surface green and nearly glabrous above but densely white-hairy below; heads 2-3 mm high; inflorescence narrow and unbranched. July-Aug. Fig. 51.

Common in stony soil of slopes and outcrops, sometimes in talus; upper montane to alpine; East, West. B.C., Alta. south to CA, UT, WY.

Artemisia tridentata Nutt., Big Sagebrush. Erect, branched shrub 40-100 cm high; herbage silvery-hairy; leaves narrow, 15-30 mm long, widening toward a blunt, shallowly 3-lobed tip; heads numerous in a branched, spreading inflorescence. Aug-Sept.

Common in the North Fork Flathead prairies. Our plants are ssp. *vasseyana* (Rydb.) Beetle. B.C., Alta. south to CA, NM, ND. Big sagebrush is killed by fire; its abundance depends on fire frequency.

Aster L., Aster

Perennial herbs (ours) with short to long rhizomes or rootstocks; leaves alternate, undivided; heads few to many, radiate; involucral bracts in several overlapping, unequal series; rays white, pink or blue, female and fertile; disk flowers bisexual; receptacle naked; pappus of numerous thin bristles.

Aster and *Erigeron* are often confused; involucral bracts of *Aster* are mostly of several different lengths and overlap each other like shingles, while bracts of *Erigeron* also overlap but are more nearly the same length. This difference is not absolute, and determinations can sometimes be difficult.

Four species of *Aster* are reported for Waterton Park (Kuijt 1982) and are expected in Glacier Park as well: *A. ciliolatus* Lindl. has arrow-shaped basal leaves; *A. lanceolatus* Willd. [=*A. hesperius* Gray] has vertical lines of hairs running down the stem from the petiole bases; *A. borealis* (T. & G.) Prov. [=*A. junciformis* Rydb.] is similar to *A. occidentalis* but has narrower leaves and occurs in fens; and *A. pansus* (Blake) Cronq. is similar to *A. falcatus* but has clustered stems and is usually found in moist habitats. *A. occidentalis*, *A. foliaceus*, and *A. eatonii* are polyploid-hybrid complex; it may be difficult to unambiguously assign all plants to a particular species (K. Chambers, pers. comm.). Reference: Dean (1966).

1. Plants glandular in the inflorescence (check just below heads) 2
1. Plants not glandular 5
2. Leaves nearly linear, largest ≤ 10 mm wide .... **A. campestris**
2. Leaves lanceolate to ovate, largest > 10 mm wide 3
3. Leaves narrowly elliptic to ovate with sharply toothed margins to the tip ......
   ....**A. conspicuus**
3. Leaves with nearly entire margins 4
4. Involucral bracts lance-shaped with a marginal fringe of hair above ....
   ....**A. engelmannii**
4. Involucral bracts linear without fringe of hair on margins .... **A. modestus**
5. Heads solitary, leaves mostly basal; stems leaves greatly reduced ....
   ....**A. alpigenus**
5. Heads usually > 1; stem leaves reduced but conspicuous 6
6. Involucral bracts purple .... **A. sibiricus**
6. Involucral bracts green to whitish 7
7. Outer involucral bracts of uppermost head as tall or taller than inner 8
7. Outer involucral bracts of uppermost head shorter than inner 9
8. Leaves strap-shaped; inflorescence leafy, with many heads .... **A. eatonii**
8. Leaves lance-shaped; inflorescence not as large .... **A. foliaceus**
9. Leaves lance-shaped .... **A. laevis**
9. Leaves strap-shaped 10
10. Tips of upper leaves and involucral bracts with cone-shaped spine ....
    ....**A. falcatus**
10. Leaf and bract tips not spine-tipped 11
11. Outer involucral bracts with a white base; plants of grasslands and moist
    meadows .... **A. ascendens**
11. Outer involucral bracts green throughout; plants of wet meadows ....
    ....**A. occidentalis**

Aster alpigenus (T. & G.) Gray. Stems 5-15 cm high, curved upward at the base; basal leaves 2-10 cm long, linear-oblong; stem leaves small and narrow;

heads solitary, 5-15 mm high; involucral bracts with long hair on margins, purplish, little overlapping; rays 10-40, violet, 7-15 mm long. July.

Usually a plant of alpine meadows and ridges; our one collection from near Lake McDonald. Our plants are var. *haydenii* (Porter) Cronq. WA to MT south to CA, NV, WY.

Aster ascendens Lindl. [*A. chilensis* Nees misapplied, *A. nelsonii* Greene]. Stems 20-60 cm tall; herbage usually hairy; leaves strap-shaped with entire margins, 2-8 cm long, petioles lacking; inflorescence open and long-branched; heads 5-7 mm high with green-tipped involucral bracts; rays 15-40, blue to pink, 5-10 mm long. Aug-Sept.

Uncommon in montane grasslands and meadows; East. B.C. to Sask. south to CA, NM.

Aster campestris Nutt. Meadow Aster. Stems 10-50 cm tall; herbage glandular above; leaves very narrowly oblong, 2-6 cm long with entire margins, mostly without petioles; inflorescence few-branched and few-headed; heads 5-8 mm high with glandular bracts; rays 15-20, purplish, 5-8 mm long. Aug-Sept.

Uncommon in montane grasslands; East, West. B.C., Alta. south to CA, UT.

Aster conspicuus Lindl. Stems glandular above, 30-70 cm tall; leaves roughened, ovate with toothed margins, 6-12 cm long, without petioles, reduced above; inflorescence open, flat-topped; heads 9-12 mm high with glandular bracts; rays 12-35, blue, 10-15 mm long. July-Aug. Fig. 52. Color plate 37.

Abundant in dry to moist montane and lower subalpine forests; East, West. B.C. to Sask., south to OR, WY. The broad, strongly toothed leaves often cover the forest floor; however, A. conspicuus rarely flowers in the shade.

Aster eatonii (Gray) Howell. Stems short-hairy above, 30-80 cm tall; leaves glabrous, linear with entire margins, 5-10 cm long without petioles; inflorescence branched, spreading, leafy; heads 4-10 mm high with green bracts; rays pink, 5-12 mm long. Aug-Sept. Fig. 53.

Uncommon in moist soil along streams and wetlands; West. B.C. to Sask., south to CA, NM.

Aster engelmannii (D. C. Eaton) Gray. Stems 50-180 cm tall; herbage glabrous to sparsely hairy and glandular; leaves lance-shaped with entire margins, 5-10 cm long, without petioles, largest at midstem; inflorescence short, unbranched; heads few, 7-15 mm high; involucral bracts purplish with white hair on the margins; rays 8-13, pink to purple, 15-25 mm long. July-Sept. Fig. 54.

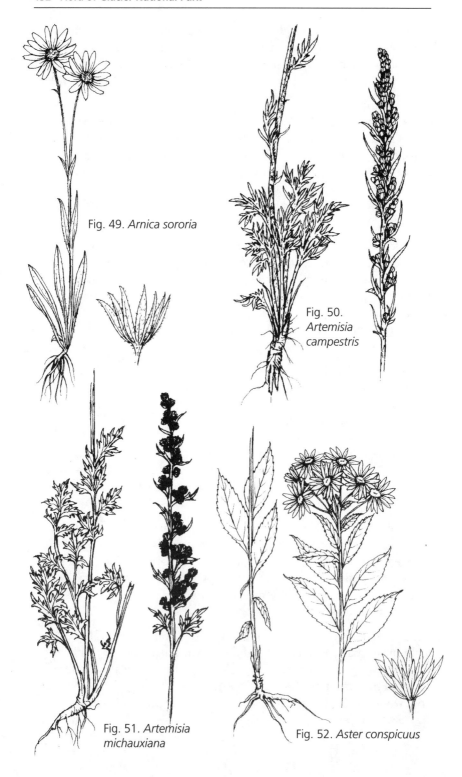

Fig. 49. *Arnica sororia*

Fig. 50. *Artemisia campestris*

Fig. 51. *Artemisia michauxiana*

Fig. 52. *Aster conspicuus*

Common in moist, open forest, aspen groves, meadows and avalanche slopes; upper montane and lower subalpine; East, West. B.C., Alta. south to WA, NV, CO.

Aster falcatus Lindl. [*A. crassulus* Rydb.]. Stems 20-50 cm tall, clustered; herbage short-hairy; leaves to 5 cm long, strap-shaped with pointed tips, petioles lacking, margins entire, the lowest withered at maturity; inflorescence leafy, narrow with heads solitary at the end of erect branches; heads 5-8 mm high with hairy, sharp-pointed, recurved bracts; rays 20-35, white, 5-8 mm long. Aug-Sept.

Uncommon, reported for grasslands along east edge of the Park (Standley 1921). AK to Man. south to AZ, TX, IA. The similar *A. pansus* has heads 2-5 mm high, numerous on the curving branches of the inflorescence. It may occur in the Park's low-elevation, grassland habitats.

Aster foliaceus Lindl. [*A. apricus* (Gray) Rydb., *A. frondeus* (Gray) Greene]. Stems to 60 cm high; herbage glabrous or hairy; leaves oblong lance-shaped, 5-12 cm long with entire margins, the lower with broad petioles; heads few with leafy outer involucral bracts mostly longer than the inner; rays 15-60, blue or purple, 1-2 cm long. Aug-Sept. Color plate 47.

Three intergradent varieties occur in the Park. Var. *parryi* (D. C. Eat.) Gray, with several heads and usually > 25 cm tall with linear outer involucral bracts, is abundant in meadows and moist, open forest in the montane and subalpine zones. Var. *apricus* Gray, less than 25 cm tall, with only 1-3 heads and purple-tipped involucral bracts, is common above treeline. Var. *foliaceus* with broad, leaf-like involucral bracts is also reported for Glacier Park (Hitchcock et al. 1955). AK to CA, NM. Reports of *A. subspicatus* Nees (=*A. umbachii* Rydb.) by Rydberg (1910) are referable here. *A. foliaceus* and *A. occidentalis* can intergrade.

Aster laevis L. Stems 40-80 cm tall; herbage glabrous; leaves lance-shaped, to 15 cm long, lower petiolate, upper clasping, margins entire or toothed; inflorescence branched, leafy, open; heads 5-10 mm high with green-tipped bracts; rays 15-30, blue, 5-8 mm long. July-Sept.

Common in open, montane forest, especially along streams, roads or other disturbed areas; East, West. B.C. to Que. south to WA, NM, GA.

Aster modestus Lindl. [*A. sayianus* Nutt.]. Stems to 30-80 cm tall; herbage glandular above; leaves lance-shaped with entire or toothed margins, 5-13 cm long, without petioles; inflorescence short, leafy; heads 7-11 mm high, involucral bracts glandular, nearly equal in length, inner ones often purplish; rays 2-45, purple, 7-12 mm long. Aug-Sept.

Common in moist thickets and avalanche slopes, especially along streams; montane and subalpine; East, West. AK to Ont. south to OR, MT, MN.

Fig. 53. *Aster eatonii*

Fig. 54. *Aster engelmannii*

Aster occidentalis Nutt., Western Aster [*A. fremontii* (T.& G.) Gray]. Stems 20-50 cm tall; herbage glabrous; leaves lance-shaped with entire margins, 3-10 cm long, the lower petiolate; heads several, 6-9 mm high; involucral bracts narrow; rays 20-50, blue or purple, 6-10 mm long. Aug-Sept.

Common in moist soil of meadows, thickets, open forest, and avalanche slopes, often associated with wetlands; montane to lower subalpine; East, West. B.C. to CA, CO. The similar *A. borealis* (T. & G.) Prov. [=*A. junciformis* Rydb.], with lines of hair running down the upper stem, could occur in montane fens.

Aster sibiricus L. [*A. meritus* A. Nels.]. Stems 10-20 cm tall; herbage short-hairy; leaves stiff, lance-shaped with toothed or entire margins, 2-5 cm long, the lower with short petioles; heads few, 8-12 mm high with purplish involucral bracts; rays 12-23, purple, 8-12 mm long. Aug-Sept.

Sparsely vegetated soil of meadows, exposed ridges, streambanks and slopes; East, West; common in subalpine and alpine zones, less common lower. Our plants are var. *meritus* (A. Nels.) Raup. Circumboreal south to OR, WY.

Balsamorhiza Nutt., Balsamroot

Balsamorhiza sagittata (Pursh) Nutt., Arrowleaf Balsamroot. Stemless perennial herb with a large taproot; leaves arrow-shaped with long petioles, the blade to 30 cm long, silvery short-hairy; heads solitary on long stalks, 20-80 cm tall; involucral bracts in several series, subequal, tapered to the tip; disk flowers yellow, ca. 1 cm long; rays ca. 13 or 21, female, fertile, yellow, 25-40 mm long; receptacle with small, leaf-like bracts between flowers; achenes glabrous, to 8 mm long; pappus lacking. May-June. Fig. 55. Color plate 1.

Common in montane grasslands; East, West. B.C., Alta. south to CA, CO, SD. The Blackfeet used the roots for food.

Bellis L., Daisy

Bellis perennis L., English Daisy. Fibrous-rooted perennial; herbage spreading hairy; leaves all basal, spoon-shaped with toothed margins, to 4 cm long; heads solitary on stalks 5-10 cm tall; involucral bracts in 1 series, lanceolate; disk flowers yellow, bisexual; ray flowers female, white, ca. 1 cm long; receptacle naked, conical; pappus lacking. May-Sept.

Locally common in the lawns of West Glacier. Introduced from Europe.

Bidens L., Beggar's-ticks

Bidens cernua L., Bur-marigold. Annual herb with stems to 50 cm tall; herbage glabrous; leaves narrowly lance-shaped, opposite, 4-10 cm long, undivided with toothed margins, petioles lacking; heads few, from stalks in upper leaf axils, often nodding; involucral bracts dimorphic; outer 6-8, strap-shaped and reflexed; inner ovate, yellowish, to 1 cm long; disk flowers bisexual, yellow; rays 6-8, yellow, to 15 mm long; receptacle with thin bracts among flowers; achenes 4-angled, prickly; pappus of 4 spines. July-Aug.

Known from the North Fork Flathead Valley. Widespread in temperate regions of N. America, Europe, Asia.

Brickellia Ell., Thoroughwort

Brickellia grandiflora (Hook.) Nutt. [*Coleosanthus grandiflorus* (Hook.) Kuntze]. Herbaceous perennial, woody at the base with stems 20-80 cm tall; herbage short-hairy; leaves alternate, triangular with toothed margins and long petioles; blades 2-8 cm long; heads nodding, discoid, on branched stalks from upper leaf axils; involucre 7-11 mm high, bracts striped, in several overlapping series; disk flowers cream-colored; receptacle naked; pappus of white bristles. Aug. Fig. 56.

Uncommon in very rocky soil of slopes and stream banks; montane and subalpine, near or east of the Divide. B.C. to Alta south to CA, CO, TX.

Carduus L., Thistle

Carduus nutans L., Musk Thistle. Biennial with winged stems to 100 cm high; leaves alternate, to 40 cm long, oblong with spine-tipped lobes and teeth, long-hairy on veins beneath; heads discoid, 4-8 cm wide, solitary on the ends of nodding stalks; outer involucral bracts 2-8 mm wide, spine-tipped; flowers purple. July.

Collected along the road in the McDonald Valley. Introduced from Europe. A serious weed of disturbed pastures.

Centaurea L., Knapweed

Perennial herbs; leaves alternate; inflorescence branched, open; heads solitary at branch tips; discoid but often appearing radiate due to enlarged, sterile marginal flowers; involucral bracts overlapping in several series, often ornamented with appendages or spines; receptacle bristly; pappus of bristles, scales or lacking. All our species are introduced from Europe.

1. Leaves divided into narrow segments .... **C. maculosa**
1. Leaves entire or shallowly toothed 2
2. Involucre 20-25 mm high; bracts with marginal pointed teeth ... **C. montana**
2. Involucre 12-18 mm high; bracts with marginal rounded lobes .... **C. jacea**

Centaurea jacea L., Brown Knapweed. Stems 20-100 cm tall; herbage glabrous or with sparse long hair; leaves lance-shaped with shallow lobes at the base, to 15 cm long, petiolate below; involucral bracts brown, ovate with papery, lobed margins; flowers rose-purple to white; pappus none. July-Aug.

One collection from near Bowman Lake campground. Meadow knapweed (*C. pratense* Thuill.), a hybrid between *C. jacea* and *C. nigra*, is similar, but the involucral bracts have fringed margins, and the pappus is of short, white hairs. It has been collected along the Middle Fork Flathead River.

Centaurea maculosa Lam., Spotted Knapweed. Stems 30-100 cm tall from a taproot; herbage sparsely hairy; leaves to 12 cm long, pinnately divided into narrowly lance-shaped segments, rosette leaves the largest; involucre 10-13 mm high, bracts striped with small, dark spines on upper half; flowers usually rose-purple (white); pappus of short hairs. July-Aug. Fig. 57.

Locally common in montane grasslands, roadsides, meadows and open forest; East, West. One of the most pernicious exotics in the Park. Knapweed seeds brought in by visitors establish along trails and roads. It then invades native plant communities, especially in disturbed areas, outcompeting or suppressing the growth of the dominant grasses. Seeds persist for many years in the soil, making it difficult to eradicate. *C. diffusa* Lam., with white flowers and short spines on the tips of the involucral bracts, has been reported for the Park, but no specimens have been seen.

Centaurea montana L. Stems to 70 cm tall from rhizomes; herbage sparsely hairy; leaves narrowly lance-shaped with entire margins, 7-18 cm long, without petioles, basal rosette lacking; involucre 20-25 mm high, bracts ovate with blackish teeth on the margin; outer enlarged flowers blue, inner disk flowers purplish. July.

Open forest near West Glacier. Cultivated for ornament and sometimes escaped.

Chrysanthemum L.

Chrysanthemum leucanthemum L., Oxeye daisy [*Leucanthemum vulgare* Lam.]. Perennial, rhizomatous herb with stems to 80 cm tall; herbage glabrous or sparsely hairy; basal leaves spoon-shaped with toothed to lobed margins, 4-15 cm long; stem leaves alternate, oblong with toothed margins; heads solitary, radiate; involucral bracts overlapping, narrow, in 2-3 series; disk flowers yellow, bisexual; rays 15-30, white, 1-2 cm long; receptacle naked; pappus none. July-Aug. Fig. 58.

Abundant along roads and in disturbed meadows, often at the edge of forest; montane; East, West. Native to Europe and Asia. This weed does not usually invade undisturbed native communities in the Park.

Cirsium Mill., Thistle

Spiny, perennial herbs with toothed or pinnately lobed, alternate leaves; heads discoid, few to several; involucral bracts, often spine-tipped, in several series of mostly unequal length; disk flowers bisexual; receptacle bristly; pappus of numerous feathery bristles.

1. Involucre ≤ 15 mm wide; plants rhizomatous, clonal .... **C. arvense**
1. Involucre > 15 mm wide; plants taprooted 2
2. Flowers white to pink .... **C. hookerianum**
2. Flowers rose to purple 3

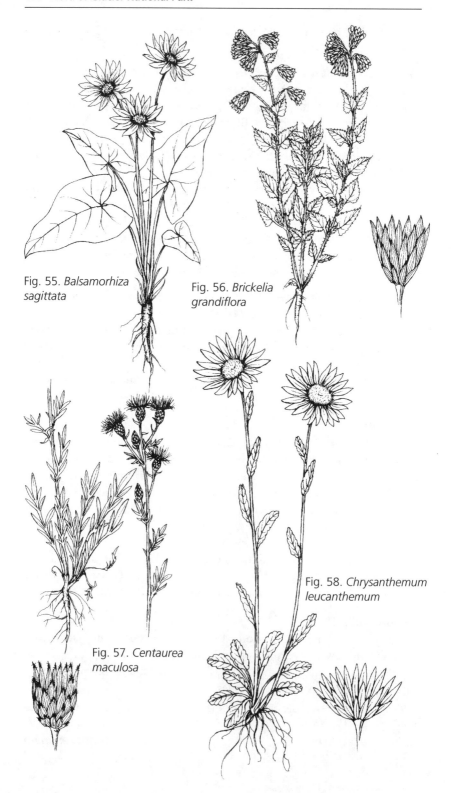

Fig. 55. *Balsamorhiza sagittata*

Fig. 56. *Brickelia grandiflora*

Fig. 57. *Centaurea maculosa*

Fig. 58. *Chrysanthemum leucanthemum*

3. Stems spreading hairy with spiny wings below leaves .... **C. vulgare**
3. Stems densely appressed white-hairy, without wings .... **C. undulatum**

Cirsium arvense (L.) Scop., Canada Thistle. Stems to 100 cm tall from deep rhizomes, tending to form colonies; leaves to 12 cm long, slenderly lance-shaped with wavy or lobed margins, hairy beneath; inflorescence open, with several heads; heads 1-3 cm high; involucral bracts unarmed or with a small spine; flowers purple (white), unisexual, only 1 sex on each stem. July-Aug.

Common in moist, usually disturbed soil of montane meadows, thickets, roadsides, and open forests, especially aspen; East, West. Native to Eurasia, introduced throughout temperate N. America. One of our worst exotic weeds; the deep rhizomes make eradication nearly impossible. Although Canada thistle is most common near roads and other human industry, the feathery seeds have carried it throughout the backcountry where it can be found on beaver dams and other moist areas of natural disturbance.

Cirsium hookerianum Nutt., White Thistle, Elk Thistle. Short-lived perennial with stems to 100 cm tall from a taproot; herbage long, white-hairy; leaves to 20 cm long, with deeply toothed margins; inflorescence compact to elongate, the uppermost head the largest; heads 2-4 cm high; involucral bracts mostly the same length with long, tangled hairs at least on the margins, the outer with short spine-tips; flowers white. July-Aug. Fig. 59.

Common in meadows, aspen groves and on open slopes at all elevations; East, West. B.C., Alta south to WA, ID, WY. Many plants, especially from higher elevations, have a compact inflorescence and appear very similar to *C. scariosum* Nutt., but the involucres usually have at least some tangled hairs typical of *C. hookerianum*.

Cirsium undulatum (Nutt.) Spreng., Wavy-leaved Thistle. Short-lived perennial with stems to 100 cm tall from a taproot; herbage densely white-hairy; leaves deeply lobed, spiny-margined; inflorescence with few heads; heads 3-4 cm high; involucral bracts with a sticky brown central ridge, outer spine-tipped; flowers light purple. July.

Uncommon in montane grasslands and meadows; East. B.C. to Man. south to OR, NM, TX, MO.

Cirsium vulgare (Savi) Tenore, Bull Thistle [*C. lanceolatum* (L.) Hill]. Biennial with winged stems to 120 cm tall; herbage spreading hairy and spiny; leaves lobed nearly to the midvein; inflorescence of several heads; heads 3-5 cm high; involucral bracts spine-tipped; flowers purple. July-Aug.

Locally common in disturbed meadows, campgrounds and roadsides; montane; East, West. Native to Eurasia, introduced throughout N. America. This short-lived weed will not persist without disturbance and is rarely a problem in native communities. Standley's (1921) report suggests it was more common 70 years ago than it is today.

Conyza Less., Horseweed

Conyza canadensis (L.) Cronq., Canada Fleabane [*Erigeron canadensis* L.]. Annual with stems rarely to 80 cm tall; herbage with spreading hairs; leaves to 10 cm long, narrowly lance-shaped with shallowly toothed margins and petiolate below; inflorescence branched, long, narrow; heads numerous, appearing discoid, 2-4 mm high; involucral bracts in 2-3 unequal series; receptacle naked; disk flowers yellow; ray flowers inconspicuous; pappus of many thin bristles. July-Aug.

Uncommon in disturbed meadows and roadsides in valleys west of the Divide. Native throughout much of temperate N. America.

Crepis L., Hawksbeard

Annual or perennial, mostly taprooted herbs with milky sap; leaves mainly basal, stem leaves alternate and reduced or absent; inflorescence open, branched, more-or-less flat-topped with several to many heads; heads ligulate; flowers yellow, bisexual; involucral bracts in 2 very unequal series; receptacle naked; pappus of many thin, white bristles.

1. Annual; basal leaves few, not much larger than stem leaves ...... **C. tectorum**
1. Perennial; leaves mainly basal 2

2. Plants of high elevations; flower heads among tufted basal leaves .... **C. nana**
2. Plants mostly montane or lower; erect stems present 3

3. Leaves pinnately lobed > halfway to the midrib 4
3. Leaves with entire or toothed margins 5

4. Leaf lobes linear, well-separated .... **C. atrabarba**
4. Leaf lobes broad at the base, close together .... **C. intermedia**

5. Basal leaves with some hair and backward-pointing teeth .... **C. runcinata**
5. Leaves entire or shallowly toothed, glabrous .... **C. elegans**

Crepis atrabarba Heller. Perennial 15-70 cm high, short-hairy when young; leaves 10-35 cm long, deeply pinnately divided into several, well-separated, linear lobes; heads 3-30, short-hairy, narrow, 8-14 mm high with 10-35 flowers; outer involucral bracts half as long as the inner; achenes narrowed at the top. June.

Rare in montane grasslands and on brushy slopes; known only from southwest of Marias Pass. B.C. to Sask., south to NV, CO.

Crepis elegans Hook. Perennial to 30 cm tall; herbage glabrous; leaves to 8 cm long, petiolate, blades ovate with entire to shallowly toothed margins; heads numerous, narrow, 6-9 mm high with 6-10 flowers; achenes with a beak 1 mm long. July-Aug. Fig. 60.

Uncommon in moist, gravelly or sandy soil along montane streams and roads, occasionally higher on moraine; East, West. AK to B.C., WY.

Crepis intermedia Gray. Perennial with stems 20-70 cm tall; herbage with short gray hairs; leaves to 30 cm long, lance-shaped, deeply lobed into narrow segments; heads 6-20, 10-15 mm high, mostly 7-12 flowered; flowers 14-30 mm long; achenes beakless. July-Aug.

Uncommon in montane grasslands, East. B.C., Alta. south to WA and CO.

Crepis nana Rich. Perennial with upward-curved stems less than 8 cm high; herbage glabrous; basal leaves numerous, to 5 cm long, petiolate; blades elliptic with entire margins; heads numerous, 8-13 mm high, 9- to 12-flowered; flowers 7-12 mm long; achenes sometimes with a short beak. July-Aug.

Locally common on moraine and talus slopes and ridges near or above treeline; East, West. AK to Newf. south to CA, UT; Asia. Flower heads occur among the mound of dark green leaves.

Crepis runcinata (James) Torr. & Gray. Perennial to 50 cm tall; herbage glabrous to sparsely hairy; basal leaves to 10 cm long, petiolate, blades elliptic, often with backward-pointing teeth; heads 1-12, 8-15 mm high, 20-50 flowered; flowers 9-18 mm long; achenes with a short, broad beak. June-July.

Common in meadows and grasslands and around wetlands; montane to subalpine; East, West. Our plants are ssp. *runcinata*. B.C. to Man. south to CA, TX, MN.

Crepis tectorum L. Annual to 50 cm tall; herbage glabrous or thinly hairy; leaves to 10 cm long, narrowly lance-shaped with toothed to lobed margins, petiolate below; heads many, 6-9 mm high, 30-70 flowered; achenes narrowed above. June-Aug.

Rare along roads and in lawns; collected once in disturbed soil at Polebridge. Native to Europe.

Erigeron L., Daisy, Fleabane

Mostly perennial herbs with alternate and basal leaves, usually with entire margins; heads radiate; involucral bracts narrow, mostly the same length; ray flowers female, white pink or blue (ours); disk flowers numerous, bisexual, yellow; receptacle naked; pappus of numerous white bristles. (Hitchcock and Cronquist 1973, Kerstetter 1994.)

Erigeron is our largest genus in the Family Asteraceae. The hair and glands of involucral bracts are important characters. Similar to *Aster* (see note under that genus). *E. divergens, E. flagellaris, E. pallens* and *E. radicatus* are reported for Waterton Park (Kuijt 1982) and may occur in Glacier as well.

1. Leaves divided into narrow segments .... ***E. compositus***
1. Leaves entire to shallowly lobed 2
2. Rays inconspicuous, ≤ 6 mm long 3
2. Rays mostly > 6 mm long 6

3. Involucral bracts deep purple .... *E. humilis*
3. Involucral bracts green or purple-tipped 4

4. Rays spreading horizontally .... *E. strigosus*
4. Rays erect 5

5. Involucral bracts glandular .... *E. acris*
5. Involucral bracts without glandular hairs .... *E. lonchophyllus*

6. At least some basal leaves 3-lobed at the tip; alpine talus .... *E. lanatus*
6. Leaves without lobed tips 7

7. Basal leaves with coarsely toothed margins; disturbed areas ....
.... *E. philadelphicus*
7. Leaf margins entire 8

8. Plants with 1 head per stem 9
8. Most plants with > 1 head per stem 13

9. Basal leaves linear, mostly < 3 mm wide, ≥ 10 times as long as wide 10
9. Basal leaves narrowly lanceolate to spoon-shaped, > 3 mm wide, < 10 times as long as wide 11

10. Rays white .... *E. ochroleucus*
10. Rays blue .... *E. lackschewitzii*

11. Stems arising from a short, mostly unbranched rhizome .... *E. peregrinus*
11. Stems from a woody, often branched crown surmounting a taproot 12

12. Involucral bracts densely long-hairy contrasting with the sparsely hairy stem
.... *E. simplex*
12. Involucre and stem about equally hairy .... *E. caespitosus*

13. Uppermost stem leaves lanceolate, > 8 mm wide 14
13. Uppermost stem leaves linear to strap shaped, < 8 mm wide 15

14. Leaves glabrous except for the margins .... *E. speciosus*
14. Leaves long-hairy on some surfaces .... *E. subtrinervis*

15. Stems usually single, fibrous-rooted; heads with > 120 rays ..... *E. glabellus*
15. Plants taprooted, usually with a woody, branched crown; heads with 30-100 rays .... *E. caespitosus*

Erigeron acris L. [*E. jucundus* Greene, *Trimorpha acris* (L.) S. F. Gray]. Short-lived perennial 5-50 cm tall; herbage sparsely hairy, sometimes glandular; leaves to 10 cm long, narrowly oblong, petiolate below; involucral bracts glandular, 5-10 mm long; rays numerous, erect, white to pink, inconspicuous, to 4 mm long. July-Aug.

Var. *kamtschaticus* (D. C.) Herder [var. *asteroides* misapplied], > 30 cm tall with many heads, is rare in open forest, on trails and slopes; East, West. Var. *debilis* Gray, to 30 cm tall with few or solitary heads, is common in moist, open soil of slopes, cliffs and streambanks; montane to alpine; East, West. Circumboreal south to CA, CO, ME.

Erigeron caespitosus Nutt. Stems 5-25 cm tall, curved at the base from a stout taproot; herbage spreading hairy; leaves narrowly oblong, rounded at the tip, to 12 cm long, petiolate below; heads 1-10 in a leafy, open inflorescence; involucral bracts silvery hairy, glandular, 4-7 mm long; rays 30-100, white to blue, 5-15 mm long. July-Aug.

Common in rocky soil of open slopes and exposed ridges at all elevations; East, West. B.C. to Man. south to AZ, NM.

Erigeron compositus Pursh., Cut-leaf Daisy. Stems to 20 cm tall from a taproot; herbage sparsely hairy and glandular; basal leaves to 6 cm long, 2-3 times divided into 3's, ultimate segments narrowly oblong; stem leaves few, linear, entire; heads solitary; involucral bracts 5-10 mm long, glandular, spreading hairy; rays 20-60, usually white, to 12 mm long. June-Aug. Fig. 61.

Common in rocky soil of grasslands, outcrops, open slopes and ridges at all elevations; East, West. Our plants are var. *glabratus* Macoun. AK to Greenl. south to CA, AZ; Que. There is a great deal of variation in leaf shape and hair and ray length in this species. Populations with short rays are thought to produce seed asexually (J. H. Beaman, personal communication).

Erigeron glabellus Nutt. [*E. asper* Nutt.]. Short-lived perennial with stems to 40 cm tall from fibrous roots; herbage sparsely hairy; leaves narrowly oblong, 4-15 cm long, petiolate below; heads few, large; involucral bracts hairy, 5-9 mm long; rays > 125, 8-15 mm long, light blue to white. June-July.

Uncommon in montane grasslands and forest openings; East, West. Var. *glabellus* has hair appressed to the stem, while var. *pubescens* Hook. has spreading hair. AK to B.C., CO, SD, WI.

Erigeron humilis Graham [*E. unalaschkensis* Vierh.]. Stems to 15 cm tall; herbage long-hairy; leaves narrowly oblong, to 5 cm long, petiolate below; heads solitary; involucral bracts purplish, 6-9 mm long, with long, purple and white hairs; rays 50-150, light purple to white, 3-6 mm long. July-Aug.

Local and uncommon in cold, open soil of moraine, wet cliffs and talus above treeline; East, West. AK to Que. south to B.C., MT.

Erigeron lackschewitzii Nesom & Weber. Taprooted perennial 3-8 cm high; herbage long-hairy; basal leaves 2-6 cm long, narrowly spoon-shaped with pointed tips; stem leaves narrow; head solitary, 10-15 mm wide; the involucral bracts green, 6-8 mm long, densely covered with long, white hairs and sessile glands; rays 30-70, blue, 8-11 mm long, notched at the tip. June-July.

Rare in gravelly, calcareous soil of turf and exposed ridges and slopes near or above treeline; known only from near Chief Mtn. Endemic to northwest Montana from Glacier Park to near Augusta.

Erigeron lanatus Hook. Stems to 10 cm high from a widely branched rootcrown; herbage long-hairy; leaves oblong, gradually narrowed to the base, to 3 cm long, some with 3 shallow lobes at the tip; heads solitary; involucral bracts 8-11 mm long, densely long-hairy, usually purple-tipped; rays 30-80, white to pink, 7-11 mm long. July-Aug.

Locally common in coarse talus above treeline, East. Endemic to B.C., Alta. south to MT; disjunct in CO. Plants are often found sprawling between the shifting rocks. The similar *E. pallens* Cronq. [*E. pupuratus* ssp. *pallens* (Cronq.) G. Dougl.] with smaller rays and involucral bracts occurs in Waterton Park (Kuijt 1982) and may occur in Glacier as well.

Erigeron lonchophyllus Hook. Short-lived perennial to 30 cm tall; herbage spreading-hairy; leaves narrowly lance-shaped, to 12 cm long, petiolate below; heads usually several in an open inflorescence; involucral bracts 5-8 mm long, hairy, usually purple-tipped; rays numerous, white, inconspicuous, 2-3 mm long. July-Aug.

Uncommon in wet soil of wetlands and along streams; montane; East, West. AK to Que. south to CA, NM, ND.

Erigeron ochroleucus Nutt. Taprooted perennial to 10 cm tall; herbage with appressed hairs; leaves linear to narrowly oblong, to 4 cm long with enlarged, whitish bases; heads solitary; involucral bracts 5-8 mm long, spreading-hairy, often purple-tipped; rays 20-80, blue to usually white, 4-10 mm long. June-July.

Rare on open slopes along east front of the Park. Our plants are var. *scribneri* (Canby) Cronq. B.C. to Alta. south to WY, NE. The similar *E. radicatus* is reported for Waterton Park (Kuijt 1982) and could occur in Glacier. It has a less branched rootcrown and < 7 stem leaves, while *E. ochroleucus* has ≥ 7 stem leaves.

Erigeron peregrinus (Pursh) Greene., Wandering Daisy [*E. salsuginosus* (Rich.) Gray]. Fibrous-rooted perennial to 70 cm tall; herbage glabrous below the inflorescence; leaves narrowly elliptic, to 15 cm long, long-petiolate below; heads usually solitary; involucral bracts 7-11 mm long, glandular; rays 30-80, pale blue to purple, 12-20 mm long. July-Aug. Fig. 62.

Abundant in moist meadows, turf, open forest and thickets; upper montane to alpine; East, West. Our plants are ssp. *callianthemus* (Greene) Cronq. AK to CA, NM. This is our common high-elevation daisy, dominating moist subalpine meadows in mid-summer. *Aster foliaceus*, which is similar and almost as common, does not have a glandular involucre.

Erigeron philadelphicus L. Short-lived, fibrous-rooted perennial 15-70 cm tall; herbage long-hairy; leaves oblong with shallowly toothed margins, to 15 cm long, petiolate below, clasping above; heads usually several; involucral bracts 4-6 mm long with a brownish midvein and white margins; rays > 150, white to light purple, 4-8 mm long. June-Aug.

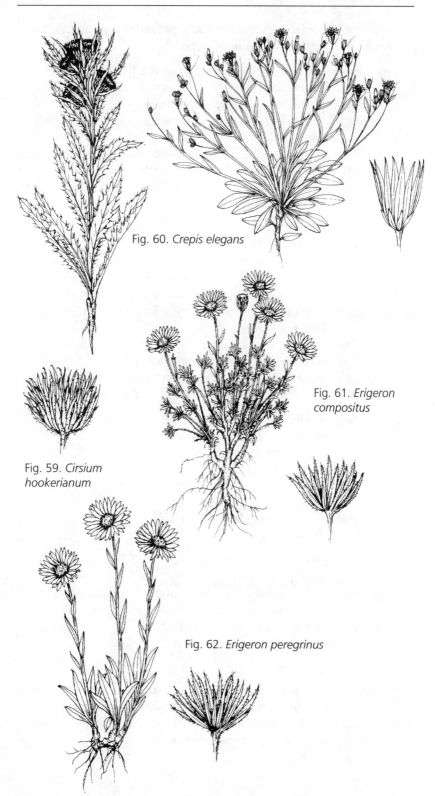

Fig. 60. *Crepis elegans*

Fig. 61. *Erigeron compositus*

Fig. 59. *Cirsium hookerianum*

Fig. 62. *Erigeron peregrinus*

Uncommon in disturbed meadows and banks and along roads and trails; montane, West. Most of temperate N. America.

Erigeron simplex Greene [*E. uniflorus* L. misapplied]. Stems to 15 cm tall from fibrous roots; herbage with long, sticky hairs; leaves narrowly oblong, to 5 cm long, petiolate below; heads solitary; involucral bracts 5-10 mm long, densely white-wooly; disk flowers 3-4 mm long; rays 50-125, pink to blue, 5-10 mm long. July.

Common in moist soil of open, alpine slopes and turf near or east of the Divide. WA, MT south to AZ, NM. Similar plants immediately north of the Canadian border are referred to *E. grandiflorus* Hook.; they occur in more exposed sites and have hairier herbage and disk corollas > 4 mm long.

Erigeron speciosus (Lindl.) D. C. [*E. macranthus* Nutt.]. Stems to 50 cm tall from a woody rootcrown; leaves oblong to ovate, to 10 cm long, the lower petiolate, the upper broader and sessile with long hair on the margin and midvein; heads usually several in an open, nearly flat-topped inflorescence; involucral bracts 5-8 mm long, glandular; rays 65-150, blue, 10-15 mm long. July-Aug. Color plate 29.

Common in montane grasslands, meadows, open forest and thickets; East, West. Our plants are var. *speciosus*. B.C.. Alta. south to CA, NM, NE. Standley (1921) reports var. *macranthus* (Nutt.) Cronq. for Glacier Park; however, Cronquist (1955) states that this variety is not in our area.

Erigeron strigosus Muhl. ex Willd., Daisy Fleabane [*E. ramosus* (Walt.) B. S. P.]. Annual or biennial with stems 20-70 cm tall; herbage finely hairy; leaves to 8 cm long, narrowly lance-shaped with mostly entire margins, petiolate below; heads several in an open, nearly flat-topped inflorescence; involucral bracts 2-3 mm long, hairy; rays 50-100, white, to 6 mm long. July-Aug.

Uncommon in disturbed areas, on banks and open slopes; montane, West. Our plants are var. *strigosus*. Much of temperate N. America.

Erigeron subtrinervis Rydb. [*E. conspicuus* Rydb.]. Similar to *E. speciosus* but the herbage is long-hairy. July-Aug.

Reported to be uncommon in lower montane grasslands, East by Standley (1921). Our plants are var. *conspicuus* (Rydb.) Cronq. B.C., Alta. south to NM, NE. Cronquist (1955) and Packer & Moss (1983) suggest that this may be no more than a variety of *E. speciosus*.

Filago L., Filago

Filago arvensis L. Annual with mostly simple stems to 40 cm tall; herbage white-wooly; leaves alternate, linear, to 2 cm long, margins entire; heads appearing discoid, small, in numerous clusters resembling small lint balls, in a narrow, leafy inflorescence; true involucral bracts minute; bracts of receptacle 3-5 mm long, appearing like involucral bracts; flowers inconspicuous; pappus of fine bristles. June-July.

Uncommon in disturbed areas of grasslands and around buildings in the valleys west of the Divide. Introduced from Europe. Could be confused with *Gnaphalium*, which has papery-tipped involucral bracts.

Gaillardia Foug.

Gaillardia aristata Pursh., Blanketflower. Taprooted perennial with stems to 40 cm tall; herbage rough-hairy; leaves alternate, lance-shaped, petiolate, to 15 cm long, lower with entire or toothed margins, upper lobed; heads radiate, large, mostly solitary; involucral bracts in 2-3 series, spreading; rays sterile, ca. 13, yellow with purple bases, 10-25 mm long, 3-lobed at the tip; disk flowers bisexual, purple with long hairs at the top; receptacle with bristles between flowers; pappus of 6-10 stiff bristles. June-Aug. Fig. 63. Color plate 4.

Common in grasslands and meadows, mostly montane but sometimes to near treeline; East, West. B.C. to Man. south to OR, CO, SD.

Gnaphalium L., Cudweed

Annual to perennial, taprooted herbs with white-wooly herbage; leaves alternate with entire margins; heads appearing discoid, small; involucral bracts in several series, papery at least at the tip; flowers yellow or whitish; receptacle naked; pappus of fine bristles.

Similar in overall appearance to *Anaphalis* (pearly everlasting) and *Antennaria* (pussytoes), which are perennial.

1. Plants <15 cm tall; involucral bracts wooly to the tip .... **G. palustre**
1. Plants >15 cm tall; involucral bracts wooly only at the base 2

2. Stem and upper leaf surfaces sticky-glandular .... **G. macounii**
2. Stem and leaves not glandular .... **G. canescens**

Gnaphalium macounii Greene [*G. viscosum* H. B. K. misapplied]. Annual or biennial with stems to 70 cm tall; herbage glandular, sticky; leaves numerous, narrowly oblong, 4-10 cm long, basal portion attached down the stem; heads 5-7 mm high, numerous in a nearly flat-topped inflorescence; involucral bracts yellowish, wooly at the base. July-Aug. Fig. 64.

Local and uncommon in open montane forest; East, West. B.C. to N. S. south to OR, TN, Mex.

Gnaphalium canescens D. C. [*G. microcephalum* Nutt.]. Short-lived perennial 20-70 cm high; leaves numerous, linear, 3-10 cm long; heads 4-5 mm high, in hemispheric clusters on stem tips; involucral bracts whitish, wooly at the base. Aug.

Rare in montane meadows of the McDonald Valley. Our plants are ssp. *microcephalum* (Nutt.) Stebbins & Keil. B.C., Alta. south to CA, CO, Mex.

Similar *Antennaria* spp. have fibrous roots and branched rootcrown rather than a taproot.

Gnaphalium palustre Nutt. Annual with stems usually < 15 cm tall; leaves oblong, 1-2 cm long; heads clustered at the stem tips, 2-3 mm high; involucral bracts brown with whitish tips, wooly. July.

Locally common in vernally moist soil around montane wetlands and disturbed areas; East, West. B.C. to Sask. south to CA, NM, NE.

Grindelia Willd. Gumweed

Grindelia squarrosa (Pursh) Dunal. Curlycup Gumweed [*G. perennis* Nels.]. Taprooted, short-lived perennial with stems 25-50 cm tall; herbage glabrous, glandular, appearing varnished; leaves alternate, oblong, often with toothed margins, to 6 cm long, petiolate below; inflorescence open, nearly flat-topped with solitary heads on branch tips; heads radiate; involucral bracts in several series, narrow, curled back; rays 25-40, female, yellow, 7-10 mm long; disk flowers yellow, bisexual or sterile; receptacle naked; pappus of few rigid bristles. July-Aug. Fig. 65.

Local in vernally moist soil around montane prairie wetlands, along roads and other disturbed areas; East, West. Our plants are var. *quasiperennis* Lunell. B.C. to Man. south to CA, TX, MO.

Helianthus L. Sunflower

Annual or perennial herbs; heads radiate; involucral bracts in several series of nearly the same length; receptacle with scales between flowers; rays yellow, sterile; disk flowers bisexual; pappus of few scales, soon falling.

1. Taprooted annuals 2
1. Rhizomatous perennials 3
2. Leaves not heart-shaped at the base; involucral bracts not long-tapered to the tip .... **H. petiolaris**
2. Some leaves with heart-shaped bases; involucral bracts with long-tapered tips .... **H. annuus**
3. Involucral bracts ovate with long hairs on the margin .... **H. rigidus**
3. Involucral bracts linear, short-hairy .... **H. nuttallii**

Helianthus annuus L., Common Sunflower. Annual with stems mostly to 50 cm tall; herbage rough-hairy; leaves mostly alternate, lance-shaped to ovate with toothed margins and often heart-shaped bases, petiolate; heads few, large but variable in size; involucral bracts oblong to ovate with a long, tapered tip; disk flowers reddish-purple.

Disturbed ground; not collected since Standley (1921) found it along the railroad near East Glacier. Native to N. America and distributed throughout. Our plants are probably the cultivated sunflower that is closely related to the native ssp. *lenticularis* (Lindl.) Cock.; it was used by Native Americans for food and cosmetic oil.

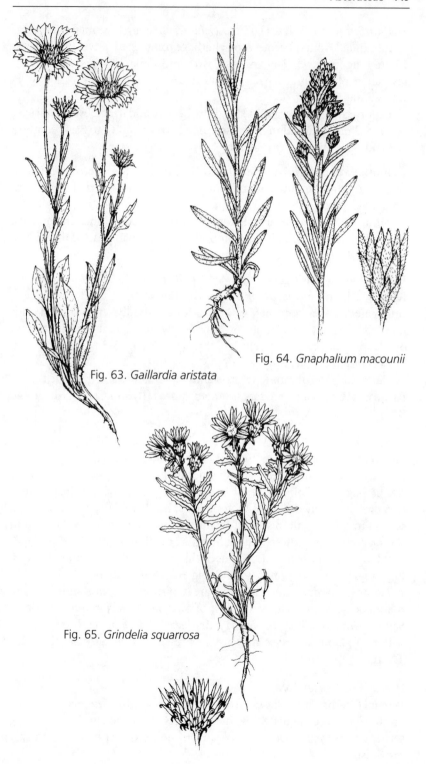

Fig. 63. *Gaillardia aristata*

Fig. 64. *Gnaphalium macounii*

Fig. 65. *Grindelia squarrosa*

Helianthus nuttallii T. & G. [*H. fascicularis* Greene]. Perennial with stems to 100 cm tall from tuberous roots; herbage roughened; leaves opposite, 8-15 cm long, narrowly lance-shaped with entire margins, petiolate; heads several; involucral bracts narrow; rays 10-16, 12-30 mm long; disk flowers yellow. Aug-Sept. Fig. 66.

Reported by Standley (1921) for disturbed ground near the Many Glacier hotel; ordinarily found in moist soil along streams. B.C. to Sask. south to NV and NM. Tubers were eaten by Native Americans.

Helianthus petiolaris Nutt. Similar to *H. annuus* but smaller; leaves lance-shaped, often with entire margins; tips of involucral bracts pointed but not long-tapered. July-Aug.

Montane grasslands and disturbed areas; collected once in the Park without specific location information, although probably east of the Divide. Native to the Great Plains, introduced elsewhere.

Helianthus rigidus (Cass.) Desf. [*H. laetiflorus* Pers., *H. subrhomboideus* Rydb.]. Rhizomatous perennial with stems 30-60 cm tall; herbage roughened; leaves opposite, to 12 cm long, broadly lance-shaped with toothed margins, petiolate below; heads few; involucral bracts ovate, in overlapping series of different lengths; rays 15-20, 15-30 mm long; disk flowers red to purple. Aug.

Rare in dry soil of montane grasslands, open slopes and disturbed areas; East, West. Our plants are ssp. *subrhomboideus* (Rydb.) Cronq. Alta. to N. B. south to GA, TX.

Heterotheca Cass., Golden Aster

Heterotheca villosa (Pursh) Shinn., Hairy Golden Aster [*Chrysopsis villosa* (Pursh) Nutt.]. Taprooted perennial herb with spreading stems to 50 cm tall; herbage long-hairy; leaves alternate, lance-shaped with entire margins, to 5 cm long, only the basal with petioles; heads radiate, 7-12 mm high, several in a short inflorescence; involucral bracts narrow, overlapping in several series; disk flowers yellow, bisexual; ray flowers 10-25, yellow, 6-10 mm long; receptacle naked; pappus of thin inner and coarse outer bristles. July-Aug. Reference: Semple (1996). Fig. 67.

Common in dry, often disturbed areas of montane grasslands or along roads and streams; East, West. Most of our plants are var. *minor* (Hook.) Semple. with glandular as well as hairy leaves. Var. *foliosa* (Nutt.) Harms, with non-glandular leaves, may also be present. B.C. to Sask. south to CA, TX, IL.

Hieracium L., Hawkweed

Perennial herbs with milky sap from rootstocks or rhizomes; leaves alternate or all basal; heads ligulate; involucral bracts in 2-3 unequal, overlapping series; flowers yellow or white; pappus of numerous fine bristles, mostly brownish.

1. Plants lacking leafy stems or with 1-2 reduced stem leaves 2
1. Plants with leafy stems 4
2. Flowers reddish orange .... *H. aurantiacum*
2. Flowers yellow 3
3. Leaves glabrous or short-hairy; subalpine or higher .... *H. triste*
3. Leaves with long, stiff hairs; low elevation weeds .... *H. caespitosum*
4. Flowers white; hairs of involucre sparse .... *H. albiflorum*
4. Flowers yellow 5
5. Basal leaves lacking; herbage nearly glabrous .... *H. umbellatum*
5. Basal rosette present; herbage with long, stiff hairs .... *H. scouleri*

Hieracium albiflorum Hook, White-flowered Hawkweed. Stems slender, up to 60 cm tall; herbage with sparse, long hairs; leaves to 15 cm long, oblong with entire or shallowly toothed leaves, the lower petiolate; heads several in a broad, open inflorescence; involucral bracts 8-11 mm long with sparse, stiff, long hairs and glands; flowers 13-34, white. June-Aug. Fig. 68.

Abundant in upper montane and subalpine forest, often with *Vaccinium* spp. and *Xerophyllum tenax*; East, West. AK to CA, CO.

Hieracium aurantiacum L., Orange Hawkweed. Nearly leafless stems to 60 cm tall from rhizomes; herbage long, black-hairy, glandular above; rosette leaves oblong, to 15 cm long, tapering to broad petioles; heads several in a tight cluster; involucral bracts 5-8 mm long with stiff, black glandular hairs; flowers reddish orange. June-July.

Locally common in lawns, disturbed meadows and roadsides; montane, West. Introduced from Europe. This exotic weed appears to be spreading.

Hieracium caespitosum Dumort, Meadow Hawkweed [*H. pratense* Tausch. misapplied]. Stems to 80 cm tall with a short rhizome and sometimes leafy stolons; herbage with long, stiff hairs; leaves mostly all basal, to 15 cm long, lance-shaped with short petioles and entire margins; heads numerous in an open, branched inflorescence; involucral bracts 6-8 mm long with black, gland-tipped hairs; flowers yellow.

Locally common along roads and in disturbed meadows and grasslands; montane, West. Native to Europe. This exotic weed appears to be increasing. *H. floribundum* Wimmer & Grab., with glabrous leaves and stolons, is a hybrid derivitive of *H. caespitosum* (Tutin et al. 1976); it has been collected on roadsides near Apgar and Polebridge.

Hieracium scouleri Hook, Wooly Weed. [*H. cynoglossoides* Arv.-Touv., *H. griseum* Rydb., *H. albertinum* Farr]. Stems to 80 cm tall; herbage with long, stiff, white hairs; leaves lance-shaped, to 20 cm long, with entire or toothed margins, those of basal rosette with broad petioles, smaller above; heads several in a short-branched inflorescence; involucral bracts 7-12 mm long; flowers 15-50, yellow. July-Aug.

Common in montane grasslands and meadows. Var. *griseum* A. Nels [*H. cynoglossoides*] has evidently glandular involucral bracts, East; var. *albertinum* (Farr) G. Dougl. & Allen has few, if any, glands, East; and var. *scouleri* lacks long, stiff, black hairs on the involucral bracts, West. B.C., Alta. south to ID, WY.

Hieracium triste Willd. ex Spreng., Alpine Hawkweed [*H. gracile* Hook.]. Stems to 20 cm tall; herbage glabrous to short-hairy; leaves all basal, to 7 cm long, ovate and sparsely toothed or entire with long petioles; heads few in an open, narrow inflorescence; involucral bracts 6-9 mm long with small star-shaped, glandular and stiff black hairs; flowers pale yellow. July-Aug. Fig. 69.

Common in meadows and open forest in the upper subalpine and lower alpine zones where snow lies late, often with *Vaccinium scoparium, Phyllodoce* spp. *Luzula hitchcockii* and *Juncus drummondii*; East, West. Our plants are ssp. *gracile* (Hook.) Calder & Taylor. AK to CA, NM.

Hieracium umbellatum L. [*H. scabriusculum* Schwein., *H. canadense* Michx.]. Stems to 100 cm tall, often spreading-hairy at the base; herbage nearly glabrous; leaves oblong with shallowly toothed margins, 3-12 cm long, petiolate below, basal rosette lacking; heads several in a broad, open inflorescence; involucral bracts 8-12 mm long, nearly glabrous with a few hairs and glands; flowers yellow, ca. 15 mm long. July-Aug.

Common in montane open forest, river banks and grasslands; East, West. AK to Newf. south to OR, CO, IA. Cronquist (1955) segregates *H. canadense* based largely on the presence of hairs at the base of the stem; plants referable to *H. canadense* have been found in grasslands east of the Divide, but are herein included in *H. umbellatum*.

Hymenoxys Cass.

Hymenoxys richardsonii (Hook.), Cockerell. Perennial herb, to 30 cm tall from a woody rootcrown; herbage with sparse hair and sunken glands; leaves alternate, and basal divided into usually 3 fleshy, linear segments; heads radiate, few, long-stalked; involucral bracts in 2 series 5-8 mm long, the outer united basally, the inner papery; receptacle naked; disks flowers bisexual, yellow; rays female, fertile, yellow, 3-lobed at the tip, 7-10 mm long; pappus of 5-6 pointed scales. June-July.

Uncommon in dry, sparsely vegetated, often clayey soil of grasslands along the east edge of the Park. Alta. to Sask. south to AZ, TX. *H. acaulis* (Pursh) Parker has been collected just south of East Glacier and may occur in the Park. It has entire, all-basal leaves and solitary heads.

Iva L., Povertyweed, Marsh Elder

Iva xanthifolia Nutt. Annual to 100 cm tall; herbage with stiff, short hairs; leaves opposite below; leaf blades arrow-shaped with toothed margins, 4-

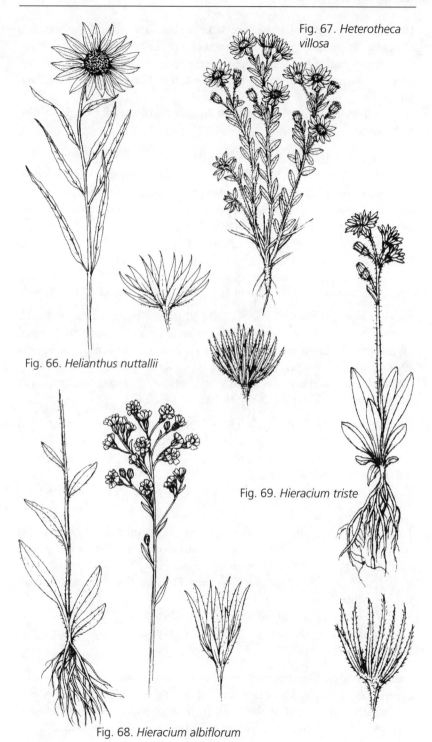

Fig. 67. *Heterotheca villosa*

Fig. 66. *Helianthus nuttallii*

Fig. 69. *Hieracium triste*

Fig. 68. *Hieracium albiflorum*

10 cm long, long-petiolate; heads discoid, densely arranged along branches of open inflorescence; involucral bracts in 2 series of 5 each, 1-3 mm long, outer green and glandular, inner papery; receptacle with scales between flowers; flowers inconspicuous, only the outer fertile; pappus lacking. Aug. Fig. 70.

Rare in disturbed ground at West Glacier and Lake McDonald. WA to Sask. south to AZ, TX, IA.

Lactuca L., Lettuce
Annual to perennial herbs with milky sap; leaves alternate, often lobed; herbage mostly glabrous; heads ligulate, in branched, open inflorescences; involucral bracts in several, overlapping, unequal series; receptacle naked; pappus of numerous long, soft bristles.

1. Flowers yellow; spines on lower side of leaf midveins .... **L. serriola**
1. Flowers blue; leaves without spines 2

2. Involucral bracts < 15 mm long; mid-stem leaves lobed .... **L. biennis**
2. Inner involucral bracts >15 mm long; mid-stem leaves not lobed **L. pulchella**

Lactuca biennis (Moench.) Fern., Tall Lettuce [*L. spicata* (Lam.) Hitchc.]. Biennial to 200 cm tall; leaves to 40 cm long with deep, sharply toothed lobes, hairy on lower midveins; involucral bracts to 12 mm long; flowers 13-35, blue; achenes nearly beakless; pappus light brown. Aug. Fig. 71.

Uncommon in moist, open, montane forest and riparian thickets; West. AK to Newf. south to CA, CO, MN, NC.

Lactuca pulchella (Pursh) D. C. Blue Lettuce. Rhizomatous perennial to 50 cm tall; leaves to 15 cm long, narrowly oblong, basal with backward-pointing lobes; involucral bracts to 15 mm long; flowers 18-50, blue; achenes beaked; pappus white. July-Aug.

Reported for open slopes near East Glacier (Standley 1921). AK to Ont. south to CA, NM, OK, WI.

Lactuca serriola L., Prickly Lettuce [*L. virosa* L.]. Annual or biennial with stems to 100 cm tall, bristly below; leaves to 20 cm long, deeply lobed with toothed margins, prickly hairs on the lower midveins, clasping the stem; involucral bracts to 10 mm long; flowers 13-27, yellow; achenes beaked; pappus white. July-Aug.

Uncommon in dry, disturbed soil around roads and habitations, West Glacier and Lake McDonald. Native of Europe, introduced in most of temperate N. America.

Liatris Schreb., Blazing Star
Liatris punctata Hook., Dotted Gay-feather. Perennial herb with stems to 30 cm tall, curved out at the base, from a woody rootcrown; herbage glabrous, gland-dotted; leaves alternate, numerous, strap-shaped, 2-7 mm

wide with stiff hairs on the margins; heads discoid, cylindric, crowded in a long, narrow inflorescence; involucral bracts to 18 mm long, awn-tipped, in several unequal, overlapping series; flowers ca. 6, purple; pappus of feathery bristles. Aug.

Rare in grasslands around St. Mary. Alta. to Man. south to NM, TX, AR, MI.

Madia Mol., Tarweed

Madia glomerata Hook., Mountain Tarweed. Annual to 30 cm tall; herbage aromatic, sticky with long, glandular hairs; leaves alternate, strap-shaped, 2-7 cm long; heads inconspicuously radiate, clustered in upper leaves; involucral bracts few, to 9 mm long, in 1 series, each enfolding a ray flower; disk flowers several, reddish-yellow; pappus lacking. Aug. Fig. 72.

Rare in vernally moist, disturbed soil in open woods or grasslands; known from Marias Pass area. B.C. to Man. south to CA, AZ, CO, MN. Dried plants were burned as incense by the Blackfeet. *M. exigua* (Sm.) Gray, with smaller heads, has been collected just west and south of the Park.

Matricaria L., Wild Chamomile

Glabrous, annual or biennial herbs; leaves alternate, 2-3 times pinnately divided into fine segments; heads hemispheric, few at ends of leafy stalks; involucral bracts brownish, in 2-3 nearly equal series, papery-margined; receptacle naked; disk flowers yellow.

1. Heads radiate; rays white; herbage unscented .... **M. maritima**
1. Heads discoid; herbage pineapple-scented .... **M. matricarioides**

Matricaria maritima L. Stems to 50 cm tall; leaves to 8 cm long; heads radiate; rays 12-25, white, to 16 mm long; pappus a low crown. July-Aug.

Rare in disturbed soil along roads in the McDonald Lake area. Native of Europe.

Matricaria matricarioides (Less.) Porter, Pineapple Weed [*M. discoidea* D. C., *Chamomilla suaveolens* Rydb.]. Stems to 20 cm high; herbage pineapple-scented; leaves to 5 cm long; heads discoid; involucral bracts ca. 4 mm long; pappus lacking. June-Aug.

Uncommon in disturbed, often compacted soil around roads and habitations; East, West. Native to the Pacific Northwest and Asia, introduced farther east. Flower heads were dried and used as perfume and insect repellent by the Blackfeet.

Microseris D. Don

Microseris nutans (Hook.) Schultz-Bip. [*Ptilocalais nutans* (Hook.) Greene]. Glabrous perennial with mostly leafless stems to 50 cm tall from fleshy roots; leaves linear, entire to lobed, up to 20 cm long, all basal or nearly so; heads ligulate, solitary; involucral bracts in 2 unequal series, the inner up to 18 mm long, black-hairy, long-pointed; flowers yellow with purple veins; pappus of 15-20 feathery bristles. June-Aug. Fig. 73.

Common in vernally moist, often stony soil of montane grasslands and meadows; East, West. B.C., Alta. south to CA, UT, CO.

Nothocalais Greene

Nothocalais cuspidata (Pursh) Greene [*Microseris cuspidata* (Pursh) Schultz-Bip.]. Taprooted perennial with milky sap and leafless stems to 30 cm tall; herbage glabrous to long-hairy; leaves linear with a wavy margin, up to 20 cm long; heads ligulate, solitary; involucral bracts up to 20 mm high, subequal, long-pointed, often purple-spotted; flowers yellow; achene tapered above but beakless; pappus of 40-80 bristles.

Montane grasslands; collected once along the North Fork Flathead River near the Canadian border. Alta. to Man. south to NM, TX, MO.

Petasites Mill., Coltsfoot

Rhizomatous, perennial, long-hairy herbs; leaves all basal, large, petiolate, arising separately from stems; stems with alternate leaf-like bracts, arising before the leaves; heads radiate, clustered in a hemispherical inflorescence; involucral bracts in 1 series; receptacle naked; disk flowers white; rays inconspicuous or lacking; pappus of numerous thin bristles.

Plants from wet spruce forest in the Belly River area appear to be intermediate between the following 2 species.

1. Leaf margins shallowly toothed .... *P. sagittatus*
1. Leaves lobed nearly 1/2-way to midvein .... *P. frigidus*

Petasites frigidus (L.) Fries [*P. nivalis* Greene]. Stems to 50 cm tall; leaf blades arrow-shaped with lobed margins, up to 15 cm long, petiolate from a wide basal sinus, stem bracts ovate, to 6 cm long, clasping the stem; involucral bracts 5-9 mm long. May.

Rare in wet soil of meadows and thickets in the North Fork Flathead Valley. Our plants are var. *nivalis* (Greene) Cronq. Circumboreal south to CA, MT, MI, MA.

Petasites sagittatus (Banks) Gray. Similar to *P. frigidus* but leaf blades shallowly toothed on the margins, up to 20 cm long. May-June. Fig. 74.

Locally common in montane boggy meadows in forest openings, often with willow and alder; East, West. AK to Que. south to CA, CO, SD, WI. The conspicuous leaves are reminiscent of balsamroot; flower stems are uncommon and ephemeral.

Prenanthes L., White Lettuce

Prenanthes sagittata (Gray) A. Nels. Glabrous perennial herb with milky sap; stems to 60 cm tall from tuberous roots; leaves alternate, winged-petiolate with arrow-shaped blades to 15 cm long, margins toothed; heads

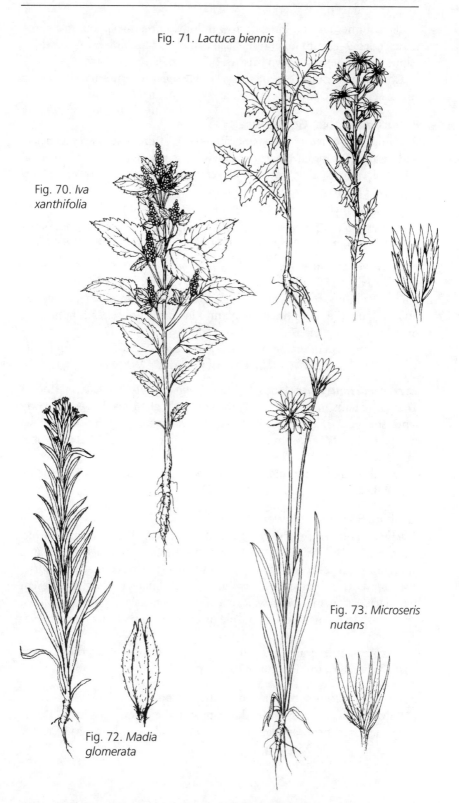

Fig. 71. *Lactuca biennis*

Fig. 70. *Iva xanthifolia*

Fig. 73. *Microseris nutans*

Fig. 72. *Madia glomerata*

ligulate, numerous in a long, narrow inflorescence; involucral bracts in 1 main series, 9-12 mm long; receptacle without scales; flowers 7-20, white; pappus of numerous brown bristles. July-Aug. Fig. 75.

Common in moist soil of thickets and open forest, especially along streams; montane and subalpine; East, West. AK to ID, MT.

Pyrrocoma Hook., Goldenweed

Perennial, taprooted herbs; stems curved out at the base; leaves alternate and basal; heads radiate, few per stem; involucral bracts in several unequal series; receptacle naked; rays yellow, female or sterile; disk flowers yellow, bisexual; pappus of fine bristles, unequal in length.

Formerly included in *Haplopappus*. Reference: Brown (1993).

1. Involucre up to 10 mm high; rays at least 5 mm long .... *P. lanceolatus*
1. Larger heads > 15 mm high; rays inconspicuous .... *P. carthamoides*

Pyrrocoma carthamoides Hook. [*Haplopappus carthamoides* (Hook.) Gray]. Stems to 50 cm tall; herbage sparsely hairy; leaves to 30 cm long, narrowly oblong, sparsely toothed to entire, the basal petiolate; heads 15-30 mm long, the outer bracts leafy; rays inconspicuous, heads appearing discoid. July-Aug. Fig. 76.

Uncommon in montane grasslands and meadows; East, West. Our plants are var. *carthamoides*. WA to MT south to CA, NV, WY.

Pyrrocoma lanceolata (Hook.) Greene [*Haplopappus lanceolatus* (Hook.) T. & G.]. Stems to 30 cm tall; herbage mostly glabrous; leaves narrowly lance-shaped, 5-15 cm long with toothed or entire margins, petiolate below; involucral bracts 5-10 mm long, green-tipped; rays 10-45, 5-12 mm long. July-Aug.

Locally common around temporary ponds near East Glacier (Standley 1921). B.C. to Sask. south to CA, UT, NE.

Ratibida Raf., Coneflower

Ratibida columnifera (Nutt.) Woot. & Standl, Prairie Coneflower. Taprooted perennial herb, 30-50 cm tall; herbage rough-hairy; leaves alternate, pinnately divided into linear segments; heads radiate, cylindric, few to several on branch tips; involucral bracts linear, in 1 series, reflexed; receptacle 2-4 cm high, with scales between flowers; rays 3-7, yellow, elliptic, 15-25 mm long; disk flowers purple or brown; pappus a low crown. Fig. 77.

Uncommon in grasslands, often in disturbed soil, collected once near East Glacier. B.C. to Man. south to AZ, TX.

Senecio L., Groundsel, Ragwort, Butterweed

Mostly perennial herbs; leaves alternate and often basal; involucral bracts in 1 series, sometimes overlapping; receptacle naked; disk flowers bisexual,

Fig. 75. *Prenanthes sagittata*

Fig. 74. *Petasites sagittatus*

Fig. 76. *Pyrrocoma carthamoides*

Fig.. 77. *Ratibida columnifera*

yellow or reddish; rays female, yellow; pappus of numerous fine, white bristles.

One of the largest genera in the Park and in the world. Some authors place many of our species in the genus *Packera*. Leaf shape and size of the flower heads are often useful characters. *S. pauperculus, S. indecorus, S. streptanthifolius,* and *S. pseudaureus* are closely related species and difficult to distinguish as are *S. cymbalarioides* and *S. cymbalaria*; habitat may serve to distinguish them as much as morphology.

1. Heads without rays 2
1. Heads radiate 3

2. Annual weed of disturbed soil .... *S. vulgaris*
2. Perennial of wet meadows .... *S. indecorus*

3. Heads solitary, occasionally 2 per stem 4
3. Heads few to many per stem 6

4. Heads large, > 12 mm high .... *S. megacephalus*
4. Heads smaller, < 12 mm high 5

5. Heads 10-20 mm high; involucral bracts purplish .... *S. cymbalaria*
5. Heads 5-9 mm high; involucre green .... *S. cymbalarioides*

6. Mid-stem leaves often largest 7
6. Mid-stem leaves smaller than basal 8

7. Stems sprawling, < 20 cm high .... *S. fremontii*
7. Stems erect, > 25 cm tall .... *S. triangularis*

8. Leaves with entire or toothed margins, not lobed 9
8. Some leaves with basal lobes 10

9. Herbage white-hairy; plants of dry soil .... *S. canus*
9. Herbage mostly glabrous; plants of moist soil .... *S. hydrophiloides*

10. Involucral bracts black-tipped .... *S. integerrimus*
10. Involucral bracts green to the tip or hairy 11

11. Plants of dry soil, grasslands, open slopes 12
11. Plants of meadows, moist forest or thickets 13

12. Herbage white-hairy .... *S. canus*
12. Herbage glabrous or nearly so .... *S. streptanthifolius*

13. Basal leaves all tapering to petiole .... *S. pauperculus*
13. Some basal leaf blades with a blunt or indented base .... *S. pseudaureus*

Senecio canus Hook. Stems to 40 cm tall with clusters of basal leaves; herbage densely or sometimes sparsely white hairy; leaves narrowly elliptic to oblong, to 8 cm long, entire or basally lobed, petiolate below; heads 4-10 mm high, several in an open inflorescence; rays 6-13 mm long. June-Aug. Fig. 78.

Common in rocky soil of grasslands, meadows, outcrops and disturbed areas at all elevations; East, West. B.C. to Sask. south to CA, CO, NE. A report of *S. werneriaefolius* Gray by Bamberg & Major (1968) is probably referable here. A relatively tall, large-flowered form of *S. canus* with fewer basal leaves occurs in deeper soil of montane grasslands, East. It can occur in close proximity to the more common, shallow-soil form without apparent intergradation.

Senecio cymbalaria Pursh [*S. conterminus* Greenm., *S. resedifolius* Less., *S. ovinus* Greene]. Stems to 15 cm tall from a thick, short rhizome; herbage glabrous or wooly below; leaf blades ovate with toothed margins, up to 25 mm long, long-petiolate; those above narrow, without petioles; heads solitary, 10-20 mm high; involucral bracts purplish; rays to 14 mm long. July-Aug.

Wet or rather dry, sparsely vegetated, stony soil of open slopes, moraine, and rock outcrops and along streams near or above treeline; near or east of the Divide. AK to Newf. south to WA, ID, MT. See note under *S. cymbalarioides.*

Senecio cymbalarioides Buek [*S. subnudus* D. C.]. Stems to 20 cm tall from short rootstocks; herbage mostly glabrous; leaves to 4 cm long, blades oblong with toothed margins long-petiolate below, narrow and lobed above; heads 5-9 mm high, 1-few; rays 6-12 mm long. July-Aug.

Common in wet, often organic soil of alpine and subalpine meadows and along small streams; East, West. B.C. to Alta. south to CA, ID, WY. *S. cymbalarioides* and *S. cymbalaria* are similar, but the former is usually in soil with ample organic matter, while the latter is mostly found in poorly developed soil.

Senecio fremontii T. & G. Stems mostly sprawling, to 25 cm long from a branched rootcrown; herbage glabrous; leaves thick, oblong with scalloped margins, 1-3 cm long, largest at mid-stem; heads several, 7-12 mm high; rays 5-10 mm long. July-Aug. Fig. 79.

Common on rocky slopes, talus and moraine near or above treeline; East, West. B.C., Alta. south to CA, UT, and CO.

Senecio hydrophiloides Rydb. [*S. foetidus* Howell]. Stems to 100 cm tall from rhizomes; herbage mostly glabrous; leaves thick, to 30 cm long, lance-shaped, with toothed margins, petiolate below, smaller above; heads 5-10 mm high, numerous in a crowded, flat-topped inflorescence; rays few, ca. 5 mm long or absent. July.

Common in wet meadows and thickets; montane and subalpine; East, West; locally abundant in disturbed meadows in the North Fork Flathead drainage. B.C., Alta. south to CA, MT.

Senecio indecorus Greene [*S. burkei* Greenm.]. Stems to 70 cm tall from a short rootcrown; herbage mostly glabrous; leaf blades lance-shaped with toothed margins, to 10 cm long, basally lobed and sessile above, the basal petiolate; heads 8-12 mm high, several in an open inflorescence; rays lacking. July-Aug.

Uncommon in montane fens and wet meadows and along streams; East, West. AK to Lab. south to WA, WY. Similar to *S. pauperculus* but lacking rays. A collection of *S. pauciflorus* Pursh from wet meadows north of Apgar is probably referable here.

Senecio integerrimus Nutt. Stems to 70 cm tall from a short rootcrown; herbage with thinning, long, white hair; leaf blades to 12 cm long, oblong to lance-shaped with toothed margins, the lower long-petiolate, the upper nearly linear; heads 5-15 mm high, several in a flat-topped cluster; rays 6-10 mm long. June-Aug.

Common in grasslands and dry forest openings; montane and subalpine; East, West. Our plants are var. *exaltatus* (Nutt.) Cronq. B.C. to Man. south to CA, CO, MN. The similar *S. lugens* Rich., which is rhizomatous and occurs at higher elevations, could occur in the Park as well.

Senecio megacephalus Nutt. Stems to 50 cm tall from a stout rootcrown; herbage thinly covered with long, white hair; leaves narrowly oblong with entire or finely toothed margins, up to 20 cm long, petiolate below; heads 12-20 mm high, usually solitary; rays 10-15 mm long. June-Aug.

Common on rocky soil of slopes and meadows, upper montane to above treeline; East, West. B.C. to Alta. south to ID, MT.

Senecio pauperculus Michx. Stems to 80 cm tall from a shallow rootcrown; herbage glabrous to lightly hairy; leaf blades to 10 cm long, lance-shaped with toothed margins, the basal long-petiolate, the upper sessile with basal lobes; heads several, 6-9 mm high; rays 5-10 mm long. June-Aug.

Moist soil of montane meadows, fens, thickets and forest openings, often along streams; mostly West. AK to Newf. south to WA, NM, GA. Similar to *S. cymbalarioides*, which has thicker leaves, is smaller and occurs at higher elevations.

Senecio pseudaureus Rydb. Stems to 80 cm tall from a short rootcrown; herbage with scattered long hair; basal leaf blades to 6 cm long, heart-shaped, petiolate, becoming lance-shaped and deeply toothed above; heads several, 7-12 mm high; rays 4-10 mm long. June-Aug.

Common in moist forest and thickets; montane and lower subalpine; East, West. B.C. to Man. south to CA, NM, MO. *S. pauperculus* is similar but not nearly as common in the shade.

Senecio streptanthifolius Greene. Similar to *S. canus* but the leaf blades green and more glabrous. June-July.

Uncommon in stony soil of open montane slopes and grasslands; East. Yuk. to CA, NM.

Senecio triangularis Hook. Stems clustered, up to 120 cm tall from short rhizomes; herbage glabrous; leaf blades narrowly triangular with toothed margins, up to 15 cm long, petiolate, only gradually reduced upward; heads

8-15 mm high, numerous in a flat-topped inflorescence; rays few, ca. 6 mm long. July-Aug. Fig. 80.

Moist soil of forests, meadows, thickets and avalanche slopes, often along streams; abundant in the montane and subalpine zones, less common higher; East, West. AK to Sask. south to CA, NM.

Senecio vulgaris L., Old Man in the Spring. Annual up to 35 cm tall; herbage nearly glabrous; leaves oblong, deeply lobed, to 5 cm long, petiolate below; heads discoid, 7-10 mm high, several in a branched, open inflorescence. June-Aug.

Collected twice at West Glacier before 1910 but not since (Jones 1910). Introduced from Eurasia. Our only annual *Senecio.*

Solidago L., Goldenrod

Fibrous-rooted perennial herbs; leaves alternate, without lobes or divisions; heads radiate, several to numerous; involucral bracts in several overlapping, unequal series; receptacle without scales; disk flowers bisexual, yellow; ray flowers female, yellow; pappus of many fine, white bristles.

S. graminifolia (L.) Salisb. is reported for wet meadows along streams in Waterton Park and may occur in Glacier as well.

1. Stems > 40 cm tall, leafy in upper half 2
1. Stems < 40 cm tall, leaves of upper half reduced 4

2. Stems white-hairy below inflorescence .... *S. canadensis*
2. Stems glabrous below inflorescence 3

3. Basal rosettes of leaves present; plants < 50 cm tall .... *S. missouriensis*
3. Basal rosette lacking; stems usually > 50 cm tall .... *S. gigantea*

4. Long hairs on margins of basal leaf petioles ....*S. multiradiata*
4. Basal leaves without marginal hairs 5

5. Inflorescence branched, spreading .... *S. missouriensis*
5. Inflorescence narrow .... *S. simplex*

Solidago canadensis L., Canada Goldenrod [*S. elongata* Nutt.]. Stems mostly 30-100 cm tall from long rhizomes; upper stem and sometimes leaves sparsely hairy; leaves narrowly lance-shaped with shallowly toothed margins, 4-10 cm long, only gradually reduced above; heads 3-6 mm high, crowded along spreading branches of pyramidal inflorescence; rays ca. 13, 4-5 mm long. July-Aug. Fig. 81.

Common in moist soil of montane meadows, thickets, open forest and avalanche slopes; East, West. Our plants are var. *salebrosa* (Piper) Jones. AK to Newf. south to CA, TX, FL. This species forms large clones and responds positively to disturbance.

Solidago gigantea Aiton [*S. serotina* Ait.]. Similar to *S. canadensis*; stems often clustered; herbage glabrous below the inflorescence; inflorescence with more strongly recurved branches. July-Aug.

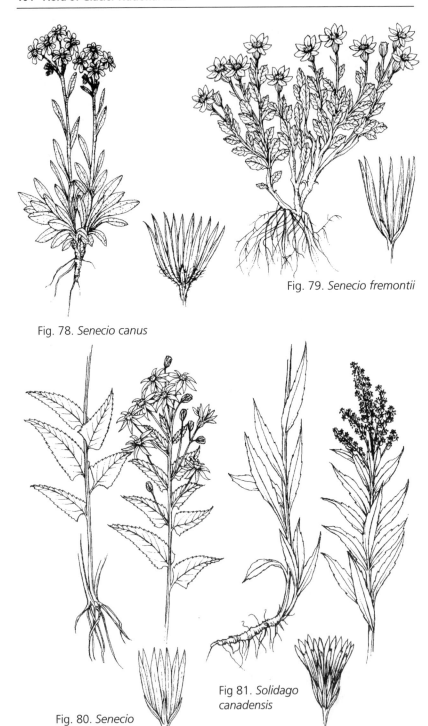

Fig. 79. *Senecio fremontii*

Fig. 78. *Senecio canus*

Fig. 80. *Senecio triangularis*

Fig 81. *Solidago canadensis*

Uncommon in montane meadows and thickets; East, West. Our plants are var. *serotina* (Ait.) Cronq. B.C. to N. S. south to OR, NM, TX, GA.

Solidago missouriensis Nutt., Missouri Goldenrod [*S. concinna* A. Nels.]. Stems to 50 cm tall from stout rootstocks; herbage glabrous below; leaves narrowly oblong with entire to shallowly toothed margins, to 30 cm long, petiolate; heads 3-5 mm high, crowded along spreading branches of inflorescence; rays 7-13. July-Aug.

Var. *missouriensis,* with heads 3-4 mm high, is common in montane grasslands and meadows, mostly East. Var. *extraria* Gray, with larger heads, is uncommon in montane meadows, open forest, and along streambanks; East. B.C. to Ont. south to AZ, NM, TX, and TN.

Solidago multiradiata Aiton [*S. ciliosa* Greene, *S. dilatata* Nels., *S. purshii* Porter, *S. scopulorum* (Gray) Nels.]. Stems to 40 cm tall from a branched rootcrown; herbage sparsely hairy; leaves oblong with toothed or entire margins, up to 15 cm long, long hairs on margins of lower leaf petioles; heads 7 mm high, several in a short, narrow inflorescence; rays ca. 13, 3-5 mm long. June-Aug. Fig. 82.

Mostly dry, stony soil of grasslands, meadows, dry open forest, and exposed ridges; abundant near or above treeline, less common in the montane zone; East, West. Our plants are var. *scopulorum* Gray. AK to Que. south to CA, NM; Asia.

Solidago simplex Humb., Bonpl. & Kunth.[*S. decumbens* Greene, *S. spathulata* D. C. misapplied]. Stems to 50 cm tall from a stout rootcrown; herbage sparsely hairy; leaves narrowly oblong with a toothed apical portion, to 12 cm long, long-petiolate below; heads 5-7 mm high, numerous in a dense, narrow inflorescence; rays 5-10. July-Aug.

Uncommon in montane grasslands and dry, open forest; East. Our plants are var. *simplex.* AK to Man. south to CA, NM.

Sonchus L., Sow Thistle

Annual or perennial herbs with milky sap; leaves alternate, lance-shaped to oblong with lobed and toothed margins, the upper clasping the stem; heads ligulate, narrow, few to several in open inflorescences; involucral bracts in several unequal series; receptacle naked; flowers yellow; achenes beakless; pappus of many fine, white bristles.

Weedy species introduced from Eurasia. *S. uliginosus* Bieb., a serious weed of alkaline meadows, may eventually be found in the Park.

1. Taprooted annual; heads < 14 mm high .... **S. asper**
1. Rhizomatous perennial; heads > 14 mm high .... **S. arvensis**

Sonchus arvensis L. Rhizomatous perennial with stems to 100 cm tall; herbage glabrous below inflorescence; leaves 6-40 cm long, prickly-margined; heads 12-20 mm high; involucral bracts with gland-tipped hairs.

Rare in moist disturbed soil.

Sonchus asper (L.) Hill. Taprooted, glabrous annual to 100 cm tall; leaves 6-30 cm long with spiny margins; heads 9-13 mm high; involucral bracts glabrous. July-Aug.

Rare in disturbed soil around roads and habitations at West Glacier.

Tanacetum L., Tansy

Tanacetum vulgare L., Common Tansy. Rhizomatous perennial; stems leafy, 40-150 cm tall; herbage aromatic, glabrous; leaves alternate, pinnately divided and sharply toothed, to 20 cm long; heads discoid, button-like, numerous in a flat-topped inflorescence; involucral bracts papery-margined in 2-3 unequal series; receptacle naked; disk flowers yellow; pappus an inconspicuous crown. Aug-Sept.

Moist, disturbed meadows, often along streams; collected near West Glacier and St.. Mary. Introduced from Eurasia.

Taraxacum Wigg., Dandelion

Taprooted perennial herbs with milky sap; herbage mostly glabrous; leaves oblong, all in basal rosettes; heads ligulate, solitary on naked stalks; involucral bracts in 2 unequal series; receptacle naked; flowers yellow; achenes tipped with a slender beak; pappus of fine bristles.

The introduced species are told by the sharply reflexed outer involucral bracts.

1. Outer involucral bracts erect or nearly so; near treeline or above 2
1. Outer involucral bracts sharply reflexed; mostly montane 3
2. Mature achenes olive or brownish; inner involucral bracts swollen at the tip . ... ***T. ceratophorum***
2. Mature achenes black; inner bracts pointed at the tip .... ***T. lyratum***
3. Mature achenes reddish; terminal lobe of leaves no bigger than ones below .... ***T. laevigatum***
3. Mature achenes olive or brownish; terminal leaf lobe larger than those below .... ***T. officinale***

Taraxacum ceratophorum (Ledeb.) D. C. [*Leontodon ceratophorum* Ledeb.]. Similar to *T. officinale*; flower stalks to 15 cm tall; leaves up to 10 cm long with large-toothed margins; heads to 20 mm high; outer involucral bracts spreading slightly, the inner often with swollen tips. July-Aug. Fig. 83.

Uncommon in moist, stony soil of open slopes and meadows near or above treeline near the Divide. Circumboreal to CA, NM, NH.

Taraxacum laevigatum (Willd.) D. C. [*Leontodon laevigatum* Willd.]. Similar to *T. officinale*, flower stalks to 30 cm tall; leaves lobed almost to the tip; heads to 20 mm high; outer involucral bracts usually reflexed, inner often swollen at the tip. May-Aug.

Montane grasslands and open, disturbed areas; reported for West Glacier (Standley 1921) but probably along the east edge of the Park as well. Introduced from Eurasia. In somewhat drier habitats than *T. officinale*.

Taraxacum lyratum (Ledeb.) D. C. [*Leontodon lyratum* Ledeb.]. Flower stalks to 10 cm tall; leaves deeply lobed, < 10 cm long; heads 7-18 mm high; involucral bracts blackish, outer erect, the inner with pointed tips. July-Aug.

Uncommon in alpine wet meadows to rocky slopes near the Divide. AK to NV, AZ, CO; Asia.

Taraxacum officinale Weber, Common Dandelion [*Leontodon taraxacum* L.]. Flower stalks to 40 cm tall; leaves to 40 cm long with lobed margins, the terminal lobe larger than those below; heads to 20 mm high; outer involucral bracts sharply reflexed; inner bracts pointed at the tip. May-Aug.

Common in moist soil of meadows, thickets, aspen groves and disturbed areas; montane, rarely higher; East, West. Introduced from Eurasia. See note under *T. laevigatum.*

Townsendia Hook., Townsendia

Taprooted biennial or perennial herbs; leaves alternate with entire margins; heads radiate, solitary or few; involucral bracts narrow, in 2-3 unequal, overlapping series; receptacle naked; rays female, white to blue; disk flowers bisexual, yellow, numerous; pappus of white bristles.

T. hookeri Beaman, a common grassland species, occurs in Waterton Park (Kuijt 1982) and may be found along the east edge of Glacier as well.

1. Rays pink; flower heads stemless .... *T. condensata*
1. Rays blue or violet; plants with leafy stems .... *T. parryi*

Townsendia condensata Parry ex D. C. Eat. Low, stemless perennial; herbage long-hairy; leaves narrowly oblong, to 25 mm long, crowded in small mats; heads 12-20 mm high, borne among the leaves; rays to 15 mm long, pink. July-Aug.

Rare in barren shingle of exposed ridges and slopes above treeline; East. Alta. south to CA, UT, WY.

Townsendia parryi D. C. Eat. Biennial or perennial to 20 cm tall; herbage with stiff, short hairs; leaves narrowly spoon-shaped, 2-6 cm long, mostly near the base; heads 9-20 mm high; involucral bracts with papery-white margins; rays blue to violet, 1-2 cm long. June-Aug. Fig. 84.

Common in sparsely vegetated soil of montane grassland or dry, open forest, rarely higher; East. B.C., Alta. south to OR, ID, WY. The size of the flower heads varies greatly.

Tragopogon L., Goatsbeard

Tragopogon dubius Scop., Salsify, Oyster-root. Taprooted biennial up to 70 cm tall; herbage glabrous with milky sap; leaves alternate, linear with entire margins, up to 20 cm long, the basal rosette grass-like; heads solitary or few, ligulate; involucral bracts narrow, in 1 series as long as flowers, glabrous; receptacle naked; flowers light yellow; achene 25-35 mm long, beaked; pappus of feathery bristles. June-Aug. Fig. 85.

Uncommon in montane grasslands and along roads and disturbed areas; East, West. Native of Europe, introduced in much of N. America.

BERBERIDACEAE: BARBERRY FAMILY

Berberis L., Oregon Grape

Berberis repens Lindl. [*B. aquifolium* Pursh var. *repens* (Lindl.) Scoggan, *Mahonia repens* (Lindl.) Don]. Low shrub with trailing stems 5-30 cm long; leaves alternate, evergreen, to 20 cm long, pinnately divided into 5-7 ovate leaflets with spiny-toothed margins; flowers clustered on stem tips, each composed of 6 yellow petals, 6 larger, yellow sepals, and 3 yellowish bracts, the largest ca. 6 mm long; stamens 6; fruit a waxy, blue, few-seeded berry. May-June. Fig. 86. Color plate 33.

Dry to moist forest and brushy slopes; abundant under montane mixed conifers, less common subalpine; East, West. B.C., Alta. south to CA, AZ, TX. Blackfeet used the bark and roots for a dye and to aid stomach and kidney afflictions.

BETULACEAE: BIRCH FAMILY

Deciduous shrubs or trees; leaves alternate, broadly lance-shaped to ovate with toothed margins, petiolate; flowers unisexual, clustered in ovate female or drooping cylindric male catkins; male flowers of 2-4 stamens, subtended by brownish bracts; female flowers of ovary with 2 styles subtended by bracts; fruit a winged or wingless nutlet. Reference: Furlow (1997).

1. Female catkins elliptic, hardened, persistent, conelike .... **Alnus**
1. Female catkins cylindric, disintegrating at maturity .... **Betula**

Alnus Ehrh., Alder

Shrubs; female bracts fleshy, becoming woody; female catkins 1-2 cm long, erect, at the base of male catkins, persistent; stamens 4; nutlet with wings or membranous margin.

A. incana is more common at lower elevations, especially west of the Divide, while *A. viridis* is more common at higher elevations. Taxonomy of N. American and Eurasian taxa is not agreed upon.

1. Fruit with a wing-margin half as wide as nutlet; winter buds sharply pointed; catkins develop on new twigs .... ***A. viridis***
1. Fruit with only a narrow papery margin; winter buds blunt or rounded; catkins on twigs of previous summer .... ***A. incana***

Alnus incana (L.) Moench, Thin-leaved Alder [*A. tenuifolia* Nutt.]. Shrubs to 4 m high with grayish-brown bark; twigs hairy; winter buds blunt or rounded at the tip; leaves 4-10 cm long, pale and hairy below; fruit with a thin margin. Apr-May.

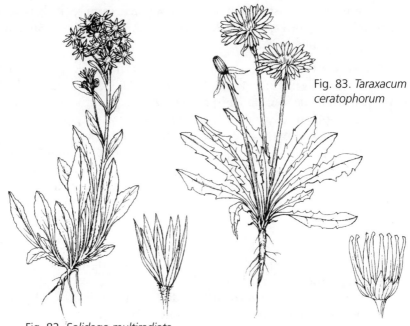

Fig. 83. *Taraxacum ceratophorum*

Fig. 82. *Solidago multiradiata*

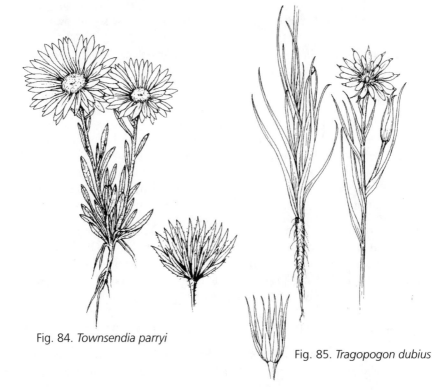

Fig. 84. *Townsendia parryi*

Fig. 85. *Tragopogon dubius*

Common in moist to wet soil of montane swamps, thickets, and streambanks; East, West. Our plants are ssp. *tenuifolia* (Nutt.) Breitung. AK to Sask. south to CA, NM.

Alnus viridis (Chaix) D. C., Green Alder, Sitka Alder [*A. crispa* (Ait.) Pursh ssp. *sinuata* (Regel) Hult, *A. sinuata* (Regel) Rydb.]. Shrubs to 3 m high with grayish bark; stems often sprawling; young twigs glandular; winter buds sharp-pointed; leaves to 10 cm long, lower surface slightly paler than upper; fruit with wings half as broad as nutlet. May-June. Fig. 87.

Abundant on avalanche slopes and in wet, open forest; montane and subalpine; East, West. Our plants are ssp. *sinuata* (Regel) Löve & Löve. Boreal Asia and N. America south to CA, WY, WI, NY. Stems often bend downhill with the pressure of snow, making walking uphill through them difficult.

Betula L., Birch

Trees or shrubs; twigs with elliptical raised glands; female bracts 3-lobed; female catkins ascending from below male catkins, disintegrating upon maturity; stamens 2; fruits winged.

B. occidentalis and *B. papyrifera* hybridize extensively in eastern WA and adjacent ID, OR (Hitchcock and Cronquist 1964).

1. Shrub usually < 2 m tall; leaves < 25 mm long; peaty soil of fens and swamps .... ***B. glandulosa***
1. Tree (often multi-stemmed) or large shrub usually > 2m tall; some leaves > 25 mm long; moist forest 2
2. Bark of older trees dark; leaves lacking tufts of brown hair beneath; twigs densely glandular .... ***B. occidentalis***
2. Bark of older trees white; leaves with tufts of brown hair in axils of veins underneath; twigs with few glands .... ***B. papyrifera***

Betula glandulosa Michx., Bog Birch. Shrub to 2 m high with brown to black bark; twigs glandular; leaves 10-25 mm long, glandular, paler beneath; mature female catkins 10-25 mm long; wings of fruit up to half as wide as nutlet. May-June.

Common in wet, organic soil of montane swamps and fens; East, West. Our plants are var. *glandulosa*. AK to Greenl. south to CA, CO, WI, Newf.

Betula occidentalis Hook., Water Birch, Black Birch [*B. fontinalis* Sarg.]. Small tree or large shrub to ca. 8 m tall; bark gray to brown, peeling little; twigs glandular; leaves 2-5 cm long, glandular when young, variously hairy, rounded to pointed at the tip; mature female catkins 2-4 cm long; wings of fruit ca. as wide as nutlet. May-June.

Common along montane streams and swamps; East, West. Our plants are var. *occidentalis*. AK to Man. south to CA, NM, and SD.

Betula papyrifera Marsh. Paper Birch. Small tree to 15 m tall; bark gray or brown when young but white and peeling when older; twigs sparsely hairy and glandular; leaves 4-7 cm long with tufts of brownish hair in axils of veins beneath; mature female catkins 2-5 cm long; wings of fruit ca. as wide as nutlet. May-June. Fig. 88.

Common in moist, open forest, along streams and lakes, and on rocky lower slopes; montane, West. Our plants are var. *subcordata* (Rydb.) Sarg. AK to Lab. south to WA, CO, MN, PA. Plants with peeling gray bark may be hybrids between *B. occidentalis* and *B. papyrifera*. The bark was used by Native Americans for making canoes.

BORAGINACEAE: BORAGE FAMILY

Annual or perennial herbs; leaves alternate, undivided with mostly entire margins; flowers bisexual, aligned on 1 side of inflorescence branches that uncoil and elongate with maturity; corolla tubular with 5 flaring lobes; calyx 5-lobed; stamens 5; style 1; ovary superior, 4-lobed, 1 or more lobes each maturing into a hard 1-seeded nutlet.

The ornamentation of mature fruits is often important for identification.

Report of *Eritrichium nanum* by Bamberg & Major (1968) is probably based on *Myosotis sylvatica*, and Standley (1921) incorrectly attributes a report of *E. howardii* to Jones (1910).

1. Flowers white, yellow or orange 2
1. Flowers blue to reddish purple 5

2. Flowers white 3
2. Flowers yellow or orange 4

3. Herbage with stiff, erect, spiny hairs .... **Cryptantha**
3. Herbage with sparse, appressed hairs .... **Plagiobothrys**

4. Annual; nutlets 2-3 mm long .... **Amsinckia**
4. Perennial; nutlets 4-6 mm long .... **Lithospermum**

5. Nutlets smooth, wrinkled or bumpy but not prickly 6
5. Nutlets with prickles 8

6. Corolla funnel-shaped with a wide mouth and exserted stamens ...... **Echium**
6. Corolla tubular, the narrow mouth enclosing the stamens 7

7. Herbage glabrous or only sparsely hairy; corolla >10 mm long ..... **Mertensia**
7. Herbage hairy; corolla < 10 mm long .... **Myosotis**

8. Flowers reddish-purple .... **Cynoglossum**
8. Flowers blue 9

9. Plants annual, mostly < 15 cm tall .... **Lappula**
9. Plants biennial or perennial, usually > 20 cm tall .... **Hackelia**

Amsinckia Lehm., Fiddleneck, Tarweed

Amsinkia menziesii Lehm. Annual to 50 cm tall; herbage with stiff, long hairs; leaves narrowly oblong, to 10 cm long, crowded below; sepals 5-10 mm long; corolla yellow, 4-7 mm long; nutlets ovate, 2-4 mm long, densely bumpy.

Reported as *A. barbata* Greene for disturbed grasslands at East Glacier by Standley (1921) where it may have been introduced and may no longer persist. B.C. to MT south to CA.

Cryptantha Lehm. ex G. Don, Cryptantha

Annual to perennial herbs with stiff long hairs, often arising from low bumps; biennials and perennials with basal rosette of leaves, annuals without; flowers small, white with yellow throats; nutlets smooth or bumpy but not spiny.

C. *celosioides*, C. *sobolifera*, and C. *spiculifera* are closely related and difficult to distinguish. C. *celosioides* usually has an unbranched rootcrown and dies after flowering once. The latter 2 species are true perennials with branched rootcrowns; C. *sobolifera* is found at higher elevations than C. *spiculifera*.

1. Plants annual; basal rosette of leaves absent 2
1. Plants biennial or perennial with basal rosettes 3
2. Longitudinal crease down the center of back side of nutlet .... **C. torreyana**
2. Longitudinal crease off-center, closer to edge .... **C. affinis**
3. Plants usually flowering 1-2 years then dying; rootcrown mostly few-branched at the base, usually only 1 main stem .... **C. celosioides**
3. Plants with branched rootcrown; > 1 main stem; old flower stems often present 4
4. Inside face of nutlet smooth; montane to alpine .... **C. sobolifera**
4. Inside face of nutlet with low ridges; valley and lower montane...................
.... **C. spiculifera**

Cryptantha affinis (Gray) Greene. Slender annual to 30 cm tall; leaves strap-shaped; corolla 1-2 mm across; nutlets 2 mm long, ovate, smooth.

Montane grasslands and open forest; collected by Umbach along the railroad near East Glacier (Standley 1921). B.C. south to CA, WY.

Cryptantha celosioides (Eastw.) Pays., Miner's Candle [*C. nubigena* (Greene) Pays. var. *celosioides* (Eastw.) Boivin, *Oreocarya glomerata* (Pursh) Greene]. Biennial or short-lived perennial to 30 cm tall; basal leaves clustered, long spoon-shaped, 2-6 cm long; stem leaves narrowly oblong, smaller; corolla 6-10 mm across; nutlets narrowly ovate, 3-5 mm long, bumpy with low ridges on outside surface. June-July.

Uncommon in montane grasslands, often in sparsely vegetated soil; East. B.C., Alta. south to OR, WY, NE.

Cryptantha sobolifera Pays. [*C. nubigena* (Greene) Pays. misapplied]. Perennial to 15 cm tall from a branched rootcrown; basal leaves spoon-shaped, to 8 cm long; corolla 4-8 mm across; nutlets lance-shaped, 3-4 mm long, bumpy with low ridges on outside surface. June-July. Fig. 89.

Locally common in sparsely vegetated soil of exposed slopes and ridges; montane to alpine; East. OR to MT south to CA.

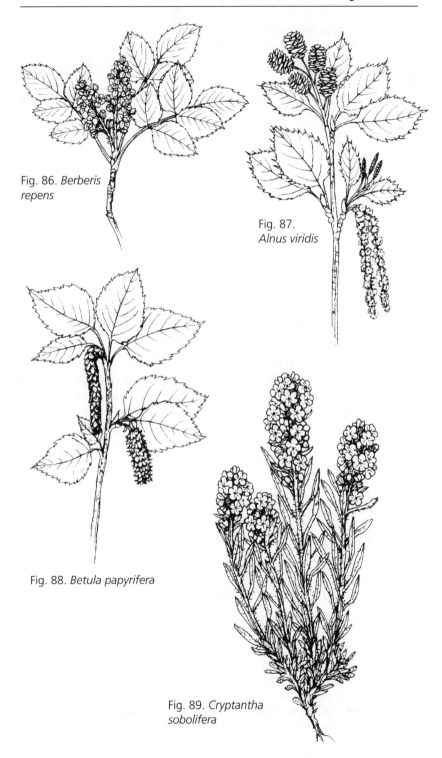

Fig. 86. *Berberis repens*

Fig. 87. *Alnus viridis*

Fig. 88. *Betula papyrifera*

Fig. 89. *Cryptantha sobolifera*

Cryptantha spiculifera (Piper) Pays., Miner's Candle. Perennial to 30 cm tall from a branched rootcrown; basal leaves narrowly oblong to 8 cm long; stem leaves strap-shaped; corolla 4-8 mm across; nutlets lance-shaped, 2-4 mm long, bumpy with low ridges on outside surface. June-July.

Uncommon in sparsely vegetated areas of montane grasslands; East. WA to MT south to CA, ID, WY.

Cryptantha torreyana (Gray) Greene. Annual to 30 cm tall; leaves strap-shaped; corolla ca. 1 mm across; nutlets ovate 1-2 mm long, smooth and shiny. May-July.

Uncommon in sparsely vegetated soil of open montane forest; West. B.C. to MT, south to CA, UT, WY.

Cynoglossum L., Hound's Tongue
Cynoglossum officinale L. Biennial to 100 cm tall; herbage hairy; leaf blades lance-shaped, to 25 cm long, petiolate below, smaller and sessile above; corolla reddish-purple, ca. 1 cm across; sepals up to 6 mm long in fruit; nutlets flattened-ovate, 5-7 mm long, with barbed prickles. July-Aug.

Local on roadsides and other disturbed areas in the valleys; East, West. Introduced from Europe.

Echium L., Viper's Bugloss
Echium vulgare L., Blueweed. Taprooted biennial 20-80 cm high; herbage bristly-hairy; basal leaves broadly strap-shaped, 6-20 cm long, petiolate; stem leaves shorter, without petioles; flowers showy in small, short-stalked clusters in a long, narrow, leafy-bracted inflorescence; corolla blue, 15-20 mm long, funnel-shaped with 5 unequal lobes at the mouth; stamens unequal, exserted; style deeply 2-lobed, hairy; nutlets bullet-shaped with a rough surface.

Rare along roads and around buildings; collected one along St. Mary Lake. Introduced from Europe.

Hackelia Opiz, Stickseed
Biennial or perennial; herbage soft-hairy; corollas blue; nutlets with barbed prickles.

1. Stems usually 1 from unbranched taproot; nutlets with few or no prickles on the face .... ***H. floribunda***
1. Stems several from branched rootcrown; nutlets with prickles on face as well as margins .... ***H. micrantha***

Hackelia floribunda (Lehm.) Johnst. [*Lappula floribunda* (Lehm.) Greene]. Biennial or short-lived perennial with 1-few stems, 40-70 cm tall; herbage spreading hairy; leaves narrowly oblong, 4-20 cm long, petiolate, the basal withering; corolla 4-7 mm across; nutlets 3-4 mm long, prickles only on margins. June-July.

Uncommon in moist soil of open montane forest, thickets and grassland; East, West. B.C. to Sask. south to CA, NM, MN.

Hackelia micrantha (Eastw.) J. L. Gentry [*H. jessicae* (McGreg.) Brand]. Similar to *H. floribunda* but usually with many stems to 100 cm high from a branched rootcrown; nutlets with marginal prickles and smaller ones on the face. June-Aug. Fig. 90.

Common in montane grasslands, rock outcrops and open forest, especially aspen groves: East, West. B.C., Alta. south to CA, UT, CO. Standley's (1921) report of *H. diffusa* (*Lappula diffusa* (Lehm.) Greene) is referable here.

Lappula Moench, Stickseed

Annual; herbage short-hairy; corolla small, blue; nutlets with barbed prickles on the margin.

1. Marginal prickles of nutlets in 1 row, united at the base .... **L. redowskii**
1. Marginal prickles in 2-3 rows, separate to base .... **L. squarrosa**

Lappula redowskii (Hornem.) Greene [*L. occidentalis* (Wats.) Greene]. Stems simple or branched, to 40 cm tall; leaves narrowly oblong to linear, up to 6 cm long; corolla ca. 2 mm across; nutlets 3-4 mm long; marginal prickles in 1 row, united at the base. June.

Uncommon in disturbed soil of montane grasslands or along roads, East. AK to Man. south to CA, TX, MO.

Lappula squarrosa (Retz.) Dumort. [*L. echinata* Gilib.]. Similar to *L. redowskii*; corolla 2-4 mm across; marginal prickles of nutlets in 2-3 rows, separate to base. June.

Collected in disturbed soil near Many Glacier Hotel and at East Glacier. Native to Eurasia and perhaps also the Northern Rocky Mountains (Cronquist et al. 1984).

Lithospermum L., Stoneseed

Lithospermum ruderale Dougl. ex Lehm., Puccoon, Gromwell. Perennial with several ascending stems to 60 cm tall from a branched rootcrown; herbage stiff-hairy; leaves narrowly lance-shaped to 10 cm long; corolla light yellow, 7-13 mm across, nutlets smooth, shiny gray, 4-6 mm long, bullet-shaped. May-June. Fig. 91.

Uncommon in montane grasslands; East, West. B.C. to Sask. south to CA, CO. Native Americans used this plant for birth control.

Mertensia Roth, Bluebells

Perennial; herbage glabrous or with soft hair; corolla blue, funnelform, much longer than calyx; nutlets wrinkled.

M. oblongifolia, M. lanceolata (Pursh) D. C., and *M. viridis* (Nels.) Nels. are very similar and intergrade throughout their range (Hitchcock et al. 1959). *M. lanceolata* has the flared portion of the corolla longer than the tube. Most of our plants do not fit this description, but plants just across the border in Alta. are referred to *M. lanceolata* by Moss & Packer (1983). *M. viridis* has a line of hair inside the corolla tube, while *M. oblongifolia* does not; otherwise these 2 species seem morphologically and ecologically indistinguishable in our area.

1. Stems 1-2 from a short tuberous root; basal leaves usually absent ....
.... ***M. longiflora***
1. Stems usually several from a branched root; basal leaves present 2
2. Corolla tube glabrous within .... ***M. oblongifolia***
2. Corolla tube with long hair on inside .... ***M. viridis***

Mertensia longiflora Greene. Stems to 20 cm tall from a short, tuberous root; herbage glabrous; leaves oblong, to 6 cm long, basal ones usually absent; corolla 12-18 mm long; corolla tube glabrous within, 2-3 times as long as the flared lobes. May-June.

Rare, collected once in vernally moist soil of rock outcrops above McDonald Lake. B.C. to Alta. south to CA, ID, MT.

Mertensia oblongifolia (Nutt.) G. Don. Stems several from a branching root, to 30 cm tall, ascending rather than erect; herbage glabrous or sparsely hairy; basal leaves lance-shaped, petiolate, to 15 cm long; stem leaves smaller, sessile; corolla 1-2 cm long; corolla tube glabrous within, 1-2 times as long as the flared lobes. June-July. Fig. 92.

Common in grasslands and meadows; montane zone to above treeline; East, expected West. WA to MT south to NV, UT, WY.

Mertensia viridis A. Nels. Similar to *M. oblongifolia*; leaves often more elliptic; corolla tube with long hair on the inside. July.

Uncommon in alpine and subalpine grasslands and turf; East. OR to MT, south to CA, UT, CO.

Myosotis L., Forget-me-not, Scorpion-grass

Annual or perennial herbs; herbage soft-hairy, corolla blue, often with yellow or white center; nutlets smooth and shiny.

1. Calyx with appressed hair only; stems often > 15 cm long 2
1. Calyx with some spreading, often hooked hairs; stems often < 15 cm tall ... 3
2. Corolla 2-5 mm across spreading lobes .... ***M. laxa***
2. Corolla 5-10 mm across .... ***M. scorpioides***
3. Plants annual at low elevations; corolla 2-5 mm across .... ***M. micrantha***
3. Plants perennial at high elevations; corolla 4-8 mm across .... ***M. alpestris***

Myosotis alpestris Schmidt, Forget-me-not [*M. sylvatica* Hoffm. var. *alpestris* (Schmidt) Koch]. Perennial 5-20 cm tall; herbage short-hairy; leaves narrowly oblong, those of basal rosette to 5 cm long, petiolate; corolla 4-6 mm across; nutlets black. June-Aug. Fig. 93.

Locally common in meadows and fellfields near or above treeline near or east of the Divide. AK to B. C, ID, WY, SD; Asia. Small plants are sometimes mistaken for *Eritrichium*.

Myosotis laxa Lehm. Short-lived perennial with weak stems to 40 cm long; herbage sparsely hairy; leaves narrowly lance-shaped, 2-4 cm long; corolla 2-5 mm across; nutlets black. June-Aug.

Rare in mud or shallow water along montane streams and wetlands; West. B.C. to CA east to northern Atlantic.

Myosotis micrantha Pallas ex Lehm. Blue Scorpion-grass. Small annual to 15 cm high, branched at the base; herbage short hairy; leaves lance-shaped, to 2 cm long; flowers borne to near base of plant; corolla 2-5 mm across; nutlets brown. May-June.

Uncommon in shallow or disturbed, soil of montane grasslands, rock outcrops and roadsides; East, West. Introduced from Eurasia.

Myosotis scorpioides L., Scorpion-grass. Short-lived perennial to 4 dm tall; stems decumbent at the base; herbage sparsely hairy; leaves narrowly oblong, 2-8 cm long; corolla 5-10 mm across; nutlets black. May-June.

Collected once in moist cedar forest along Middle Fork Flathead River near West Glacier. Introduced from Europe.

Plagiobothrys F. & M., Popcorn Flower

Plagiobothrys scouleri (H. & A.) Johnst. [*P. scopulorum* (Greene) Johnst., *Allocarya californica* (F. & M.) Greene]. Annual with ascending to prostrate stems; herbage sparsely hairy; leaves strap-shaped, to 3 cm long; corolla 2-3 mm across; nutlets light brown, bumpy and ridged. June-July. Fig. 94.

Uncommon in drying mud around wetlands along the east edge of the Park. AK to Man. south to CA, NM, MN. Diminutive plants may also occur along roads. Standley's (1921) report of *P. leptocladus* (Greene) Johnst. (*Allocarya orthocarpa* Greene) was based on an immature specimen and was probably *P. scouleri*.

BRASSICACEAE: MUSTARD FAMILY

Annual to perennial herbs; herbage often covered with branched hairs; leaves alternate or basal; branches of inflorescence narrow, arising from leaf axils, becoming elongate in fruit; petals 4, separate; sepals 4; stamens 6- 4 long and 2 short; ovary superior, 2-celled; fruits are long and narrow (siliques) to short and capsule-like (silicles). Reference: Rollins (1993).

Mature fruit is essential for identification.

1. Fruit ovate, circular, elliptic or triangular, up to 2 times as long as wide 2
1. Fruit narrowly elliptic to linear, > 2 times as long as wide 10

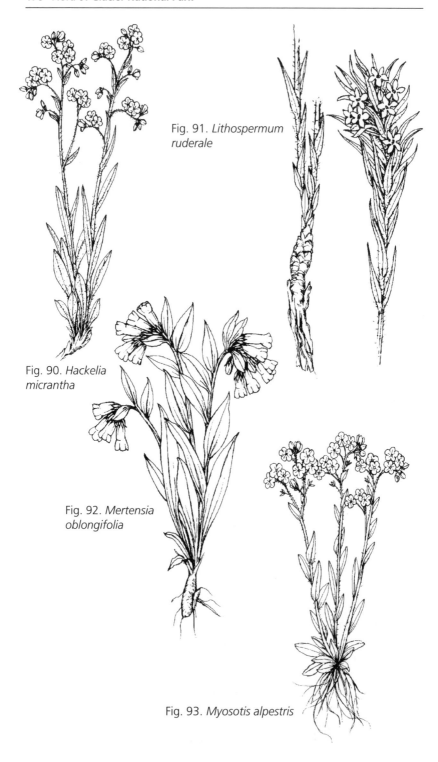

Fig. 91. *Lithospermum ruderale*

Fig. 90. *Hackelia micrantha*

Fig. 92. *Mertensia oblongifolia*

Fig. 93. *Myosotis alpestris*

2. Fruit 2-lobed above, inflated, > 1 cm across .... **Physaria**
2. Fruit < 1 cm across or not inflated 3

3. Fruit inverted-triangular, shallowly dished out on top .... **Capsella**
3. Fruit not triangular 4

4. Fruit notched or indented on top 5
4. Fruit rounded or pointed on top 7

5. Mature fruit > 10 mm across .... **Thlaspi**
5. Mature fruit < 5 mm across 6

6. Leaves with toothed margins .... **Lepidium**
6. Leaves with entire margins .... **Alyssum**

7. Style at top of fruit at least 2 mm long 8
7. Style < 2 mm long 9

8. Plants > 10 cm tall; basal leaves few or none .... **Camelina**
8. Plants < 10 cm tall; basal leaves numerous .... **Lesquerella**

9. Basal leaves conspicuously toothed or lobed .... **Rorippa**
9. Basal leaves entire to shallowly toothed .... **Draba**

10. Leaves entire or shallowly toothed 11
10. Leaves lobed or divided 13

11. Fruit 2-7 times as long as broad .... **Draba**
11. Fruit > 7 times as long as broad 12

12. Flowers white to purple, not yellow; fruit flattened .... **Arabis**
12. Flower yellow; fruit round or 4-angles in cross section .... **Erysimum**

13. Upper stem leaves with basal lobes that clasp the stem 14
13. Stem leaves absent or without clasping bases 16

14. Mature fruit with a beak-like tip > 5 mm long .... **Brassica**
14. Beak of fruit < 4 mm long 15

15. Fruits erect .... **Barbarea**
15. Fruits spreading perpendicular to the stem .... **Rorippa**

16. Leaves not lobed to near the midrib.. 17
16. At least 1 lobe of some leaves going to midrib or nearly so 18

17. Petals white; mature fruit without a distinct beak .... **Arabis**
17. Petals yellow; fruit with a beak-like tip > 1 mm long; uncommon weed ......
.... **Diplotaxis**

18. Fruit with a sterile beak or beak-like style > 1 mm long at the tip........... 19
18. Style at tip of fruit < 1 mm long 20

19. Beak > 2 mm long .... **Brassica**
19. Beak-like style < 2 mm long... **Erucastrum**

20. Herbage glabrous or nearly so .... **Cardamine**
20. Herbage hairy.. 21

21. Plant high-elevation perennial with woody rootcrown............. **Smelowskia**
21. Plant low-elevation annual... 22

22. Fruits to 3 cm long .... **Descurainia**
22. Fruits > 5 cm long .... **Sisymbrium**

Alyssum L., Alyssum

Annuals with branched stems; herbage with radially branched hairs; leaves narrowly oblong, with entire margins; flowers yellow, inconspicuous; petals light yellow, ca. 2 mm long, falling upon opening; fruits circular, ca. 3 mm long. Introduced from Europe.

1. Fruits hairy .... *A. alyssoides*
2. Fruits glabrous .... *A.. desertorum*

Alyssum alyssoides L. Stems to 15 cm tall; leaves to 2 cm long; fruits hairy. May-June.

Uncommon in disturbed soil along trails and roads in the valleys; East, West.

Alyssum desertorum Stapf. Similar to *A. alyssoides* but with glabrous fruits.

Uncommon, collected once along the road through Big Prairie.

Arabis L., Rockcress

Biennial or perennial; basal rosette of leaves present; stem leaves usually clasping; inflorescence without leafy bracts; flowers white to purple but not yellow; fruit long, glabrous, with 1 or 2 rows of seeds in each chamber. Reference: Mulligan (1995).

1. Fruits spreading to near horizontal or reflexed 2
1. Fruits erect or ascending 4
2. Fruit stalks mostly to 4 mm long; rootcrown branched .... *A. lemmonnii*
2. Fruit stalks > 4 mm long; rootcrown usually unbranched 3
3. Fruits somewhat arc-shaped, near horizontal .... *A. sparsiflora*
3. Fruits straight, hanging below horizontal or reflexed .... *A. holboellii*
4. Basal leaves shallowly lobed .... *A. lyrata*
4. Basal leaves not conspicuously lobed 5
5. Mature fruits 2-3 mm wide 6
5. Mature fruits < 2 mm wide 8
6. Plants < 20 cm tall .... *A. lyallii*
6. Plants usually > 20 cm tall 7
7. Seeds in 2 columns on each side of the fruit .... *A. drummondii*
7. Seeds in 1 series on each side of the fruit .... *A. divaricarpa*
8. Fruits 1-2 cm long; leaves with long, erect hairs on the margin .... *A. nuttallii*
8. Fruits > 3 cm long; conspicuous marginal hairs absent 9
9. Fruits ascending but not strictly erect; stem leaves < 5 mm wide ....
.... *A. divaricarpa*
9. Fruits strictly erect; some stem leaves > 1 cm wide 10
10. Fruits nearly round in cross section; upper stem waxy .... *A. glabra*
10. Fruits obviously flattened; stem not waxy................................... *A. hirsuta*

Arabis divaricarpa A. Nels. [*A. bourgovii* Rydb., *A. confinis* Wats.]. Short-lived perennial to 80 cm tall from a mostly unbranched rootcrown; leaves narrowly lance-shaped with mostly entire margins, to 4 cm long; petals pink to purple; mature fruits spreading, 2-8 cm long, ca. 2 mm wide. June-July.

Common in open montane forests and grasslands; East, West. AK to Que. south to CA, CO, IA, VT.

Arabis drummondii Gray. Perennial to 50 cm tall from a usually unbranched rootcrown; herbage mostly glabrous; leaves narrowly oblong with entire margins, to 7 cm long; petals 5-7 mm long, white or tinged with purple; mature fruits 4-8 cm long, 2-3 mm wide, clustered-erect. June-July. Fig. 95.

Common in sparsely vegetated soil of rocky slopes, moraine, meadows, and grasslands; montane to above treeline; East, West. AK to Newf. south to CA, CO, OH, DE. Both *A. drummondii* and *A. glabra* are both glabrous with erect fruits, but the former has wider fruits.

Arabis glabra (L.) Bernh, Tower Mustard. Biennial or short-lived perennial to 100 cm high from an unbranched taproot; herbage glabrous above, hairy below; basal leaves oblong with wavy margins, 3-14 cm long; stem leaves lance-shaped; petals white, 4-6 mm long; mature fruits 6-10 cm long, 1-2 mm wide, erect. June-July.

Disturbed or sparsely vegetated soil of montane meadows, grasslands and open forest; East, West. AK to Que., south to CA, NM, MN, GA. See note under *A. drummondii*.

Arabis hirsuta (L.) Scop. Biennial to 60 cm tall from an unbranched taproot; herbage hairy at least below; basal leaves oblong with wavy margins, 2-3 cm long; stem leaves lance-shaped; petals white, 4-9 mm long; fruits 2-8 cm long, 1-2 mm wide, erect. June-July.

Uncommon in sparsely vegetated, often stony soil of meadows and open forest; montane and subalpine; East, West. AK to Que. south to CA, NM, MN, GA.

Arabis holboelii Hornem. [*A. lignipes* Nels., *A. retrofracta* Grah.]. Short-lived perennial to 70 cm tall from a mostly unbranched rootcrown; herbage with branched hairs; basal leaves oblong, 1-5 cm long with toothed margins; stem leaves narrowly lance-shaped; petals white to purple, 5-10 mm long; fruits 3-8 cm long pendent to reflexed. May-June.

Var. *retrofracta* (Grah.) Rydb., with sharply reflexed fruits, is common in sparsely vegetated soil of montane grasslands, open forest and rock outcrops; East, West. Var. *pinetorum* (Tides.) Rollins, with pendulous fruits and clasping stem leaves, is known from the Two Medicine Valley. Var. *pendulocarpa* (A. Nels.) Rollins, with pendulous fruits and non-clasping stem leaves, has been found on open slopes of Elk Mountain. AK to Greenl. south to CA, CO, MN, MI.

Arabis lemmonii Wats. Perennial to 20 cm high from a branched rootcrown; herbage with branched hairs; basal leaves spatulate, to 3 cm long; petals purple, 5-7 mm long; fruits 2-5 cm long, ca. 2 mm wide, mostly all on 1 side of the stem, horizontal to somewhat pendant. June-July.

Common in stony, sparsely vegetated soil of moraine, talus slopes, and ridges; subalpine and alpine, less common along montane rivers; near or east of the Divide. Our plants are var. *drepanoloba* (Greene) Rollins. B.C., Alta. south to CA, CO.

Arabis lyallii Wats. Tufted, glabrous perennial to 20 cm high from a branched-spreading rootcrown; basal leaves oblong, to 3 cm long, fleshy; stem leaves with clasping bases; flowers few; petals purple, 6-10 mm long; fruits 2-6 cm long, 2-3 mm wide, erect. June-Aug.

Common in sparsely vegetated soil of meadows, ridges and moraine near or above treeline; East, West. B.C. to Alta. south to CA, UT, WY. *A. murrayi* Mulligan is similar but does not have clasping stem leaves; it has been found both north and south of the Park.

Arabis lyrata L. Short-lived perennial, to 25 cm tall from a branched or simple rootcrown; herbage mostly glabrous; basal leaf blades oblong with shallowly lobed margins, 1-5 cm long, petiolate; stem leaves few-lobed, not clasping; petals white, 5-8 mm long; mature fruits 1-4 cm long, erect or ascending. May-July.

Rare in sparsely vegetated soil of montane and subalpine rocky slopes and open forest; known from the Many Glacier area, the slopes of Mount Brown and near Kintla Lake. Our plants are var. *kamchatica* (Fisch.) Hultén. AK to Que., south to WA, MT, MN, GA.

Arabis nuttallii Robins. Perennial to 20 cm high, often from a branched rootcrown; herbage sparsely hairy; basal leaves to 3 cm long, narrow-spoon-shaped with hairs at least on the margins; stem leaves oblong, not clasping; petals white, 5-8 mm long; fruits 1-2 cm long, erect on spreading stalks. May-June. Fig. 96.

Abundant in montane grasslands, meadows and on rock outcrops, less common near treeline; East, West. B.C., Alta. south to WA, UT, WY.

Arabis sparsiflora Nutt. Short-lived perennial to 60 cm tall from a mostly unbranched rootcrown; herbage hairy; basal leaves oblong, 1-4 cm long; petals blue, 6-8 mm long; fruits 4-14 cm long, arcuate-spreading. May-June.

Locally common in dry, open montane forest, West. Our plants are var. *columbiana* (Macoun) Rollins. B.C. to MT, south to CA, UT, WY.

Barbarea R. Br., Wintercress

Glabrous biennial herbs; basal leaves pinnately divided with large, ovate terminal lobes; stem leaves alternate, clasping; flowers yellow; fruits round or 4-angled in cross section.

1. Style at tip of fruit 2-3 mm long; petals 6-8 mm long .... **B. vulgaris**
1. Style < 2 mm long; petals 3-5 mm long .... **B. orthoceras**

Barbarea orthoceras Ledeb. [*Campe orthoceras* (Ledeb.) Heller]. Stems to 50 cm tall; herbage often purplish; basal leaves to 12 cm long; petals 3-5 mm long; fruits 15-50 mm long, 1-3 mm wide, erect or spreading. May-July.

Uncommon in moist soil of montane thickets, open forest and meadows; West, expected East. AK to Newf. south to CA, AZ, MN, NH; Asia.

Barbarea vulgaris R. Br., Yellow Rocket. Similar to *B. orthoceras*; stems somewhat taller; stem leaves becoming merely lobed upward; petals 6-8 mm long; fruits 1-3 cm long with a beak-like style 2-3 mm long. May-July.

Uncommon in moist soil of disturbed montane meadows and roadsides, West. Introduced from Europe.

Brassica L., Mustard
Annual or biennial; stems to 100 cm; leaves alternate, the lower pinnately divided or lobed, the upper sessile; flowers yellow; fruits erect or ascending with a distinct, sterile beak at the tip.

Weeds introduced from Europe.

1. Leaf bases wrapped partly around the stem .... **B. campestris**
1. Leaf bases not wrapped around the stem 2

2. Fruit 4-5 cm long with flattened beak .... **B. kaber**
2. Fruit 2-4 cm long with a conical beak .... **B. juncea**

Brassica campestris L., Rape. Stems 40-70 cm tall; herbage glabrous and waxy; stem leaves lance-shaped with bases clasping the stem; petals 6-10 mm long; fruits 3-7 mm long with a conical beak.

Disturbed forest at West Glacier.

Brassica juncea (L.) Cosson., Chinese Mustard. Herbage glabrous; leaves up to 25 cm long; petals 7-9 mm long; fruits 2-4 cm long with a conical beak.

Reported for disturbed ground along the east border of the Park by Standley (1921).

Brassica kaber (D. C.) Wheeler, Charlock [*B. arvensis* (L.) Kuntze]. Herbage with stiff hairs; petals 6-8 mm long; fruits 4-5 cm long with a flattened beak.

Reported for disturbed ground around West Glacier by Standley (1921).

Camelina Crantz, False flax
Camelina microcarpa Andrz. ex D. C. Annual to 50 cm tall; herbage with dense, branched and simple hairs; leaves lance-shaped with entire margins, 1-3 cm long, mostly without petioles; petals pale yellow, 4-5 mm long; fruits spreading, 4-6 mm long, pear-shaped, flattened with a ridge on each face. June.

Montane grasslands and disturbed areas east of the Divide (Standley 1921). Introduced from Eurasia.

Capsella Medic.

Capsella bursa-pastoris (L.) Medic., Shepherd's Purse [*Bursa bursa-pastoris* (L.) Weber]. Annual to 50 cm tall; herbage with sparse, branched hairs; basal leaves pinnately lobed, to 2-8 cm long; stem leaves narrowly lance-shaped, clasping, much smaller; flowers white; petals 2-4 mm long; inflorescence of spreading fruits; fruits inverted-triangular, flattened, ca. 5 mm long, indented on top. May-June.

Common in disturbed soil around residences and roads; East, West. Introduced from Europe.

Cardamine L., Bittercress

Annual or perennial herbs; herbage mostly glabrous; leaves basal and alternate, some pinnately lobed (ours), petiolate; flowers white; fruits flattened, erect or ascending.

1. Rhizomatous perennial; basal leaf blades with few lobes .... *C. breweri*
1. Annual or biennial; rhizomes absent; basal leaves with > 1 pair of lobes 2

2. Fruits > 1 mm wide with 15-22 seeds .... *C. oligosperma*
2. Fruits up to 1 mm wide with 24-40 seeds .... *C. pensylvanica*

Cardamine breweri Wats. Rhizomatous perennial with erect to prostrate stems up to 50 cm long; basal leaf blades undivided (often absent at flowering), up to 3 cm long; stem leaves with narrow lobes and a large, ovate terminal segment; petals 3-7 mm long; fruits 1-3 cm long. June-July.

Uncommon in wet soil of fens and along streams and lakes, montane and lower subalpine; East, West. Our plants are var. *breweri*. AK to MT, south to CA, WY.

Cardamine oligosperma Nutt. ex T. & G. Annual or biennial to 10 cm tall; basal leaves with ovate, shallowly lobed leaflets; stem leaves with lance-shaped lobes; petals 2-4 mm long, 1-2 mm wide; fruits erect, 15-25 mm long, in a flat-topped inflorescence. July-Aug.

Rare in wet, sparsely vegetated soil along streams and rock outcrops; subalpine and alpine near the Divide. Our plants are var. *kamtschatica* (Regel) Detl. B.C. to MT, south to CA, CO; Asia. Similar to the more common *C. pensylvanica*.

Cardamine pensylvanica Muhl. ex Willd. Annual or biennial to 50 cm tall; basal leaves 2-8 cm long with oblong leaflets; stem leaves with linear lobes; petals 2-4 mm long; fruits ascending, 1-3 cm long, ca. 1 mm wide. June-July. Fig. 97.

Common in wet, sparsely vegetated soil along montane streams, lakes and wetlands, mainly West. B.C. to Newf. south to CA, CO, OK, FL.

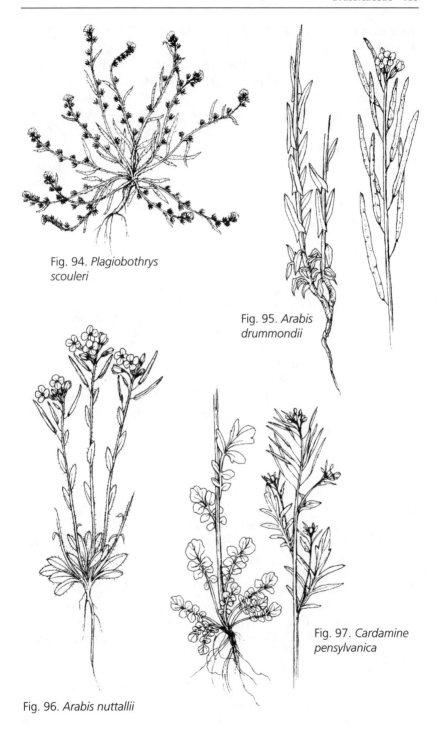

Fig. 94. *Plagiobothrys scouleri*

Fig. 95. *Arabis drummondii*

Fig. 96. *Arabis nuttallii*

Fig. 97. *Cardamine pensylvanica*

Descurainia Webb & Berth., Tansymustard

Annual or biennial herbs; stems often branched; herbage with short, grayish hair; leaves basal and alternate, pinnately divided; flowers yellow; petals 1-4 mm long; fruits on spreading or ascending stalks.

Standley's (1921) report of the Mexican species *D. hartwegiana* (Fourn.) Britt. [*Sophia hartwegiana* (Fourn.) Greene] is unlikely. *D. sophia* (L.) Webb, with finely divided leaves and narrow fruits, is widely introduced and may eventually be found in disturbed areas of the Park.

1. Fruit club-shaped with rounded tip; seeds in 2 rows in each half .. **D. pinnata**
1. Fruit linear and pointed at the tip; seeds in 1 row in each half .... **D. incisa**

Descurainia incisa (Gray) Britton [*Sophia richardsonii* (Sweet) Schulz var. *viscosa* (Rydb.) Peck]. Stems 15-80 cm tall; herbage glandular; leaves 1-2 times pinnately divided or lobed, 2-10 cm long; fruits ascending or spreading, 5-17 mm long, with stalks 5-10 mm long. June-July. Fig. 98.

Locally common in sparsely vegetated, often disturbed soil of montane grasslands and open forest; East, West. Our plants are ssp. *viscosa* (Rydb.) Rollins. B.C., Alta. south to CA, NM, TX.

Descurainia pinnata (Walter) Britton [*S. intermedia* Rydb.]. Stems 20-70 cm tall; leaves 1-2 times pinnately divided, to 10 cm long; fruits 8-12 mm long, the stalks 6-12 mm long, spreading. June-July.

Local in sparsely vegetated soil of montane grasslands and meadows; East, West. Ssp. *brachycarpa* (Rich.) Det. has glandular stems, while ssp. *intermedia* (Rydb.) Det. is without glands or glandular only in the inflorescence. B.C. to Que., south to most of U. S.

Diplotaxis D. C.

Diplotaxis muralis (L.) D. C., Wall Rocket [*D. erucoides* (L.) D. C. misapplied]. Annual or biennial with ascending stems to 40 cm tall; herbage mostly glabrous; leaves mainly basal, to 8 cm long, oblong, shallowly lobed; petals yellow; fruit 25-45 mm long with a beak 1-3 mm long.

Roadsides; reported for railroad right-of-way near East Glacier (Standley 1921). Introduced from Europe.

Draba L., Whitlow-wort

Annual to perennial herbs; herbage mostly with branched hairs; leaves alternate, undivided, basal rosette usually present; flowers in a narrow, usually unbranched inflorescence; petals white or yellow; fruit narrowly elliptic to linear.

The mat-forming perennials can be difficult to distinguish due to similar gross morphology. The white-flowered *D. lonchocarpa* and *D. nivalis* can be confused as can the yellow-flowered *D. densifolia*, *D. incerta*, *D. oligosperma*, and *D. paysonii*. Shape of mature fruit and hairs on the leaves are important characters. Many *Draba* spp. are more common in calcareous soil.

1. Stem leaves > 2, at least on larger plants 2
1. Stem leaves 0-2 7

2. Fruit glabrous 3
2. Fruit hairy 4

3. Fruit stalk > 1.5 times as long as the fruit .... *D. nemorosa*
3. Fruit stalk < 1.5 times as long as the fruit .... *D. albertina*

4. At least some fruits twisted 5
4. Fruits with flat faces, not twisted 6

5. Petals white; stem leaves with shallow teeth .... *D. breweri*
5. Petals yellow; stem leaves with entire margins .... *D. aurea*

6. Fruit stalk > 1.5 times as long as the fruit; flowers yellow .... *D. nemorosa*
6. Fruit stalk < 1.5 times as long as the fruit; flowers white .... *D. praealta*

7. Plants apparently annual with a weak taproot and unbranched rootcrown . 8
7. Plants perennial, usually with a woody, branched rootcrown, often mat-
   forming 10

8. Petals white, 2-lobed; stem leaves absent .... *D. verna*
8. Petals yellow, not 2-lobed; stem leaves present on larger plants 9

9. Upper leaf surface hairy .... *D. nemorosa*
9. Upper leaf surface glabrous, hairs only on the margins .... *D. crassifolia*

10. Fruit hairy 11
10. Fruit glabrous 15

11. Petals white; fruit > 3 times as long as wide, often twisted *D. lonchocarpa*
11. Petals yellow; fruit < 3 times as long as wide 12

12. Leaves glabrous except for straight, stiff hairs on the margins *D. densifolia*
12. Leaves with hair on the surfaces 13

13. Leaves with long hairs on the margin and long-stalked, branched hairs on
    the lower surface .... *D. paysonii*
13. Leaves without long marginal hairs; lower surface with short-stalked, comb-
    like hairs 14

14. Lower leaf surface with closely appressed comb-like hairs; leaves up to 1.5
    mm wide; fruits to 7 mm long; mostly low elevation .... *D. oligosperma*
14. Comb-like leaf hairs short-stalked; leaves > 1.5 mm wide; high elevation ...
    .... *D. incerta*

15. Fruits linear, at least some twisted .... *D. lonchocarpa*
15. Fruits narrowly elliptic to ovate, not twisted 16

16. Fruits ovate, 3-4 mm wide .... *D. macounii*
16. Fruits narrowly lance-shaped or elliptic, < 3 mm wide 17

17. Leaves nearly glabrous except for marginal hairs .... *D. crassifolia*
17. Leaves hairy on at least 1 surface .... *D. nivalis*

Draba albertina Greene [*D. nitida* Greene, *D. stenoloba* Ledeb. var *nana* (Schulz) Hitchc.]. Biennial to 20 cm tall from an unbranched rootcrown; herbage with simple and branched hairs; basal leaves 1-3 cm long, narrowly oblong with shallow marginal teeth; stem leaves 2-8, lance-shaped; petals yellow, 2-4 mm long; fruit linear, 8-12 mm long, glabrous. May-July.

Common in sparsely vegetated soil of grasslands, open forest and streambanks; montane and subalpine; East, West. AK south to WA, MT.

Draba aurea Vahl ex Hornem. [*D. mccallae* Rydb]. Short-lived perennial, 5-40 cm tall from a mostly unbranched rootcrown; herbage densely covered with simple and branched hairs; basal leaves narrowly oblong, 1-5 cm long; stem leaves several; petals yellow, 4-6 mm long; fruits lance-shaped, 8-12 mm long, hairy, often twisted. June-July. Fig. 99.

Locally common in sparsely vegetated, rocky soil of grasslands, open slopes and ridges; montane to alpine; East. AK to Greenl. south to AZ, NM.

Draba breweri Wats. [*D. cana* Rydb., *D. lanceolata* Royle misapplied]. Perennial to 20 cm tall from a few-branched rootcrown; herbage with dense, simple and star-shaped hairs; basal leaves oblong, 1-3 cm long; stem leaves several with toothed margins; petals white 3-5 mm long; fruits narrowly lance-shaped, 4-12 mm long, hairy, often twisted. June-July.

Uncommon in sparsely vegetated, stony soil of grasslands, meadows and open forest; montane to near treeline; East. Our plants are var. *cana* (Rydb.) Rollins. AK to Que., south to CA, CO, WI, VT.

Draba crassifolia Graham. Short-lived perennial to 10 cm tall from a few-branched rootcrown; herbage mostly glabrous; basal leaves narrowly oblong, to 3 cm long; stem leaves 0-2; petals yellow, 2-3 mm long; fruits narrowly ovate, glabrous, 5-12 mm long. July-Aug.

Local in moist, often late snowmelt, sparsely vegetated soil of meadows, rocky slopes and cliffs near or above treeline; East, West. AK to Greenl. south to CA, AZ, CO. *D. fladnizensis* Wulfen is similar to *D. crassifolia* but has white flowers. Reports of *D. fladnizensis* from Glacier are likely referable here because flowers of *D. crassifolia* fade to white.

Draba densifolia Nutt. Mat-forming perennial to 8 cm tall; simple and forked hairs on the stem; leaves tongue-shaped, glabrous with long, stiff hairs on the margins, stem leaves absent; petals yellow, 2-6 mm long; fruits elliptic, hairy, 3-7 mm long. June-July.

Uncommon in stony soil of exposed alpine slopes and ridges, East. AK to Alta. south to CA, UT, WY. The more common *D. paysonii, D. incerta*, and *D. oligosperma* have hair on the leaf surfaces.

Draba incerta Payson [*D. glacialis* Adams misapplied]. Mat-forming perennial to 20 cm tall; herbage with simple and branched hairs; basal leaves narrowly oblong, 5-13 mm long; stem leaves absent; petals yellow, 3-5 mm long; fruits narrowly ovate, 5-10 mm long, hairy. July-Aug.

Common in stony soil of alpine tundra, cliffs and exposed slopes near or east of the Divide. AK to ID, WY, Que.

Draba lonchocarpa Rydb. Perennial to 8 cm tall from a few-branched rootcrown; herbage with dense star-shaped hairs; basal leaves 5-10 mm long, narrowly oblong; stem leaves absent; petals white, 2-5 mm long; fruits linear, 6-15 mm long, glabrous or hairy, often twisted. July-Aug.

Common in moist soil of cliffs and cool, rocky slopes near or above treeline; East, West. Our plants are var. *lonchocarpa*. AK to OR, ID, CO.

Draba macounii Schulz. Perennial to 4 cm tall from a loosely branched rootcrown; herbage nearly glabrous; basal leaves 6-10 mm long, broadly oblong with wrinkled hairs on the tips and margins; stem leaves absent; petals white; fruits ovate, 4-8 mm long, glabrous, wrinkled. July-Aug.

Rare in wet soil of cool, open slopes, rock outcrops and along streams above treeline along the Divide. AK to MT, CO.

Draba nemorosa L. Annual to 25 cm tall; basal leaves narrowly ovate often with a few shallow teeth, 1-3 cm long, branched-hairy; stem leaves few; petals yellow, 2-4 mm long; fruits narrowly elliptic, 3-10 mm long, glabrous or hairy. May-June.

Locally common in disturbed soil of montane grasslands and open forest; East. AK to Que. south to CA, CO, MN. Often associated with pocket gopher or ground squirrel digging.

Draba nivalis Lilj. Perennial to 6 cm tall from a branched rootcrown; basal leaves oblong, 5-10 mm long, with dense, branched hairs; stem leaves 0-2; petals white; siliques narrowly elliptic, ca. 5 mm long, glabrous. June-July.

Rare in stony turf of cool alpine slopes, occasionally on exposed ridges at lower elevations; East. Circumpolar south to CO. Similar to *D. lonchocarpa* but fruits are shorter and not twisted. Standley's (1921) report of *D. oreibata* Macbr. & Pays. can probably be referred here.

Draba oligosperma Hook. [*D. andina* (T.& G.) Nels.]. Mat-forming perennial to 6 cm tall; basal leaves narrowly oblong, 3-10 mm long, with comb-like branched hairs appressed to the surface and on the margins; stem leaves absent; petals yellow, 3-5 mm long; fruits ovate, 2-7 mm long, hairy. May-June.

Uncommon in sparsely vegetated soil of montane rock outcrops and exposed alpine slopes and ridges; East. AK south to CA, CO. The upper surfaces of some leaves of *D. oligosperma* may become almost glabrous. Hairs of the similar *D. incerta* are not appressed to the leaf surface.

Draba paysonii Macbr. Mat-forming perennial to 5 cm tall; basal leaves narrowly oblong, 4-8 mm long, densely covered with tangled, branched hairs; stem leaves absent; petals yellow, 2-5 mm long; fruits ovate, hairy. May-July. Fig. 100.

Var. *treleasii* (Schulz) Hitchc., with fruits 3-5 mm long, is common in stony, sparsely vegetated soil of exposed ridges and slopes near or above treeline; East. Var. *paysonii*, with fruits 5-8 mm long, is uncommon in similar, limestone habitats at lower elevations; East. AK south to CA, UT, WY.

Draba praealta Greene. Short-lived perennial to 30 cm tall, with a mostly unbranched rootcrown; herbage with dense, simple and branched hairs; basal leaves oblong, 1-3 cm long; stem leaves several, lance-shaped with few shallow teeth on the margins; petals white, 2-4 mm long; fruits narrowly lance-shaped, 8-14 mm long, hairy. May-July.

Uncommon in sparsely vegetated soil of open slopes and stream banks; montane and subalpine; East. AK south to CA, WY.

Draba verna L. Annual to 10 cm high; basal leaves lance-shaped with entire or toothed margins, to 10 mm long, densely covered with branched hairs; stem leaves absent; petals white, 2-3 mm long, 2-lobed; fruits narrowly elliptic, 3-8 mm long, glabrous. May.

Locally common in shallow, gravelly soil along montane streams and roads; East, West. Rollins (1993) considers this native only to Europe, while Hitchcock et al. (1964) and Price (1993) consider it circumboreal. Conspicuous for only a brief time in early spring.

Erucastrum Presl

Erucastrum gallicum (Willd.) Schulz, Dog Mustard. Annual to 50 cm tall; herbage short-hairy; leaves deeply pinnately lobed and toothed, 3-15 cm long; flowers subtended by leaf-like bracts; petals pale yellow, 4-7 mm long; fruits linear, 20-45 mm long, 1-2 mm wide with a sterile tip 2-3 mm long.

Moist, disturbed, often calcareous soil at low elevations. Reported by Dorn (1984) for Glacier Park. Introduced from Eurasia.

Erysimum L., Wallflower

Annual to perennial herbs; herbage with branched hairs; both basal and stem leaves unlobed; petals yellow; fruit ascending or erect, linear with a prominent style at the tip.

Erysimum asperum (Nutt.) D. C., with larger flowers than the following species, is reported for Waterton Park (Kuijt 1982) and could be in Glacier as well.

1. Fruit 15-30 mm long with a style tip ca. 1/2 mm wide; petals < 6 mm long .. .. ***E. cheiranthoides***
1. Fruit 25-50 mm long with style tip ca. 1 mm wide; petals > 6 mm long .... ***E. inconspicuum***

Erysimum cheiranthoides L., Treacle Mustard [*Cheirinia cheiranthoides* (L.) Link]. Annual to 60 cm tall; leaves narrowly lance-shaped, to 8 cm long; petals 3-5 mm long; fruit 15-30 mm long with a style-tip ca. 1/2 mm wide. June-Aug.

Uncommon in disturbed soil of montane meadows, grasslands, avalanche slopes and along streams; East, West. AK to Newf. south to CA, CO, MO, FL.

Erysimum inconspicuum (Wats.) MacM. [*Cheirinia inconspicua* (Wats.) Rydb.]. Short-lived perennial to 70 cm tall; leaves strap-shaped, to 8 cm long; petals 7-10 mm long; fruit 25-60 mm long with a style tip ca. 1 mm wide. July-Aug. Fig. 101.

Common in grasslands, meadows and open slopes; montane to subalpine, East. AK to Newf. south to OR, CO, MN.

Lepidium L., Peppergrass

Lepidium densiflorum Schrader. Annual with branched stems to 30 cm tall; herbage minutely hairy; leaves narrowly lance-shaped with toothed margins, 2-5 cm long; sepals ca. 1 mm long; petals absent; fruit elliptic ca. 3 mm long with a notch at the tip.

Standley (1921) reported this plant to be common in disturbed soil at low elevations, but it hasn't been collected in the Park since. Native to Eurasia and much of N. America.

Lesquerella Wats., Bladderpod

Lesquerella alpina (Nutt. ex T. & G.) Wats. [*L. spathulata* Rydb.]. Tufted perennial to 10 cm high from a branched rootcrown; herbage with appressed, star-shaped hairs; leaves basal and alternate, narrowly oblong with entire margins, to 4 cm long; petals yellow, 5-7 mm long; fruit ovate, 4-5 mm long with a style ca. 2 mm long; fruit stalks spreading, 2-10 mm long, often s-shaped.

Exposed slopes of grasslands along the east boundary. Our plants are var. *alpina*. Alta. and Sask south to ID, CO, SD. *L. arenosa* (Rich.) Rydb. occurs in southwest Alta. and could also be in grasslands along the east edge of the Park.

Physaria (Nutt. ex T. & G.) Gray, Twinpod

Tufted perennial herbs with a branched rootcrown; herbage gray with appressed, star-shaped hairs; leaves unlobed (ours), basal broader than those of the stem; flowers yellow; fruit inflated, wider than high, 2-lobed at the top with style in the sinus; fruit stalks spreading.

1. Fruit 2-lobed above and below; 4 seeds in each half; lower elevation ....
.... *P. didymocarpa*
1. Fruit 2-lobed only above; 2 seeds in each chamber; near treeline ....
.... *P. saximontana*

Physaria didymocarpa (Hook.) Gray. Stems 3-10 cm long; basal leaves 2-5 cm long, spoon-shaped, often with shallow teeth, spreading and appressed to the ground; stem leaves few; petals 9-12 mm long; fruits 1-2 cm high, 2-lobed above and below, the two halves connected only in the middle with 4 seeds in each. May-June. Fig. 102.

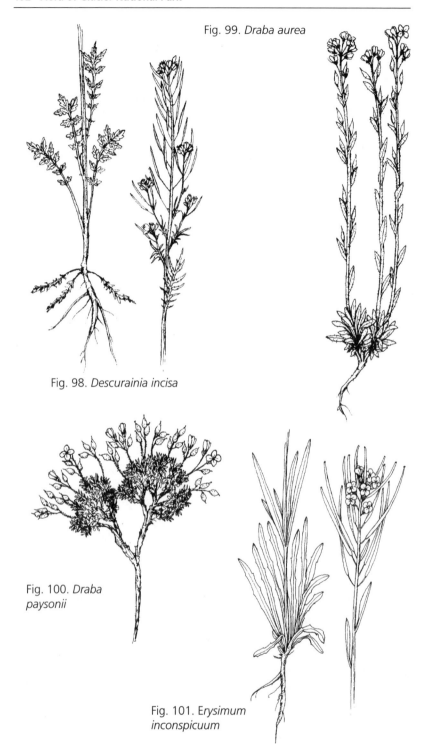

Fig. 99. *Draba aurea*

Fig. 98. *Descurainia incisa*

Fig. 100. *Draba paysonii*

Fig. 101. E*rysimum inconspicuum*

Common on banks and open slopes; montane; East. Our plants are var. *didymocarpa*. B.C., Alta. south to ID, WY. Blackfeet Indians used the juice of *P. didymocarpa* to treat sore throats and stomach troubles.

Physaria saximontana Rollins. Similar to *P. didymocarpa*; basal leaves with broad-toothed margins; fruit ca. 1 cm high, 2-lobed above but not below; seeds 2 in each chamber. June-July.

Uncommon in shallow, stony soil of exposed ridges and slopes near or above treeline east of the Divide. Our plants are var. *dentata* Rollins. MT, WY.

Rorippa Scop., Yellowcress

Annual, biennial or short-lived perennial; herbage glabrous to sparsely hairy; leaves toothed to lobed, alternate and in a basal rosette; flowers yellow; fruits linear to narrowly ovate, plump, on spreading stalks.

1. Fruit linear, often curved, > 6 mm long .... **R. curvisiliqua**
1. Fruit 3-6 mm long, ovate to cylindrical 2
2. Stems erect or ascending; fruit stalks as long as fruits; petals > 1 mm long ...
.... **R. palustris**
2. Stems mostly prostrate; fruit stalks shorter than fruit; petals ≤ 1 mm long ....
.... **R. tenerrima**

Rorippa cuvisiliqua (Hook.) Bessey ex Britt. [*Radicula curvisiliqua* (Hook.) Greene, *Radicula lyrata* (Nutt.) Greene]. Perennial; stems ascending to prostrate, to 40 cm long; leaves 2-10 cm long; petals 1-2 mm long; fruits linear, 6-12 mm long, often gently curved. July-Aug.

Local in muddy soil around montane and subalpine streams and temporary ponds; East, West. Both var. *curvisiliqua* and var. *lyrata* (T. & G.) Peck may occur in Glacier Park; they often grow together and intergrade. B.C. to CA, CO.

Rorippa palustris (L.) Besser [*R. islandica* (Oeder) Borbas, *Radicula palustris* (L.) Moench]. Perennial; stems erect or ascending, 20-60 cm tall; leaves 6-15 cm long; petals ca. 2 mm long; fruits narrowly ovate, 2-5 mm long. June-Aug. Fig. 103.

Common in wet, often disturbed soil along montane streams and wetlands; East, West. Var. *hispida* (Desv.) Rydb. has stiff hairs on the lower leaf surfaces, while var. palustris has glabrous lower leaf surfaces. Native to most of N. America.

Rorippa tenerrima Greene. Annual; stems ascending to prostrate, 5-20 cm long; leaves 2-7 cm long; petals to 1 mm long; fruits ovoid, 3-6 mm long. Aug.

Mud around montane ponds and lakes; known only from Two Medicine Lake. B.C. to Sask. south to CA, NM, Mex.

Sisymbrium L., Tumble Mustard

Sisymbrium altissimum L., Jim Hill Mustard [*Nortia altissima* (L.) Britt.]. Annual to 100 cm high, branched above; herbage sparsely hairy below, glabrous above; leaves to 10 cm long, pinnately divided into lance-shaped, toothed leaflets; petals pale yellow, 6-8 mm long; fruits narrowly linear, 5-10 cm long, round in cross section, ascending to spreading. May-July.

Common in disturbed soil of montane grasslands and along roads; East, West. Introduced from Eurasia.

Smelowskia C. A. Meyer

Smelowskia calycina (Stephen) C. A. Mey. [*S. lobata* Rydb., *S. americana* Rydb.]. Tufted perennial to 15 cm tall from a branched rootcrown; herbage with simple and branched hairs; leaves basal and alternate, to 5 cm long, petiolate, deeply divided into narrowly oblong lobes; flowers congested; petals white, 4-7 mm long; fruit erect, narrowly elliptic, 5-9 mm long, nearly round in cross section. June-July. Fig. 104. Color plate 58.

Common in stony soil of moraine, cliffs, talus slopes, and exposed ridges near or above treeline; East, West. Our plants are var. *americana* (Reg. & Herd.) Drury & Rollins. AK to OR, CO; Asia. Flowering soon after the snow melts.

Thlaspi L., Pennycress

Thlaspi arvense L., Fanweed. Glabrous annual, 20-50 cm tall; leaves to 6 cm long, narrowly lance-shaped with shallowly toothed margins, clasping the stem above; petals white, 3-4 mm long; fruit stalks spreading; fruit broadly elliptic, strongly flattened, 8-17 mm long, deeply notched at the tip, erect. June-July. Fig. 105.

Common in disturbed soil of montane grasslands and around roads and habitations; East, West. Introduced from Europe. The report of *T. montanum* L. [*T. fendleri* Gray) for Siyeh Pass (Bamberg and Major 1968) is probably erroneous.

CALLITRICHACEAE: WATER-STARWORT FAMILY

Callitriche L., Water-starwort

Delicate, glabrous, annual, aquatic plants with numerous stems from fibrous roots; leaves opposite with entire margins; flowers 1-3 at the base of leaves, small, green, unisexual; male flower of 1 stamen; female flower an ovary with 2 styles; fruit 4-lobed separating into 4 single-seeded achenes.

1. Floating leaves oblong; fruits with small bracts at the base .... **C. verna**
1. Floating leaves linear; fruits without bracts .... **C. hermaphroditica**

Callitriche hermaphroditica L. [*C. autumnalis* L.]. Stems 5-20 cm long; leaves linear, 5-10 mm long with 2-lobed tips; fruit 1-2 mm long, circular with wing-margins, without bracts. July-Sept. Fig. 106.

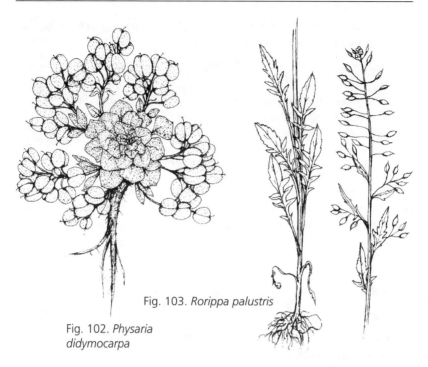

Fig. 103. *Rorippa palustris*

Fig. 102. *Physaria didymocarpa*

Fig. 104. *Smelowskia calycina*

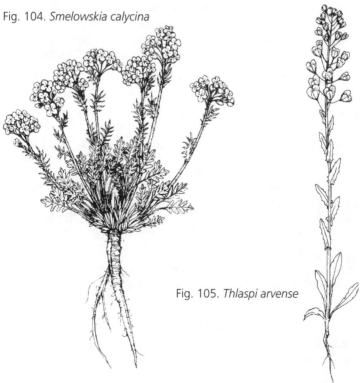

Fig. 105. *Thlaspi arvense*

Common in mud beneath shallow water of montane ponds, lakes, and slow streams; East, West. Circumboreal south to CA, NM, NE, MN, VT.

Callitriche verna L. [*C. palustris* L.]. Stems 5-30 cm long; submerged leaves linear, 5-20 mm long; floating leaves oblong, to 4 mm wide; fruit ca. 1 mm wide and somewhat longer; subtended by small bracts. July-Sept.

Common in mud beneath shallow water of montane ponds and lakes; East, West. Circumboreal through much of N. America.

CAMPANULACEAE: BELLFLOWER FAMILY

Perennial herbs; leaves basal and alternate; flowers bisexual; corolla united at least at the base, 5-lobed; sepals 5; stamens 5; ovary below base of corolla (inferior) with 1 style; fruit a capsule with numerous seeds.

Lobelia and related genera are sometimes placed in a separate family, the Lobeliaceae.

1. Corolla bell-shaped with 5 equal lobes .... **Campanula**
1. Corolla united below with 2 lobes above and 3 below .... **Lobelia**

Campanula L., Bellflower
Perennial herbs with milky sap; flowers blue; corolla conical to bell-shaped; style 1 with 3 stigmas.

1. Flowers without stalks, clustered; introduced weed .... **C. glomerata**
1. Flowers stalked; native 2
2. Corolla with lobes ≥ than lower tubular portion; flowers solitary .. **C. uniflora**
2. Corolla lobes < the tube; flowers often > 1 .... **C. rotundifolia**

Campanula glomerata L., Clustered Bellflower. Stems 30-50 cm tall; herbage fine-hairy; leaves lance-shaped with small-toothed margins, petiolate below; flowers conical, 2-3 cm long, clustered in upper leaf axils. June-July.

Garden plant escaped into disturbed forest margins at West Glacier. Introduced from Eurasia.

Campanula rotundifolia L., Harebell. Stems 10-50 cm high from slender rootstocks; herbage glabrous; basal leaves long petiolate with broadly spade-shaped blades to 2 cm long, often absent at flowering; stem leaves linear, 15-50 mm long; flowers few on spreading stalks; corolla 10-25 mm long, bell-shaped. July-Aug. Fig. 107.

Common in grasslands, meadows, open forest, cliffs, exposed slopes, and ridges at all elevations; East, West. Circumboreal south to CA, TX, IA, PA.

Campanula uniflora L., Arctic Bellflower. Stems to 10 cm tall from slender rootstocks; lower leaves narrowly oblong, 2-3 cm long; upper leaves linear; flowers solitary, erect; corolla 6-12 mm long, the long lobes spreading. Aug.

Rare in moist meadows and turf of cool slopes near or above treeline near or east of the Divide. Circumboreal south to CO.

Lobelia L., Lobelia

Lobelia kalmii L. Biennial or short-lived, glabrous perennial with fibrous roots and stems to 30 cm high; basal leaves spoon-shaped, 1-3 cm long; stem leaves strap-shaped; stalked flowers few, 8-15 mm long, solitary in axils of uppermost, reduced leaves; corolla blue with a white center, 2-lipped with 3 larger lobes below and 2 above; capsule 4-8 mm long. July-Aug.

Rare in montane fens; known only from near Lee Creek in the northeast corner of the Park. Boreal Canada south to WA, MT, MN, PA.

CAPPARACEAE: CAPER FAMILY

Cleome L., Bee Plant

Cleome serrulata Pursh., Rocky Mountain Bee Plant. Glabrous annual 20-50 cm tall, branched below; leaves alternate, of 3 narrowly lance-shaped leaflets, 2-7 cm long; inflorescence compact at first, elongating in fruit; flowers rose-purple; petals 4, separate, 8-11 mm long; calyx 5-lobed, cup-shaped, stamens 6, longer than corolla; fruit a long-stalked, nodding, pod-like capsule, 25-50 mm long.

Collected along the railroad near East Glacier where perhaps introduced (Standley 1921). Native to grasslands B.C. to Man. south to CA, TX, MO.

CAPRIFOLIACEAE: HONEYSUCKLE FAMILY

Erect or trailing shrubs or vines; leaves opposite; flowers bisexual; corolla 5-lobed; stamens usually 5; ovary below base of corolla (inferior); fruit usually a berry.

1. Plants shrubs with erect stems 2
1. Plants trailing subshrubs or climbing vines 5

2. Leaves divided or obviously lobed 3
2. Leaves of older stems with mostly entire margins 4

3. Leaves divided into 5-7 leaflets .... **Sambucus**
3. Leaves 3-lobed, maple-like .... **Viburnum**

4. Flowers to 7 mm long; berries white .... **Symphoricarpos**
4. Flowers > 10 mm long; berries red or black .... **Lonicera**

5. Trailing subshrub; leaves shallowly toothed on the tip .... **Linnaea**
5. Vine; leaves with entire margins .... **Lonicera**

Linnaea L.

Linnaea borealis L., Twinflower. Stems slender, trailing, with short erect shoots to 8 cm high; herbage sparsely hairy, sometimes glandular; leaves ovate with shallow teeth on the upper half, petiolate, 7-20 mm long; flowers

2, nodding on stem tips; pinkish corolla narrowly bell-shaped, hairy within, 9-16 mm long; fruit a small, round, glandular-hairy achene. June-July. Fig. 108. Color plate 44.

Abundant in moist, cool forest; montane to lower subalpine; East, West; often with *Vaccinium* spp. and *Clintonia uniflora*. Circumboreal south to CA, NM, IN, WV.

Lonicera L., Honeysuckle

Leaves usually with entire margins; pairs of flowers from leaf axils (ours); corolla funnelform, bulged at the base, hairy within; stigma globose; fruit a several-seeded berry.

1. Vining plants; flowers orange .... *L. ciliosa*
1. Erect shrubs; flowers yellow or white 2
2. Flowers yellow with purplish bracts; berry black .... *L. involucrata*
2. Flowers whitish without bracts; berry red .... *L. utahensis*

Lonicera ciliosa (Pursh) D. C. Stems weak, prostrate or twining, hollow; leaves elliptic, 4-10 cm long, white-waxy beneath, hairy on the margins, short-petiolate; clusters of several short-stalked flowers subtended by the united uppermost pair of leaves; corolla orange, 2-4 cm long, the flared lobes nearly equal; berries red, ca. 1 cm across. June-July.

Rare in montane forest openings; known only from near West Glacier. B.C. to MT south to CA.

Lonicera involucrata (Rich.) Banks ex Spreng., Bearberry, Black Twinberry. Stems erect, 1-3 m high with 4-angled twigs; leaves oblong-ovate, 5-14 cm long, sparsely hairy beneath; pairs of flowers subtended by 2 purplish bracts reflexed in fruit; corolla yellow, 10-20 mm long, glandular; berry black, ca. 1 cm wide. June-July. Fig. 109.

Common on avalanche slopes and in moist forest, especially along streams; montane and subalpine; East, West. Our plants are var. *involucrata*. B.C. to Que. south to CA, NM, WI. The berries are so bitter it is hard to imagine they are eaten by bears as the common name implies.

Lonicera utahensis Wats. Stems branched, 1-2 m tall; leaves ovate to elliptic with rounded tips, 2-8 cm long, glabrous or sparsely hairy beneath; pairs of flowers bractless, on a stalk elongating in fruit; corolla cream-colored, 1-2 cm long; berry red, 5-8 mm wide. June-July.

Common in moist montane and subalpine forest both sides of the Divide. B.C. to Alta. south to CA, UT, WY.

Sambucus L., Elderberry

Shrubs; stems with soft pith, to 3 m high; twigs waxy; leaves pinnately divided into 5-9 ovate to lance-shaped, toothed, glabrous leaflets with long-pointed tips; flowers numerous in a much-branched inflorescence on stem tips; calyx absent; corolla saucer-shaped; fruit a few-seeded berry.

1. Inflorescence pyramidal in outline; fruits shiny black .... *S. racemosa*
1. Inflorescence flat-topped; fruits blue and white-waxy .... *S. cerulea*

Sambucus cerulea Raf., Blue Elderberry. Leaflets 5-9, 5-15 cm long; flowers in a flat-topped, umbrella-shaped inflorescence 4-20 cm across; corollas white, 4-6 mm across; berry blue with a white-waxy coating; 4-6 mm wide.

Rare in open, montane forest and thickets; common west of the Divide just south of the Park and collected once near St. Mary, where it may have been introduced. B.C. and MT south to CA, AZ, NM. The fruits are edible.

Sambucus racemosa L., Black Elderberry [*S. melanocarpa* Gray]. Leaflets 5-7, 4-17 cm long; flowers tightly clustered in a pyramidal inflorescence; corollas cream-white, ca. 3 mm across; berry black, 4-6 mm wide. June-July. Fig. 110.

Common in moist, open forest, thickets, avalanche slopes and along streams; montane to subalpine; East, West. Our plants are var. *melanocarpa* (Gray) McMinn., but var. *pubens* (Michx.) Koehne, with red or yellow berries, is reported for Waterton Park and could be in Glacier as well. Circumboreal south to CA, NM, IL, GA. The fruits of *S. racemosa* are said to be poisonous.

Symphoricarpos Duhamel, Snowberry, Buckbrush

Stems from spreading rootstocks; leaves petiolate with entire margins (juvenile stem leaves may have lobed margins); flowers in small clusters at stem tips and solitary in leaf axils; corolla white or pinkish, hairy within; fruit a white, 2-seeded berry.

An infusion from the leaves was used by Blackfeet for sore eyes. Chipmunks are the only animal I have observed collecting the fruit. *S. albus* is found in forest, while *S. occidentalis* is generally in more open habitat.

1. Style glabrous; corolla bell-shaped; twigs glabrous .... *S. albus*
1. Style hairy at mid-length; corolla bowl-shaped; twigs hairy .... *S. occidentalis*

Symphoricarpos albus (L.) Blake, Snowberry. Stems to 1 m high with glabrous twigs; leaves ovate, 2-5 cm long, sparsely hairy beneath; corolla bell-shaped, 4-7 mm long; style hairless, 2-3 mm long; berry 10-15 mm long. July-Aug. Fig. 111.

Abundant in moist to dry, montane to subalpine forest; East, West. Our plants are var. *laevigatus* (Fern.) Blake. AK to N. S. south to CA, CO, NE, VT. Plants bloom over a period of many weeks.

Symphoricarpos occidentalis Hook., Buckbrush. Similar to *S. albus* but spreading by rhizomes more readily; twigs and lower leaf surfaces hairy; flowers bowl-shaped, 3-5 mm long; style 3-8 mm long, hairy at mid-length; berry 6-9 mm long. July.

Common in montane grasslands and meadows or dry, open forest; East, West. B.C. to Ont. south to WA, NM, OK, MI.

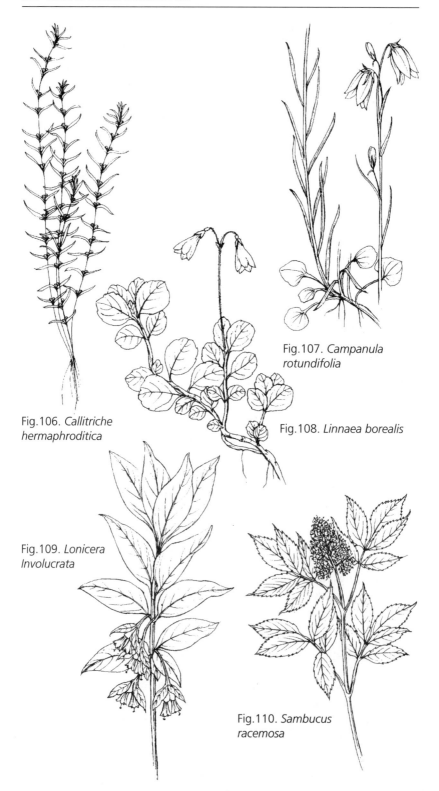

Fig.106. *Callitriche hermaphroditica*

Fig.107. *Campanula rotundifolia*

Fig.108. *Linnaea borealis*

Fig.109. *Lonicera Involucrata*

Fig.110. *Sambucus racemosa*

Viburnum L.

Viburnum edule (Michx.) Raf., Mooseberry, Low-bush Cranberry [*V. pauciflorum* Pylaie]. Stems to 2 m tall; leaves maple-leaf-like with 3 pointed lobes and toothed margins, 3-10 cm long, hairy and glandular beneath; flowers in a flat-topped cluster on stem tips; corolla bowl-shaped, deeply lobed, white, 3-6 mm across; fruit 1-seeded, red or orange, 8-10 mm long. June. Fig. 112.

Uncommon in moist forest, often near streams; montane and lower subalpine, mainly west of the Divide. AK to Newf. south to OR, CO, MN, PA. Birds eat the berries; leaves turn red in autumn.

CARYOPHYLLACEAE: PINK FAMILY

Annual or perennial herbs; stems often swollen at the nodes; leaves with entire margins, mostly opposite; flowers with 5 separate petals, 5 separate or united sepals and 10 stamens; ovary with 2-5 styles; fruit a capsule. Reference: Hartman (1993).

Minuartia and *Arenaria* are similar; in our area *Arenaria* is usually erect and ≥ 10 cm high or diminutive annuals, while *Minuartia* spp. are short or lax perennials.

1. At least the lower half of the sepals united 2
1. Sepals separate to the base 3

2. Styles 2; petals pink, ca. 2 cm long; annual or biennial.................. **Dianthus**
2. Styles 3-5; perennial .... **Silene**

3. Petals absent 4
3. Petals present 5

4. Plants clumped; stems < 5 cm long.. **Minuartia**
4. Stems not clumped; often > 5 cm long .... **Stellaria**

5. Papery bracts (stipules) on the stem at the leaf bases .... **Spergularia**
5. Stipules absent 6

6. Petals deeply 2-lobed... 7
6. Petals with rounded to shallowly indented tips 8

7. Styles mostly 3; fruit capsule ovoid .... **Stellaria**
7. Styles mostly 5; capsule cylindric.. **Cerastium**

8. Styles 4-5 .... **Sagina**
8. Styles 3.. 9

9. Leaves narrowly oblong, > 3 mm wide .... **Moehringia**
9. Leaves linear or ovate, < 3 mm wide.. 10

10. Leaves linear, < 10 mm long; capsule splitting into 3 sections ..... **Minuartia**
10. Leaves ovate or > 10 mm long; capsule splitting into 6 sections .... **Arenaria**

Arenaria L., Sandwort

Annual or perennial herbs; leaves without petioles; flowers cup-shaped, several to many; sepals separate; petals white with blunt tips; styles 3; capsule egg-shaped, opening by 6 slits in the top.

Many species formerly in *Arenaria* are now placed in *Minuartia* and *Moehringia*.

1. Plant annual; petals smaller than sepals; leaves ovate .... **A. serpyllifolia**
1. Plant perennial; petals longer than sepals; leaves needle-like 2
2. Inflorescence glabrous, congested; sepals lance-shaped; capsule ca. as long as sepals .... **A. congesta**
2. Inflorescence glandular, open; sepals ovate; capsules twice as long as sepals .... **A. capillaris**

Arenaria capillaris Poir. [*A. formosa* Fisch.]. Tufted perennial with numerous stems 5-25 cm tall; leaves glabrous, needle-like, 2-4 cm long; inflorescence glandular, open; petals 5-7 mm long; sepals ovate with membranous margins and barely-pointed tips; capsule ca. twice as long as sepals. July-Aug. Fig. 113.

Sparsely vegetated, rocky soil of grasslands, meadows, open forest, and cliffs; abundant alpine and subalpine, less common lower; East, West. Our plants are var. *americana* (Maguire) Davis. AK to Alta. south to OR, NV, MT.

Arenaria congesta Nutt. Perennial 10-20 cm high from a widely branched rootcrown; leaves glabrous, 1-6 cm long, needle-like with smaller leaves in the axils; inflorescence congested-hemispheric; petals 5-9 mm long; sepals lance-shaped with papery margins and definitely pointed tips; capsule ca. as long as sepals. July.

Rare in montane grasslands, sometimes higher; East, West. Our plants are var. *lithophila* Rydb. Alta. and Sask. south to CA, UT, CA.

Arenaria serpyllifolia L. Annual 5-15 cm tall, branched at the base; leaves short-hairy, ovate, 3-5 mm long; inflorescence open but narrow; petals ca. 2 mm long; sepals lanceolate, hairy, longer than petals; capsule ca. as long as calyx. May-June.

Locally common in sparsely vegetated, often disturbed soil of montane grasslands and rock outcrops and along roads and trails; East, West. Introduced from Eurasia. Plants are inconspicuous and ephemeral.

Cerastium L., Mouse-ear Chickweed

Perennial (ours); herbage hairy; flowers several in a branched inflorescence; sepals united below; petals white, 2-lobed; styles mostly 5; capsule cylindrical, often curved at maturity, opening by 10 slits.

1. Petals ca. as long as sepals; disturbed areas .... **C. fontanum**
1. Petals at least 1.5 times as long as sepals; grasslands to turf 2
2. Main leaves with smaller leaves in the axils; all elevations .... **C. arvense**
2. Leaves lacking smaller axillary leaves; high elevation .... **C. beeringianum**

Cerastium arvense L., Field Chickweed [*C. strictum* L.]. Mat-forming, 5-30 cm tall, with trailing stems, branched above; herbage often glandular;

leaves narrowly lance-shaped, 1-3 cm long with smaller leaves in lower axils; sepals 5-7 mm long; petals 2-3 times as long as calyx; capsule 1-2 times as long as calyx. June-July. Fig. 114.

Common in dry, open montane forest and grassland, as well as in alpine and subalpine meadows and turf; East, West. Circumboreal south to CA, NM, NE, MN, GA.

Cerastium beeringianum Cham.& Schlecht. [*C. alpinum* L.]. Similar to *C. arvense* but somewhat smaller with fewer flowers; leaves narrowly oblong, 1-2 cm long; petals 6-8 mm long; capsule ca. twice as long as calyx. July-Aug.

Common in alpine turf, limestone talus, exposed ridges, and moraine; East. AK to Newf. south to CA, AZ, CO; Eurasia. Plants intermediate between *C. arvense* and *C. beeringianum* were referred to *C. alpinum* by Standley (1921).

Cerastium fontanum Baumg. [*C. vulgatum* L. misapplied]. Stems 10-40 cm long, trailing and rooting at the nodes; herbage glandular; leaves narrowly ovate, 1-2 cm long; sepals ca. 4 mm long; petals 4-6 mm long; capsule twice as long as calyx. June.

Uncommon in disturbed soil of lawns, trails and roadsides; montane; West. Our plants are ssp. *vulgare* (Hartm.) Greut. & Burd. Introduced from Europe.

Dianthus L., Pink

Dianthus armeria L., Deptford Pink. Annual or biennial to 50 cm tall; stems glandular-hairy near nodes; leaves glabrous, strap-shaped, 4-10 cm long; stem leaves erect; inflorescence of few several-flowered, tight clusters; sepals narrow, pointed, united below, ca. 15 mm long; petals deep pink, 20-25 mm long with wavy, rounded tips; capsule as long as calyx. July-Aug.

Rare in grasslands; collected once near Marias Pass. Introduced from Europe; a common weed in other areas of northwest MT.

Minuartia L., Sandwort

Tufted or mat-forming perennials; leaves linear; flowers solitary to several; sepals separate; petals white with blunt tips; styles 3; capsule ovoid, opening by 3 slits in the top. Reference: Porsild & Cody (1980).

These species have previously been placed in *Arenaria*.

1. Tip of sepals blunt, hood-like .... **M. obtusiloba**
1. Sepals with sharp-pointed tips 2
2. Herbage glabrous .... **M. rossii**
2. Herbage glandular 3
3. Stem leaves 5-10 mm long; plants loosely spreading over ground ....
.... **M. nuttallii**
3. Stem leaves < 5 mm long; plants in tight cushions .... **M. rubella**

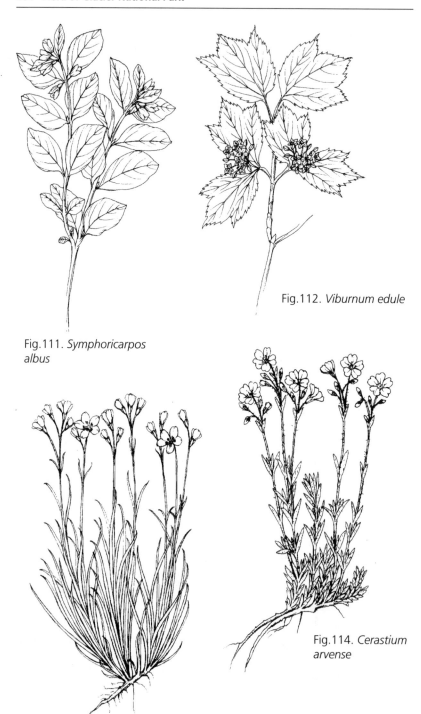

Fig.111. *Symphoricarpos albus*

Fig.112. *Viburnum edule*

Fig.113. *Arenaria capillaris*

Fig.114. *Cerastium arvense*

Minuartia nuttallii (Pax) Briq. [*Arenaria nuttallii* Pax]. Loose, mat-forming with ascending or trailing stems, 4-12 cm long; herbage glandular-hairy; leaves 5-10 mm long; inflorescence open but small; sepals 3-5 mm long, lance-shaped; petals shorter than sepals; capsule shorter than calyx. July-Aug.

Common on rocky alpine slopes and talus, especially limestone, rarely lower; East. Our plants are var. *nuttallii*. B.C., Alta. south to CA, UT, WY.

Minuartia obtusiloba (Rydb.) House [*Arenaria obtusiloba* (Rydb.) Fern., *A. sajanensis* Willd., *A. laricifolia* L.]. Tufted perennial 3-10 cm tall; herbage glandular-hairy; leaves 5-8 mm long; flowers solitary; sepals 4-6 mm long, lance-shaped with blunt tips; petals 5-8 mm long; capsules as long or longer than calyx. July-Aug. Fig.115.

Stony soil of exposed ridges and slopes; abundant near or above timberline; East, West, occasionally lower, East. AK to Greenl. south to OR, NM. Widely spreading plants with glabrous sepals occurring in moist, cool, alpine sites have been called *M. biflora* (L.) Schinz & Thell. [=*A. sajannensis* Willd.], but Weber (1987) considers it simply an ecotype of *M. obtusiloba*.

Minuartia rossii (R. Br.) Graebn. [*Arenaria rossii* R. Br., *M. elegans* (Cham. & Schlecht., *M. austromontana* Wolf & Packer]. Tufted or loose mat-forming with stems to 5 cm tall; herbage glabrous; leaves 4-6 mm long, with smaller leaves in the axils; flowers solitary; sepals lance-shaped, 2-3 mm long; capsules equal to or shorter than calyx. July-Aug.

Ssp. *elegans* (Cham. & Schlecht.) Rebr., with petals and a loose habit, is rare in moist soil of cool alpine slopes; ssp. *rossii*, small-mat-forming and lacking petals, is common in stony soil of exposed alpine slopes and ridges; East, West. AK to Greenl. south to WA, ID, CO; Asia.

Minuartia rubella (Wahl.) Graebn. [*Arenaria rubella* (Wahl) Smith, *A. propinqua* Rich.]. Tufted with stems to 10 cm tall; herbage glandular-hairy; leaves 3-10 mm long; flowers few in an open inflorescence; sepals lance-shaped with papery margins, pointed, 3-4 mm long; petals ca. as long as sepals; capsule longer than calyx. June-Aug.

Common in sparsely vegetated, often stony soil of grasslands, river banks, cliffs, open slopes, and ridges at all elevations; East, West. Circumpolar south to CA, NV, CO.

Moehringia L., Sandwort

Moehringia lateriflora (L.) Fenzl. [*Arenaria lateriflora* L.]. Rhizomatous perennial 5-20 cm tall; herbage finely hairy; leaves thin, narrowly oblong, 1-3 cm long; flowers few on long stalks from upper leaf axils; sepals separate, rounded at the tip, 2-3 mm long; petals white with rounded tips, 4-6 mm long; styles 3; capsule globose, shorter than the sepals, opening by 6 slits on top. June-July. Fig. 116.

Common in moist soil of grasslands, meadows, open forest, and thickets; montane and subalpine; East, West. AK to Newf. south to CA, NM, Man., PA.

Sagina L., Pearlwort

Small, tufted, glabrous, short-lived perennial herbs; leaves linear; flowers few, on stalks from leaf axils; sepals separate; petals white or absent; styles 4-5; capsule opening by 4-5 slits on top.

1. Sepals mostly 5, usually green; petals mostly 5 .... *S. saginoides*
1. Sepals mostly 4, purple-margined; petals mostly 4 .... *S. nivalis*

Sagina nivalis (Lindlbl.) Fries. Stems numerous, to 5 cm tall; leaves rolled; sepals 4, purplish, 1-2 mm long, erect when mature; petals mostly 4, 1-2 mm long; capsule 2-3 mm long.

Rare in wet, sparsely vegetated soil of shaded alpine cliffs along the Divide. Circumpolar south to MT.

Sagina saginoides (L.) Karst. Stems 2-8 cm cm tall; leaves 5-10 mm long; sepals 5, 2-3 mm long, usually green; petals 5, shorter than sepals; capsules 3-5 mm long. July. Fig. 117.

Common but inconspicuous in moist, sparsely vegetated soil of meadows, cliffs and rock outcrops; montane to alpine; East, West. Circumpolar south to CA, NM.

Silene L., Campion, Catchfly

Perennial herbs; flowers often unisexual with sexes on different plants (dioecious); sepals united; calyx tubular; petals mostly white, 2- or 4-lobed mostly with 2 appendages where it widens to the flared portion; styles 3-5; ovary mounted on a short, thick stalk; capsule opening by 6-10 slits on top.

Some authorities split the genus (e.g., *Melandrium, Lychnis*) based on the number of styles, but this character often varies within the same population of a single species.

1. Flowers pink; plants forming cushions to 2 cm high .... *S. acaulis*
1. Flowers white to purplish; stems > 2 cm 2

2. Flowers usually solitary, nodding .... *S. uralensis*
2. Flowers mostly > 1, erect or ascending 3

3. Calyx glabrous 4
3. Calyx glandular 5

4. Calyx 15-20 mm long in fruit, inflated, mottled-veiny .... *S. cucubalus*
4. Calyx 8-12 mm long, not inflated, not mottled .... *S. cserei*

5. Calyx < 10 mm long .... *S. menziesii*
5. Calyx ≥ 10 mm long 6

6. Plants with unisexual flowers; disturbed areas .... *S. latifolia*
6. Plants with bisexual flowers; grasslands .... *S. parryi*

Silene acaulis L., Moss Campion. Glabrous, cushion-forming, often unisexual; leaves needle-like, mostly glabrous, 4-12 mm long, all basal; flowers solitary, nearly sessile or on stems to 2 cm high; calyx 4-8 mm long; petals rose-purple, 6-9 mm long, 2-lobed; styles 3, capsule opening by 3 slits. June-Aug. Fig. 118. Color plate 56.

Abundant in stony soil of tundra and exposed ridges and slopes near or above treeline; East, West. Our plants are var. *exscapa* (All.) D. C. Circumpolar south to OR, AZ, NM. On cool, sheltered slopes, plants are more diffuse than the bright green cushions of exposed fellfields.

Silene cserei Baumg., Bladder Campion. Glabrous, short-lived perennial, 20-80 cm tall; leaves lance-shaped; flowers numerous in a narrow inflorescence; calyx 8-11 mm long; petals 2-lobed; styles 3; capsules slightly longer than calyx. July.

Rare, along montane roads and banks; East, West. Introduced from Europe.

Silene cucubalus Wibel, Bladder Campion. Similar to *S. cserei*; leaves narrowly lance-shaped, 3-8 cm long; inflorescence branched, open; calyx inflated and up to 2 cm long in fruit. July.

Collected once in disturbed soil along the Camas Road. Introduced from Europe.

Silene latifolia Poir., White Campion [*S. pratensis* (Rafn) Godron & Gren. *Lychnis alba* Mill.]. Plants unisexual, stems 40-70 cm tall; herbage hairy, glandular above; leaves narrowly oblong, 2-8 cm long; inflorescence open; calyx 15-20 mm long; petals 2-lobed with a 4-lobed appendage, 2-4 cm long; styles 5; capsule ovoid, 10-15 mm long, opening by 5 slits. July-Aug.

Uncommon in disturbed soil of montane meadows, lawns, and roadsides; East, West. Introduced from Europe. The male calyx has 10 veins, the female 20. Flowers open in evening and close in late morning.

Silene menziesii Hook. Rhizomatous, mostly dioecious herbs with trailing stems 5-30 cm long; herbage hairy, glandular above; leaves narrowly lance-shaped, 2-5 cm long; flowers several in a small inflorescence; calyx 5-8 mm long; petals 6-8 mm long, 2-lobed; styles 3; capsule as long as calyx. June-Aug.

Uncommon in open montane forest, aspen groves, and along streams; East, West. Both var. *menziesii* with eglandular lower stems and var. *viscosa* (Greene) Hitchc. & Mag. with glandular stems occur in the Park. AK to CA, NM.

Silene parryi (Wats.) Hitchc. & Mag. Stems 20-50 cm tall from a branched rootcrown; herbage short-hairy, glandular above; leaves narrowly oblong, 3-8 cm long; inflorescence narrow, 3- to 7-flowered; calyx purple-veined,

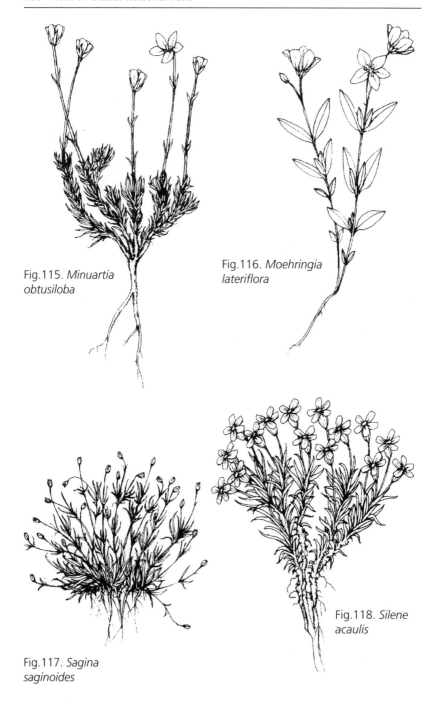

Fig.115. *Minuartia obtusiloba*

Fig.116. *Moehringia lateriflora*

Fig.117. *Sagina saginoides*

Fig.118. *Silene acaulis*

12-16 mm long; petals often purple-tinged, ca. 16 mm long, 4-lobed; styles 3; capsule, longer than calyx. July-Aug. Fig. 119.

Grasslands, meadows, and open slopes at all elevations; common East, less common West. B.C., Alta south to WA, ID, WY. Reports of *S. douglasii* Hook. and *S. repens* Pers. are referable here.

Silene uralensis (Rupr.) Bocq. [*Lychnis apetala* L., *Melandrium apetalum* (L.) Fenzl]. Stems to 10 cm tall from a woody, branched rootcrown; herbage short-hairy, glandular above; leaves linear-oblong, 2-4 cm long; flowers usually solitary, nodding; calyx 10-15 mm long, inflated; petals purplish, ca. 13 cm long, 2-lobed with a tooth on each side; styles 5; capsules erect, opening by 5 slits. July.

Rare in moist, stony soil of moraine, cliffs, and talus above treeline near or east of the Divide. Circumpolar south to UT, CO.

Spergularia J. & C. Presl, Sand Spurry

Spergularia rubra (L.) J. & C. Presl. Glandular-hairy annual with prostrate stems to 15 cm long; leaves 3-5 mm long, linear with sharp-pointed tips and small leaves in the axils; flowers few, long-stalked from upper leaf axils; sepals separate, 3-4 mm long, white-margined; petals ca. 4 mm long, light purple; styles 3; capsule ovoid, ca. as long as calyx. May-July.

Uncommon in disturbed soil along roads and trails in the montane zone both sides of the Divide. Introduced from Europe.

Stellaria L., Chickweed

Small annual or perennial herbs; leaves mostly without petioles; sepals separate; petals white and 2-lobed or absent; styles usually 3, capsule ovoid, opening by 6 slits. References: Chinnappa & Morton (1991), Morton & Rabeler (1989).

1. Herbage glandular; plants of shifting limestone talus .... **S. americana**
1. Plants mostly glabrous, usually not in talus 2

2. Lower leaves with a petiole nearly 1/2 as long as the blade; annual of disturbed areas .... **S. media**
2. Perennial; petioles very short or lacking 3

3. Flowers solitary in leaf axils 4
3. Flowers in a compact or open, terminal inflorescence 6

4. Leaves linear to narrowly lance-shaped, bluish-waxy .... **S. longipes**
4. Leaves ovate to broadly lance-shaped, not bluish-waxy 5

5. Leaves glabrous with wavy margins ....**S. crispa**
5. Leaf margins not wavy but with hairs near base............................ **S. obtusa**

6. Small leaf-like bracts of inflorescence papery-membranous 7
6. Bracts of inflorescence green 9

7. Petals minute or absent .... **S. umbellata**
7. Petals as long or longer than sepals 8

8. Lower branches of inflorescence spreading; capsule yellow-green when ripe
.... *S. longifolia*
8. Branches of inflorescence erect or ascending; mature capsule purplish-black
.... *S. longipes*
9. Petals minute or absent 10
9. Petals longer than sepals 11
10. Leaves narrowly elliptic, to 25 mm long; near or above treeline ....
.... *S. calycantha*
10. Leaves narrowly lance-shaped, 2-4 cm long; montane or subalpine ....
.... *S. borealis*
11. Plants freely branched at the base with prostrate stems often forming mats
.... *S. crassifolia*
11. Stems erect or ascending, not mat-forming .... *S. longipes*

Stellaria americana (Porter) Standley. Glandular-hairy perennial with trailing stems, 5-15 cm long from rhizomes arising from a taproot; leaves ovate, 1-3 cm long; flowers several in a compact inflorescence; sepals ovate, 3-5 mm long; petals 4-10 mm long; capsule ca. as long as calyx. July-Aug. Fig. 120.

Common in limestone talus near or above treeline; East. Endemic to MT and adjacent Alta. The long underground stems (rhizomes) extend upslope, covered by the shifting talus.

Stellaria borealis Bigelow [*S. alpestris* Fries, *S. calycantha* (Ledeb.) Bong. var. *sitchana* (Steud.) Fern.]. Glabrous perennial with weak stems to 40 cm long; leaves narrowly lance-shaped, 2-4 cm long; flowers few in an open inflorescence; sepals lance-shaped, 2-4 mm long with transparent margins; petals short or absent; capsule ovoid, longer than sepals. July-Aug.

Moist montane forests and stream margins; common West, less so East. Ssp. *sitchana* (Steud.) Piper has reflexed fruit stalks. Ssp. *borealis* with spreading fruit stalks is also reported for the Park (Morton and Rabeler 1989). Circumboreal south to CA, UT, CO.

Stellaria calycantha (Ledeb.) Bong. Glabrous perennial with weak stems to 15 cm long; leaves narrowly elliptic, 10-25 mm long; flowers few in an open inflorescence; sepals lance-shaped, 1-3 mm long, papery-margined; petals shorter or absent; capsule globose, longer than sepals. July-Aug.

Common but inconspicuous in wet soil of meadows and cliffs, subalpine and alpine; East, West. Circumpolar south to CA, UT, WY.

Stellaria crassifolia Ehrh. Glabrous perennial with weak stems 5-15 cm long, forming loose mats; leaves narrowly lance-shaped, 6-16 mm long; flowers few; sepals 2-4 mm long, lance-shaped; petals shorter or absent; capsules to 2 times as long as sepals.

Reported for wet thickets near St. Mary (Standley 1921). Circumpolar south to B.C., ID, CO.

Stellaria crispa Cham. & Schlecht. Mostly glabrous perennial with weak, mat-forming stems 8-30 cm long; leaves narrowly ovate, 10-20 mm long with wavy margins; 1-2 stalked flowers in leaf axils; sepals 2-4 mm long, lance-shaped with membranous margins; petals minute or absent; capsule ca. twice as long as calyx. June-July.

Common but inconspicuous in wet, sparsely vegetated soil of thickets and open forest, especially along trails; montane and subalpine; East, West. AK to CA, ID, MT.

Stellaria longifolia Muhl. Glabrous perennial with lax stems to 50 cm long; leaves strap-shaped, 15-60 mm long; flowers several in an open inflorescence with papery bracts; sepals lance-shaped, 3-4 mm long; petals and capsules longer than sepals. July.

Uncommon in wet soil along montane streams and wetlands; East, West. Circumpolar south to CA, NM, MO, SC.

Stellaria longipes Goldie [*S. laeta* Rich., *S. monantha* Hult.]. Glabrous, rhizomatous perennial, 5-30 cm tall; leaves narrowly lance-shaped, 1-3 cm long, often bluish-waxy; flowers 1-2 per stem; sepals narrowly ovate, 3-5 mm long; petals ca. as long as sepals; capsule longer than calyx. July. Fig. 121.

Var. *longipes*, > 15 cm high with papery-margined inflorescence bracts, is common in moist soil of meadows and along streams; alpine and subalpine. Var. *altocaulis* (Hult.) Hitchc. [=*S. monantha*] is mostly < 10 cm tall without papery-margined inflorescence bracts and occurs in moist soil or talus near or above treeline; East, West. Circumpolar south to CA, NM, SD, MN, NY. Chinnappa & Morton (1991) believe that these varieties do not merit recognition, but they seem morphologically distinct and to have ecological significance in Glacier Park.

Stellaria media (L.) Cyrill. Common Chickweed. Annual with trailing stems to 50 cm long, short-hairy in longitudinal lines; leaves 1-3 cm long, petiolate, ovate, glabrous with marginal hairs; flowers several in a compact terminal inflorescence and solitary from leaf axils; sepals 4-6 mm long, lance-shaped, hairy; petals shorter or absent; capsule longer than calyx. June-July.

Standley (1921) reports it around Many Glacier Hotel. Introduced from Eurasia.

Stellaria obtusa Engelm. Glabrous, mat-forming perennial with prostrate stems to 15 cm long; leaves 4-8 mm long, ovate with long hairs on lower margins; 1-2 stalked flowers from leaf axils; sepals narrowly ovate with blunt tips, 2-3 mm long; petals absent; capsule as long as calyx.

Rare in moist soil, especially near streams; montane to subalpine; East, West. B.C. to Alta. south to CA, CO.

Stellaria umbellata Turcz. Glabrous perennial with slender stems, 4-10 cm long; leaves lance-shaped, 5-20 mm long; inflorescence open; sepals 2-3 mm long, lance-shaped with membranous margins; petals absent or tiny; capsule 4-5 mm long. Aug.

Uncommon in moist, often mossy soil of cliffs and along streams near or above treeline near the Divide. AK to CA, CO; Eurasia.

CELASTRACEAE: BITTERSWEET FAMILY

Paxistima Raf. [Pachistima]
Paxistima myrsinites (Pursh) Raf., Mountain Lover, Mountain Box. Glabrous, evergreen shrub with prostrate to ascending, 4-ridged stems 20-80 cm high; leaves shiny, deep green, opposite, elliptic with toothed margins, short-petiolate; flowers dish-shaped, in leaf axils; sepals 4, petals 4, 1-2 mm long, maroon; stamens 4; style 1; ovary imbedded in a central disk; fruit an ovoid capsule, 3-4 mm long. May-June. Fig. 122. Color plate 42.

Montane and subalpine forests; abundant West, common East. B.C., Alta. south to CA, AZ, Mex.

CHENOPODIACEAE: GOOSEFOOT FAMILY

Annual or perennial herbs (ours); herbage sometimes mealy, covered with tiny white flakes; leaves undivided, alternate, petiolate; flowers small, greenish; sepals usually 5; petals absent; stamens 1-5; styles 1-5; ovary enclosed in the calyx; fruit thin-walled, 1-seeded.

1. Leaves linear, fleshy, round in cross section, ≤ 2 mm wide 2
1. Leaves strap-, lance-, or arrow-shaped, > 2 mm wide 3

2. Leaves and bracts spine-tipped .... **Salsola**
2. Leaves not spine-tipped .... **Suaeda**

3. Seed enclosed in a triangular, envelope-like pair of bracts .... **Atriplex**
3. Seed partially exposed or enclosed only by the calyx 4

4. Calyx of 1 sepal; seed mostly exposed .... **Monolepis**
4. Sepals 5; seed mostly enclosed by the calyx .... **Chenopodium**

Atriplex L., Orache
Atriplex patula L., Fat-hen, Spear Oracle [*A. hastata* L.]. Annual with prostrate to erect stems 5-30 cm long; herbage mealy when young; leaves lance-shaped with few-toothed margins, the basal teeth often lobe-like; flowers unisexual, sexes mixed in distinct clusters in a narrow inflorescence; female flowers with 2 triangular bracts sandwiching the fruit, 4-12 mm long at maturity.

Standley (1921) collected a few small plants around ponds near East Glacier. Our plants are var. *hastata* (L.) Gray. Much of N. America, Eurasia.

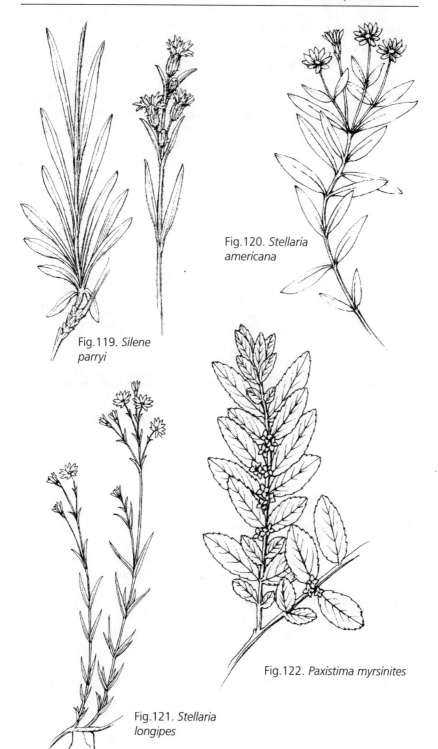

Fig.119. *Silene parryi*

Fig.120. *Stellaria americana*

Fig.121. *Stellaria longipes*

Fig.122. *Paxistima myrsinites*

Chenopodium L., Goosefoot

Annuals (ours); herbage often mealy; leaves mostly lance- to arrow-shaped, petiolate, often with toothed margins; flowers bisexual, densely clustered in narrow inflorescences from leaf axils; fruit lense-shaped, ca. 1 mm long.

    C. fremontii and *C. leptophyllum*, our native, dry grassland species, are similar, but the former has lance-shaped leaves, while those of the latter are linear. *C. simplex* (Torr.) Raf. [=*C. gigantospermum* Ael.], with leaves > 5 cm long, is reported to be just west of the Park.

1. Leaves with entire margins, linear to narrowly lance-shaped 2
1. Some leaves with toothed or lobed margins, lance- to arrow-shaped 3

2. Leaves strap-shaped, > 4 times as long as wide .... *C. leptophyllum*
2. Leaves lance-shaped, < 4 times as long as broad .... *C. fremontii*

3. Herbage glandular, aromatic... *C. botrys*
3. Herbage glabrous to mealy, not glandular.. 4

4. Flowers in red, berry-like clusters .... *C. capitatum*
4. Flowers not in red, berry-like clusters 5

5. Leaves green above but mealy-white beneath; stems usually prostrate ....
.... *C. glaucum*
5. Both leaf surfaces about the same color; stems usually erect...................... 6

6. Seeds horizontal in the inflorescence .... *C. album*
6. Seeds vertical (on edge) in the inflorescence .... *C. rubrum*

Chenopodium album L., Lamb's Quarters. Stems erect, to 60 cm high, often reddish; leaves somewhat mealy below, 3-10 cm long; flower clusters small, in narrow inflorescences from leaf axils; seed black, borne horizontally. July-Aug.

    Uncommon in disturbed ground near roads, trails, campgrounds and buildings; montane; East, West. Introduced from Eurasia.

Chenopodium botrys L., Jerusalem Oak. Stems to 50 cm tall; herbage glandular and aromatic; leaves pinnately lobed and toothed, 1-6 cm long; flowers numerous in a long, branched, narrow inflorescence; seed black, borne horizontally. Aug-Sept.

    Uncommon, collected in disturbed soil at West Glacier and gravel bars of North Fork Flathead River. Introduced from Eurasia.

Chenopodium capitatum (L.) Asch., Strawberry Blite. Stems to 50 cm tall; herbage glabrous; leaf blades triangular, to 6 cm long, petiolate; flowers red, in juicy, globose clusters, 5-10 mm wide; seeds black, borne vertically. July-Aug.

    Uncommon but conspicuous in sparsely vegetated soil of montane meadows, roadsides, and gravel bars; East, West. AK to N. S. south to CA, NM, MN, NJ; Eurasia.

Chenopodium fremontii Wats. [*C. atrovirens* Rydb.]. Stems to 50 cm tall; leaves mealy below, 1-6 cm long with mostly entire margins; inflorescences slender, not long; seeds borne horizontally. Fig. 123.

Reported for open slopes in grasslands near East Glacier (Standley 1921). Our plants are var. *atrovirens* (Rydb.) Fosberg. B.C. to Sask. south to CA, NM, TX.

Chenopodium glaucum L. [*C. salinum* Standley]. Stems prostrate to ascending, 10-30 cm long; leaf blades mealy beneath, 2-3 cm long, margins toothed; seeds brown, borne horizontally or vertically. June-July.

Drying mud around ponds near East Glacier (Standley 1921). Our plants are ssp. *salinum* (Standl.) Aellen. AK to Ont. south to CA NM, KS; Eurasia. Sometimes mistaken for *Monolepis nuttalliana*, the leaves of which are entire above and green on both surfaces.

Chenopodium leptophyllum (Moq.) Wats. Stems to 40 cm tall; herbage glabrous or mealy; leaves 1-3 cm long, strap-shaped with entire margins; seeds black, borne horizontally.

Sandy, sparsely vegetated soil of grasslands, reported for East Glacier (Standley 1921). Our plants are var. *leptophyllum*. B.C. to Man. south to CA, NM.

Chenopodium rubrum L. [*C. humile* Hook.]. Stems erect, often reddish, 20-80 cm tall; herbage glabrous; leaves 3-10 cm long; seeds brown, borne vertically.

Sparsely vegetated, saline soil around ponds near East Glacier (Standley 1921). Circumboreal through much of N. America. Similar to the more common *C. album*, which has horizontal seeds.

Monolepis Schrad., Poverty Weed

Monolepis nuttalliana (Schultes) Greene. Glabrous annual, branched from the base with prostrate or ascending, often reddish stems 10-25 cm long; leaf blades lance-shaped with 2 basal lobes, 1-5 cm long, petiolate; flowers clustered in upper leaf axils, bisexual or female; calyx of 1 sepal; ovary with 2 stigmas; seed disk-like, ca. 1 mm wide.

Uncommon in vernally moist, sparsely vegetated soil of grasslands and wetlands; montane, East. B.C. to Man. south to CA, NM, and MO. See note under *Chenopodium glaucum*.

Salsola L.

Salsola pestifer Nels., Tumbleweed, Russian Thistle [*S. kali* L., *S. australis* R. Br., *S. iberica* Sennen & Pau]. Spiny annual branched from the base; stems to 80 cm high, purple-striped; lower leaves linear, 1-3 cm long, sharp-pointed; upper leaves shorter, stiff, spine-like; flowers bisexual, solitary in leaf axils with 2 spine-like bracts; sepals 5, winged on back, becoming purplish; fruit cup-shaped, 2-5 mm across.

Sporadic and not persistent in disturbed soil near roads and buildings. Introduced from Eurasia.

Suaeda Forsk., Sea Blite.

Suaeda calceoliformis (Hook.) Moq. [*S. depressa* (Pursh) Wats., *Dondia depressa* (Pursh) Britt.]. Annual; stems erect, 10-30 cm tall; herbage glabrous; leaves linear, fleshy, 1-3 cm long; flowers bisexual, in small clusters in leaf axils; 5 sepals unequal, 1-2 mm long; styles 2; seeds black, most born horizontally.

Sparsely vegetated, saline soil around wetlands, reported for the East Glacier area (Standley 1921). Yuk. to Newf. south to CA, TX, MO.

CONVOLVULACEAE: MORNING-GLORY FAMILY

Convolvulus L., Bindweed

Convolvulus arvensis L., Field Bindweed. Rhizomatous perennial with prostrate or twining stems up to 1 m or more long; leaves usually hairy, arrow-shaped, 2-6 cm long; 1-2 stalked flowers from leaf axils, each subtended by 2 ovate, leaf-like bracts; corolla funnel-shaped with 5 united lobes, 15-25 mm long, white or pink; calyx 5-lobed; stamens 5; fruit an ovoid capsule, 5-7 mm long. Fig. 124.

Disturbed meadows and open forest. Introduced from Europe.

CORNACEAE: DOGWOOD FAMILY

Cornus L., Dogwood

Shrub or subshrub; leaves opposite or whorled, elliptic with entire margins and prominent veins; flowers small, white, clustered in hemispheric inflorescences; petals 4 separate; sepals 4, inconspicuous; stamens 4; ovary superior, with 1 style; fruit a berry.

1. Creeping subshrub to 20 cm tall; leaves whorled .... *C. canadensis*
1. Shrub > 1 m tall; leaves opposite .... *C. sericea*

Cornus canadensis L., Bunchberry. Glabrous, rhizomatous subshrub to 20 cm high; leaves 2-6 cm long, 4-7 whorled at stem tips, usually smaller pair below; flowers in a dense head subtended by 4 white, ovate, petal-like bracts, 1-2 cm long; petals 1-2 mm long; fruit red, 6-8 mm long. June-July. Fig. 125. Color plate 14.

Abundant in deep, moist, montane and lower subalpine forest; West, local in protected valleys East. AK to Greenl. south to CA, NM, OH, NJ; Asia.

Cornus sericea L., Red Osier Dogwood [*C. stolonifera* Michx.]. Shrub with prostrate to erect, reddish stems 1-3 m tall; leaf blades to 10 cm long, hairy beneath, petiolate; flowers clustered in a branched, nearly flat-topped inflorescence; petals 2-4 mm long; fruit white, 1-seeded, 5-7 mm long. May-July. Fig. 126.

Common in moist, montane forest, thickets and avalanche slopes, especially along streams; East, West. Our plants are ssp. *stolonifera* (Michx.) Fosb. AK to Newf south to Mex. Leaves turn red in autumn. Blackfeet used the inner bark in a tobacco mixture.

CRASSULACEAE: STONECROP FAMILY

Sedum L., Stonecrop

Glabrous perennial herbs; leaves alternate, succulent with entire margins; flowers in compact, branched inflorescences; petals 5, separate, erect or ascending; sepals 5, united below; stamens 8-10; ovaries 4-5, united below; fruit a group of dry, many-seeded capsules.

Lower leaves of *S. stenopetalum* have often fallen by flowering time.

1. Flowers red or purple; leaves oblong, flattened .... *S. roseum*
1. Flowers yellow; leaves linear, round in cross section 2
2. Leaves with ridge-like lower midvein; plantlets in upper axils ....
.... *S. stenopetalum*
2. Leaves without prominent midveins; plantlets lacking .... *S. lanceolatum*

Sedum lanceolatum Torrey. Stems 5-15 cm high from rhizomes; leaves 5-20 mm long, nearly linear, many clustered in basal rosettes; flowers bisexual; petals yellow, lance-shaped, 5-7 mm long; capsules 5-6 mm long. June-Aug.

Common in rocky, often shallow soil of cliffs and open slopes in the montane to alpine zones; East, West. AK to Sask. south to CA, NE.

Sedum roseum (L.) Scop., King's Crown, Red Orpine, Roseroot [*S. integrifolium* (Raf.) Nels., *Tolmachevia integrifolia* (Raf.) Löve & Löve, *Rhodiola integrifolia* Raf.]. Stems to 10 cm tall from a branched rootcrown; leaves oblong, 7-15 mm long; flowers red, unisexual, different sexes on separate plants; petals ca. 3 mm long; capsules ca. 4 mm long. July-Aug.

Moist soil of cliffs, exposed ridges, talus, and turf near or above treeline; East, West. Our plants are ssp. *integrifolium* (Raf.) Hult. AK to Greenl. south to CA, CO, NY; Asia.

Sedum stenopetalum Pursh [*S. douglasii* Hook.]. Similar to *S. lanceolatum* but leaves with prominent midrib below; upper leaves with small plantlets in the axils. June-July. Fig. 127.

Uncommon in vernally moist, stony, shallow soil of cliffs and grasslands; montane and lower subalpine; East, West. B.C., Alta south to CA, MT. Locally common in the North Fork prairies.

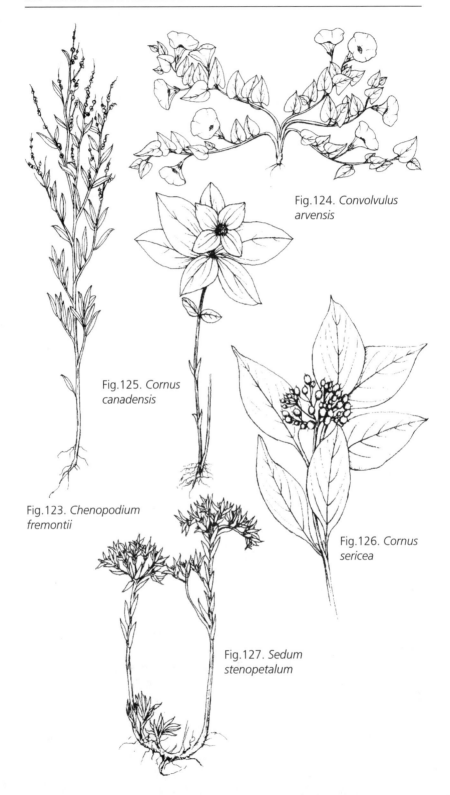

Fig.124. *Convolvulus arvensis*

Fig.125. *Cornus canadensis*

Fig.123. *Chenopodium fremontii*

Fig.126. *Cornus sericea*

Fig.127. *Sedum stenopetalum*

DROSERACEAE: SUNDEW FAMILY

Drosera L., Sundew

Insectivorous, perennial herbs; leaves basal, petiolate, the upper surface with long, purple, gland-tipped hairs; flowers bisexual, in a narrow, 1-sided inflorescence; petals white, usually 5, separate; sepals 5; stamens 4-20; styles 3-5, deeply 2-lobed; fruit a many-seeded capsule.

These plants occur only in montane sphagnum fens. *D. linearis* Goldie, with longer, more linear leaves than *D. anglica*, occurs just south of the Park.

1. Leaf blades oblong to linear, 1-3 cm long .... ***D. anglica***
1. Leaf blades orbicular, to 12 mm long .... ***D. rotundifolia***

Drosera anglica Huds. Leaves erect to prostrate; blades narrowly oblong, 1-3 cm long; petiole 2-6 cm long; flower stem 6-20 cm tall; petals ca. 6 mm long; stamens slightly shorter; styles 4-5. July-Aug.

Rare West. Circumboreal south to CA, WY, Sask.

Drosera rotundifolia L. Similar to *D. anglica* but leaf blades orbicular, 6-10 mm long; petioles 1-4 cm long; styles mostly 3. July-Aug. Fig. 128.

Locally common West and in the Waterton Valley. Circumboreal south to CA, MT, ND, and FL.

ELAEAGNACEAE: OLEASTER FAMILY

Shrubs (ours) with star- or flake-shaped hairs; leaves lance-shaped to narrowly elliptic with entire margins, petiolate; flowers bisexual or unisexual; calyx 4-lobed, petals absent; style 1; fruit 1-seeded.

1. Leaves opposite; berry orange, juicy .... **Shepherdia**
1. Leaves alternate; berry silvery, dry .... **Elaeagnus**

Elaeagnus L.

Elaeagnus commutata Bernh. ex Rydb., Silverberry, Wolf Willow. Plants to 4 m high from extensively branched rootstocks; herbage silvery; leaves alternate, 2-7 cm long, petiolate; flowers bisexual, 1-3 nodding in leaf axils; calyx funnel-shaped, 6-14 mm long, yellowish; stamens 4; fruit a dry, mealy berry 9-12 mm long. June-July. Fig. 129.

Common along montane streams and rivers; East, West. AK to Que. south to ID, UT, ND, MN. Flowers are sweetly scented. Blackfeet used the fruits as decorative beads.

Shepherdia Nutt., Buffaloberry

Shepherdia canadensis (L.) Nutt. Canada Buffaloberry, Soapberry [*Lepargyrea canadensis* (L.) Greene]. Plants to 3 m high with brown-scurfy branches; leaves opposite, 2-7 cm long, brown and white mealy-scurfy beneath, short-petiolate; flowers yellowish, clustered below leaves, unisexual with different sexes on separate plants; calyx 1-2 mm long, lobes spreading; stamens 8; fruit a juicy orange, elongate berry, 4-6 mm long. May-June. Fig. 130.

Common in montane and lower subalpine forest and higher grasslands both sides of the Divide. AK to Newf. south to OR, NM, SD, OH. Most grasslands supporting *S. canadensis* are probably maintained by fire.

ERICACEAE: HEATH FAMILY

Perennial herbs or shrubs; leaves mostly alternate, not lobed or divided; flowers dish- to urn-shaped; sepals and petals mostly 5; stamens 5-10; ovary superior with 1-12 cells; fruit a berry or capsule.

This family is often split into three: Monotropaceae (non-green genera), Pyrolaceae (*Chimaphila, Moneses, Orthilia, Pyrola*), and Ericaceae with the woody genera. Members of the Monotropa group obtain their nourishment indirectly from canopy trees by parasitizing their mycorrhizal fungi. *Pyrola uniflora* and *P. secunda* are herein placed in *Moneses* and *Orthilia* respectively. *Rhododendron albiflorum* Hook., with nodding white flowers ca. 2 cm across, has never been collected in Glacier National Park although it occurs to the south, west, and north. It should be looked for in the North Fork Flathead drainage on cool, subalpine slopes with *Menziesia ferruginea*. Reference: Haber & Cruise (1974).

1. Stems and leaves white, yellow or brown, without chlorophyll (leafless forms of Pyrola could key here, but stems are usually somewhat green) 2
1. Plants with chlorophyll, at least in the stems 3
2. Petals united into an urn-shaped corolla; stems glandular, mostly > 25 cm tall .... **Pterospora**
2. Petals separate; stems < 25 cm tall, glabrous .... **Monotropa**
3. Shrubs or subshrubs, at least lower stems woody; stigma ball- or disk-shaped 4
3. Perennial herbs; stigma 4- or 5-lobed 11
4. Petals separate nearly to the base .... **Ledum**
4. Petals united at least halfway, forming an urn-, cup- or dish-shaped corolla 5
5. Ovary inferior; fruit a juicy berry .... **Vaccinium**
5. Ovary superior; fruit a capsule or dry berry 6
6. Plants mat-forming, heather-like; leaves needle-like, up to 2 mm wide 7
6. Not mat-forming; leaves wider 8
7. Leaves opposite, appressed to the stem; flowers white .... **Cassiope**
7. Leaves alternate, spreading; flowers yellow to purple .... **Phyllodoce**
8. Flowers dish-shaped; leaves with down-rolled margins .... **Kalmia**
8. Flowers urn- to cup-shaped 9
9. Erect shrubs; fruit a capsule .... **Menziesia**
9. Plants low, spreading; fruit a dry berry 10

10. Flowers in leaf axils; leaves ovate with sharp-pointed tip .... **Gaultheria**
10. Flowers in a terminal cluster; leaves oblong, rounded at the tip ....
.... **Arctostaphylos**
11. Flowers solitary .... **Moneses**
11. Flowers > 1 12
12. Leaves whorled .... **Chimaphila**
12. Leaves basal or alternate, not whorled 13
13. Flowers bell-shaped in a 1-sided inflorescence .... **Orthilia**
13. Flowers dish-shaped, not on just 1 side of the inflorescence .... **Pyrola**

Arctostaphylos Adans., Manzanita

Arctostaphylos uva-ursi (L.) Spreng., Bearberry, Kinnikinnick [*A. adenotricha* (Fern. & Macbr.) Löve et al.]. Shrub to 15 cm tall with prostrate or trailing, hairy and sometimes glandular stems; leaves leathery, shiny, oblong, 1-3 cm long; flowers in small, nodding clusters at branch tips; corolla pink, urn-shaped, ca. 5 mm long; fruit a mealy, dry red berry, 6-8 mm wide. May-June. Fig. 131.

Abundant in rather dry montane forest dominated by pine, larch, or Douglas fir; also common in grasslands, open slopes, and exposed ridges, occasional in fens; found at all elevations; East, West. Circumboreal south to CA, NM, SD, MN, VA. A number of segregate species and subspecies have been proposed based on the presence of hairs and glands on the stems. Native Americans used the leaves in tobacco smoking mixtures.

Cassiope D. Don, White Mountain Heather

Cassiope tetragona (L.) Don. Trailing, often mat-forming shrubs to 30 cm tall; leaves opposite, needle-like, evergreen, 2-4 mm long, sparsely hairy with a prominent groove on the back, pressed to the stem and overlapping; flowers nodding, solitary on long stalks from upper leaf axils; corolla 4-6 mm long, white to rose, cup-shaped with spreading lobes; stamens 8 or 10; fruit a 4- or 5- chambered capsule. July. Fig. 132.

Uncommon in moist, mossy, often organic soil where snow lies late on cliffs, meadows, and among dwarfed trees near or above treeline; East, West. Our plants are ssp. *saximontana* (Small) Porsild. Circumpolar south to WA, MT, Que. In Montana known only from the Lewis, Cabinet and Bitterroot ranges. Reports of *C. mertensiana* are referable here.

Chimaphila Pursh, Pipsissewa

Chimaphila umbellata (L.) Bart., Prince's Pine. Rhizomatous herb, woody at the base, 10-25 cm high; leaves whorled, glabrous, shiny, evergreen, 2-6 cm long, narrowly lance-shaped with toothed margins, short-petiolate; flowers dish-shaped, long-stalked, few at branch tips; petals separate, elliptic, ca. 6 mm long, pink; stamens 10, enlarged at the base; ovary 5-lobed, hemispheric; fruit a 5-celled capsule 5-7 mm wide. July-Aug. Fig. 133. Color plate 41.

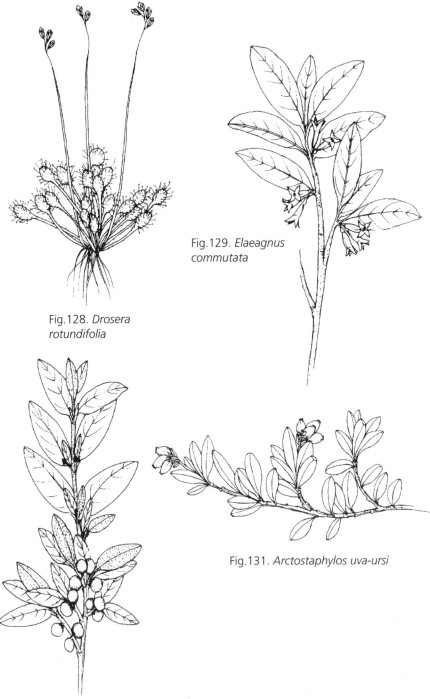

Fig.129. *Elaeagnus commutata*

Fig.128. *Drosera rotundifolia*

Fig.131. *Arctostaphylos uva-ursi*

Fig.130. *Shepherdia canadensis*

Moist, usually deep, montane and lower subalpine forests; abundant West, less common East. Our plants are ssp. *occidentalis* (Rydb.) Hult. Circumboreal south to CA, NM, SD, OH, GA. Native Americans used the leaves in tobacco mixtures. Reported to be an ingredient in commercial soft drinks. *C. menziesii* (Don) Spreng., with 1-3 flowers, occurs just to the south and west of the Park.

Gaultheria L., Creeping Wintergreen

Creeping, often mat-forming shrubs; leaves ovate, evergreen; flowers solitary in leaf axils; corolla cup-shaped; stamens 8 or 10; fruit a bright red berry-like capsule.

1. Leaves 1-2 cm long with entire margins; subalpine .... **G. humifusa**
1. Leaves 2-4 cm long with toothed margins; montane .... **G. ovatifolia**

Gaultheria humifusa (Graham) Rydb. Stems appressed to the ground; herbage mostly glabrous; leaves 5-15 mm long with entire margins; corolla ca. 3 mm long, whitish; fruit 5-7 mm wide. July-Aug. Fig. 134.

Local, uncommon and inconspicuous in moist, organic, often mossy soil of turf and open forest, often along streams and ponds in the subalpine zone near or west of the Divide. B.C., Alta. south to WA, CO.

Gaultheria ovatifolia Gray. Stems 5-20 cm long, hairy, leaves 2-4 cm long with toothed margins; corolla 3-5 mm long, white to pink; fruit 6-8 mm wide. May-June.

Uncommon and local in openings of moist montane forest, West and Waterton Valley. B.C. to CA, ID, MT.

Kalmia L., Laurel

Kalmia polifolia Wang. [*K. occidentalis* Small, *K. microphylla* (Hook.) Heller]. Shrubs with ascending stems to 60 cm tall; leaves opposite, mostly evergreen, 1-3 cm long, lance-shaped with turned-under margins, shiny green above, grayish below; flowers saucer-shaped, stalked, several at stem tips; petals united, pink; stamens 10; fruit a 5-celled capsule. May-July. Fig. 135.

Ssp. *polifolia,* > 20 cm tall with leaves 2-3 cm long and corollas ca. 15 mm across, is uncommon in wet organic soil of montane swamps and fens; West; ssp. *microphylla* (Hook.) Calder & Taylor, to 15 cm tall with smaller leaves and corollas 10-12 mm across, is abundant in wet turf, especially around streams and ponds; subalpine and alpine; East, West. AK to Newf. south to CA, CO, PA, NJ.

Ledum L., Labrador Tea

Ledum glandulosum Nutt., Trapper's Tea. Evergreen shrub 30-60 cm high with reddish, hairy twigs; leaves 1-8 cm long, elliptic, petiolate, shiny above, glandular beneath; flowers dish-shaped, long-stalked, several on branch

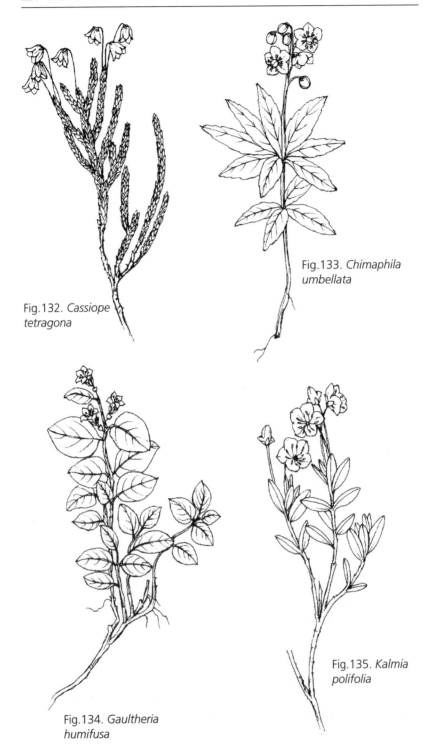

Fig.132. *Cassiope tetragona*

Fig.133. *Chimaphila umbellata*

Fig.134. *Gaultheria humifusa*

Fig.135. *Kalmia polifolia*

tips; petals separate, white, ca. 5 mm long; stamens 8-12, longer than the style; fruit an ovoid capsule 3-5 mm long. May-June. Fig. 136. Color plate 21.

Locally common in montane swamps, fens, thickets, and wet spruce forest; West. Our plants are var. *glandulosum*. B.C., Alta. south to CA, WY.

Menziesia Smith, Menziesia

Menziesia ferruginea Smith, Fool's Huckleberry [*M. glabella* Gray]. Shrub to over 2 m high with shredding bark; herbage glandular-hairy; leaves 2-6 cm long, narrowly elliptic with finely toothed margins; flowers long-stalked, several from base of new twigs; corolla urn-shaped, 4-lobed, ca. 7 mm long, dull yellow and bronze; stamens 8; fruit an ovoid capsule 5-7 mm long. June-Aug. Fig. 137. Color plate 40.

Abundant in moist montane and subalpine forest and swamps; East, West; often with *Vaccinium* spp., *Clintonia uniflora*, and *Xerophyllum tenax*. Our plants are ssp. *glabella* (Gray) Calder & Taylor. AK to CA, WY. The sticky stems distinguish this from *Vaccinium*.

Moneses Salisb., One-flowered Wintergreen

Moneses uniflora (L.) Gray, Woodnymph [*Pyrola uniflora* L.]. Glabrous perennial herb, 5-10 cm tall; leaves all basal, ovate with finely toothed margins, petiolate, 1-3 cm long; flower dish-shaped, solitary, nodding on stem tip; petals separate, ovate, waxy-white, 5-10 mm long; stamens 10; ovary 5-lobed, hemispheric; fruit a 5-chambered globose capsule 6-7 mm wide. July-Aug. Fig. 138.

Widespread but sparse in mossy soil of moist, deep forest; montane and lower subalpine; East, West. Circumboreal south to CA, NM, SD, PA. Often found under spruce.

Monotropa L.

Non-green, perennial herbs; stems fleshy, nodding, turning black with drying; leaves scale-like; petals erect, pouched at the base, separate; fruit a globose capsule.

1. Stems white; flowers solitary, > 14 mm long .... **M. uniflora**
1. Stems yellow to pink; flowers several, < 14 mm long .... **M. hypopitys**

Monotropa hypopitys L., Pinesap (*Hypopitys monotropa* Crantz, *H. latisquama* Rydb.]. Stems 10-25 cm high, yellow or pink, hairy; leaves 5-15 mm long; petals 4-5, yellowish, hairy, 10-12 mm long; stamens 8; capsule 5-8 mm wide. July-Aug. Fig. 139.

Rare in moist montane forest both sides of the Divide. Circumboreal south to CA, NM, MN, MO, FL.

Monotropa uniflora L., Indian Pipe. Stems 10-25 cm tall, white, glabrous, nodding in flower but erect in fruit; petals 5, 15-20 mm long; stamens 10; capsule ca. 6 mm wide. Aug.

Uncommon in deep, moist, montane forest; West. AK to Newf. south to CA, AZ, CO, ND, FL. Old stems often persist into the following summer.

Orthilia Raf., Sidebells Wintergreen
Orthilia secunda (L.) House [*Pyrola secunda* L.]. Glabrous, rhizomatous perennial herb, 5-20 cm high; leaves ovate with toothed margins, petiolate, blades 1-5 cm long; flowers bell-shaped, short-stalked, several nodding on 1 side of the long, narrow inflorescence; petals greenish, elliptic, 4-5 mm long; stamens 10; fruit a globose capsule ca. 5 mm wide. July-Aug. Fig. 140.

Abundant in moist, usually well shaded, montane and lower subalpine forest both sides of the Divide. Circumboreal south to CA, NM, SD, OH, VA.

Phyllodoce Salisb., Mountain Heather
Low, spreading shrubs 10-30 cm high; leaves evergreen, crowded, alternate, needle-like with down-rolled margins; flowers nodding on long stalks, several at stem tips; corolla bell-shaped; stamens 10; fruit a globose, 5-chambered capsule.

P. intermedia (Hook.) Camp., with a pale pink, sparsely glandular corolla, is a hybrid between the following two species and can sometimes be found where they occur together.

1. Corolla rose, glabrous, lobes recurved .... ***P. empetriformis***
1. Corolla yellow, glandular, lobes spreading .... ***P. glanduliflora***

Phyllodoce empetriformis (Smith) Don. Leaves 5-12 mm long; flower stalks reddish, glandular-hairy; corolla glabrous, rose, 5-8 mm long, lobes recurved; sepals ca. 2 mm long. July-Aug.

Abundant in moist subalpine and alpine turf on open slopes or among dwarf trees where snow lies late; East, West. AK to CA, ID, WY. May occur at lower elevations than the following species.

Phyllodoce glanduliflora (Hook.) Coville [*P. aleutica* (Spreng.) Heller]. Similar to *P. empetriformis* but with smaller leaves averaging a little shorter; flower stalks densely glandular; corolla dull yellow, glandular with spreading lobes. July-Aug. Fig. 141.

Abundant in turf and stony meadows near or above treeline; East, West. AK to OR, WY. Often found in drier or more exposed habitats than the former species.

Pterospora Nutt.
Pterospora andromedea Nutt., Pinedrops. Non-green perennial herb; stems reddish-brown, glandular, unbranched, 20-120 cm tall; leaves scale-like, to 15 mm long, purple; stalked flowers numerous, nodding in a long, narrow inflorescence; corolla urn-shaped, 7-8 mm long, yellowish, glabrous; stamens 10; fruit a globose capsule 8-12 mm long. Aug. Fig. 142. Color plate 36.

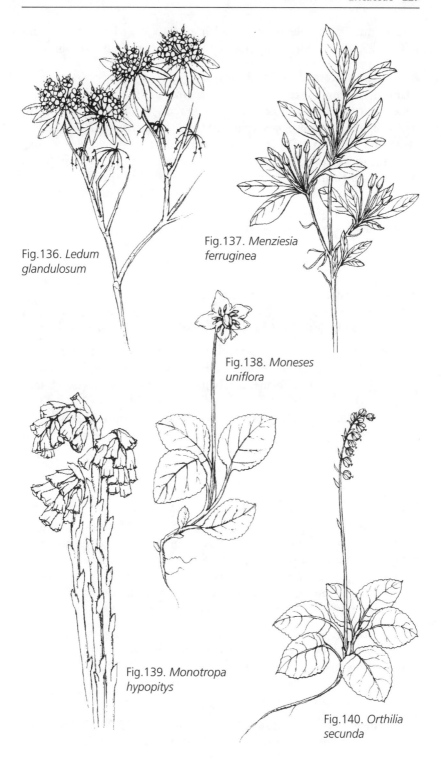

Fig.136. *Ledum glandulosum*

Fig.137. *Menziesia ferruginea*

Fig.138. *Moneses uniflora*

Fig.139. *Monotropa hypopitys*

Fig.140. *Orthilia secunda*

Uncommon in rather dry, montane forest; East, West. B.C. to Que. south to CA, CO, MI, NY. Most common in the North Fork Flathead valley.

Pyrola L., Wintergreen, Shinleaf

Glabrous perennial herbs from shallow rhizomes; leaves mainly basal, petiolate with entire to obscurely toothed margins; flowers in a narrow, terminal inflorescence; petals separate; stamens 10; style usually bent down at the base and up at the tip; ovary 5-lobed; fruit a 5-lobed, globose capsule.

Plants form strong symbiotic or parastic relationships with soil fungi. In some cases, they may become almost entirely dependent on this relationship and fail to form leaves, producing only flowering stems. These aphyllous plants may be difficult to identify to species. In the past they have been mistakenly combined into *P. aphylla* Smith.

1. Leaves with white-margined veins above ....*P. picta*
1. Leaves not variegated with white 2

2. Style straight, 1-2 mm long ....*P. minor*
2. Style curved down at the base and upward at the tip 3

3. Flowers pink; some leaf blades > 3 cm long ....*P. asarifolia*
3. Flowers whitish; leaf blades mostly < 3 cm long.......................*P. chlorantha*

Pyrola asarifolia Michx., Pink Wintergreen [*P. bracteata* Hook.]. Stems 15-30 cm high; leaf blades elliptic to heart-shaped, 3-7 cm long, shiny; petiole shorter than the blade; petals pink, 5-7 mm long; style 5-8 mm long. June-July. Fig. 143.

Common in moist forest and thickets, around streams, in swamps or fens; montane and subalpine; East, West. Ssp. *asarifolia* has sepals ca. 3 mm long and round-toothed leaf margins, while ssp. *bracteata* (Hook.) Haber has sepals > 4 mm long and sharply toothed leaves. AK to Newf. south to CA, NM, SD, MN. The two varieties occur together and do not appear to have ecological significance in the Park.

Pyrola chlorantha Sw. [*P. virens* Schweigg. & Koerte]. Stems 8-30 cm high; leaf blades ovate to elliptic, 1-4 cm long, dull green below, deep green above; petiole longer than the blade; flowers 2-8; petals greenish-white to pinkish, 5-7 mm long; style 4-7 mm long, curved down. July-Aug.

Moist, deep, montane or subalpine forest or thickets; common West, less common East. Circumboreal south to CA, NM, SD, PA.

Pyrola minor L. Stems 10-20 cm high; leaf blades ovate to orbicular, 1-4 cm long, deep green; petioles as long as the blade; flowers 5-20; petals white to pink, 4-5 mm long; style straight, 1-2 mm long. July.

Rare in moist to wet montane and subalpine forest; East, West. Circumboreal south to CA, NM, MN.

Pyrola picta Smith, Variegated Wintergreen. Stems 10-20 cm high; leaf blades ovate to oblong, 1-6 cm long, deep green and white-veined; flowers

10-25; petals yellowish or greenish white, 6-8 mm long; style ca. 5 mm long. July.

Rare in deep montane forest both sides of the Divide. B.C., Alta. south to CA, CO.

Vaccinium L., Huckleberry, Whortleberry, Blueberry

Branched shrubs; leaves petiolate, mostly deciduous; flowers 1to several in leaf axils; corolla urn-shaped; stamens 8-10; ovary below corolla (inferior); fruit a many-seeded berry.

The leaves of many species turn brilliant red in autumn. Reference: Vander Kloet (1988).

1. Leaves densely soft-hairy on veins and margins .... *V. myrtilloides*
1. Leaves glabrous to sparsely hairy 2
2. Young twigs nearly round in cross section; leaves broadest above the middle .... *V. caespitosum*
2. Twigs distinctly angled; leaves broadest at or below middle 3
3. Most plants > 30 cm high; berry usually > 8 mm wide ... *V. membranaceum*
3. Plants < 30 cm high; berry < 8 mm wide 4
4. Branches stiffly erect or ascending; twigs glabrous .... *V. scoparium*
4. Branches spreading; twigs short-hairy .... *V. myrtillus*

Vaccinium caespitosum Michx., Dwarf Huckleberry, Dwarf Bilberry. Stems spreading, to 30 cm high from rhizomes; leaves 1-4 cm long, oblong with finely toothed margins; corolla pinkish, 4-6 mm long, half as wide; berry blue, waxy, 5-8 mm wide. May-June.

Common in open, montane and lower subalpine forest, usually on gentle slopes, often under spruce; less common in grasslands and high-elevation meadows; East, West. AK to Newf. south to CA, CO, MN. Plants flower early and rarely set fruit.

Vaccinium membranaceum Dougl. ex Hook., Huckleberry [*V. globulare* Rydb.]. Stems 50-100 cm high; twigs angled; leaves 2-5 cm long, glabrous, broadly lance-shaped with toothed margins; flowers solitary in leaf axils; corolla yellow-green or pinkish, 5-6 mm long, not as wide; berry deep purple, 7-10 mm wide. May-June. Fig. 144.

Abundant in moist to rather dry, montane and subalpine forest; East, West. B.C. to Alta. south to CA, ID, WY. Berries are eaten by many animals and are an important staple for bears. Size of the crop varies from year to year, and size of the berries is also highly variable. Bushes bear more fruit where the overstory canopy is opened, perhaps by fire or avalanche. This species is harvested commercially outside the Park. *V. globulare* has often been recognized as separate from *V. membranaceum*, but characters that separate these two putative taxa merge in the Park.

Vaccinium myrtilloides Michx. Stems 10-60 cm tall with hairy twigs; leaves lance-shaped to oblong, 1-4 cm long, hairy on at least the margins and midribs; flowers in small clusters at branch tips; corolla white, tinged with pink, 4-5 mm long, just as wide; berry blue, 4-8 mm wide. June.

Locally common in moist montane forest in the West Glacier-Apgar area. B.C. to Newf. south to MT, Man., VA.

Vaccinium myrtillus L. Stems spreading, widely branched to 30 cm high; twigs green, angled; leaves 2-3 cm long with toothed margins; flowers solitary in leaf axils; corolla 10-25 mm long; berry dark blue, 5-8 mm wide. June. Fig. 145.

Abundant in montane and subalpine spruce-fir or lodgepole pine forest; East, West. B.C. to Alta. south to WA, NM; Eurasia. Our most common low *Vaccinium*.

Vaccinium scoparium Leiberg ex Cov., Grouse Whortleberry. Stems and branches erect, to 20 cm high; twigs angled; leaves 6-20 mm long, ovate with toothed margins; flowers solitary in leaf axils; corolla ca. 3 mm long; berry red, 3-5 mm wide. June-July.

Common in open subalpine forest both sides of the Divide; less common in drier montane forest; West. B.C. to Alta. south to CA, CO, and SD. The erectly branched, broom-like growth form is distinctive. This species is much more common in southwest Montana.

EUPHORBIACEAE: SPURGE FAMILY

Euphorbia, Spurge

Euphorbia esula L. Leafy Spurge. Glabrous perennial with milky sap spreading by horizontal roots; stems 20-80 cm tall, branched above; leaves alternate, linear with entire margins, 2-6 cm long; inflorescence flat-topped, umbrella-like with opposite, ovate, leaf-like bracts 12-16 mm long; flowers (actually cyathia) apparently bisexual, yellowish, cup-shaped, 2-3 mm high, 4-lobed and subtended by 2 ovate, yellow-green bracts; ovary hemispheric, 3-lobed; stamens 5; fruit a 3-lobed capsule, ca. 4 mm wide. June-July. Fig. 146.

Locally common in the North Fork prairies; also present near St. Mary. Introduced from Europe. Potentially one of the worst weeds in the Park, capable of invading undisturbed native prairie.

FABACEAE: PEA FAMILY

Annual to perennial herbs; leaves alternate, mostly pinnately or palmately (leaflets originate at 1 point) divided into distinct leaflets; leafy or membranous stipules often at the base of petiole; flowers bisexual, in unbranched inflorescences; corolla of 5 separate dissimilar petals: upper

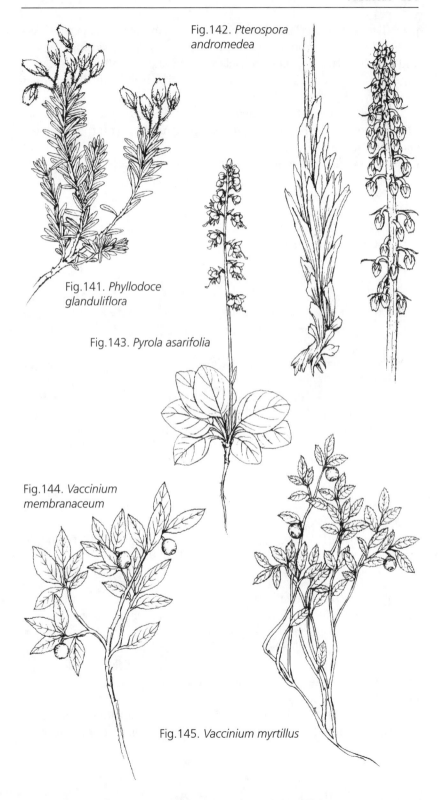

Fig.142. *Pterospora andromedea*

Fig.141. *Phyllodoce glanduliflora*

Fig.143. *Pyrola asarifolia*

Fig.144. *Vaccinium membranaceum*

Fig.145. *Vaccinium myrtillus*

petal (banner) usually largest and bent up in the middle, lower petals (keel) united along the bottom to form a canoe-like envelope enclosing the style and stamens; lateral petals are called wings; stamens 10; style 1; ovary superior; fruit a pod (legume) or a linear series of individually enclosed seeds (loment).

Identification may be difficult without both flowers and mature fruits. Members of this family form symbiotic relationships with root bacteria, allowing them to use atmospheric nitrogen and thrive in infertile soils of banks, gravel bars and moraine. References: Barneby 1989, Isely 1998.

1. Terminal leaflet reduced to a bristle or curling tendril 2
1. Terminal leaflet similar to lateral ones 3

2. Leaflets 2-8, not including stipules at base of petiole; if 8, then flowers yellowish .... **Lathyrus**
2. Leaflets 8-14; flowers blue-purple .... **Vicia**

3. Leaflets 3, not including stipules at base of petiole 4
3. Leaflets > 3 7

4. Pod > 4 cm long, curved; grassland native .... **Thermopsis**
4. Pod < 1 cm long; introduced in disturbed meadows and roadsides 5

5. Inflorescence 4-12 cm long, loosely flowered .... **Melilotus**
5. Inflorescence < 3 cm long 6

6. Pod curved, much longer than calyx .... **Medicago**
6. Pod ovoid, mostly hidden by the calyx .... **Trifolium**

7. Leaves palmate; leaflets all attached at a single point....................... **Lupinus**
7. Leaves pinnate; leaflets attached along the midrib 8

8. Pod with hooked prickles; leaflets gland-dotted below; stems leafy ....
.... **Glycyrrhiza**
8. Pod without prickles; leaves without glands or stems leafless 9

9. Pod constricted between seeds, flattened; keel petal squared off in front .....
.... **Hedysarum**
9. Pod not constricted between seeds; keel rounded or pointed 10

10. Keel petal with beak-like point at the tip; pod with sharp-pointed tip at least 1 mm long .... **Oxytropis**
10. Keel petal rounded at the tip; pod often without sharp-pointed tip ....
.... **Astragalus**

Astragalus L., Milkvetch, Locoweed

Perennial herbs; leaves pinnately divided into an odd number of leaflets; flowers in stalked clusters from upper leaf axils; corolla with banner longer than wings or keel; fruit a legume usually longer than wide.

Having mature fruit is essential for positive identification. This genus has many narrowly endemic species in western N. America, but all of ours except *A. bourgovii* are widespread. Roots of *A. australis* and *A. canadensis* were eaten by Native Americans. *A. crassicarpus* Nutt., *A. eucosmus* Robins., *A. gilviflorus* Sheld. (with 3 leaflets) and *A. miser* Dougl. have been collected at low elevations in Waterton Park (Kuijt 1982) and may occur in Glacier as well.

1. Fruit erect 2
1. Fruit spreading, nodding or pendulous 4

2. Leaves mostly > 12 cm long; pods mostly > 10 mm long .... **A. canadensis**
2. Leaves < 12 cm long; pods mostly < 10 mm long 3

3. Stems arising together from a closely branched rootcrown ..... **A. adsurgens**
3. Stems arising singly from rhizomes ....**A. agrestis**

4. Pod nearly round in cross section with a groove on 1 side 5
4. Pod flattened or round in cross section, but not grooved 6

5. Stems prostrate to ascending, often mat-forming, < 30 cm long ..**A. alpinus**
5. Stems erect, > 30 cm high ....**A. drummondii**

6. Mature pod with a basal stalk as long or longer than the calyx 7
6. Mature pod without an obvious basal stalk 8

7. Pod linear-elliptic, flattened; leaflets linear ....**A. tenellus**
7. Pod crescent-shaped, inflated; leaflets narrowly elliptic ....**A. australis**

8. Mature pod < 12 mm long ....**A. vexilliflexus**
8. Mature pod > 12 mm long 9

9. Pod obviously black-hairy.. 10
9. Pod glabrous or sparsely gray-hairy 11

10. Pod 12-15 mm long; calyx lobes > 1/2 as long as tube........... **A. bourgovii**
10. Pod 16-18 mm long; calyx lobes < 1/2 as long as tube............. **A. robbinsii**

11. Pod 2-4 mm wide; grasslands ....**A. flexuosus**
11. Pod > 5 mm wide; open forest or thickets ....**A. americanus**

Astragalus adsurgens Pallas [*A. striatus* Nutt. ex T. & G.]. Stems 15-30 cm high, numerous from a tightly branched rootcrown, closely hairy; leaflets 13-25, narrowly elliptic; inflorescences above the leaves, densely flowered; flowers erect or ascending; corolla 12-16 mm long, purple; pod 7-10 mm long, 3-4 mm wide, erect, appressed hairy, grooved on 1 side. June-Aug.

Grasslands along the east edge of the Park. Our plants are var. *robustior* Hook. AK to Man. south to WA, NM, NE, IA; Asia.

Astragalus agrestis Dougl. ex Don [*A. dasyglottis* Fisch. ex D. C., *A. goniatus* Nutt.]. Stems weak, 5-20 cm long, erect to nearly prostrate from rhizomes; herbage glabrous to sparsely hairy; leaflets 15-19, narrowly elliptic; inflorescence head-like, as high as leaves; flowers erect, 13-20 mm long; corolla purple; pod ca. 1 cm long, erect, long-ovate, black-hairy, grooved on 1 side. June-July.

Uncommon in moist soil of montane grasslands and thickets, often near streams; East. AK to Sask. south to NM, KN, IA; Asia.

Astragalus alpinus L. Stems 5-30 cm long, prostrate or ascending, often mat-forming from a widely branched rootcrown; herbage sparsely hairy; leaflets 13-25, elliptic; inflorescence 10-30 flowered, above the leaves; flowers spreading to nodding, 7-12 mm long; calyx black-hairy; corolla purple with white wings; pod pendant with a stalk as long as calyx, 8-12 mm long, 2-4 mm wide, grooved on 1 side. June-Aug. Fig. 147.

Common in moist, stony soil of tundra and open slopes near or above treeline and on montane stream gravel bars; East, West. Circumpolar south to OR, NM, Man. Gravel bar plants may form large mats as much as 60 cm across.

Astragalus americanus (Hook.) Jones. Large herbs with erect stems 50-100 cm high; herbage mostly glabrous; leaflets 7-17, narrowly elliptic; inflorescence loosely 15- to 40-flowered, ca. as high as leaves; flowers nodding, 13-15 mm long; corolla yellow-white; pod pendant, 15-25 mm long, ca. 7 mm wide, round in cross section, glabrous. July.

Uncommon in open, montane forest and thickets; East, West. AK to Ont. south to B.C., CO, Man. Often beneath aspen.

Astragalus australis (L.) Lam. [*A. aboriginum* Richards., *A. forwoodii* Wats.]. Stems 10-35 cm long, spreading from a woody, branched rootcrown; herbage long-hairy; leaflets 7-15, narrowly elliptic; inflorescence above the leaves, 10- to 20-flowered; calyx black-hairy; corolla 7-12 mm long, white with purple keel tip; pod 2-3 cm long, flattened, glabrous to sparsely hairy, pendant, crescent-shaped with a stalk protruding from the calyx. June-July.

Uncommon in sparsely vegetated, stony, often calcareous soil of grasslands and open slopes, montane to above treeline; East. Our plants are var. *glabriuscula* (Hook.) Isely. AK to Que., south to OR, CO, ND; Eurasia.

Astragalus bourgovii Gray. Stems to 30 cm long, prostrate to ascending, numerous from a branched rootcrown; herbage finely hairy; leaflets 13-23, lance-shaped; inflorescence above the leaves, loosely 5- to 10-flowered; calyx black-hairy; corolla 8-11 mm long, purple; pod black-hairy, flattened, 12-17 mm long, narrowly elliptic with a pointed tip. June-Aug. Fig. 148.

Abundant in meadows, moraine and open slopes near or above treeline, occasional in stream gravels lower, near or east of the Divide. B.C., Alta south to ID, MT.

Astragalus canadensis L. Stems erect, 30-60 cm tall from short rhizomes; herbage sparsely appressed-hairy; leaflets 13-29, narrowly elliptic, blunt; inflorescence above the leaves, crowded, to 10 cm long in flower; flowers horizontal to declined, 12-15 mm long; corolla light yellow with purple-tipped keel; calyx with black and white hairs; pods erect, crowded, 10-20 mm long, 4-5 mm wide, pointed, grooved on 1 side. July.

Uncommon in open, montane forest or thickets in the West Glacier-McDonald Lake area. Our plants are var. *mortonii* (Nutt.) Wats. B.C. to Que. south to CA, CO, TX, AR, VA.

Astragalus drummondii Dougl. ex Hook. Stems erect, 30-60 cm high from a tightly branched rootcrown; herbage long-hairy; leaflets 15-33, linear-elliptic; inflorescence above the leaves with 20-50 flowers; calyx black-hairy; flowers white, 18-25 mm long; pod pendant, linear, 2-4 cm long with a stalk as long as the calyx, grooved on 1 side. June.

Uncommon in grasslands along the east edge of the Park (Standley 1921). Alta. to Sask. south to ID, UT, NM, KS.

Astragalus flexuosus (Hook.) Don. Stems erect or ascending, 30-60 cm long from a widely branched rootcrown; herbage short-hairy; leaflets 15-21, linear-oblong with blunt tips; inflorescence above leaves, loosely 10- to 30-flowered; flowers spreading, 8-10 mm long, pink to purple; calyx mostly gray-hairy; pod 10-20 mm long, 2-4 mm wide, sharp-pointed, sparsely hairy. June-July.

Grasslands near East Glacier (Standley 1921). B.C. to Man. south to CO, KS.

Astragalus robbinsii (Oakes) Gray [*A. macounii* Rydb.]. Stems ascending, 20-50 cm high from a loosely branched rootcrown; herbage nearly glabrous; leaflets 7-17, lance-shaped; inflorescence above leaves, 7- to 12-flowered; flowers pale purple, spreading, 8-10 mm long; calyx black-hairy; pods linear-elliptic, stalked, 10-18 mm long, spreading or nodding, black-hairy. June-July.

Common in open, montane forest, especially along rivers and streams, occasional in subalpine and alpine meadows; East, West. Our plants are var. *minor* (Hook.) Barneby. AK to Newf. south to WA, ID, CO, VT.

Astragalus tenellus Pursh. Stems erect, 20-50 cm high, numerous from a branched rootcrown; herbage sparsely hairy; leaflets 9-19, linear; inflorescence within leaves, loosely 7- to 20-flowered; flowers spreading; corolla dull white with purple-tipped keel, 6-9 mm long; pods glabrous, pendant, flattened, 7-15 mm long, 3-5 mm wide with a stalk as long as calyx. June-July.

Uncommon in gravelly soil of grasslands and roads along the east edge of the Park. Yuk. to Man. south to NV, NM, NE.

Astragalus vexilliflexus Sheldon. Stems numerous, 10-25 cm long, prostrate to ascending from a loosely branched rootcrown; herbage close, short-hairy; leaflets 7-11, narrowly lance-shaped; inflorescence 5- to 10-flowered, among leaves; corolla usually purple, 6-8 mm long; pods spreading, white-hairy, 7-10 mm long, 2-3 mm wide, flattened except around seeds. May-July.

Common in stony, barren ground of montane gravel bars, open slopes and exposed ridges in the montane zone, less common near or above treeline; East, West. Our plants are var. *vexilliflexus*. B.C. to Sask. south to ID, WY, SD.

Glycyrrhiza L., Wild Licorice

Glycyrrhiza lepidota Pursh. Rhizomatous perennial herb 30-100 cm high; herbage glandular; leaves pinnately compound; leaflets 7-15, lance-shaped; flowers 10-15 mm long, in a dense, elongate cluster on long stalks from leaf axils; corolla off-white; banner only slightly bent up; wings as long as the keel; pods ovoid, 10-15 mm long, covered with hooked spines, brown at maturity. Aug. Fig. 149.

Uncommon in moist meadows and thickets along streams in the valleys; East, West. Our plants are var. *lepidota*. B.C. to Ont. south to CA, AZ, TX, MO. Native Americans used boiled roots as a tonic for the throat. The licorice of commerce is a European member of the same genus.

Hedysarum L., Sweetvetch

Perennial herbs; leaves pinnately compound; leaflets lance-shaped, with minute glands above; flowers spreading or nodding in long, narrow, stalked inflorescences from upper leaf axils; banner reflexed, as long as the wings; keel longer than either, squarish; pod flat, veiny, constricted between seeds (loment).

The constricted pods and blunt keel separate *Hedysarum* from *Astragalus* and *Oxytropis*.

1. Flowers yellow or whitish; common .... ***H. sulphurescens***
1. Flowers pink to magenta; uncommon 2

2. Calyx lobes linear, nearly equal; loment mostly cross-veined .... ***H. boreale***
2. Calyx lobes triangular, shorter above; loment net-veined .... ***H. alpinum***

Hedysarum alpinum L. [*H. americanum* (Michx.) Britt.]. Stems 20-70 cm high; leaflets 15-21, sparsely hairy; flowers pink, 11-15 mm long; calyx lobes triangular, upper shorter than lower; loment with 1-5 seeds, veins net-like.

Uncommon on gravel bars and in moist meadows of aspen parkland; East. Our plants are ssp. *americanum* (Michx.) Fedtsch. AK to Ont. south to B.C., WY, SD, VT.

Hedysarum boreale Nutt. [*H. cinerascens* Rydb.]. Stems erect to nearly prostrate, 20-50 cm high; leaflets 9-13, hairy; flowers pink or magenta, 12-15 mm long; calyx lobes linear, equal in length; loment with 2-6 seeds, veins mostly perpendicular to the axis. June-July.

Uncommon in sparsely vegetated soil of montane banks and open slopes; East. Our plants are ssp. *boreale*. AK to Newf. south to OR, NM, TX.

Hedysarum sulphurescens Rydb. Stems 20-60 cm tall from a woody, branched rootcrown; leaflets 9-17, sparsely hairy; flowers light yellow,, 12-18 mm long; calyx lobes shorter above than below; loment with 1-4 seeds, lightly net-veined. June-Aug. Fig. 150.

Common in montane grasslands and gravel bars; abundant in stony, especially calcareous soil of moraine, fellfield and turf near or above treeline, near or east of the Divide. B.C., Alta. south to WA, WY.

Fig.147. *Astragalus alpinus*

Fig.146. *Euphorbia esula*

Fig.148. *Astragalus bourgovii*

Fig.149. *Glycyrrhiza lepidota*

Lathyrus L., Sweet Pea

Rhizomatous perennial herbs with erect or weak, twining stems; leaves pinnately compound; prominent leaf-like wings (stipules) where leaves attach to the stem; terminal leaflet replaced by a curling tendril or short bristle; flowers in small, stalked clusters from leaf axils; banner strongly bent upward, shorter than the wings; pod linear, somewhat flattened, glabrous.

1. Leaflets linear, terminal one replaced by a bristle .... **L. bijugatus**
1. Leaflets elliptic, terminal one replaced by curling tendrils 2

2. Leaflets 2; flowers pinkish; uncommon weed .... **L. latifolius**
2. Leaflets 6-8; flowers yellowish; common .... **L. ochroleucus**

Lathyrus bijugatus White [*L. lanszwertii* Kellog in part]. Stems ascending, 10-30 cm long; herbage glabrous; leaflets 2-4, linear to lance-shaped, terminal one replaced by a bristle; flowers blue, 10-13 mm long, 2-3 per cluster; pod 3-4 cm long.

Local in open, park-like forest in North Fork Flathead Valley. Endemic to ID, adjacent WA and MT.

Lathyrus latifolius L. Stems 60-100 cm long, climbing; herbage glabrous; leaflets 2, elliptic, terminal one replaced by tendrils; flowers pinkish, 15-20 mm long, 5-15 per cluster; pod 6-10 cm long. July-Aug.

Collected once in open forest near West Glacier. Introduced from Europe.

Lathyrus ochroleucus Hook. Stems to 80 cm long, climbing; herbage glabrous; leaflets 6-8, elliptic, terminal one replaced by tendrils; flowers yellowish-white, 10-15 mm long, 5-10 per cluster; pod 4-7 cm long. June-July. Fig. 151.

Common in montane thickets and moist, open forest, especially aspen; East, West. AK to Que. south to WA, WY, NE, OH, PA.

Lupinus L., Lupine

Perennial herbs (ours) with branched taproots; leaves palmately compound; leaflets narrowly oblong; flowers in narrow inflorescences at stem tips; calyx 2-lipped, the upper lip 2-lobed; corolla blue with strongly upturned banner; keel crescent-shaped; pod flattened, 2-chambered, hairy.

Our species are all similar in appearance. Reports of *L. lepidus* are all referable to *L. argenteus* var. *depressus*. Standley's (1921) report of *L. leucophyllus* Dougl. ex Lindl., a more southern and western species, was probably based on *L. sericeus*. *L. laxiflorus* Lindl., with a conical-based calyx, occurs just south of the Park.

1. Banner hairy on the back, especially around and under the calyx . **L. sericeus**
1. Banner glabrous or with a few hairs on the back 2

2. Plants of moist or wet meadows; lower surface of leaves glabrous or nearly so; uncommon .... ***L. polyphyllus***
2. Plants of grassland, dry forest or open slopes; lower leaf surface hairy ....
.... ***L. argenteus***

Lupinus argenteus Pursh. Plants 10-60 cm tall, branched at the base; herbage hairy; leaflets 7-8, 2-3 cm long, more hairy below; corolla 9-11 mm long; banner glabrous or sparsely hairy on back, upturned only at the tip; pod 15-25 mm long. June-Aug. Fig. 152.

Var. *argenteus*, with an open inflorescence and > 25 cm tall, is common in grasslands and open forest in the montane zone both sides of the Divide. Var. *depressus* (Rydb.) Hitchc., with dense flower clusters and 10-20 cm tall, is common in sparsely vegetated, stony soil of meadows and exposed ridges and slopes near or above treeline and occasionally lower near or east of the Divide. B.C. to Sask. south to CA, NM, SD. Plants of var. *depressus* have been mistaken for *L. lepidus* Dougl. ex Lindl. [=*L. minimus* Dougl ex Hook.]; however, the latter has persistent flower bracts while *L. argenteus* does not. Reports of *L. wyethii* Wats are also referable here.

Lupinus polyphyllus Lindl. Stems 50-100 cm tall; herbage glabrous to sparsely hairy; leaflets 9-13, 4-10 cm long; corolla 10-16 mm long, glabrous, strongly upturned; pod 3-5 cm long. July.

Moist to wet, montane meadows; collected once near Lake McDonald. Our plants are var. *burkei* (Wats.) Hitchc. AK to Alta. south to CA, CO. This is the wild progenitor of our garden lupine.

Lupinus sericeus Pursh, Silky Lupine. [*L. flexuosus* Lindl., *L. leucopsis* Agar.]. Stems 20-60 cm tall; herbage long-hairy; leaflets 5-10, equally hairy above and below; corolla 10-12 mm long, banner hairy on basal portion, strongly upturned; pod 2-3 cm long. June-July. Color plate 6.

Common in montane grasslands, meadows and grassy, open forest; East, West. Our plants are var. *sericeus*. B.C., Alta. south to CA, NM.

Medicago L., Medick

Annual or perennial herbs; leaflets 3, with toothed margins; flowers clustered in stalked, head-like inflorescences; calyx nearly as long as corolla with equal lobes; banner erect, longer than wings and keel; pod coiled. Introduced from Europe.

1. Leaflets < 2 cm long; flowers < 5 mm long, yellow .... ***M. lupulina***
1. Leaflets > 2 cm long; flowers ca. 10 mm long, blue or white .... ***M. sativa***

Medicago lupulina L., Black Medick. Sparsely long-hairy annual with prostrate stems 10-40 cm long; leaflets broadly oblong, 5-20 mm long; flowers 10-40, yellow, 2-3 mm long; pod nearly round, black, 2-3 mm long. June-Aug.

Locally common on gravelly montane roadsides; East, West. Becoming common in grasslands near Polebridge.

Medicago sativa L., Alfalfa, Lucerne. Sparsely hairy perennial 30-80 cm tall; leaflets oblong, 2-4 cm long; flowers 20-100, purple or white, ca. 1 cm long; pod round in 2-3 spirals, 3-4 mm long. July.

Collected around horse corrals at the head of Lake McDonald. An important component of hay crops.

Melilotus Mill., Sweet Clover

Sweet-scented, branched biennials (in our area); herbage nearly glabrous; leaflets 3, oblong with toothed margins; flowers in narrow inflorescences from upper leaf axils; calyx lobes nearly equal; pod ovoid, mostly 1-seeded. Introduced from Eurasia.

1. Flowers white .... *M. alba*
1. Flowers yellow .... *M. officinalis*

Melilotus alba Medikus, White Sweet Clover. Stems 50-100 cm tall; leaflets 15-30 mm long; flowers white, 4-5 mm long; pod ca. 4 mm long. July.

Locally common along roads and other disturbed areas; East, West.

Melilotus officinalis (L.) Pallas, Yellow Sweet Clover. Similar to *M. alba*; flowers yellow, 4-6 mm long; pod ca. 3 mm long. July. Fig. 153.

Locally common along montane roads; East, West.

Oxytropis D. C., Crazyweed

Tufted perennial herbs from a taproot with a branched crown; leaves pinnately divided, all basal or nearly so; leaflets mostly lance-shaped; flowers in a narrow inflorescence on naked stems; banner erect; keel with a pointed beak at the tip, hidden by the wings; pod hairy, with a groove on 1 side and a sharp-pointed tip, mostly without a stalk.

Report of *O. parryi* Gray for the Park (Booth and Wright 1966) was based on a few-flowered collection of either *O. campestris* or *O. sericea* (Barneby, pers. comm.). *O. campestris* and *O. sericea* can be difficult to distinguish east of the Divide; mature pods or old pods from the previous season are required.

1. Leaflets in whorls of 3-4 on leaf axis .... *O. splendens*
1. Leaflets opposite each other on leaf axis 2

2. Pods pendant, flowers spreading .... *O. deflexa*
2. Pods and flowers erect or ascending 3

3. Herbage glandular as well as hairy .... *O. borealis*
3. Herbage hairy but not glandular 4

4. Flowers and pods 1-3; pods papery, inflated; flowers purple ... *O. podocarpa*
4. Flowers and pods usually > 3; pods not inflated; flowers yellowish 5

5. Pod thin-walled, yielding to pressure of fingertip when mature; calyx mostly 6-9 mm long .... *O. campestris*
5. Pod thick-walled, not yielding to pressure when mature; calyx mostly 10-13 mm long .... *O. sericea*

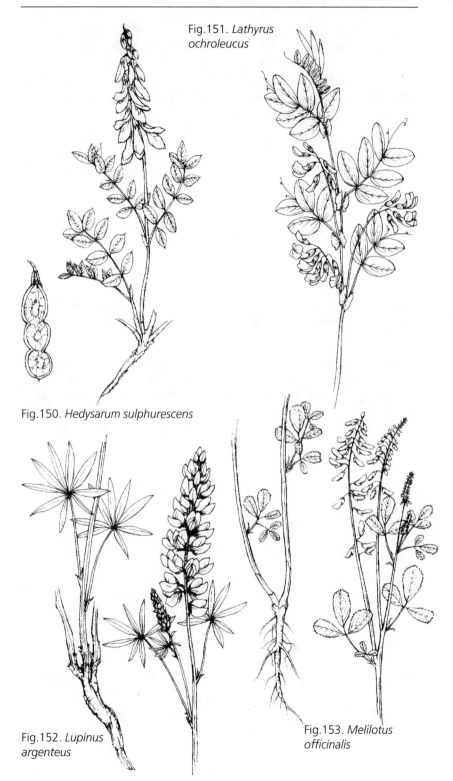

Fig.151. *Lathyrus ochroleucus*

Fig.150. *Hedysarum sulphurescens*

Fig.152. *Lupinus argenteus*

Fig.153. *Melilotus officinalis*

Oxytropis borealis D. C. [*O. viscida* Nutt.]. Stems 5-25 cm high; herbage glandular-hairy; leaflets 15-35, 8-15 mm long; flowers 3-20, purple, 11-15 mm long; pod 10-15 mm long. June-July.

Common in sparsely vegetated, gravelly soil of montane grasslands and exposed ridges; East. Our plants are var. *viscida* (Nutt.) Welsh. AK to CA, CO.

Oxytropis campestris (L.) D. C. [*O. cusickii* Greenm., *O. gracilis* (Nels.) Jones, *O. monticola* Gray, *O. alpicola* (Rydb.) Jones]. Stems 5-40 cm high; herbage long-hairy; flowers 10-20 mm long, yellowish; pods crowded, erect, 10-25 mm long, thin-walled. June-July. Fig. 154.

Var. *gracilis* (Nels.) Barneby [=*O. monticola, O. gracilis, O. c.* var. *spicata* Hook.], > 15 cm high with > 17 leaflets, is common in montane grasslands and on gravel bars; East, West. Var. *cusickii* (Greenm.) Barneby, mostly < 15 cm high with < 17 leaflets, is uncommon in stony soil of exposed ridges and slopes, montane to alpine; East. Circumboreal south to WA, CO, ND. Ellisens & Packer (1980) reported var. *columbiana* (St. John) Barneby from North Fork Flathead River, and, although these plants usually have a purple-spotted keel, they are better referred to var. *gracilis* (S. Welsh, pers. comm.). See *O. sericea.*

Oxytropis deflexa (Pall.) D. C. Stems spreading, 10-35 cm long with 1-2 leaves below; herbage long-hairy; leaflets 6-12 mm long; flowers 10-40, pale blue, 7-10 mm long; pods pendant, widely separated, 10-18 mm long. June-July.

Rare in thickets and moist, open forest, especially aspen, often near streams, montane; East. Our plants are var. *sericea* T. & G. AK to Newf. south to OR, NM, ND.

Oxytropis podocarpa Gray. Stems spreading, 1-5 cm long; herbage long-hairy; leaflets 9-25, linear, 2-8 mm long; flowers 1-3, purple, 11-14 mm long; pods erect, inflated, thin-walled, ovoid, 15-25 mm long. June.

Rare in stony, barren, usually calcareous soil of exposed ridges and slopes above treeline; East. B.C. to Lab. south to CO.

Oxytropis sericea Nutt. [*O. spicata* (Hook.) Standl., *O. sericea* var. *speciosa* (Torr. & Gray) Welsh]. Stems 5-30 cm high; herbage long-hairy; leaflets 7-17, 10-25 mm long; flowers 6-25, yellowish, 15-22 mm long; pods crowded, erect, 15-25 mm long, thick-walled and rigid. May-July.

Common in grasslands and montane to alpine exposed ridges and slopes; East, West. Our plants are var. *spicata* (Hook.) Barneby. Yuk. south to NV, NM, and TX. Barneby (pers. comm.) believes this is the common species at high elevations in the Park; Welsh as cited in Isely (1998) suggests that *O. campestris* var. *cusickii* is a high-elevation extension of *O. sericea* var. *spicata.*

Oxytropis splendens Dougl. ex Hook. Stems 10-30 cm high; herbage densely long-hairy; leaflets numerous in 7-15 whorls of 3-4, 5-20 mm long; flowers 12-35, magenta, 10-15 mm long; pods erect, 10-17 mm long, sessile, thin-walled. July.

Common in montane grasslands and river banks, East. AK to Ont. south to NM, ND, MN. The leaves with whorled, silky-hairy leaflets are distinctive.

Thermopsis R. Br., Golden Pea

Thermopsis rhombifolia (Nutt.) Richards. Rhizomatous herb 10-40 cm high, sparsely hairly to glabrous; leaves with 3 elliptic leaflets, 1-3 cm long, and 2 leaf-like wings at the base of the petiole; flowers yellow, 15-20 mm long, in a short, narrow inflorescence; pod long-hairy, flattened around the seeds, 4-8 cm long, curled. Fig. 155.

Clayey soil of grasslands along the east edge of the Park, also sparingly introduced along the road near Lake McDonald. B.C. to Man. south to CO, NE.

Trifolium L., Clover

Annual or perennial herbs; herbage glabrous to sparsely hairy; leaves with 3 oblong leaflets with toothed margins; flowers in stalked, globose clusters; banner slightly upturned, longer than wings and keel; pod often shorter than calyx.

Although native clovers are a conspicuous part of the montane flora in southern Montana and farther south in the Rocky Mountains, all of our species are introduced from Europe.

1. Flowers yellow 2
1. Flowers white or pink to purple 3
2. Flower clusters < 12 mm wide; leaf-like stipules at base of petiole, ovate ......
   .... **T. procumbens**
2. Some flower heads > 12 mm wide; stipules linear .... **T. agrarium**
3. Flowers 12-20 mm long, red to purple; heads > 25 mm wide ..... **T. pratense**
3. Flowers 7-11 mm long, white to pink; heads < 25 mm wide 4
4. Calyx glabrous; flowers mainly white .... **T. repens**
4. Calyx with hair between lobes; flowers usually pink .... **T. hybridum**

Trifolium agrarium L., Yellow Clover [*T. aureum* Poll.]. Annual, stems erect or ascending, 15-40 cm high; leaflets narrow, 1-3 cm long; inflorescence 1-2 cm long, 30- to 100-flowered; flowers yellow, 5-7 mm long; calyx mostly glabrous, upper lobes longer than lower. July-Aug. Fig. 156.

Abundant along roads and in disturbed meadows in the lower montane zone; West, uncommon East.

Trifolium hybridum L., Alsike Clover. Perennial; stems ascending or prostrate, 10-50 cm long; leaflets broad, 1-4 cm long; flowers white and pink, 7-10 mm long; calyx lobes narrow, equal, with a few hairs between. July.

Common in disturbed meadows and open forest in the montane zone; West, East.

Trifolium pratense L., Red Clover. Perennial, stems ascending or spreading, 10-60 cm long; leaflets elliptic, 2-4 cm long; flowers 50-200, purplish-red, 13-20 mm long; calyx long-hairy; lobes with stiff hairs, upper lobes shorter. July.

Common along roads and in disturbed meadows in the lower montane zone; West, East.

Trifolium procumbens L., Hop Clover. Annual; stems prostrate to ascending, 10-30 cm long; leaflets with indented tips, 1-3 cm long; flowers yellow, 4-6 mm long; calyx glabrous, upper lobes shorter.

Collected by Standley (1921) along the railroad near West Glacier. Could be confused with *Medicago lupulina*, but the latter has smaller flowers and curled pods.

Trifolium repens L., White Clover, Dutch Clover. Perennial; stems prostrate, 10-50 cm long; leaflets broad with indented tips, 1-2 cm long; flowers white, sometimes pinkish-tinged, 6-9 mm long; calyx glabrous, lobes nearly equal. July.

Disturbed meadows, trails, and roadsides; common West, rare East.

Vicia L., Vetch

Vicia americana Muhl. ex Willd. Perennial herb with twining stems; leaves with 8-14 linear to elliptic leaflets 1-3 cm long, terminated by 1-3 curling tendrils; flowers 4-10 in a loose, short-stalked cluster from leaf axils; corolla blue to purple, 15-25 mm long; banner strongly upturned; pod glabrous, flattened, 25-35 mm long. June-July. Fig. 157.

Common in montane meadows, grasslands, thickets, and open forests; East, West. Our plants are var. *truncata* (Nutt.) Brew. AK to Ont. south to CA, NM, TX, MO, OH.

FUMARIACEAE: FUMITORY FAMILY

Glabrous, annual to perennial herbs; leaves alternate, pinnately divided, waxy; flowers bisexual, sepals 2, falling early; petals 4 in 2 different pairs, separate, 1 or 2 sac-like at the base; stamens 6; style 1; stigma 2-lobed; fruit a many-seeded capsule.

1. Flowers solitary; both outer petals with recurved spurs .... **Dicentra**
1. Flowers > 1; 1 outer petal with sac-like spur .... **Corydalis**

Corydalis Medic., Corydalis

Annual to biennial herbs; leaves ovate in outline, 2-4 times pinnately divided and lobed into narrow, lance-shaped or oblong segments; flowers stalked, in a narrow inflorescence from leaf axils; corolla 2-lipped with 2 lateral

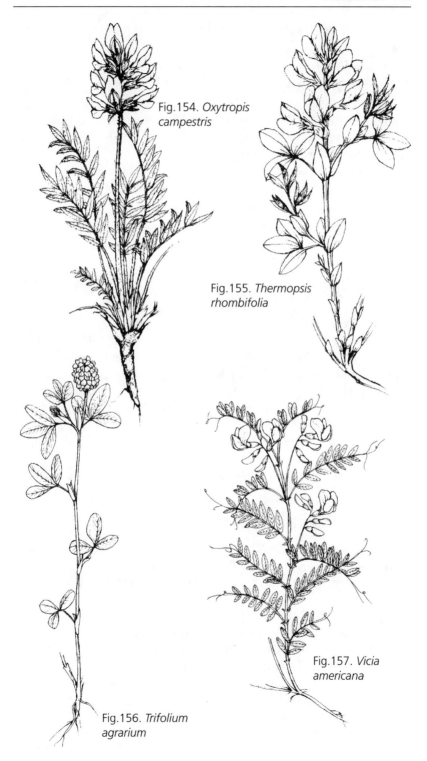

Fig.154. *Oxytropis campestris*

Fig.155. *Thermopsis rhombifolia*

Fig.156. *Trifolium agrarium*

Fig.157. *Vicia americana*

wings; upper petal with cone-shaped sac (spur) at the base; fruit a long, pod-like capsule.

1. Flowers yellow; capsules nodding, > 2 mm wide .... **C. aurea**
1. Flowers mainly pink; capsules erect, < 2 mm wide .... **C. sempervirens**

Corydalis aurea Willd., Golden-smoke [*Capnoides aureum* (Willd.) Kuntze]. Stems ascending to prostrate, 10-30 cm long; leaves to 12 cm long; flowers yellow, 12-15 mm long; pods nodding, 20-25 mm long, curved, constricted between seeds. June-Sept.

Uncommon in moist, sparsely vegetated soil of montane stream banks and steep slopes; East, West. AK to Que. south to CA, TX, LA, OH, VT.

Corydalis sempervirens (L.) Pers., Rock Harlequin [*Capnoides sempervirens* (L.) Borkh.]. Stems erect, 30-70 cm high; leaves petiolate; flowers deep pink with yellow tips, ca. 15 mm long with a short spur; capsules erect, 25-50 mm long. June-July. Fig. 158.

Rare in rocky, disturbed or eroding soil of steep slopes in open, montane forest; East, West. AK to Newf. south to B.C., MT, Man., GA.

Dicentra Bernh.

Dicentra uniflora Kell., Steer's-head. Glabrous perennial with waxy, leafless stems 4-8 cm high from tuberous roots; leaves 2-3 times pinnately divided into narrowly oblong segments, as long as the stem; flowers solitary, 12-15 mm long, white to pink; both outer petals with recurved spur; inner petals narrowly arrow-shaped, purple-tipped; capsule 20-35 mm long, 3-4 mm wide. May-June. Fig. 159.

Rare in sparsely vegetated, often shallow, vernally moist soil of open forest in the Blacktail Hills. WA to MT, south to CA, UT, WY. Rarely seen because it flowers very early in habitat that is usually still surrounded by snow.

GENTIANACEAE: GENTIAN FAMILY

Annual to perennial, usually glabrous herbs; leaves mostly opposite, undivided with entire margins; flowers bisexual; corolla dish- to funnel- or bell-shaped, 4- to 5-lobed; stamens 4-5; style 1; ovary superior; fruit a capsule.

Gentianella and Gentianopsis were previously placed in *Gentiana*. *Frasera speciosa* Griseb., with tall spikes of greenish, saucer-shaped flowers, and *Swertia perennis* L. with purple, saucer-shaped flowers, occur just south of the Park, but reports by Bamberg & Major (1968) have not been confirmed. *Halenia deflexa* (Sm.) Griseb., an annual with greenish-blue, spurred petals, occurs just west of the Park.

1. Corolla with conspicuous folds (plaits) between the lobes .... **Gentiana**
1. Corolla lacking plaits 2

2. Corolla > 2 cm long; corolla lobes with fringed margins .... **Gentianopsis**
2. Corolla < 2 cm long; corolla lobes entire, although sometimes fringed within
.... **Gentianella**

Gentiana L., Gentian

Annual to perennial herbs; corolla with folds (plaits) between the 5 lobes; capsule with a short stalk.

1. Stems > 10 cm tall; basal cluster of leaves absent 2
1. Stems < 10 cm tall; basal leaves conspicuous 3

2. Calyx lobes pointed; flowers usually > 1; montane........................ *G. affinis*
2. Calyx lobes rounded; flowers usually 1; subalpine and above ...... *G. calycosa*

3. Leaves mostly > 5 mm long; stem leaves 2-3 pairs .... *G. glauca*
3. Leaves mostly < 5 mm long; stem leaves > 3 pairs .... *G. prostrata*

Gentiana affinis Griseb., Prairie Gentian [*Pneomonanthe affinis* (Griseb.) Greene]. Perennial with clustered ascending stems 10-30 cm high from a branched rootcrown; leaves lance-shaped to ovate, 1-4 cm long; usually several flowers clustered in a leafy inflorescence; calyx lobes unequal, pointed; corolla deep blue, funnel-shaped, 2-3 cm long with fringed plaits between spreading lobes. Aug.

Moist soil of montane meadows and grasslands; reported for the East Glacier area (Standley 1921). B.C. to Man. south to CA, AZ, CO, SD.

Gentiana calycosa Griseb., Explorer's Gentian [*Pneomonanthe calycosa* (Griseb.) Greene]. Perennial with clustered ascending stems, 10-30 cm high; leaves ovate, 1-3 cm long; flowers mostly solitary on stem tips; calyx lobes low, rounded; corolla deep blue, 25-50 mm long, funnel-shaped with fringed plaits between the lobes. July-Aug. Fig. 160. Color plate 53.

Common in moist to wet soil of meadows, cliffs and open forest, often where snow lies late in the subalpine and alpine zones. Our plants are var. *obtusiloba* (Rydb.) Hitchc., the type locality of which is near Sperry Glacier. B.C. to Alta. south to CA, ID, WY.

Gentiana glauca Pallas. Small, mat-forming perennial 3-10 cm high; leaves ovate, fleshy, 5-15 mm long, clustered at the base; few to several flowers clustered in a leafy inflorescence; corolla tubular, blue, 15-20 mm long. July-Aug.

Wet soil of rock ledges near or above treeline. AK to WA, MT; Asia. The only known population in the U. S. Rocky Mountains is near Sperry Glacier, a great distance from the closest site in Banff Park. Marcus Jones (1910) first discovered our small population, and it was still extant in 2000.

Gentiana prostrata Haenke, Moss Gentian. Biennial (or short-lived perennial?) with erect to prostrate stems 2-5 cm high; leaves 3-5 mm long, ovate below, lance-shaped above; flowers solitary on stem tips; corolla light purple, funnel-shaped 1-2 cm long with spreading, triangular lobes and lobed pleats; ovary becoming longer than withered corolla when mature. Aug.

Uncommon and inconspicuous in moist, low turf above treeline near the Divide. AK to CA, UT, CO. Flowers open only on sunny afternoons and must be searched for among the taller grasses and sedges.

Gentianella Moench, Gentian

Annual or biennial herbs; corolla funnel-shaped with 4 lobes; capsule globose. Reference: Gillett 1957.

1. Corolla lobes fringed within; all elevations .... **G. amarella**
1. Corolla lobes not fringed; alpine .... **G. propinqua**

Gentianella amarella (L.) Boerner (*Gentiana amarella* L., *G. acuta* Michx.]. Stems to 40 cm tall; leaves lance-shaped, 1-4 cm long; short-stalked flowers clustered among upper leaves; corolla pale violet, 10-20 mm long, lobes fringed on the inside. June-Aug. Fig. 161.

Locally common in grassland, wet meadows, open forest, and thickets, often where disturbed; montane to alpine; East, West. Our plants are ssp. *acuta* (Michx) Gillett. Circumboreal south to CA, NM, SD, MN, VT. Sizes of populations vary widely among years.

Gentianella propinqua (Richards.) Gillett [*Gentiana propinqua* Richards.]. Stems angled, 2-10 cm high; leaves 5-20 mm long, oblong below, ovate and sessile above; stalked flowers in the leaf axils; corolla purple, 1-2 cm long, lower ones smaller, rarely fully open. Aug.

Uncommon and inconspicuous in moist turf above treeline near the Divide. AK to Newf. south to MT, WY.

Gentianopsis Ma, Fringed Gentian

Gentianopsis macounii (Holm) Iltis [*Gentianella crinita* (Froel.) Don ssp. *macounii* (Holm) Iltis, *G. procera* ssp. *macounii* (Holm) Iltis]. Annual, few-branched herb 5-30 cm high; basal leaves oblong; stem leaves linear, 1-3 cm long; long-stalked flowers from branch tips; corolla deep blue, 2-4 cm long, 4 broad lobes with fringed margins. Aug.

Known from a single calcareous fen in the northeast corner of the Park. Alta. to Que. south to MT, SD. A report of *Gentiana detonsa* Rottb. from above treeline (Bamberg and Major 1968) is unlikely. Reference: Iltis (1965).

GERANIACEAE: GERANIUM FAMILY

Annual or perennial herbs; herbage hairy; leaves deeply lobed; flowers bisexual, bowl-shaped, long-stalked in open inflorescences; sepals and petals 5, separate; stamens 5 or 10; ovary 5-lobed; 5 styles united into a column; fruit a 5-parted, 5-seeded capsule.

1. Leaf blade oblong, pinnately divided .... **Erodium**
1. Leaf blade orbicular in outline; palmately divided .... **Geranium**

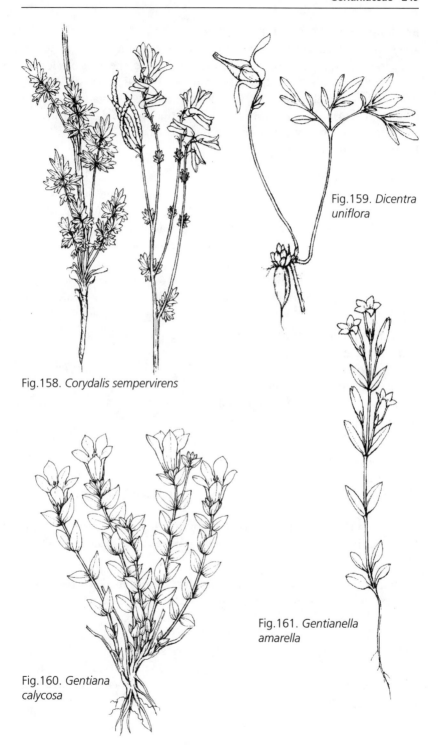

Fig.159. *Dicentra uniflora*

Fig.158. *Corydalis sempervirens*

Fig.161. *Gentianella amarella*

Fig.160. *Gentiana calycosa*

Erodium L'Her., Crane's-bill

Erodium cicutarium (L.) L'Her. Annual with spreading stems to 30 cm long; leaves mostly basal, opposite, pinnately divided, the divisions lobed; flowers in long-stalked, small clusters from leaf axils; sepals bristle-tipped, 3-7 mm long; petals pink, somewhat longer than sepals; stamens 5; style column 2-3 cm long at maturiity; styles separating, becoming twisted. June.

Uncommon in disturbed areas around roads and buildings; West. Introduced from Europe.

Geranium L., Geranium

Annual to perennial herbs; leaves nearly orbicular in outline, palmately lobed; flowers long-stalked, mostly paired in leaf axils; sepals glandular, bristle-tipped; petals broad at the tip but narrow at the base; stamens 10; capsules splitting from the base into 5 recurved sections.

1. Annual or biennial; petals < 1 cm long .... **G. bicknellii**
1. Perennial; petals > 1 cm long 2
2. Petals rose; herbage below inflorescence glandular .... **G. viscosissimum**
2. Petals white; glandular in the inflorescence only .... **G. richardsonii**

Geranium bicknellii Britt. Freely branched annual to biennial 10-50 cm high; leaves alternate below, opposite above, long-petioled; leaf blade 2-7 cm wide with 5-7 toothed lobes; petals 5-7 mm long, deep pink, obovate; mature capsule ca. 2 cm long. June-July.

Locally abundant following fire in montane forest; uncommon in disturbed soil of banks and along roads; East, West. AK to Newf. south to CA, UT, CO, SD, IN, MA. Seeds require heat or abrasion to germinate.

Geranium richardsonii Fisch. & Trautv., White Geranium. Perennial 40-90 cm high; leaves 6-14 cm wide with 3-7 lobed divisions; inflorescence glandular; petals white, 10-18 mm long, purple-veined; capsule 20-25 mm long. July-Aug.

Common in montane thickets and moist open forest, especially aspen; East, occasional in Middle Fork Flathead drainage. Yuk. south to CA, NM, SD.

Geranium viscosissimum Fisch & Mey., Sticky Geranium. Perennial 20-90 cm high; herbage long-hairy and glandular; leaves 5-14 cm wide with 5-7 sharply toothed divisions; petals 14-20 mm long, rose, purple-veined; capsule 25-30 cm long. Aug. Fig. 162.

Common in montane grasslands, meadows, and open forest; East, West. B.C. to Sask. south to CA, CO, SD. Occasional hybrids between G. *richardsonii* and G. *viscisissimum* can be found in the aspen parkland along the east edge of the Park.

GROSSULARIACEAE: GOOSEBERRY FAMILY

Ribes L., Currant, Gooseberry

Shrubs up to 150 cm high; leaves alternate, 3- to 5-lobed like a maple leaf, often clustered, petiolate; flowers bisexual, saucer- to tube-shaped, stalked, few to several in leaf axils; calyx 5-lobed, united below; petals 5, separate, arising from between calyx lobes; stamens 5; ovary below base of calyx (inferior); styles 2; fruit a globose berry.

Ribes cereum Dougl. is reported for high elevations (Bamberg and Major 1968), a very unlikely habitat; the plant is common farther south in Montana, but is not known from Alberta (Kuijt 1982, Moss and Packer 1983); it may occur at low elevations in the southwest portion of the Park. Bamberg and Major's report of R. montigenum McClat., a more southern species, is probably based on R. oxycanthoides ssp. hendersonii. Reference: Sinnott (1985).

1. Stems with 1 to many spines at some or all nodes (gooseberries) 2
1. Stems lacking spines at the nodes (currants) 4

2. Flowers saucer-shaped; berry glandular-hairy .... **R. lacustre**
2. Flowers bell- to tube-shaped; berry glabrous 3

3. Stamens as long as petals; leaves glandular .... **R. oxyacanthoides**
3. Stamens longer than petals; leaves without glands .... **R. inerme**

4. Leaves and berries without glands .... **R. inerme**
4. Leaves and berries glandular 5

5. Leaves densely soft-hairy; berry glandular-hairy .... **R. viscosissimum**
5. Leaves sparsely hairy; berries glandular but without hairs ... **R. hudsonianum**

Ribes hudsonianum Richards., Stinking Currant. Stems erect, unarmed, 50-150 cm high; leaves 3-11 cm across, the lobes toothed, crystalline-glandular, aromatic, hairy; flowers cup-shaped, 6-12 in long, narrow, ascending inflorescences; calyx lobes lance-shaped, white, ca. 4 mm long, spreading; petals white, ca. 2 mm long; berry black, 7-12 mm long, often glandular, unpalatable. May-July.

Uncommon in moist montane forest, often along streams; West. Our plants are var. petiolare (Dougl.) Jancz. AK to Ont. south to WA, MT, MN.

Ribes inerme Rydb. [Grossularia inermis (Rydb.) Cov. & Britt.]. Stems spreading to erect, to 150 cm high, with an occasional nodal spine; leaves 2-6 cm across, the lobes toothed and lobed, mostly glabrous; flowers 2-4, funnel-shaped; calyx lobes green to purplish, lance-shaped, 3-4 mm long spreading; petals 1-2 mm long, white to pink; berry reddish-purple, 7-9 mm long, glabrous, edible. May-June.

Common in open montane forest, thickets, rocky slopes and stream banks; East, West. B.C. to Alta. south to CA, NM.

Ribes lacustre (Pers.) Poir, Prickly Swamp Currant. Stems erect, 50-100 cm high, spiny at the nodes and bristly between; leaves 1-7 cm across, lobes sharply toothed, nearly glabrous and shiny above, glabrous to sparsely hairy and glandular on the veins beneath; flowers saucer-shaped, 7-15 in drooping inflorescences; calyx lobes ovate, 2-3 mm long, greenish to reddish; petals pink, broad, 1-2 mm long; berries 6-8 mm long, black, bristly-glandular. May-July. Fig. 163.

Abundant in moist to wet forest, avalanche slopes and cliffs, often along streams in the montane and subalpine zones; East, West. AK to Newf. south to CA, CO, SD, PA. Density of spines is quite variable.

Ribes oxyacanthoides L., Wild Gooseberry [*R. setosum* Lindl., *R. irriguum* Dougl., *R. hendersonii* Hitchc.]. Stems prostrate to erect, to 150 cm tall, spiny at the nodes and usually bristly between; leaves 1-4 cm wide with rounded teeth, hairy and sometimes glandular; flowers 1-3, funnel-to bell-shaped, 2-6 mm long; calyx lobes whitish, oblong, glabrous; petals white, 1-3 mm long; stamens as long as petals; berries 7-16 mm long, green to purple, glabrous, edible. May-June.

Ssp. *setosum* (Lindl.) Sinnott, with glabrous, nearly tubular flowers > 10 mm long with sepals extended, purple berries and bristly stems, is uncommon on open, rocky, montane slopes; East. Ssp. *hendersonii* (Hitchc.) Sinnott, with mostly prostrate stems, reddish berries, and bell-shaped flowers < 8 mm long, is uncommon in subalpine and alpine boulder fields and exposed slopes; East. Ssp. *irriguum* (Doug.) Sinnott, with mainly erect, non-bristly stems, purple berries, and bell-shaped flowers 8-11 mm long, is uncommon on rocky slopes in open, lower montane forest; West. AK to Ont. south to NV, WY, NE, MN.

Ribes viscosissimum Pursh, Sticky Currant. Stems erect, unarmed, 50-100 cm high; leaves shallowly toothed, glandular and soft-hairy, 2-10 cm across; flowers bell-shaped, 1-2 cm long, 3-7 in a short inflorescence; calyx lobes pointed, ascending, greenish-white, tinged with pink; petals 3-4 mm long, white; stamens as long as petals; berry black, 10-12 mm long, glandular-hairy. May-July. Fig. 164.

Common in montane and lower subalpine forest and avalanche slopes; East, West. Our plants are var. *viscosissimum*. B.C. to Alta. south to CA, AZ, CO.

HALORAGACEAE: WATER-MILFOIL FAMILY

Myriophyllum L., Water-milfoil
Myriophyllum exalbescens Fern. [*M. spicatum* L. var. *exalbescens* (Fern.) Jeps]. Aquatic perennial, overwintering as tight balls of unexpanded leaves (turions); leaves 4-whorled, 1-3 cm long, divided comb-like into filiform segments; flowers small, subtended by entire bracts, sessile in terminal

spikes, mostly unisexual, male above female; sepals and petals 4; stamens 8; stigmas 2-4, feathery; ovary below sepals (inferior); fruit globose, splitting into 4 seeds 2-3 mm long. Fig. 165.

Common in shallow to rather deep water of montane ponds and lakes; East, West. AK to Newf. south to CA, AZ, TX, IL, WV; Europe. M. verticillatum L., with deeply divided floral bracts, may also occur in the Park. These plants are superficially similar to *Utricularia vulgaris* but lack leaf bladders. Reference: Aiken 1981.

HIPPURIDACEAE: MARE'S-TAIL FAMILY

Hippuris L., Mare's-tail
Hippuris vulgaris L. Rhizomatous, at least partly submerged perennial 10-60 cm tall; leaves 6-to 12-whorled, strap-shaped, 1-4 cm long; flowers bisexual, green, small, sessile in leaf axils; calyx tubular; petals absent; style and stamen 1; ovary elliptic, ca. 1 mm long; fruit 1-seeded, ca. 2 mm long. July-Sept. Fig. 166.

Locally common in shallow water of montane ponds and lakes; East, West. Circumboreal south to CA, NM, NE, MN, NY. Submerged leaves are flaccid, while those above water are more rigid.

HYDRANGEACEAE: HYDRANGEA FAMILY

Philadelphus L., Mock Orange, Syringa
Philadelphus lewisii Pursh. Shrub to 3 m high; twigs reddish, glabrous; leaves opposite, elliptic, 2-6 cm long with mostly entire margins, petiolate; flowers bisexual, fragrant, short-stalked, 3-11 in open, terminal clusters; calyx 4-lobed, tubular portion united with the ovary; petals 4, white, ovate, 1-2 cm long; stamens 20-40; stigmas 4; fruit an elliptic, 4-chambered capsule, 6-10 mm long. July. Fig. 167.

Locally common on dry, rocky, montane slopes in the Middle Fork Flathead Valley and near St. Mary Lake. B.C. to Alta. south to CA, ID, MT. One of our finest native shrubs for horticultural purposes.

HYDROPHYLLACEAE: WATERLEAF FAMILY

Annual or perennial herbs; leaves mostly alternate, mostly pinnately lobed; flowers bisexual, mostly borne in stalked clusters that elongate with maturity; calyx 5-lobed; corolla 5-lobed, cup- to funnel-shaped; stamens 5; ovary superior, 1-chambered; style 2-lobed.

1. Leaf blades fan-shaped, palmately lobed .... **Romanzoffia**
1. Leaf blades ovate to lance-shaped in outline, entire to pinnately lobed 2
2. Flowers solitary in leaf axils; calyx with reflexed lobes between sepals ....
.... **Nemophila**
2. Flowers in expanding clusters; calyx without lobes between sepals 2

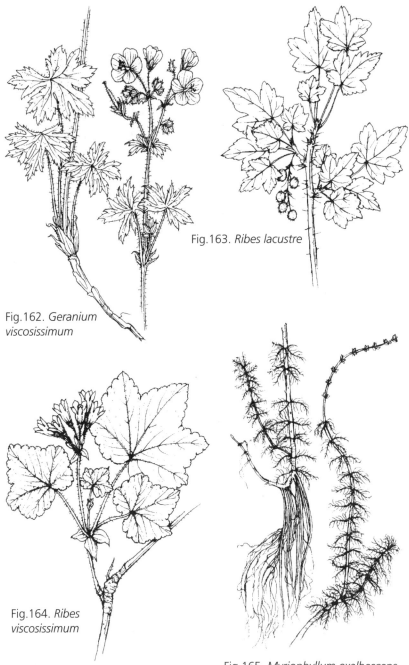

Fig.163. *Ribes lacustre*

Fig.162. *Geranium viscosissimum*

Fig.164. *Ribes viscosissimum*

Fig.165. *Myriophyllum exalbescens*

3. Flower clusters shorter than the leaves.................................. **Hydrophyllum**
3. Flower clusters among and above the leaves.................................... **Phacelia**

Hydrophyllum L., Waterleaf

Hydrophyllum capitatum Dougl. ex Hook., Ballhead Waterleaf. Fleshy-rooted perennial, 10-25 cm high; leaves to 20 cm long, petiolate, pinnately divided with lobed leaflets; flowers in stalked, globose clusters below leaves; corolla lavender, 5-10 mm long, cup-shaped; stamens longer than corolla; capsule globose with 1-4 seeds. May-July. Fig. 168. Color plate 30.

Locally common, often in disturbed or slumping, vernally moist soil of open forest, grassland, and thickets; montane and lower subalpine; East, West. B.C. to Alta. south to CA, CO.

Nemophila Nutt.

Nemophila breviflora Gray. Annual with weak, prostrate to ascending stems, 5-20 cm long; herbage fine-prickly; leaves 2-4 cm long, pinnately divided into ca. 5 lance-shaped lobes; stalked flowers nodding, solitary at leaf bases; calyx long-hairy with 5 reflexed lobes between sepals, enlarging at maturity; corolla light-purplish, cup-shaped, ca. 2 mm long, smaller than calyx; stamens shorter than corolla; capsule 3-5 mm long, 1-seeded. May-June.

Uncommon and inconspicuous in vernally moist, often disturbed soil of grasslands and open forest; East, West. B.C. to Alta. south to CA, CO.

Phacelia Juss., Phacelia

Annual or perennial herbs; flowers in axillary and terminal, branched clusters that elongate with maturity; calyx without lobes between sepals; corolla bell-shaped; stamens longer than petals.

A report of *P. heterophylla* Pursh, a short-lived perennial with mostly solitary, erect stems, from above treeline (Bamberg and Major 1968) was probably based on *P. hastata*.

1. Leaves unlobed or with 1-2 pairs of lobes 2
1. Leaves with > 2 pairs of lobes 3

2. Annual; flowers > 8 mm across; leaves strap-shaped, green .... *P. linearis*
2. Perennial; flowers < 8 mm across; leaves lance-shaped, silvery ...... *P. hastata*

3. Leaves with linear lobes; stamens 2-3 times as long as corolla .... *P. sericea*
3. Leaves with triangular lobes; stamens 1-2 times as long as corolla .... *P. lyallii*

Phacelia hastata Dougl. ex Lehm. [*P. leucophylla* Torr., *P. leptosepala* Rydb.]. Perennial with stems 10-50 cm high, curved at the base from a branched rootcrown; herbage silver-hairy; basal leaves narrowly lance-shaped with entire margins, petiolate, 4-8 cm long; stem leaves reduced, sometimes lobed at the base; corolla white to blue, 4-7 mm long. June-Aug.

Common in stony, sparsely vegetated soil of grasslands, open slopes, and exposed ridges at all elevations; East, West. B.C. to Alta. south to CA,

CO, NE. A number of varieties have been proposed, but variation in our area seems continuous rather than discrete.

Phacelia linearis (Pursh) Holz. Short-hairy annual, 10-25 cm high; leaves strap-shaped, 1-4 cm long, often with 1-2 pairs of lobes near the base; calyx stiff-hairy; corolla blue, ca. 1 cm across. June.

Rare in sparsely vegetated soil of montane grassland and dry, open forest; collected once near the head of Kintla Lake. B.C. to Alta. south to CA, UT, WY.

Phacelia lyallii (Gray) Rydb. Short-hairy perennial, 10-25 cm high from a branched rootcrown; leaves clustered near the base, oblong, 5-10 cm long, pinnately lobed into toothed, triangular segments; corolla blue, 5-9 mm long. July-Aug. Fig. 169.

Common in meadows, along streams, and on rocky slopes near or above treeline; East, West. Endemic to B.C., Alta., MT. Named for David Lyall, doctor and botanist for the International Boundary survey of 1861, who first collected the plant.

Phacelia sericea (Graham) Gray. Long-hairy perennial, 10-40 cm high from a branched rootcrown; leaves to 15 cm long, pinnately divided into numerous linear lobes, petiolate below, reduced above; corolla purple, 5-7 mm long; stamens > 2 times as long as petals. July-Aug.

Common on rocky slopes, talus and stream gravels; montane to alpine, East. B.C. to Alta. south to CA, CO. Standley (1921) states that when *P. sericea* occurs with *P. lyallii*, the former flowers earlier.

Romanzoffia Cham., Mist-maiden

Romanzoffia sitchensis Bong. Weak-stemmed perennial to 20 cm high with bulbs among the leaf bases; leaves mostly basal, petiolate, the blade fan-shaped with shallow, rounded lobes, glabrous, 1-3 cm wide; flowers in an open inflorescence; corolla white, funnel-shaped, 5-10 mm long; stamens shorter than petals; capsule pod-like, many-seeded. July-Aug.

Common in wet, shallow soil of cliffs and streambanks near or above treeline, uncommon lower; East, West. AK to CA, ID, MT.

HYPERICACEAE: ST. JOHN'S-WORT FAMILY

Hypericum L., St. John's-wort

Perennial herbs; leaves opposite, with entire margins, without petioles; flowers bisexual, in branched, open inflorescences; sepals 5; petals 5, separate, yellow; stamens numerous; styles 3-4; ovary superior; fruit a many-seeded capsule.

1. Leaves lance-shaped; sepals with pointed tips; montane ..... ***H. perforatum***
1. Leaves broadly elliptic; sepals rounded; mostly higher .... ***H. formosum***

Plants of grassland

1. Grassland with Balsamorhiza sagitata

2. Castilleja hispida

3. Galium boreale

4. Gaillardia aristata

*5. Dodecatheon
conjugens*

6. Lupinus sericeus

7. Eriogonum flavum

8. Monarda fistulosa

Plants of warm montane forest

9. Warm montane forest

10. Diphasiastrum complanatum

11. Lysichiton americanus

12. Holodiscus discolor

13. Athyrium filix-femina

14. Cornus canadensis

15. Viola orbiculata

Plants of wetlands

16. Wetland

17. Carex vesicaria

18. Platanthera hyperborea

19. Menyanthes trifoliata

20. Spiranthes romanzoffiana

21. Ledum glandulosum

22. Heracleum sphondylium

Plants of aspen woodlands

23. Aspen woodlands

24. Prunus emarginata

25. Allium cernuum

26. Rosa woodsii

27. Smilacina racemosa

28. Amelancier alnifolia

29. Eriogeron speciosus

30. Hydrophyllum capitatum

31. Cold montane forest

32. Corallorhiza striata

33. Berberis repens

Plants of cold montane forest

35. Calypso bulbosa

34. Rubus parviflorus

36. Pterospora andromeda

37. Aster conspicuus

38. Arnica cordifolia

Plants of subalpine forest

*39. Subalpine forest with
Xerophyllum tenax*

42. Paxistima myrsinites

40. Menziesia ferruginea

41. Chimaphila umbellata

*43. Chamerion
angustifolium*

44. Linnaea borealis

45. Spiraea betulifolia

Plants of subalpine meadow

46. Subalpine meadow

47. Aster foliaceus

48. Castilleja rhexifolia

49. Mimulus lewisii

50. Erythronium grandiflorum

51. Spiraea densiflora

52. Aquilegia flavescens

53. Gentiana calycosa

Plants of alpine
tundra

54. Alpine tundra with Dryas octopetala

55. Eriogonum androsaceum

56. Silene acaulis

57. Potentilla uniflora

58. Smelowskia calycina

59. Ranunculus eschscholtzii

60. Aquilegia jonesii

All photographs by Peter Lesica

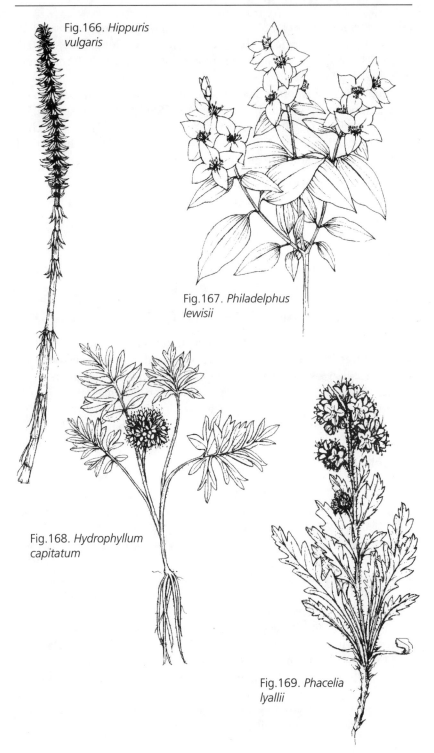

Fig.166. *Hippuris vulgaris*

Fig.167. *Philadelphus lewisii*

Fig.168. *Hydrophyllum capitatum*

Fig.169. *Phacelia lyallii*

Hypericum formosum H. B. K. [*H. scouleri* Hook.]. Stems glabrous, 5-20 cm tall from rhizomes; leaves broadly elliptic, 1-2 cm long, black-dotted; petals 6-12 mm long; sepals ovate, amber-dotted; capsule 6-9 mm long. July-Aug. Fig. 170.

Abundant in wet, often shallow soil of meadows, streambanks, and ledges; upper montane to alpine; East, West. Our plants are ssp. *scouleri* (Hook.) Hitchc. B.C., Alta. south to CA, WY. Var. *nortoniae* (Jones) Hitchc., the high-elevation ecotype, was named for Gertrude Norton who studied the botany of the Flathead region in the early part of the twentieth century.

Hypericum perforatum L., Goatweed, Klammath Weed. Taprooted and rhizomatous with stems 30-70 cm tall; leaves lance-shaped, 1-3 cm long, translucent-dotted; sepals narrowly lance-shaped; petals 10-14 mm long; capsule 5-8 mm long. July-Aug.

Locally common in montane grassland, open forest and roadsides; East, West. Introduced from Europe. This pernicious weed is especially common on old burns along Middle Fork Flathead River.

LAMIACEAE: MINT FAMILY

Annual or perennial, often aromatic herbs; stems usually square in cross section; leaves opposite, mostly petiolate, often toothed but mostly not lobed or divided; flowers bisexual; corolla tubular, 2-lipped, lower lip 3-lobed; stamens 2 or 4; ovary superior, 4-lobed; style 2-lobed; fruit 4 nutlets (hard-coated seeds).

Physostegia parviflora Gray, with purplish, funnel-shaped corollas, 8-16 mm long, has been collected along the Flathead River near Columbia Falls and could occur in the Park as well.

1. Flowers cup-or funnel-shaped, barely 2-lipped 2
1. Flowers distinctly 2-lipped, usually somewhat tubular 4

2. Corolla funnel-shaped, > 10 mm long .... **Satureja**
2. Corolla cup-shaped, < 10 mm long 3

3. Stamens 2; corolla white; herbage not strongly aromatic .... **Lycopus**
3. Stamens 4; corolla blue; herbage aromatic .... **Mentha**

4. Stamens 2 .... **Monarda**
4. Stamens 4 5

5. Flowers in upper leaf axils without a terminal cluster 6
5. Terminal cluster of flowers present 7

6. Leaves fan-shaped; lawn or roadside weed .... **Glecoma**
6. Leaves lance-shaped; wetlands and riparian areas .... **Scutellaria**

7. Leaf-like bracts of the inflorescence spine-tipped .... **Dracocephalum**
7. Bracts of inflorescence not spine-tipped 8

8. Stamens exserted beyond mouth of corolla .... **Agastache**
8. Stamens included in the corolla 9

9. Leaf margins entire or with indistinct teeth .... **Prunella**
9. Leaf margins distinctly toothed 10
10. Calyx glandular, with > 14 veins .... **Nepeta**
10. Calyx not glandular, with 5-10 veins .... **Stachys**

Agastache Clayton, Giant Hyssop, Horsemint

Agastache urticifolia (Benth.) Kuntze. Perennial with numerous erect stems 50-100 cm high from a branched rootcrown; herbage glabrous or finely hairy; leaf blades to 8 cm long, arrow-shaped with toothed margins, short-hairy below; calyx purplish; flowers in dense, elongate clusters at stem tips; corolla white, ca. 10 mm long with short lobes; stamens exserted. July-Aug.

Meadows and thickets; montane and lower subalpine; collected once on an avalanche slope near Elizabeth Lake. B.C. to MT south to CA, CO.

Dracocephalum L., Dragonhead

Dracocephalum parviflorum Nutt. [*Moldavica parviflora* (Nutt.) Britt.]. Annual or biennial, 10-60 cm high; leaf blades to 8 cm long, lance-shaped with toothed margins; flowers clustered above leaf-like, spiny-margined bracts in a dense, terminal inflorescence; calyx 10-14 mm long with spine-tipped lobes, the upper larger; corolla light blue slightly longer than calyx; stamens not exserted. June-Aug. Fig. 171.

Montane forest and along streams, roads, and trails; East, West. AK to Newf. south to OR, NM, NE, WI, NY. Abundant following fire, the plant is otherwise uncommon; sporadic small colonies appearing with disturbance.

Glecoma L., Hemp Nettle

Glecoma hederacea L., Creeping Charlie, Gill-over-the-Ground. Perennial with prostrate stems 20-50 cm long; leaves with blunt-toothed, fan-shaped blades 1-4 cm wide; flowers whorled in leaf axils; calyx 5-6 mm long, upper teeth longer; corolla blue, purple-spotted, 13-23 mm long, the middle lobe of lower lip large; stamens not exserted.

Moist, disturbed soil of lawns or trails; collected once near Lake McDonald. Introduced from Eurasia.

Lycopus L., Water Horehound, Bugleweed

Rhizomatous or stoloniferous perennial, non-aromatic herbs; leaves with toothed margins; flowers in dense clusters in upper leaf axils; corolla white, cup-shaped, barely 2-lipped, hairy within; fertile stamens 2, not exserted; nutlets 3-angled.

1. Calyx lobes as long as nutlets, not awn-tipped; leaves toothed .. **L. uniflorus**
1. Calyx lobes longer than nutlets, awned; leaves lobed 1/2-way to middle ....
.... **L. americanus**

Lycopus americanus Muhl. Stems 20-80 cm tall; herbage glabrous; leaves lobed halfway to midvein, 2-7 cm long; calyx lobes awned; corolla 2-3 mm long.

Montane marshes, meadows, and along streams and wetlands; collected once near West Glacier. Native to much of N. America.

Lycopus uniflorus Michx. Stems 20-50 cm tall from tuberous roots; herbage glabrous or short-hairy; leaves sessile, 2-6 cm long; calyx lobes not awned; corolla 2-4 mm long. July-Aug. Fig. 172.

Locally common on the margins of montane fens and wet thickets; East, West. AK to Newf. south to CA, WY, NE, AK, NC.

Mentha L., Mint

Mentha arvensis L., Field Mint [*M. canadensis* L.]. Perennial 20-60 cm high from rhizomes; herbage hairy, glandular; leaf blades lance-shaped with toothed margins, 2-6 cm long; flowers clustered around leaf axils; corolla cup-shaped, 4-lobed, blue, 4-7 mm long; stamens 4, exserted. July-Sept. Fig. 173.

Common in montane wet meadows and fens and mud along streams and shores; East, West. Circumboreal south to CA, NM, TX, MO, WV. This plant is the source of the familiar mint odor when walking along shores of ponds and streams. Native Americans used it for flavoring meat.

Monarda L., Bee Balm

Monarda fistulosa L., Wild Bergamot, Horsemint [*M. menthaefolia* Benth.]. Rhizomatous perennial 30-70 cm high; herbage short-hairy; leaf blades lance-shaped, 2-8 cm long with toothed margins; flowers in dense, hemispheric terminal clusters; corolla lavender, tubular, 15-30 mm long, strongly 2-lipped, the lower perpendicular to the tube; stamens 2, exserted beyond upper lip. July-Aug. Fig. 174. Color plate 8.

Common in montane grasslands, meadows, and open forest; East, West. Our plants are var. *menthaefolia* (Grah.) Fern. B.C. to Que., south to AZ, TX, GA. Blackfeet use an infusion of the flowers for eyewash and a tea of the leaves for stomach pain.

Nepeta L.

Nepeta cataria L., Catnip. Perennial 30-80 cm high from a taproot; herbage soft, short-hairy; leaf blades triangular, 2-7 cm long with toothed margins; flowers in a dense, elongate, terminal cluster; corolla white, purple-dotted, 8-12 mm long, 2-lipped, the upper hood-like, the lower reflexed with a wavy margin; stamens 4, shorter than upper lip. July.

Uncommon around buildings and along roads; collected near the head of Lake McDonald. Introduced from Eurasia.

Prunella L., Self-heal

Prunella vulgaris L., Heal-all. Perennial 10-30 cm high from a short, creeping root crown; herbage sparsely hairy; leaf blades 2-7 cm long, lance-shaped with entire or shallowly toothed margins; flowers subtended by

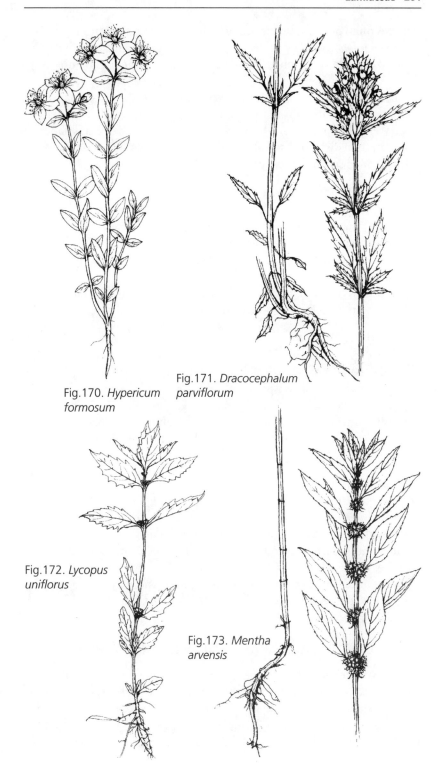

Fig.170. *Hypericum formosum*

Fig.171. *Dracocephalum parviflorum*

Fig.172. *Lycopus uniflorus*

Fig.173. *Mentha arvensis*

leaf-like, often purplish bracts in a terminal cluster; calyx with 1 broad and 2 narrow lobes; corolla purple, 10-15 mm long, 2-lipped, the upper lip hood-like, the lower 3-lobed; stamens 4, included in upper lip. June-Aug. Fig. 175.

The native var. *lanceolata* (Bart.) Fern. with leaf blades at least 3 times as long as wide, is common in montane forest, meadows and streambanks; East, West. The introduced var. *vulgaris* with broader leaf blades is uncommon in disturbed soil along streams and roads and in lawns; West. Circumboreal south to most of U. S.

Satureja L., Savory

Satureja vulgaris (L.) Fritsch, Wild Basil. Rhizomatous perennial 20-40 cm tall; herbage sparsely hairy; leaf blades 1-3 cm long, lance-shaped with scalloped margins; flowers in a dense, terminal head; corolla purplish, ca. 10 mm long, funnel-shaped, indistinctly 2-lipped; stamens 4, included in corolla. July.

Collected once along road near Lake McDonald. Native to Europe and eastern N. America but introduced in Glacier.

Scutellaria L., Skullcap

Scutellaria galericulata L. [*S. epilobifolia* Ham.]. Rhizomatous, short-hairy perennial 10-60 cm high; leaf blades 2-5 cm long, lance-shaped with scalloped margins; flowers paired at upper leaf nodes; calyx shallowly 2-lipped, cup-shaped; corolla 12-20 mm long, blue, tubular, 2-lipped, the upper lip short; stamens 4, included in corolla. July. Fig. 176.

Uncommon in montane fens, marshes, and wet meadows; West. Circumboreal south to CA, AZ, TX, IN, WV.

Stachys L., Hedge Nettle

Stachys palustris L., Woundwort [*S. scopulorum* Greene]. Rhizomatous perennial 30-70 cm high; herbage soft-hairy; leaf blades 4-8 cm long, lance-shaped with toothed margins; flowers in a dense but interrupted, elongate, terminal inflorescence; corolla purple, spotted, 10-15 mm long, 2-lipped, the lower 3-lobed; stamens 4, included in corolla. July-Aug. Fig. 177.

Uncommon in moist or wet soil along montane streams and ponds; East, West. Our plants are ssp. *pilosa* (Nutt.) Epling. Circumboreal south to OR, AZ, OH, NY.

LENTIBULARIACEAE: BLADDERWORT FAMILY

Carnivorous perennials; leaves basal or alternate; flowers 1 to few on naked stalks; calyx 4- or 5-lobed; corolla tubular, 2-lipped with a spur at base of the tube; stamens 2; ovary superior; fruit a cpasule with numerous seeds.

1. Terrestrial plants with all leaves basal and undivided .... **Pinguicula**
1. Aquatic plants with alternate, divided leaves .... **Utricularia**

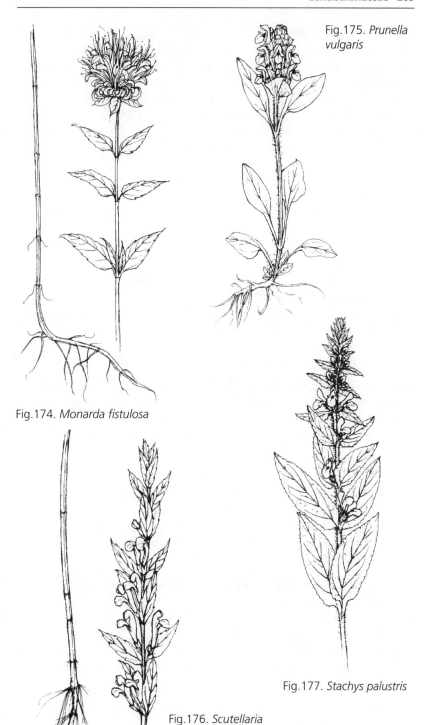

Fig.175. *Prunella vulgaris*

Fig.174. *Monarda fistulosa*

Fig.177. *Stachys palustris*

Fig.176. *Scutellaria galericulata*

Pinguicula L., Butterwort

Pinguicula vulgaris L. Stems leafless, 4-15 cm high from fibrous roots; leaves succulent, 2-5 cm long, oblong with inrolled margins, the sticky surface capturing small insects; flower solitary; corolla purple, funnel-shaped, 5-lobed, barely 2-lipped, ca. 2 cm long. July-Aug. Fig. 178.

Wet soil of fens, ledges and along streams; locally common subalpine and alpine, occasional lower; East, West. Circumboreal south to CA, MT, MI, NY. Our only alpine carnivorous plant; the flowers superficially resemble violets.

Utricularia L., Bladderwort

Aquatic, submersed, non-rooted plants; leaves alternate, palmately divided into linear segments; small, bladder-like insect traps borne on leaf segments or separate branches; flowers few on naked, emergent stalks; corolla yellow, 2-lipped, snapdragon-like with a spur at the base of the tube; plants overwintering as large vegetative buds (turions). July-Aug.

Plants flower only sporadically.

1. Bladders on separate, leafless branches .... *U. intermedia*
1. Bladders on leaf segments 2
2. Bladders few, ca. 2 mm long; flowers 4-8 mm long .... *U. minor*
2. Bladders numerous, 3-5 mm long; flowers 14-20 mm long .... *U. vulgaris*

Utricularia intermedia Hayne. Stems slender; leaves 5-15 cm long, the segments flattened, toothed; bladders ca. 5 mm long, on separate, leafless branches; flowers 10-15 mm long.

Rare in sphagnum fens of the McDonald Valley. Circumboreal south to CA, MT, OH, DE. Plants are inconspicuous, creeping on peat beneath shallow water among emergent plants.

Utricularia minor L. Stems slender; leaves 3-10 mm long, segments with entire margins; bladders few, ca. 2 mm long, attached to leaf segments but sometimes also on separate branches; flowers 4-8 mm long. July.

Rare in shallow water of montane fens; East, West. Circumboreal south to CA, CO, IA, NJ.

Utricularia vulgaris L. [*U. macrorhiza* LeConte]. Stems coarse; leaves crowded, 1-5 cm long, segments entire-margined; bladders numerous, 3-5 mm long, attached to leaf segments; flowers 14-20 mm long. July-Aug. Fig. 179.

Locally common in water of montane lakes and sloughs; West, expected East. Circumboreal south to most of U. S. In autumn the turions float and resemble dark green brussel sprouts.

LINACEAE: FLAX FAMILY

Linum L., Flax

Annual or perennial herbs; leaves linear, alternate; flowers in an open, terminal inflorescence; sepals 5, separate; petals 5, blue (ours), narrowly fan-shaped; stamens and styles 5; ovary superior; fruit a 10-seeded capsule.

1. Annual with simple stems; inner sepals with marginal hairs **L. usitatissimum**
1. Perennial, branched at the base; sepals hairless .... **L. lewisii**

Linum lewisii Pursh, Blue Flax [*L. perenne* L. var. *lewisii* (Pursh) Eat. & Wright]. Glabrous perennial 20-60 cm high, branched at the base; leaves 5-20 mm long; petals ca. 1 cm long, quickly falling; capsule longer than the sepals. July. Fig. 180.

Common in montane grasslands and on open, alpine slopes; East. AK to Que., south to CA, TX, WI.

Linum usitatissimum L., Common Flax. Annual 20-80 cm high, mostly unbranched; leaves 1-3 cm long; petals 10-15 mm long; capsule as long as sepals.

Standley (1921) reported a few plants near East Glacier, but not collected since. Introduced from Europe, cultivated for fiber and seed oil.

LOASACEAE: BLAZING-STAR FAMILY

Mentzelia L., Blazing Star

Mentzelia dispersa Wats. Annual 10-40 cm high; herbage covered with hooked hairs; leaves alternate, 3-10 cm long, lance-shaped with toothed margins, petiolate below; inflorescence terminal, compact; calyx joined to ovary, 5-lobed; petals 5, separate, yellow, 3-6 mm long; stamens numerous; fruit a linear capsule 1-3 cm long. Fig. 181.

Sparsely vegetated soil of grasslands and open slopes, collected once near East Glacier (Standley 1921). WA to MT, south to CA, CO. *M. decapetala* (Pursh ex Sims) Urb. & Gilg, with large white flowers, has been collected just east of the Park and could be expected on sparsely vegetated, gravelly slopes.

MALVACEAE: MALLOW FAMILY

Annual to perennial; herbage often with star-shaped hairs; leaves alternate, palmately lobed, petiolate; flowers axillary; calyx 5-lobed; petals 5, separate, wedge-shaped; stamens numerous, united into a column around the solitary, branched style; ovary superior; fruit a wheel-like capsule with numerous radial sections like an orange.

1. Perennial, stems > 50 cm tall; petals ca. 2 cm long .... **Iliamna**
1. Annual, stems < 50 cm long; petals ca. 1 cm long .... **Malva**

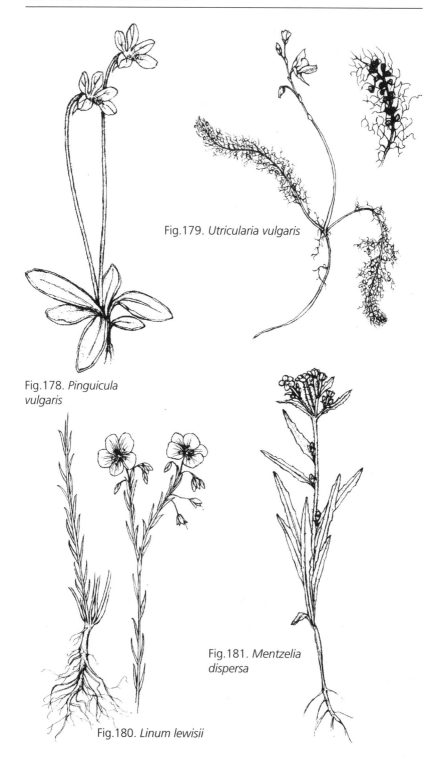

Fig.179. *Utricularia vulgaris*

Fig.178. *Pinguicula vulgaris*

Fig.181. *Mentzelia dispersa*

Fig.180. *Linum lewisii*

Iliamna Greene

Iliamna rivularis (Dougl. ex Hook.) Greene, Wild Hollyhock [*Sphaeralcea rivularis* Hook.]. Perennial 50-200 cm high; leaves 5-20 cm wide, maple-like with 5-7 lobes; flowers in narrow, dense clusters in axils of upper leaves; petals 2-3 cm long, pink to rose; fruit segments 2- to 3-seeded. June-Aug. Fig. 182.

Common along streams and roads and in open montane forest; East, West. B.C. to Alta. south to OR, CO. The hard seeds germinate readily after scouring or fire.

Malva L., Mallow

Malva neglecta Wallr., Cheeses. Annual with prostrate to ascending stems 20-50 cm long; leaves nearly round in outline with shallow lobes, 1-4 cm long; flowers stalked, few in leaf axils; petals white to light blue, ca. 10 mm long. July-Aug.

Rare in lawns and around buildings; collected once at Polebridge. Introduced from Europe.

MENYANTHACEAE: BUCK-BEAN FAMILY

Menyanthes L., Buck-bean

Menyanthes trifoliata L. Rhizomatous, glabrous perennial; leaves succulent, alternate with 3 oblong leaflets, 3-8 cm long, clustered at base of ascending flowering stems 10-30 cm high; short-stalked flowers in a narrow inflorescence; calyx 5-lobed; corolla white, funnel-shaped, 5-lobed, the lobes with long, thick hairs; stamens 5; stigma 2-lobed; ovary superior; fruit an ovate capsule. June-July. Fig. 183. Color plate 19.

Common in montane fens; East, West. Circumboreal south to CA, CO, MO, PA. The bean-like leaves in the inundated peat are unmistakeable.

NYMPHAEACEAE: WATER-LILY FAMILY

Nuphar J. E. Smith, Yellow Water Lily

Nuphar luteum (L.) Sibth. & Smith [*N. polysepalum* Engelm, *N. variegatum* Engelm, *Nymphaea polysepala* (Engelm.) Greene]. Aquatic perennial from a thick, short rhizome; leaf blades 10-30 cm long, heart-shaped, petiolate, arising from rhizome, floating; flowers solitary, floating on long stalks; sepals 5-12, yellow, 3-6 cm long; petals 10-20, inconspicuous; stamens numerous; ovary superior with a large, plate-like stigma; fruit a many-seeded capsule, ca. 4 cm long. June-Aug. Fig. 184.

Common in a few small, montane lakes; West. Ssp. *variegatum* (Engelm.) Beal has 6 sepals and yellow stamens, while ssp. *polysepalum* (Engelm.) Beal has ca. 9 sepals and reddish stamens. AK to Newf. south to CA, CO, OH, NJ.

ONAGRACEAE: EVENING-PRIMROSE FAMILY

Annual or perennial; leaves undivided, opposite or alternate; flowers bisexual; sepals usually 4; petals usually 4, separate; stamens mostly 4 or 8; ovary below the base of the calyx (inferior); style 1; fruit a many-seeded capsule.

1. Petals and sepals 2; fruit with hooked hairs .... **Circaea**
1. Petals and sepals 4; fruit without hooked hairs 2

2. Capsule hard, nut-like, 7-9 mm long, not stalked .... **Gaura**
2. Capsule mostly > 9 mm long or stalked 3

3. Stem branches fine and hair-like; ovary with 2 chambers .... **Gayophytum**
3. Branches not so slender; ovary 4-chambered 4

4. Seeds with a tuft of hair at the tip 5
4. Seeds without tuft of hair 6

5. Petals > 1 cm long, completely separate, unlobed...................... **Chamerion**
5. Petals < 1 cm long, united at the very base, lobed at the tip .... **Epilobium**

6. Stigma with 4 linear lobes 6
6. Stigma round or slightly lobed 7

7. Petals deeply lobed, pink .... **Clarkia**
7. Petals notched but not deeply lobed .... **Oenothera**

8. Stems lacking; petals 6-8 mm long, yellow .... **Camissonia**
8. Leafy stems present; petals 2-4 mm long; petals white or pink ..... **Epilobium**

Camissonia Link, Sun Cups

Camissonia breviflora (T. & G.) Raven [*Oenothera breviflora* T. & G., *Taraxia breviflora* (T. & G.) Nutt.]. Stemless, taprooted perennial; herbage short-hairy; leaves 5-15 cm long, narrowly lance-shaped with lobed margins, petiolate; flowers with a long, stalk-like hypanthium between the ovary and petals, borne among the basal rosette; petals yellow, 6-8 mm long; stigma round; capsules among leaf bases, oblong, hairy, 10-25 mm long. Aug.

Locally common in drying mud around ponds along the south border of the Park. B.C. to Sask. south to CA, WY.

Chamerion Raf. ex Holub, Fireweed

Perennial; leaves sessile or short-petiolate, mostly alternate; petals unequal, the lower slightly narrower; stigma 4-long-lobed; stamens 8; capsule elongate; seeds with a tuft of fine, white hair at the tip.

The fireweeds were formerly placed in *Epilobium*. Reference: Cronquist et al. (1997).

1. Inflorescence leafy with mostly 1-7 flowers .... ***C. latifolium***
1. Inflorescence not leafy; flowers numerous .... ***C. angustifolium***

Chamerion angustifolium (L.) Holub, Fireweed [*Epilobium angustifolium* L.]. Rhizomatous with unbranched stems 15-120 cm high; herbage glabrous below; leaves 5-15 cm long, lance-shaped with mostly entire margins; flowers numerous in a long, narrow, terminal inflorescence, nodding in bud; petals magenta 1-2 cm long; capsule 2-10 cm long; seeds cross-corrugated. July-Sep. Fig. 185. Color plate 43.

Abundant on open slopes and stream banks and in avalanche chutes and open forest at all but the highest elevations; East, West. Our plants are ssp. *angustifolia.* Circumboreal south to CA, NM, SD, OH, NC. The windborne seeds quickly colonize burned-over forest, and the plant persists until the new tree canopy closes over, a process that may take centuries in some subalpine areas.

Chamerion latifolium (L.) Holub, Alpine Fireweed [*Epilobium latifolium* L.]. Stems 4-60 cm high from a branched rootcrown; herbage glabrous; leaves 2-6 cm long, lance-shaped with entire margins, opposite below; flowers several in a short, terminal inflorescence; petals rose, 1-3 cm long; capsule 3-10 cm long, glabrous or short-hairy; seeds glabrous or hairy. July-Aug.

Common on gravel bars in the montane zone; uncommon on stony, especially limestone, slopes near or above treeline; East, West. Our plants are ssp. *latifolium.* AK to Greenl. south to CA, CO; Asia. One of the first plants, along with *Dryas drummondii,* to colonize and stabilize fresh river gravels. Low-elevation plants often have narrower leaves than those near treeline.

Circaea L., Enchanter's Nightshade
Circaea alpina L. Perennial 10-40 cm tall from slender rhizomes; herbage hairy above; leaves opposite; blades 2-5 cm long, petiolate, broadly lance-shaped; flowers in terminal, branched inflorescences; sepals 2, reflexed; petals 2, 1-2 mm long, white; stamens 2; fruit ovate, ca. 2 mm long, with white, hooked hairs. July. Fig. 186.

Common in moist forest and thickets. Ssp. *alpina,* with toothed leaves with a heart-shaped base, is found in mesic montane forest, mainly West. Ssp. *pacifica* (Asch & Magnus) Raven, with leaf blades with entire margins and a rounded base, is generally found higher and in less shaded sites; East, West. Circumboreal south to CA, CO, GA. The fruits adhere to fur and clothing, and the plant thrives in moderately disturbed soil.

Clarkia Pursh, Clarkia
Clarkia pulchella Pursh, Elk Horns, Ragged Robin. Annual 10-50 cm tall; herbage finely hairy; leaves alternate, 2-7 cm long, linear with entire margins; flowers few, terminal; petals lavender, 15-25 mm long, 3-lobed; stamens 4; stigma oblong, white; capsule linear ca. 2 cm long. July. Fig. 187.

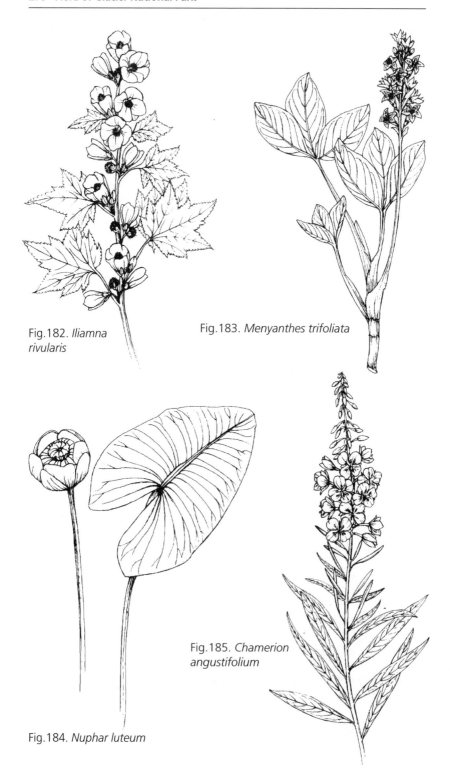

Fig.182. *Iliamna rivularis*

Fig.183. *Menyanthes trifoliata*

Fig.185. *Chamerion angustifolium*

Fig.184. *Nuphar luteum*

Rare in montane grasslands and cliffs; collected once near West Glacier. B.C. to MT south to OR, ID.

Epilobium L., Willow Herb

Mostly perennials, sometimes forming bulb-like turions at the base; leaves sessile or short-petiolate, usually opposite below but often alternate above; petals 2-lobed; stigma 4-lobed; stamens 8; capsule elongate; seeds mostly with a tuft of fine, white hair at the tip.

Large-flowered species formerly placed in *Epilobium* are now placed in *Chamerion*. The treatment herein follows that of Peter Hoch as presented in Moss & Packer (1983) and Cronquist et al. (1997).

1. Stigma of 4 long lobes; petals yellow .... *E. suffruticosum*
1. Stigma round or slightly lobed; petals white to purple 2

2. Plants annual; leaves often alternate at least below .... *E. brachycarpum*
2. Plants perennial; leaves mostly opposite 3

3. Inflorescence, especially fruits, white-hairy with few glands .... *E. palustre*
3. Inflorescence glandular-hairy or glabrous, not white-hairy 4

4. Stems glabrous and waxy-coated below inflorescence .... *E. glaberrimum*
4. Stems hairy, often in lines from leaf bases 5

5. Stems curved or angled upward at the base, usually < 30 cm high, branched at the base or unbranched; basal rosette or bulb-like turions lacking; rhizomes often present 6
5. Stems erect at the base, often branched above middle, sometimes > 30 cm high; basal rosette or turions usually present; rhizomes short or lacking 9

6. Petals white to pale pink, 2-5 mm long; seeds smooth .... *E. lactiflorum*
6. Petals pink to purple or > 5 mm long; seeds sometimes bumpy 7

7. Plants mostly 15-30 cm high; petals 5-13 mm long; leaves broadly lance-shaped .... *E. hornemannii*
7. Plants mostly < 15 cm high; petals 3-6 mm long; leaves narrowly lance-shaped 8

8. Mature fruit ca. 1 mm wide .... *E. anagallidifolium*
8. Mature fruit 1.5-2.0 mm wide .... *E. clavatum*

9. Inflorescence without glands .... *E. leptocarpum*
9. Inflorescence, especially fruits, at least sparsely glandular 10

10. Leaves mostly clasping .... *E. saximontanum*
10. Leaves petiolate, not clasping 11

11. Basal rosette or large turions present .... *E. ciliatum*
11. Compact turions present; basal rosette lacking .... *E. halleanum*

Epilobium anagallidifolium Lam. [*E. alpinum* L. var. *alpinum*]. Often mat-forming perennial, mostly 3-15 cm high; stems sparsely hairy; leaves 8-25 mm long, elliptic with mostly entire margins, short-petiolate; inflorescence nodding in bud; sepals mostly glabrous; petals 3-4 mm long, pink to purple; capsule 17-36 mm long; seeds cross-corrugated. July-Sept. Fig. 188.

Common along streams and on gravelly slopes; mainly alpine and subalpine; East, West. Circumboreal south to CA, CO, and NH.

Epilobium brachycarpum Presl [*E. paniculatum* Nutt. ex T. & G.]. Annual 5-80 cm high, branched above; herbage glabrous below glandular above; leaves narrowly lance-shaped with entire to finely toothed margins, 15-50 mm long, short-petiolate, often with small leaves in the axils; inflorescence branched, spreading; petals 2-15 mm long, white to purple; capsule 15-32 mm long, glabrous or glandular-hairy; seeds bumpy. June-Aug.

Common in disturbed soil of montane meadows, grasslands and open slopes; East, West. B.C. to Que. south to CA, NM, SD, MN. Plants from Glacier Park previously identified as *E. minutum* Lindl. ex Lehm. are referred here.

Epilobium ciliatum Raf. [*E. glandulosum* Lehm. in part]. From turions or rosettes, 5-70 cm high; herbage hairy and glandular above; leaves 3-12 cm long, lance-shaped with finely toothed margins, petiolate; inflorescence erect in bud; petals white to purple, 2-12 mm long; capsule 4-10 cm long, hairy; seeds ridged. June-Aug.

Uncommon in moist soil along streams or lakes in the montane and subalpine zones; East, West. Ssp. *ciliatum,* has leaves < 15 mm wide and a branched inflorescence, while ssp. *glandulosum* (Lehm.) Hoch & Raven has some leaves > 20 mm wide and mostly unbranched inflorescences. N. America, Asia, S. America.

Epilobium clavatum Trel. [*E. alpinum* L. var. *clavatum* (Trel.) Hitchc.]. Stems 5-15 cm high, spreading by wiry runners; stems sparsely hairy, glandular above; leaves elliptic with mostly entire margins, 5-15 mm long, short-petiolate or sessile; inflorescence erect in bud; sepals glabrous or sparsely glandular; petals pink to purple, 4-6 mm long; capsule 2-4 cm long, mostly glabrous, somewhat club-shaped; seeds cross-corrugated or bumpy. July-Aug.

Common in wet meadows, on rocky slopes and along streams, mostly near or above treeline; East, West. AK to CA, CO.

Epilobium glaberrimum Barbey [*E. platyphyllum* Rydb.]. Clustered, glabrous stems 10-40 cm high; leaves opposite, clasping the stem, ovate, 10-34 mm long, waxy; inflorescence erect in bud; sepals glabrous; petals pink to purple, 3-7 mm long; capsule 20-55 mm long, glabrous; seeds with lines of bumps. July.

Common in moist, sparsely vegetated soil along streams and trails; montane and subalpine; East, West. Our plants are ssp. *fastigiatum* (Nutt.) Hoch & Raven. B.C., Alta. south to CA, UT, WY.

Epilobium halleanum Hausskn. [*E. watsonii* Barbey in part]. Stems 20-80 cm high with fleshy, compact, underground bulbs; herbage hairy and glandular above; leaves mostly sessile, narrowly elliptic, 2-8 cm long with finely toothed margins; inflorescence nodding in bud; sepals glabrous or sprasely glandular-hairy; petals white to pink, 2-6 mm long; capsule 24-60 mm long; seeds bumpy. July-Aug.

Common in moist soil of meadows and along streams; subalpine and alpine; East, West. B .C. to Sask. south to CA, AZ, CO.

Epilobium hornemannii Reichb. [*E. alpinum* L. var. *nutans* (Hornem.) Hook.]. Stems 10-25 cm high, glandular above; leaves 15-30 mm long, petiolate below, often with finely toothed margins; inflorescence erect in bud; sepals sparsely glandular; petals 3-9 mm long; mature capsule 40-50 mm long with a stalk 5-25 mm long, sparsely hairy; seeds bumpy. July-Aug.

Common in wet soil of meadows and along streams; subalpine and alpine, less common montane; East, West. Our plans are ssp. *hornemannii*. Circumpolar south to CA, NM, NH.

Epilobium lactiflorum Hausskn. [*E. alpinum* L. var. *lactiflorum* (Hausskn.) Hitchc.]. Stems 5-25 cm high with lines of hair from the leaf bases; leaves 2-5 cm long, the lower petiolate; inflorescence nodding in the bud; sepals sparsely glandular-hairy; petals white or pinkish, 2-5 mm long; capsule 5-10 cm long, glandular-hairy; seeds finely ridged. July-Aug.

Uncommon in moist soil of meadows, cliffs and along streams; alpine and subalpine; East, West. AK to Newf. south to CA, CO, NH.

Epilobium leptocarpum Hausskn. [*E. glandulosum* Lehm. var. *macounii* Trel. Hitchc.]. Similar to *E. halleanum*, 8-30 cm high, glabrous below, hairy above; leaves petiolate and entire below, sessile and toothed above; inflorescence noding to erect; sepals hairy; petals white; capsule 25-55 mm long, sparsely hairy; seeds bumpy. Aug.

Rare on moist, montane and lower subalpine cliffs; East. AK to OR, ID, MT.

Epilobium palustre L. [including *E. leptophyllum* Raf.]. Perennial 5-50 cm high, often with stolons; stems hairy; leaves narrowly lance-shaped with entire margins, 15-50 mm long, without petioles, glabrous or short-hairy; sepals hairy; petals white to pink, 2-9 mm long; capsule 3-9 cm long, densely hairy; seeds, finely bumpy. June-July.

Uncommon in montane fens; West, expected East. Circumboreal south to CA, CO, SD, MI, PA. *E. leptophyllum* has been segregated on the degree of leaf pubescence, a character that varies within populations and even as individual plants age.

Epilobium pygmaeum (Speg.) Hoch & Raven [*Boisduvalia glabella* (Nutt.) Walp.]. Annual 5-15 cm high, branched at the base; leaves alternate, without petioles, 5-15 mm long, lance-shaped with toothed margins; flowers solitary in crowded upper leaf axils; petals pink, 2-4 mm long; capsule club-shaped, 6-8 mm long; short-hairy.

Locally common around shallow ponds near East Glacier (Standley 1921). B.C. to Sask. south to CA, UT, WY, SD.

Epilobium saximontanum Hausskn. [*E. watsonii* Barbey in part]. Stems 40-80 cm high from underground turions; herbage hairy and glandular above; leaves 2-8 cm long, lance-shaped with minutely toothed margins, mostly clasping the stem; inflorescence erect or nodding in the bud; sepals sparsely hairy; petals white or purplish, 2-5 mm long; capsule 20-55 mm long, hairy and glandular; seeds bumpy in lines. July-Aug.

Common in meadows or along streams; montane to lower subalpine; East, West. Alta. to Newf. south to CA, NM.

Epilobium suffruticosum Nutt. Prostrate to ascending stems, 10-25 cm long, woody at the base; herbage finely hairy; leaves mostly opposite, 1-2 cm long, lance-shaped with entire margins; flowers in upper leaf axils; petals yellow, 7-9 mm long, 1 slightly larger; stigma 4-lobed; capsule 10-25 mm long. Aug.

Collected on gravel bars of the North Fork Flathead River just outside the Park. ID, MT, WY.

Gaura L., Butterfly-weed

Gaura coccinea Pursh. Perennial with numerous ascending stems 10-30 cm long; herbage mostly short-hairy; leaves alternate, lance-shaped, 1-3 cm long with shallow teeth; flowers dense in a narrow, terminal inflorescence; petals pink-red, 4-7 mm long, unequal; stamens 8; capsule club-shaped, 1-seeded, 4-angled, short-hairy.

Uncommon in grasslands along the eastern edge of the Park. B.C. to Man. south to CA, TX, MO.

Gayophytum Juss., Groundsmoke

Slender annuals; leaves mostly alternate, linear with entire margins; flowers in axils of upper, reduced leaves; petals white or pink; calyx lobes reflexed, stamens 8; stigma globose; capsules linear; seeds without long hair.

1. Capsules with stalks at least 2 mm long .... **G. diffusum**
1. Capsules sessile or nearly so 2
2. Seeds end-to-end in capsules .... **G. racemosum**
2. Seeds stacked horizontally or nearly so .... **G. humile**

Gayophytum diffusum T. & G. [*G. intermedium* Rydb., *G. nuttallii* T. & G.]. Stems 15-60 cm high, open-branched; herbage glabrous or hairy; leaves 15-50 mm long; petals 1-5 mm long; capsules erect, stalked, glabrous, 4-12 mm long. Aug. Fig. 189.

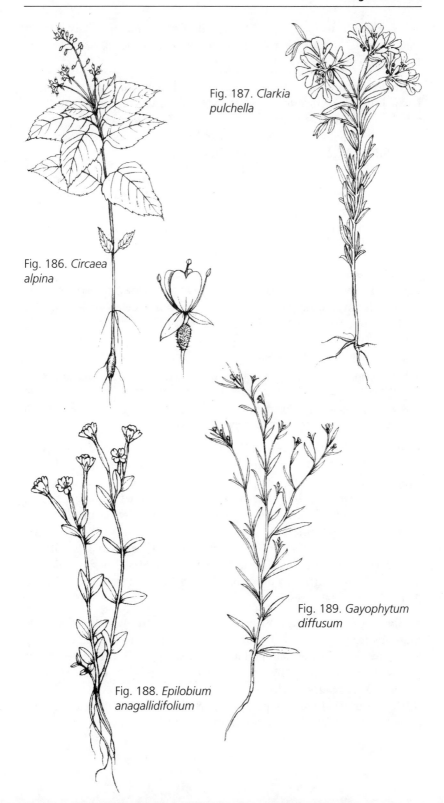

Fig. 187. *Clarkia pulchella*

Fig. 186. *Circaea alpina*

Fig. 189. *Gayophytum diffusum*

Fig. 188. *Epilobium anagallidifolium*

Uncommon in sparsely vegetated soil of grasslands near East Glacier (Standley 1921). Our plants are ssp. *parviflorum* Lewis & Szweykowski. B.C. to SD south to CA, NM.

Gayophytum humile Juss. Stems 5-20 cm high, leafy; herbage glabrous or hairy; leaves 5-30 mm long; petals ca. 1 mm long; capsules sessile, 7-15 mm long, glabrous or hairy.

Rare in vernally moist soil of grassland and open forest in the southern half of the Park; East, West. WA to MT south to CA.

Gayophytum racemosum T. & G. Stems 10-20 cm high, branched at the base; herbage mostly glabrous; leaves 1-2 cm long; petals < 1 mm long; capsules 8-14 mm long, without stalks or nearly so, glabrous.

Uncommon in sparsely vegetated soil of grasslands near East Glacier (Standley 1921). WA to Alta. south to CA, UT, CO.

Oenothera L., Evening Primrose

Perennial or biennial; leaves alternate or in basal rosettes; flowers yellow or white with a hypanthium tube between the ovary and petals; sepals reflexed; stigma 4-lobed; stamens 8; capsule 4-celled.

1. Plants without stems; leaves all in a rosette 2
1. Leafy stems at least 40 cm high 3
2. Petals white, > 2 cm long; capsules bumpy .... ***O. caespitosa***
2. Petals yellow, < 2 cm long; capsules 4-angled, not bumpy .... ***O. flava***
3. Flowers white; hypanthium tube as long as petals .... ***O. nuttallii***
3. Flowers yellow; hypanthium tube longer than petals...................... ***O. villosa***

Oenothera caespitosa Nutt., Rock Rose, Gumbo Lily [*Pachylophus caespitosus* (Nutt.) Raim.]. Stemless, taprooted; herbage mostly glabrous; leaves 5-15 cm long, narrowly oblong with toothed margins, petiolate; flowers among leaf rosettes; petals 3-4 cm long, white but pink with age; hypanthium tube 4-8 cm long; capsules often at or below ground level, 2-3 cm long, bumpy. June-July.

Sparsely vegetated soil of banks and open slopes near East Glacier (Standley 1921). WA to Sask., south to CA, NM, TX, Mex. The showy flowers are open at night and early morning. The crushed root was used by Native Americans to treat sores.

Oenothera flava (A. Nels.) Garrett [*Lavauxia flava* A. Nels.]. Stemless, taprooted; herbage nearly glabrous; leaves 5-20 cm long, the blades oblong with lobes along lower half, petiolate; flowers among leaf rosettes; petals yellow, 12-18 mm long; hypanthium tube 4-7 cm long; capsule among leaves, 2-3 cm long, 4-angled.

Drying mud of shallow ponds; reported for the Esat Glacier area by Standley (1921). WA to Sask., south to CA, CO, ND.

Oenothera nuttallii Sweet [*Anogra nuttallii* (Sweet) A. Nels]. Stems whitish, 40-100 cm high from rhizomes; herbage short-hairy, glandular above; leaves 5-10 cm long, narrowly oblong with mostly entire margins, without petioles, reduced above; flowers scented, in upper leaf axils; petals white, ca. 2 cm long; hypanthium tube as long as petals; capsule linear, 2-3 cm long.

Sandy soil of grasslands; reported for East Glacier area (Standley 1921). B.C. to Ont. south to CO, KS, WI.

Oenothera villosa Thunb. [*O. strigosa* (Rydb.) Mack. & Bush, *O. biennis* L. misapplied]. Rosette-forming, taprooted biennial, 40-70 cm high; herbage short-hairy, glandular above; leaves narrowly lance-shaped with wavy margins, 5-10 cm long, petiolate below; flowers in upper leaf axils; petals yellow, 12-25 mm long; hypanthium tube 2-3 cm long; capsule club-shaped, hairy, 2-3 cm long. July-Aug. Fig. 190.

Uncommon around roads and buildings and on dry slopes in the montane zone; East, West. Our plants are ssp. *strigosa* (Rydb.) Dietr. & Raven. B.C. to Man. south to CA, AZ, OK.

OROBANCHACEAE: BROOMRAPE FAMILY

Orobanche L., Broomrape

Non-green annual or perennial herbs; leaves alternate, scale-like; flowers bisexual, terminal on long, erect stalks; sepals 5; corolla tubular, 5-lobed, slightly 2-lipped; stamens 4; ovary superior; fruit a many-seeded capsule.

Roots attach to neighboring green plants and obtain all their water and nutrition parasitically.

1. Stem apparent, at least 2 cm above ground; flowers 3-10 ...... *O. fasciculata*
1. Stems mostly below ground; flowers mostly 1-4 .... *O. uniflora*

Orobanche fasciculata Nutt. Perennial with brownish stems 2-7 cm high; flowers 3-10; corolla purplish, 15-30 mm long. June-July.

Rare in montane grasslands; East and Blacktail Hills. AK to Ont. south to CA, NM, TX, IN. Usually parasitic on *Artemisia* or other members of Asteraceae.

Orobanche uniflora L. [*O. sedi* (Suks.) Fern.]. Annual with underground stems; few flowers apparently solitary on stalks 5-20 cm long; corolla curved 15-25 mm long, purple marked with yellow and white. June-July. Fig. 191.

Uncommon in stony, often moist soil of open, montane, and subalpine slopes; East, West. Our plants are var. *minuta* (Suks.) Beck. Much of temperate N. America. Parasitic on many hosts including Saxifragaceae, Asteraceae, Crassulaceae and Ranunculaceae (McLaughlin 1935).

PAPAVERACEAE: POPPY FAMILY

Papaver L., Poppy

Papaver pygmaeum Rydb. [*P. radicatum* Rottb. var. *pygmaeum* (Rydb.) Welsh]. Alpine glacier poppy. Tufted, taprooted perennial 5-12 cm high; herbage with stiff hairs and milky sap; leaves all basal, the blades ovate, 1-2 cm long with deep, rounded lobes, petiolate; flowers bisexual, solitary on long stalks, nodding in bud; sepals 2, falling in flower; petals 4, ca. 1 cm long, yellowish orange; ovary superior, stamens numerous; fruit a bristly, many-seeded, conical capsule, 10-15 mm long. July-Aug. Fig. 192.

Uncommon in stony soil of exposed, alpine ridges and slopes near or east of the Divide. Plants can be found in many alpine areas throughout the Park but are never very abundant. Endemic to the Waterton-Glacier area; a segregate of the circumpolar *P. radicatum* complex. Welsh et al. (1987) believe the same plant occurs in Utah.

PLANTAGINACEAE: PLANTAIN FAMILY

Plantago L., Plantain

Perennial (ours); leaves with prominent nerves and entire margins, all basal; flowers small, bisexual, each subtended by an ovate bract, in a congested terminal spike borne on a naked stalk; sepals 4; corolla white, 4-lobed, the tube enclosed by sepals; ovary superior; stamens 4, exserted from corolla; fruit a few-seeded capsule.

1. Leaf blades ovate, 1-2 times as long as broad, glabrous .... *P. major*
1. Leaf blades lance-shaped, > 2 times as long as wide, often hairy 2

2. Flower stalks long-hairy; sepals often hairy on the tip .... *P. canescens*
2. Flower stalks short-hairy; sepals glabrous .... *P. lanceolata*

Plantago canescens Adams [*P. septata* Morris]. Herbage long-hairy; leaves lance-shaped; flower stalk 10-20 cm high, the spike ca. 2-5 cm long; sepals ovate with papery margins. June. Fig. 193.

Rare in grasslands along the east margin of the Park. AK to B.C., Alta., MT; Asia.

Plantago lanceolata L., English Plantain. Herbage sparsely long-hairy, wooly at the petiole bases; leaves narrowly lance-shaped, 6-15 cm long; flower stalks 15-45 cm high; spikes congested, 1-2 cm long, conical, becoming cylindric; sepals hairy on the tip. June-July.

Uncommon at low elevations along roads and trails; West. Introduced from Eurasia.

Plantago major L. Herbage mostly glabrous; leaf blades ovate, 5-15 cm long, petiolate; flower stalks 10-40 cm high; spikes congested to somewhat sparse, 5-15 cm long, narrowly cylindric; sepals glabrous. June-July.

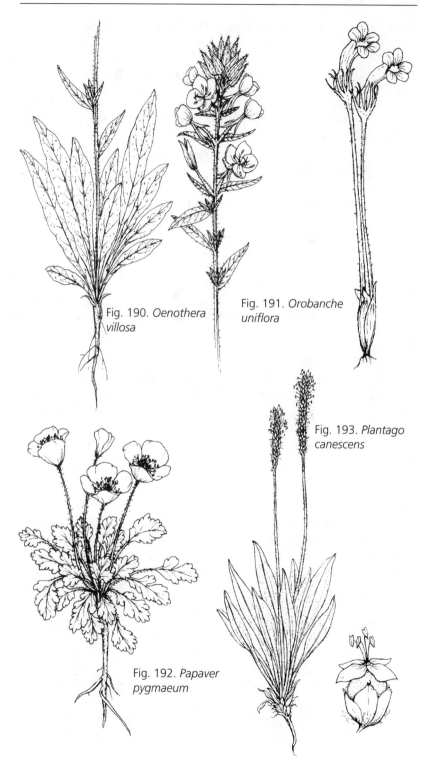

Fig. 190. *Oenothera villosa*

Fig. 191. *Orobanche uniflora*

Fig. 193. *Plantago canescens*

Fig. 192. *Papaver pygmaeum*

Uncommon along roads and around buildings and in lawns; East, West. Introduced from Eurasia.

POLEMONIACEAE: PHLOX FAMILY

Annual or perennial herbs; flowers bisexual; calyx 5-lobed; corolla tubular, 5-lobed; stamens 5; ovary superior; style mostly 3-lobed; fruit a 3-chambered capsule.

1. Leaves with entire margins, sometimes needle-like 2
1. Leaves divided; leaflets sometimes needle-like 4
2. Leaves < 5 mm wide, needle- or awl-like .... **Phlox**
2. Some leaves > 5 mm wide, lance-shaped but not awl-like 3
3. Lower leaves opposite .... **Microsteris**
3. Lower leaves alternate .... **Collomia**
4. Leaves opposite, the linear divisions appearing whorled .... **Linanthus**
4. Leaves alternate 5
5. Leaves lobed into needle-like segments, inflorescence not glandular ....
.... **Navarretia**
5. Leaflets elliptic to ovate; inflorescence glandular .... **Polemonium**

Collomia Nutt.
Collomia linearis Nutt. Annual 10-40 cm high; herbage short-hairy, glandular above; leaves alternate, narrowly lance-shaped with entire margins, 2-5 cm long, without petioles; flowers in a congested, leafy, hemispheric, terminal inflorescence; corolla pink, 7-12 mm long. June-July. Fig. 194.

Common in grasslands, open forest, thickets, rock outcrops, and along streams in the montane zone; East, West. B.C. to Ont. south to CA, NM, MO, IL.

Linanthus Benth.
Linanthus septentrionalis Mason. Annual 5-25 cm high, often branched above; leaves opposite, glabrous or short-hairy, palmately divided to the base into 5-7 linear lobes, 5-20 mm long; flowers long-stalked in an open inflorescence; corolla white, cup-shaped, 3-4 mm long. June-July. Fig. 195.

Uncommon in sparsely vegetated soil of grasslands and warm, open slopes; montane and lower subalpine near or east of the Divide. B.C. to Sask. south to CA, CO. Standley's (1921) report of the similar but more southern *S. harknessii* (Curran) Greene is referable here.

Microsteris Greene
Microsteris gracilis (Hook.) Greene. Annual 10-25 cm high, short-hairy and sometimes glandular above; leaves narrowly lance-shaped with entire margins, 2-3 cm long, opposite below; flowers in small clusters in upper leaf axils; corolla 8-12 mm long with a yellow tube and pink lobes. June. Fig. 196.

Uncommon in montane grasslands and on open slopes and stream banks; East, West. B.C., Alta. south to CA, NM.

Navarretia Ruiz & Pavon

Navarretia intertexta (Benth.) Hook. Annual to 10 cm high; leaves alternate, 5-20 mm long, 2 times pinnately divided into spine-like segments, long-hairy at the base; inflorescence open with spiny-bracted, hemispheric flower clusters at branch tips; corolla pale blue, 4-8 mm long; stigmas 2. June-July.

Rare in drying mud around ponds; Standley's (1921) report of *N. minima* Nutt. for East Glacier is referred here. Our plants are ssp. *propinqua* (Suksd.) Day. WA to Sask. south to CA, AZ.

Phlox L., Phlox

Mat-forming (ours) perennials; leaves opposite, firm, awl-shaped; flowers sessile, 1-3 at branch tips; calyx papery between lobes above; corolla tubular with conspicuous spreading lobes.

1. Leaves 2-5 mm wide with white margins .... *P. alyssifolia*
1. Leaves < 2 mm wide without white margins 2
2. Leaves with cobwebby hairs at the base; East .... *P. hoodii*
2. Leaves without cobwebby hairs; West .... *P. caespitosa*

Phlox alyssifolia Greene. Stems mostly prostrate, to 8 cm high; herbage spreading-hairy; leaves oblong-linear, 5-15 mm long with white margins; calyx glandular; corolla usually white, 15-18 mm long. June-July.

Uncommon in stony, sparsely vegetated soil of montane grassland and exposed alpine slopes along east edge of the Park. B.C. to Sask. south to CO, SD.

Phlox caespitosa Nutt. Stems woody at the base, ascending to 10 cm tall; herbage glandular-hairy; leaves linear, 5-13 mm long; calyx long-hairy and glandular; corolla ca. 10 mm long, usually white. June.

Rare in stony soil of montane grasslands and dry, open forest; West. B.C. to MT, south to OR, ID.

Phlox hoodii Richardson. Stems prostrate and cushion-forming; leaves linear, 2-8 mm long with long, tangled hair at the base; calyx with tangled hair; corolla usually white, ca. 6 mm long. May-June. Fig. 197.

Uncommon in grasslands and on banks and exposed slopes, montane to treeline; East. AK to CA, UT, CO, NE.

Polemonium L.

Tufted perennials from a branched rootcrown; herbage glandular; leaves alternate, pinnately divided; flowers in moderately congested, hemispheric, terminal inflorescence; calyx cup-shaped, corolla blue, funnel-shaped.

Plants, especially of *P. viscosum*, often have a skunk-like odor, but the intensity is variable among populations.

1. Corolla > 15 mm long; leaflets deeply divided, appearing whorled ....
.... *P. viscosum*
1. Corolla < 10 mm long; leaflets with entire margins .... *P. pulcherrimum*

Polemonium pulcherrimum Hook., Jacob's Ladder [*P. parvifolium* Nutt.].
Stems 5-30 cm high; leaves mostly basal with 11-23 entire-margined leaflets
to 1 cm long; corolla 5-10 mm long with yellow tube. May-July.

Common in stony soil of montane cliffs, outcrops, grasslands, and open
slopes; less common on exposed ridges and slopes near or above treeline;
East, West. Our plants are var. *pulcherrimum*. AK to CA, CO.

Polemonium viscosum Nutt., Sky Pilot. Stems 5-15 cm high; herbage
densely glandular; leaves to 10 cm long, mainly basal; leaflets deeply divided
into 2-5 ovate lobes, 1-6 mm long, appearing partly whorled on the leaf
axis; corolla 15-20 mm long. July-Aug. Fig. 198.

Common in rocky slopes and talus above treeline near or east of the
Divide. B.C., Alta. south to OR, NV, NM.

POLYGONACEAE: BUCKWHEAT FAMILY

Annual or perennial herbs; leaves mostly alternate with entire margins and
membranous stipules that sheath the stem where it joins the petiole; flowers
bisexual, dish- to funnel-shaped; petals lacking; sepals 4-6, small, often petal-
like, separate or sometimes united at the base; stamens 4-9; ovary with 2-3
styles; fruit a hard-coated seed (achene).

1. Flowers in flat-topped or hemispheric inflorescences, mostly yellow or white,
   sometimes reddish .... **Eriogonum**
1. Flowers in elongate inflorescences, mostly green or reddish 2

2. Basal leaves nearly orbicular in outline, as wide as long .... **Oxyria**
2. Basal leaves linear to elliptic or arrow-shaped, longer than wide 3

3. Sepals 5; achene enclosed by unswollen sepals .... **Polygonum**
3. Sepals 6; the inner 3 becoming swollen, wing-like and often net-veined with
   the achene at the base .... **Rumex**

Eriogonum Michx., Wild Buckwheat

Perennial herbs; leaves basal, petiolate with entire margins, lacking stipules;
flowers funnel-shaped, stalked, several arising from a bell-shaped, stalked
involucre, several of these arising from the stem tip (umbel) and subtended
by a whorl of leaf-like bracts; sepals 6; stamens 9; styles 3; achene 3-sided,
enclosed by the calyx.

Often very showy and excellent rock garden plants. *E. heracleoides* Nutt.,
with a whorl of leaves at mid-stem, occurs just south and west of the Park.

1. Flowers glabrous 2
1. Flowers hairy, at least at the base 3

2. Inflorescence a solitary globose umbel; leaf blades orbicular .. ***E. ovalifolium***
2. Inflorescence of many globose umbels; leaf blades narrowly elliptic ....
.... ***E. umbellatum***

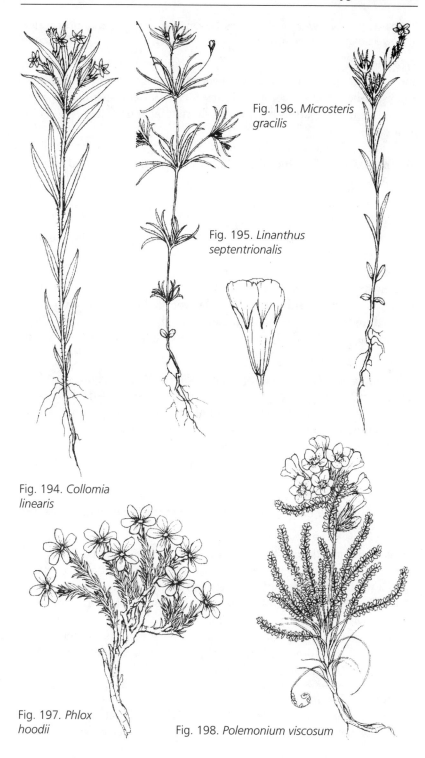

Fig. 196. *Microsteris gracilis*

Fig. 195. *Linanthus septentrionalis*

Fig. 194. *Collomia linearis*

Fig. 197. *Phlox hoodii*

Fig. 198. *Polemonium viscosum*

3. Flowers pale yellow or cream; leaves 1-2 cm long .... **E. androsaceum**
3. Flowers bright sulfur-yellow; leaves > 2 cm long .... **E. flavum**

Eriogonum androsaceum Benth. Mat-forming with stems to 8 cm high and old leaves at the base; leaves crowded, narrowly oblong, 1-2 cm long, appressed-hairy; umbel compact; bracts linear; flowers 3-5 mm long, light yellow to pinkish, sparsely hairy with a short, tubular base. July-Aug. Color plate 55.

Common in stony soil of exposed ridges and slopes near or above treeline; East and along the Divide. Endemic to northwest MT and adjacent B.C., Alta. Leaves turn pink in late summer.

Eriogonum flavum Nutt. [*E. piperi* Greene]. Cushion-forming with white-hairy stems 10-30 cm high; leaves petiolate, the blades 2-4 cm long, oblong; bracts spreading, leaf-like; flowers yellow, 4-5 mm long, hairy with a tubular base. June-Aug. Fig. 199. Color plate 7.

Ssp. *piperi* (Greene) Stokes, with longer tubular corolla bases, is abundant in stony soil of dry meadows, exposed ridges, and talus slopes, montane to alpine; ssp. *flavum*, with a short tubular corolla base, is uncommon on exposed, montane slopes along the east margin of the Park; East, West. B.C. to Man. south to OR, CO, ND. An extreme form of ssp. *piperi* with very short, green leaves occurs in the North Fork Flathead prairies.

Eriogonum ovalifolium Nutt. [*E. depressum* (Blank.) Rydb.]. Cushion-forming with stems 5-15 cm high; leaves spoon-shaped, closely white-hairy, the blades to 15 mm long; umbel hemispheric, bracts inconspicuous; flowers white, 4 mm long, glabrous, without a tubular base. July-Aug.

Var. *ovalifolium*, with stems > 10 cm high, is uncommon in sparsely vegetated soil of montane grasslands along the east edge of the Park; var. *depressum* Blank. with stems < 10 cm high, is common in stony soil of upper montane to alpine exposed ridges and slopes; East, West. B.C., Alta. south to CA, CO. Early prospectors believed this plant indicated silver deposits.

Eriogonum umbellatum Torrey [*E. subalpinum* Greene]. Stems 10-40 cm high from a branched rootcrown; leaf blades narrowly elliptic, 1-3 cm long, densely white-hairy beneath; umbel with spreading rays, subtended by oblong, leaf-like bracts; flowers yellowish or greenish white, 4-6 mm long, glabrous with a tubular base. June-Aug.

Common in stony soil of dry meadows, open forest, rock outcrops and open slopes; montane and subalpine; East, West. Our plants are var. *subalpinum* (Greene) Jones. B.C., Alta south to CA, NV, and CO.

Oxyria Hill, Mountain Sorrel
Oxyria digyna (L.) Hill. Glabrous perennial; stems 5-40 cm tall; leaves basal, long-petiolate, the blade 1-6 cm wide, nearly orbicular with a round-

lobed base; flowers short-stalked, whorled in a narrow inflorescence, each whorl subtended by an cup-shaped, membranous bract; sepals 4, red or green, 1-2 mm long; stamens 6; styles 2; achene oval, 4-6 mm wide with broad, flattened edges. July-Aug. Fig. 200.

Abundant in stony soil of moraine, rock outcrops, moist meadows, and talus slopes near or above treeline; East, West. Circumpolar south to CA, AZ, NM. The edible leaves have an acidic taste reminiscent of rhubarb.

Polygonum L., Knotweed, Smartweed

Annual or perennial; leaves with stipules sheathing the stem at the petiole base; sepals usually 5 (4-6), the outer ones often somewhat larger; stamens 8; styles mostly 3; fruit an achene enclosed in the calyx. Reference: Hickman (1993).

Mature achenes are often needed for positive identification. Annuals are predominately self-pollinating, sometimes creating an array of locally distinctive forms.

1. Flowers and fruits numerous in terminal, elongate clusters 2
1. Flowers and fruits 1-4 in axils of leaves or leaf-like bracts 5

2. Plants aquatic or stems prostrate and rooting at the nodes ..... *P. amphibium*
2. Plants not aquatic and rooting at the nodes; stems mostly erect 3

3. Fibrous-rooted annual; montane .... *P. lapathifolium*
3. Perennial from bulb-like rhizome; upper montane and above 4

4. Lower flowers converted to bulbs; stems 15-20 cm high .... *P. viviparum*
4. Bulbs lacking in the inflorescence; stems > 15 cm high .... *P. bistortoides*

5. Stems twining or creeping; leaf blades arrow-shaped .... *P. convolvulus*
5. Stems erect to prostrate but stiff; leaf blades linear to elliptic 6

6. Stems with 8-16 longitudinal ribs 7
6. Stems 3- to 5-angled but not ribbed 9

7. Leaves elliptic, 1/3-1/2 as wide as long .... *P. achoreum*
7. Leaves more linear, ≤ 1/4 as wide as long 8

8. Stems prostrate .... *P. aviculare*
8. Stems erect or nearly so .... *P. ramosissimum*

9. Leaves linear 10
9. Leaves elliptic 11

10. Mature achene black, shiny .... *P. douglasii*
10. Mature achene light brown to dull black .... *P. polygaloides*

11. Flowers erect or spreading; plants upper subalpine and alpine . *P. minimum*
11. Older flowers nodding; plants montane to lower subalpine ...... *P. douglasii*

Polygonum achoreum Blake [*P. erectum* L. in part]. Annual; stems branched at the base, ascending, 10-30 cm high; leaves ovate, 10-25 cm long; flowers 1-3 in leaf axils; sepals 3 mm long, green with white margins, the outer hood-like at the tip; achene 3-angled, yellow-brown, 2-3 mm long. July.

Uncommon in disturbed areas of montane grasslands or along roads and trails; East. WA to Que. south to OR, CO, NY.

Polygonum amphibium L. Water Smartweed [includes *P. coccineum* Muhl.]. Rhizomatous, aquatic perennial; stems prostrate or floating, rooting at the nodes; leaves lance-shaped, the blades to 10 cm long; foliage glabrous to hairy; flowers clustered in a terminal spike 1-10 cm long; sepals pink, 4-5 mm long; achene flattened, 2-3 mm long, shiny dark brown. July-Aug.

Uncommon in shallow water of montane ponds; we have var. *emersum* Michx., with hairy leaves, East, and var. *stipulaceum* Coleman, with glabrous leaves, West. Circumboreal south to most of U. S.

Polygonum aviculare L., Dooryard Knotweed. Annual; stems prostrate to ascending, 5-30 cm long; leaves narrowly lance-shaped, 1-2 cm long, short-petiolate; flowers 1-3 in leaf axils; sepals 2-3 mm long, greenish with white or pink margins; achene 3-sided, brown, 2-3 mm long. June-Aug.

Locally common in disturbed soil of grasslands and especially along roads and trails, montane to lower subalpine; East, West. Introduced from Europe and widespread in N. America. Some of our plants may be better referred to *P. arenastrum* Jord. ex Boreau.

Polygonum bistortoides Pursh, American Bistort. Perennial from a bulb-like rhizome; stems erect, 15-50 cm high, unbranched; basal leaves petiolate, the blades 5-10 cm long, narrowly lance-shaped; stem leaves smaller, sessile; flowers clustered in a thimble-shaped spike 1-5 cm long; sepals white to pink, 3-5 mm long; stamens exserted; achene 3-sided, shiny pale brown. June-Aug. Fig. 201.

Common in moist grasslands, meadows and turf where wet early but drying later in the season; montane to alpine; East, West. B.C., Alta. south to CA, NM. The root was eaten by Native Americans.

Polygonum convolvulus L., Black Bindweed. Glabrous annual; stems twining or trailing; leaves long-petiolate, the blades arrow-shaped, 1-4 cm long; flowers in leaf axils and short, narrow terminal clusters; sepals greenish, ca. 4 mm long; achene 3-sided, 3-4 mm long, black. July-Aug.

Uncommon in disturbed ground along roads and trails; East, West. Introduced from Europe.

Polygonum douglasii Greene [*P. austiniae* Greene, *P. engelmannii* Greene]. Annual; stems erect or ascending, 4-30 cm high; leaves sessile, 5-40 mm long, linear to elliptic, reduced above; flowers 1-4 in upper leaf axils; sepals pink to white, 2-5 mm long; achene 3-sided, 2-3 mm long, shiny black. July-Aug. Fig. 202.

Ssp. *douglasii*, with linear leaves, is common in sparsely vegetated soil of stream banks, grasslands and open slopes, montane to subalpine; East, West. Ssp. *austiniae* (Greene) Murray, with elliptic leaves, is uncommon in shaly, open soil of montane grasslands and stream banks along the south margin of the Park. B.C. to Que. south to CA, NM, NY.

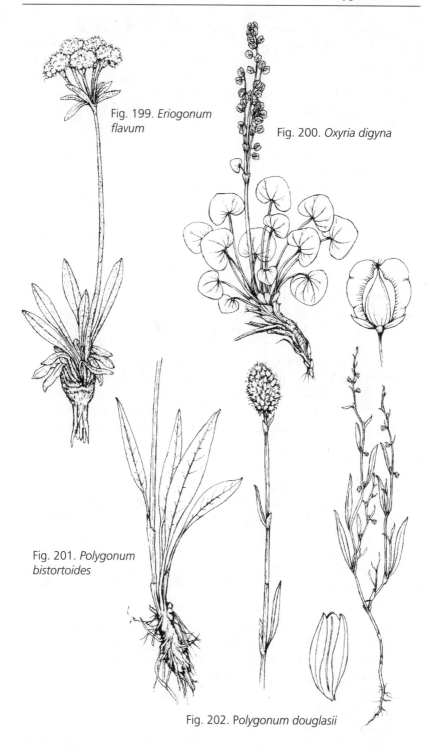

Fig. 199. *Eriogonum flavum*

Fig. 200. *Oxyria digyna*

Fig. 201. *Polygonum bistortoides*

Fig. 202. P*olygonum douglasii*

Polygonum lapathifolium L. Annual; stems 15-70 cm high, usually branched; leaves lance-shaped, 2-10 cm long, short-petiolate; flowers in a series of glandular spikes from upper leaf axils; sepals 2-3 mm long, green to white; achene flattened, 2-3 mm long, shiny dark brown. July-Sept.

Rare in moist soil along montane streams or ponds; collected once along North Fork Flathead River. Much of N. America, Europe. Hickman (1993) considers this species native, while Hitchcock et al. (1964) believe it is introduced in the Pacific Northwest.

Polygonum minimum Watson. Annual; stems branched, ascending, 1-10 cm high; leaves ovate, nearly sessile, 3-10 mm long; flowers 1-4, concealed in leaf axils; sepals 2 mm long, greenish with pink to white margins; achene 3-sided, ca. 2 mm long, shiny black. July-Sept.

Common in sparsely vegetated stream banks, moist rock outcrops and vernally moist open slopes, subalpine to alpine; East, West. B.C., Alta. south to CA, CO. This is one of our few alpine annuals.

Polygonum polygaloides Meissner [*P. kelloggii* Greene, *P. watsonii* Sm.]. Glabrous annual; stems erect, often branched, 1-10 cm high; leaves linear, 5-15 mm long; flowers 1-4 in axils of closely overlapping leaf-like bracts at stem tips; sepals 2-3 mm long, white to pink with green midrib; achene 3-sided, ca. 2 mm long, light brown to black. Aug.

Ssp. *confertiflorum* (Piper) Hickman, with white-margined floral bracts broader than the leaves, is uncommon in vernally wet soil in montane grasslands, rock outcrops, and around ponds; East and near Marias Pass; ssp. *kelloggii* (Greene) Hickman, with narrow, spreading, green-margined floral bracts, is uncommon in moist, open soil of subalpine ponds and meadows; East.

Polygonum ramosissimum Michx. Annual; stems erect, branched, 10-20 cm high; leaves stap-shaped, 1-3 cm long, short-petiolate; flowers 1-3 in leaf axils; sepals ca. 3 mm long, green with yellow margins; achene dark brown, 3-sided.

Collected once along railroad tracks near West Glacier by Standley (1921). Native in eastern N. America but thought to be introduced west of the Continental Divide (Hickman 1993). Similar to *P. aviculare* but with erect stems.

Polygonum viviparum L., Alpine Bistort. Perennial from a bulb-like rhizome; stems erect, 5-25 cm high, unbranched; basal leaf blades narrowly oblong, 2-8 cm long, petiolate; stem leaves smaller, sessile; flowers unisexual, clustered in a narrow spike 2-6 cm long, the lower replaced by purplish, ovoid bulbs; sepals white to pink, 3-4 mm long; achene 3-sided, brown. July-Aug. Fig. 203.

Common in moist alpine turf, often with *Dryas octopetala* and dwarf willows; East, West. Circumpolar south to OR, NM, MN, NH. Male flowers have exserted stamens, while they are short in female flowers; viable seed is rarely produced.

Rumex L., Dock, Sorrel

Annual or perennial herbs; leaves with membranous stipules sheathing the stem at the base of the petiole; flowers stalked in small whorls in a branched inflorescence; sepals 6, outer 3 reflexed, inner 3 colored and becoming enlarged and veiny in fruit, sometimes with a bump-like swelling toward the base; stamens 6, styles 3; achene 3-angled, enclosed at base of inner sepals.

Dates given correspond to time of mature fruit, which may be needed for positive identification.

1. Blades of some leaves with back- or outward-pointing lobes at the base, arrow-shaped 2
1. Leaf blades strap-, lance-, or ellipse-shaped 3
2. Stem ca. 1 mm wide; leaf blades ≤ 4 cm long .... *R. acetosella*
2. Stem ≥ 2 mm wide; largest leaf blades > 4 cm long .... *R. acetosa*
3. Inner sepals enclosing the seed with long-toothed margins at maturity; uncommon 4
3. Margins of mature inner sepals without teeth or lobes 5
4. Taprooted perennial .... *R. obtusifolius*
4. Fibrous-rooted annual or biennial .... *R. maritimus*
5. Blades of some basal leaves > 6 cm wide .... *R. occidentalis*
5. Basal leaf blades ≤ 5 cm wide 6
6. Basal leaf blades with wavy (crisped) margins .... *R. crispus*
6. Basal leaf blades with entire margins 7
7. Stem leaves nearly as large as basal ones .... *R. salicifolius*
7. Stem leaves greatly reduced upward .... *R. paucifolius*

Rumex acetosa L. Glabrous perennial; stems 20-70 cm high, ribbed; leaf blades long arrow-shaped, 2-12 cm long, the basal long-petiolate; flowers unisexual on separate plants (dioecious); inflorescence 10-25 cm long; outer sepals 2-3 mm long; inner sepals of female flowers 3-5 mm long in fruit, net-veined, without a swelling. July-Aug.

Uncommon in moist meadows and talus slopes, upper montane to alpine; East, West. Our plants are ssp. *alpestris* (Scop.) Löve. Circumboreal south to OR, WY. A closely related ssp. is introduced from Europe and grown as a garden herb that tastes much like rhubarb.

Rumex acetosella L., Sheep Sorrel. Glabrous annual or perennial; stems 10-40 cm high; leaves mostly basal, the blades lance- to elongate arrow-shaped, 1-3 cm long; flowers unisexual on separate plants (dioecious); inflorescence narrow; outer sepals ca. 1 mm long; inner sepals of female flowers ca. 2 mm long in fruit, without a swelling. July.

Common in sparsely vegetated, often disturbed soil of grasslands and meadows, often along trails and roads, montane and subalpine; East, West. Introduced from Europe into most of N. America.

Rumex crispus L., Curly Dock. Perennial; stem 30-80 cm tall, unbranched; leaf blades broadly strap-shaped, 5-20 cm long with wavy margins; flowers in numerous dense spikes in a branched inflorescence; inner sepals 3-6 mm in fruit with an ovate swelling at the base. July-Aug.

Uncommon in moist, often disturbed soil along montane ditches or streams, sometimes in aspen groves; East, West. Introduced from Europe into most of N. America.

Rumex maritimus L. Annual or biennial; stems 5-40 cm high, branched; leaf blades narrowly lance-shaped, 2-10 cm long, reduced above; flowers in dense whorls in axils of leaf like-bracts on branch tips; sepals 1-2 mm long; inner sepals swelling to 3 mm long, net-veined with several bristle-like lobes on the margins and a cylindrical swelling at the base.

Rare in drying mud around ponds and lakes on the east margin of the Park. Circumboreal south to S. America.

Rumex obtusifolius L., Bitter Dock. Perennial; stems 60-100 cm high from a tap root; leaf blades ovate, the lower 10-20 cm long; flowers in dense clusters subtended by reduced leaves in a spike-like inflorescence at branch tips; sepals ca. 3 mm long, the inner 5 cm long at maturity, net-veined with long-toothed margins and an ovoid swelling at the base.

Standley (1921) reported a few plants at St. Mary. Introduced from Europe.

Rumex occidentalis Watson. Mostly glabrous perennial; stems 50-120 cm high; leaves mostly basal, the blades ovate to arrow-shaped, 5-25 cm long; inflorescence branched, leafy, 20-40 cm long; sepals ca. 2 mm long, the inner 5-10 mm long at maturity, net-veined, with mostly entire margins and no swelling at the base. July-Aug.

Uncommon in montane fens, wet meadows, and thickets; East, West. AK to Que. south to CA, NV, NM, SD.

Rumex paucifolius Nutt. Glabrous, taprooted perennial; stems 20-60 cm high; leaves mostly basal, the blades narrowly elliptic, 6-10 cm long; flowers unisexual on separate plants (dioecious), in dense spike-like inflorescence branches subtended by reduced leaves; inner sepals 3-4 mm long without a swelling at the base. July-Aug.

Rare in vernally moist grasslands and meadows; reported for East Glacier by Standley (1921). B.C., Alta. south to CA, CO.

Rumex salicifolius Weinm. [*R. triangulivalvis* (Dans.) Rech., *R. mexicanus* Meisn.]. Glabrous, taprooted perennial; stems 30-70 cm high; leaf blades narrowly elliptic, 5-15 cm long, little reduced above; flowers in dense, spike-like branches of a leafy-bracted inflorescence, 10-30 cm long; sepals green, ca. 2 mm long, the inner ca. 3 mm long at maturity with an ovoid swelling at the base. July-Aug. Fig. 204.

Uncommon in moist montane meadows and thickets, often along streams or around ponds; East, West. Our plants are var. *triangulivalvis* (Dans.) Hickman. AK to Que. south to CA, NM, IN, NY; Europe. Decoction from boiled roots was used by Native Americans to reduce swelling.

PORTULACACEAE: PURSLANE FAMILY

Annual or perennial herbs; leaves often succulent, undivided with entire margins; flowers bisexual, dish- to funnel-shaped; sepals usually 2, separate or united at the base; petals 5 (sometimes more), separate or united at the base; stamens usually 5; ovary and style 1, above base of sepals (superior); fruit a capsule.

1. Sepals > 2 or appearing so, completely separate; petals ≥ 5 .... **Lewisia**
1. Sepals 2, united at the base; petals 5 2

2. Stem leaves 2, opposite .... **Claytonia**
2. Stem leaves alternate, usually > 2 .... **Montia**

Claytonia L.
Glabrous annuals or perennials; basal leaves petiolate; stem leaves 2, sessile, opposite; flowers several, stalked in an open inflorescence; petals 5, notched at the tip; stamens 5; style divided in 3; capsule 3-chambered. Reference: Chambers (1993).

1. Basal leaves usually 1-2; stems from a globose rhizome .... *C. lanceolata*
1. Basal leaves > 2; plants taprooted or fibrous-rooted 2

2. Stem leaves linear; alpine perennial .... *C. megarhiza*
2. Stem leaves wider; plants montane; annual 3

3. Stem leaves broadly united into a disk with the stem through the center ......
.... *C. perfoliata*
3. Stem leaves ovate, not broadly united .... *C. sibirica*

Claytonia lanceolata Pursh, Spring Beauty. Perennial; stems 5-15 cm tall from a globose root; basal leaves mostly 1-2, the blade lance-shaped; stem leaves broadly lance-shaped, 10-60 mm long; flower stalks 1-3 cm long, spreading; petals white with pink lines, 7-14 mm long; capsule 4 mm long. May-July. Fig. 205.

Common in vernally moist, often shallow soil of grasslands, meadows and open forest, montane to alpine; East, West. B.C., Alta. south to CA, NM. Plants seem to prefer warm slopes where snow accumulates and flower shortly after it melts. Native Americans ate the roots (corms) fresh or roasted.

Claytonia megarhiza (Gray) Parry. Perennial; stems 5-15 cm high from a thick taproot; foliage often reddish; basal leaves numerous, spoon-shaped, 1-8 cm long; stem leaves linear; flower stalks 5-40 mm long, ascending; petals pink, 6-12 mm long; capsule 4-6 mm long. July-Aug.

Uncommon in very stony, often moist soil of alpine open slopes and ridges; East, West. B.C., Alta. south to NV, UT, NM. Grizzly bears will dig the roots when they are common enough.

Claytonia perfoliata Donn ex Willd. [*Montia perfoliata* (Willd.) Howell]. Annual; stems 3-15 cm tall; basal leaf blades linear to elliptic; stem leaves united, appearing like a disk with the stem through the center; flower stalks spreading; petals white, 2-4 mm long; capsule 2-3 mm long. May-June.

Rare in vernally moist, sparsely vegetated soil of montane open forest and grasslands, often along streams; known only from a mossy talus slope near West Glacier. B.C. to ND south to CA, UT, AZ.

Claytonia sibirica L. [*Montia sibirica* (L.) Howell]. Annual to short-lived perennial; stems 10-30 cm high; basal leaf blades elliptic, 1-4 cm long; stem leaves broadly ovate, 1-7 cm long; flower stalks spreading, 5-20 mm long; petals white, 6-12 mm long; capsules 3-5 mm long. July.

Rare in moist, montane forest openings; collected once in the upper McDonald Creek Valley. AK to MT south to CA, UT.

Lewisia Pursh

Glabrous, succulent perennials; leaves basal or opposite (ours); petals 5-18; stamens 5-50; style 1 with 3-8 stigmas.

1. Petals > 10 mm long; sepals 6-9 .... **L. rediviva**
1. Petals < 9 mm long; sepals 2-4 2
2. Basal leaves numerous on flowering plants .... **L. pygmaea**
2. Flowering plants without basal leaves .... **L. triphylla**

Lewisia pygmaea (Gray) Robinson [*Oreobroma pygmaea* (Gray) Howell]. Taprooted; stem 1-4 cm high with a pair of narrow, opposite, leaf-like bracts; basal leaves numerous, linear, to 8 cm long; flowers deep pink, 1 per stem; sepals 2, with gland-toothed margins; petals 6-8, 4-10 mm long. July-Aug. Fig. 206.

Common in gravelly, shallow, vernally moist soil of banks, rock ledges and open slopes, montane to alpine; East, West. Yuk to Alta. south to CA, NM. Abundant in the Logan Pass "hanging gardens."

Lewisia rediviva Pursh, Bitterroot. Taprooted; stem 1-3 cm tall with a whorl of membranous bracts below the solitary flower; basal leaves numerous, tubular, 15-50 mm long, appearing and sometimes disappearing before flowering; sepals 6-9, unequal; petals 12-18, pink (ours) 18-35 mm long; stamens 30-50; style branches 4-8. June.

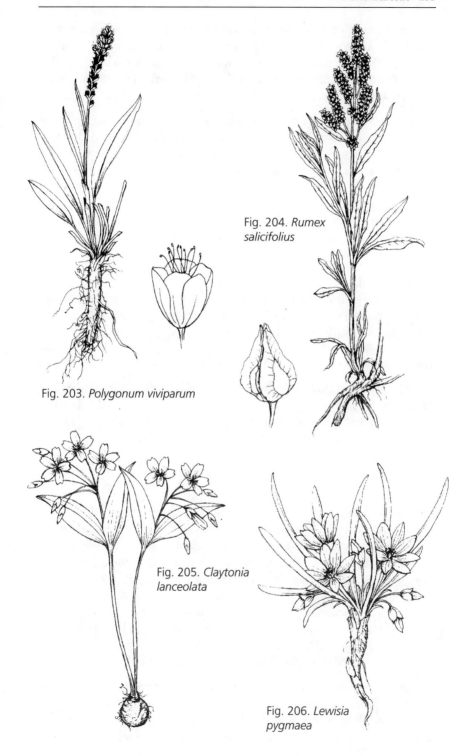

Fig. 203. *Polygonum viviparum*

Fig. 204. *Rumex salicifolius*

Fig. 205. *Claytonia lanceolata*

Fig. 206. *Lewisia pygmaea*

Rare in gravelly, well-drained, usually sparsely vegetated soil of montane grasslands; known only from the Marias Pass area. B.C. to MT south to CA, AZ, CO. The state flower of Montana. Native Americans ate the boiled or steamed roots, and may have introduced the plant into western Glacier County (DeSanto 1993).

Lewisia triphylla (Watson) Robinson. Stems 2-6 cm long from a globose root; basal leaf blades 1-3 cm long; stem leaves 2-3, linear, 1-3 cm long, whorled; flowers several, stalked, in an open inflorescence; sepals 2 but appearing 4; petals 5-9, white to pink, 4-7 mm long; stamens 5; style branches 3-5. June-Aug.

Rare in gravelly, shallow, vernally moist, sparsely vegetated soil of rock outcrops and open grasslands, montane to subalpine; East, West. WA to MT south to CA, CO. This inconspicuous plant may be more common than records indicate.

Montia L., Miner's Lettuce

Annual or short-lived perennial; leaves basal and alternate; flowers stalked in a narrow inflorescence; sepals 2; petals 5; styles 3, stamens 3-5.

1. Basal leaves mostly > 3 mm wide; small plantlets in lower leaf axils .... ***M. parvifolia***
1. Basal leaves 1-3 mm wide; stem leaves without plantlets .... ***M. linearis***

Montia linearis (Dougl. ex Hook.) Greene [*Claytonia linearis* Hook., *Montiastrum lineare* (Hook.) Rydb.]. Annual; stems 3-10 cm high, branched at the base; leaves linear, 1-4 cm long; flowers several, all on 1 side of inflorescence; petals white, 3-5 mm long; stamens 3. May-July.

Common in vernally moist, sparsely vegetated soil of montane grasslands and rock outcrops and around ponds; East, West. B.C. to Sask.. south to CA, UT.

Montia parvifolia (Moc. ex D. C.) Greene [*Claytonia parvifolia* Moc. ex D. C.]. Perennial with stolons (runners); stems 5-30 cm long, erect or lax; basal leaves petiolate, 15-30 mm long; the blades narrowly to broadly elliptic; stem leaves small, lance-shaped, often with plantlets in the axils; flowers few; petals pink, 6-10 mm long; stamens 5. June-Aug. Fig. 207.

Common in shallow, often mossy soil of stream banks, rock outcrops and cool talus slopes, montane to near treeline; East, West. AK to Alta. south to CA, UT. The small axillary plantlets establish readily in moist moss.

PRIMULACEAE: PRIMROSE FAMILY

Annual or perennial herbs; leaves unlobed and undivided, mostly basal or opposite; flowers bisexual, stalked, dish- to vase-shaped or with petals reflexed; petals mostly 5, united at the base or into a basal tube; sepals 5, united; stamens 5; style 1; ovary above the base of the sepals (superior); fruit a capsule.

1. Leaves primarily on stems 20-80 cm high .... **Lysimachia**
1. Leaves all basal; naked stems to 25 cm high 2
2. Flowers nodding; corolla lobes sharply reflexed upward .... **Dodecatheon**
2. Flowers and petals erect 3
3. Flowers mostly solitary on stalks .... **Douglasia**
3. Flowers several, the stalks arising from stem tip in an open or congested
   umbel .... **Androsace**

Androsace L., Fairy Candelabra

Annual or perennial; leaves all basal; flowers several, the stalks all arising
from the stem tip (umbel), subtended by inconspicuous bracts; corolla
tubular, 5-lobed.

1. Corolla with a yellow center; umbel congested .... **A. chamaejasme**
1. Corolla pure white; umbel open 2
2. Bracts at the base of the flower stalks strap-shaped .... **A. septentrionalis**
2. Flower bracts narrowly elliptic or ovate .... **A. occidentalis**

Androsace chamaejasme Host, Rock-jasmine [*A. lehmanniana* Spreng.].
Perennial forming loose cushions from a branched rootcrown; stems 2-10
cm high; leaves narrowly oblong, hairy; umbel congested, hemispheric;
corolla white with a yellow center, 5-7 mm across. June-Aug.

Locally common in calcareous soil of grasslands and rocky slopes along
the east front of the mountains, montane to alpine. Our plants are ssp.
lehmanniana (Spreng.) Hult. Circumpolar south to CO.

Androsace occidentalis Pursh. Similar to *A. septentrionalis*; bracts at the
base of the umbel elliptic to ovate; corolla shorter than calyx. June.

Rare in montane grasslands or open forest, sometimes in disturbed soil;
collected once at West Glacier. B.C. to Ont. south to CA, NM, TX, AR.

Androsace septentrionalis L. [*A. puberulenta* Rydb., *A. subumbellata* (Nels.)
Small]. Annual; stems 2-12 cm high, often several from a rosette; leaves
narrowly lance-shaped, 5-20 mm long with entire or shallowly toothed
margins; flower stalks long in an open umbel subtended by linear bracts;
corolla white, the tube 4-5 mm long, as long as the calyx. May-July. Fig.
208.

Common but inconspicuous in sparsely vegetated soil of montane to
alpine grasslands, meadows and rocky slopes; East, West. Circumboreal
south to CA, AZ, NM.

Dodecatheon L., Shooting Star

Perennial; leaves all basal, petiolate, the margins entire; flowers 1-several,
nodding, the stalks arising from the stem tip (umbel); petals pink to
magenta, basally united, reflexed upward; anthers exserted, the stalks short,
united into a tube; capsules erect.

Although it is synonymous with *D. pulchellum*, *D. cusickii* Greene reported by Standley (1921) is probably referable to *D. conjugens.*

1. Dark base of anther tube horizontally wrinkled; capsule opening by a hole on top; leaves glandular-hairy; montane grasslands .... **D. conjugens**
1. Dark base of anther tube smooth or wrinkled up-and-down; capsule opening by 5 slits on top; leaves glabrous; wet montane to alpine habitats .... **D. pulchellum**

Dodecatheon conjugens Greene [*D. acuminatum* Rydb.]. Stem 8-25 cm high; leaves mostly glandular-hairy, oblong, 3-12 cm long; corolla lobes 8-20 mm long; base of anther tube purple, horizontally wrinkled. May-June. Color plate 5.

Common in grasslands, montane to near treeline; our plants are var. *viscidum* (Piper) Mason; East, West. B.C. to Sask. south to CA, WY.

Dodecatheon pulchellum (Raf.) Merr. [*D. pauciflorum* (Durand) Greene]. Stem 5-35 cm high; leaves 2-15 cm long, the blades narrowly lance-shaped to oblong; corolla lobes 10-15 mm long; base of anther tube smooth or wrinkled up-and-down. June-Aug. Fig. 209.

Abundant in perennially moist to wet soil of meadows and along streams, montane to alpine; East, West. Smaller alpine plants with few flowers have been called var. *watsonii* (Tieds.) Hitchc. AK to Man. south to CA, AZ, CO, NE.

Douglasia Lindl.

Douglasia montana Gray. Cushion-forming, moss-like perennial; rootcrown long-branched, terminating in rosettes; leaves basal, linear, 4-8 mm long; flowers solitary, erect on short stalks 5-10 mm high; corolla 6-8 mm long, magenta, tubular with 5 lobes ca. 4 mm long; capsule longer than the calyx. May-June.

Uncommon in stony soil of exposed slopes and ridges, montane grasslands and alpine fellfields; East, West. Alta. south to ID, WY.

Lysimachia L., Loosestrife

Rhizomatous perennials; leaves opposite with entire margins; flowers yellow, stalked; calyx and corolla 5- or 6-lobed; petals separate to near the base; stamens 5-6.

1. Flowers in congested spikes paired at nodes; leaves sessile ...... **L. thyrsiflora**
1. Flowers long-stalked, paired at nodes; leaves with hairy petioles ...... **L. ciliata**

Lysimachia ciliata L., Fringed Loosestrife [*Steironema ciliatum* (L.) Raf.]. Stems 20-50 cm high; leaf blades ovate, 3-6 cm long, the short petioles with long hairs; flowers dish-shaped, 15-20 mm wide, long-stalked and paired from each upper leaf node; stamens 5. July-Aug.

Rare in montane thickets and aspen groves along the east margin of the Park; reported for Lower St. Mary Lake (Maguire 1934). B.C. to N. S. south to most of N. America.

Lysimachia thyrsiflora L. Stems 20-60 cm high; leaves sessile, narrowly lance-shaped, 5-10 cm long, dark-dotted, the lower small; flowers bell-shaped in dense thimble-shaped, long-stalked spikes, paired at leaf nodes; petals linear, 4-6 mm long, purple-dotted; style and stamens exserted. July. Fig. 210.

Uncommon in montane fens and marshes; East, West. Circumboreal south to CA, CO, MO, WV.

RANUNCULACEAE: BUTTERCUP FAMILY

Mostly perennial herbs or woody vines; leaves mostly basal and/or alternate; flowers mostly bisexual, dish-shaped (except *Aquilegia*); sepals mostly 5-20, separate, sometimes petal-like; petals mostly 5-10 or lacking, separate; ovaries 1-many, above the base of sepals (superior); stamens 5-many.

1. Fruits dry, 1-chambered, many-seeded, opening at the top and along the edges (follicles), in radial clusters of 3-20 2
1. Fruit a berry or hard-coated seed (achene), often with a beak or feather-like style on top 4
2. Leaves 1-2 times pinnately divided into leaflets; flower with 5 backward-pointing spurs .... **Aquilegia**
2. Leaves deeply palmately lobed; flower with 1 or no spurs 3
3. Flowers dish-shaped, without a spur; follicles 10-20 .... **Trollius**
3. Flowers with a backward-pointing spur; follicles 3 .... **Delphinium**
4. Plants twining or trailing vines, woody at least toward the base ...... **Clematis**
4. Plants herbaceous with erect, creeping or floating stems 5
5. Fruit a red or white berry; petals and sepals inconspicuous; stamens white ...
.... **Actaea**
5. Fruit an achene with a persistent style; flowers not as above 6
6. All leaves basal, linear .... **Myosurus**
6. Plants with divided or lobed leaves or with stem leaves 7
7. Stem leafless except for a whorl of 3 divided leaves at the middle . **Anemone**
7. Leaves alternate or all basal 8
8. Leaves divided into > 10 stalked leaflets; male and female flowers on separate plants; sepals and petals inconspicuous or greenish .... **Thalictrum**
8. Leaves deeply lobed or divided into 3-5 stalked leaflets; flowers white or yellow, bisexual .... **Ranunculus**

Actaea L., Baneberry
Actaea rubra (Ait.) Willd., Doll's-eyes. Stems 30-80 cm high; leaves 2-3 times divided into 3; the ultimate leaflets ovate, 3-12 cm long, with lobed or toothed margins; flowers on ascending stalks in a short, hemispheric inflorescence becoming narrower with age; sepals 3-5, white, petal-like, quickly falling; petals white, narrow, 2-3 mm long; stamens many; ovary 1; fruit a several-seeded red or white berry. June-July. Fig. 211.

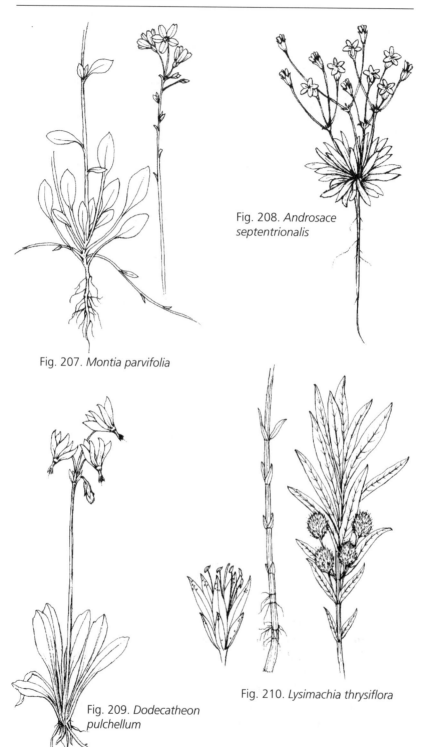

Fig. 207. *Montia parvifolia*

Fig. 208. *Androsace septentrionalis*

Fig. 209. *Dodecatheon pulchellum*

Fig. 210. *Lysimachia thrysiflora*

Common in moist montane and lower subalpine forest and thickets, often with spruce; East, West. AK to Newf. south to CA, NM, SD, OH, NJ. Fruits are poisonous; roots were used by Native Americans to treat coughs and colds. Red- and white-fruited plants may occur in the same population.

Anemone L., Windflower, Anemone

Leaves basal and 1 whorl of 3 near mid-stem, divided or deeply lobed, the basal petiolate; flowers 1-few, long-stalked; sepals 5-9, petal-like; petals absent; ovaries and stamens numerous; fruit a hard-coated seed (achene) with hairs around the base clustered in a head. Reference: Boraiah & Heimburger (1964).

A. cylindrica Gray, similar to A. multifida but with a longer seed head, occurs in Waterton Park (Kuijt 1982). The rhizomatous A. piperi Britt. occurs just south of the Park.

1. Styles of mature achenes > 15 mm long; seed heads shaggy; sepals ≥ 2 cm long 2
1. Styles < 5 mm long; seed heads sometimes cottony but not shaggy; sepals < 2 cm long 3
2. Sepals blue on the outside; mature styles spreading; montane ...... **A. patens**
2. Sepals white; mature styles reflexed; subalpine to alpine .... **A. occidentalis**
3. Leaf blades once divided into 3 wedge-shaped, lobed segments .... **A. parviflora**
3. Leaf blades 2-4 times divided into linear segments 4
4. Stems mostly > 20 cm high; flowers often > 1 per stem; montane .... **A. multifida**
4. Stems mostly < 20 cm high; flowers 1 per stem; subalpine, alpine 5
5. Styles often hooked, 1-1.5 mm long; ultimate leaf segments long-tapered ... . **A. tetonensis**
5. Styles straight, > 2 mm long; ultimate leaf segments oblong, not long-tapered .... **A. lithophila**

Anemone lithophila Rydb. [A. drummondii Wats. var. lithophila (Rydb.) Hitchc.]. Stems 10-20 cm high; foliage sparsely long-hairy; basal leaves 3-4 times divided and lobed into 3's, the ultimate lobes oblong; flowers mostly solitary; sepals white, bluish on the outside, 9-12 mm long; styles straight, 2-4 mm long; seed head globose, ca. 1 cm wide. July-Aug. Fig. 212.

Common in meadows and stony slopes near or above treeline; East, West. B.C., Alta. south to WA, ID, WY.

Anemone multifida Poiret [A. globosa Nutt.]. Stems 20-70 cm high; foliage long-hairy; basal leaves 2-3 times divided and lobed into 3's, ultimate segments long-tapered to the tip; flowers mostly 2-4, some stalks with a whorl of leaf-like bracts; sepals white to magenta, 9-14 mm long; styles 1-2 mm long, straight; seed head globose, ca. 1 cm wide. June-July.

Common in montane grasslands and open forest; East, West. AK to Newf. south to CA, NM, MN, ME; S. America. Smoke from burning ripe seed heads was inhaled by Blackfeet to cure headache.

Anemone occidentalis Watson [*Pulsatilla occidentalis* (Wats.) Freyn]. Stems 15-40 cm high, increasing with maturity; foliage long-hairy; leaf blades 2-3 times divided into short, narrowly triangular segments 1-3 mm wide; flowers 1 per stem; sepals 5-7, white, 2-3 cm long; styles feather-like, expanding to 2-4 cm long at maturity, reflexed down. June-Aug. Fig. 213.

Common in subalpine meadows and open forest where there is good snow cover; East, West. B.C., Alta. south to CA, ID, MT.

Anemone parviflora Michx. Stems 5-25 cm tall; foliage nearly glabrous; leaf blades divided into three wedge-shaped, shallowly to deeply lobed leaflets; flowers 1 per stem; sepals 5-6, white or bluish, 8-12 mm long; style straight, 1-2 mm long. July-Aug.

Common in moist meadows and stony, open slopes near or above treeline; rare on hummocks in calcareous, montane fens; East, West. AK to Que. south to OR, ID, CO. Plants from dry limestone-derived soil of Divide Mtn. have sepals 15-20 mm long.

Anemone patens L., Pasqueflower, Wild Crocus [*A. nuttalliana* D. C., *Pulsatilla ludoviciana* (Nutt.) Heller, *P. patens* (L.) Miller]. Stems 5-25 cm high, increasing with maturity; foliage long-hairy; leaf blades 3-8 cm wide, twice divided into linear segments 1-2 mm wide; flowers 1 per stem; sepals 5-7, blue or whitish on the inside, 2-4 cm long; styles feather-like, expanding to 2-4 cm long at maturity, spreading. May-June.

Common in montane grasslands, uncommon in alpine meadows; East. Circumboreal south to NM, TX, and MO. Plants bloom early, close to the ground; stems and leaves expand as the seeds mature.

Anemone tetonensis Porter ex Britt. [*A. multifida* Poir. var. *tetonensis* (Britt.) Hitchc.]. Similar to *A. multifida*; stems 10-25 cm high; ultimate leaf divisions strap-shaped and tapered to a point; flowers solitary; sepals white with blue on the back; styles hooked, 1-1.5 mm long. July-Aug.

Common in suabalpine and alpine meadows and open slopes; East, West. ID, MT south to NV, NM.

Aquilegia L., Columbine

Leaves long-petiolate, 2-3 times divided and lobed; flowers few; sepals 5, petal-like; petals erect, smaller than sepals but expanded behind point of attachment into a nectar-bearing tube (spur); stamens numerous; fruit with 5 cylindical chambers divergent and opening above.

Putative hybrids between *A. jonesii* and *A. flavescens* are occasionally found near treeline in the Many Glacier area; these have been called *A. jonesii* ssp. *elatior* by Standley (1921).

1. Flowers blue, solitary, erect; stems < 20 cm high .... *A. jonesii*
1. Flowers several, yellow, nodding; stems > 20 cm high .... *A. flavescens*

Aquilegia flavescens Watson. Stems 20-70 cm high; foliage glabrous or sparsely hairy, glandular; leaf blades twice divided into wedge-shaped, lobed segments 15-40 mm long; flowers several, nodding; sepals yellow, 15-25 mm long, spreading; petals yellow, the spurs 6-10 mm long; fruits glandular, ca. 2 cm long. June-Aug. Fig. 214. Color plate 52.

Common in meadows and open forest, often along streams and on rocky slopes, montane to lower alpine; East, West. B.C., Alta. south to OR, UT, CO.

Aquilegia jonesii Parry. Stems 5-10 cm high; foliage sparsely short-hairy and glandular; leaves basal, the blades 5-10 mm wide, twice divided into crowded oblong segments; flowers solitary, erect; sepals blue, 15-20 mm long; petals light blue with spurs ca. 10 mm long; fruits glabrous, 15-25 mm long. June-July. Color plate 60.

Common in stony, calcareous soil of exposed alpine ridges and slopes, often growing in cushion plant communities; East and along the Divide. Endemic from southern Alta. to northern WY.

Clematis L., Virgin's-bower

Vines, trailing or climbing on trees and shrubs; stems limber but somewhat woody; leaves petiolate, opposite, divided into 3-7 stalked leaflets; sepals 4, petal-like; petals absent or inconspicuous; stamens and ovaries numerous; achenes in a dense head, hairy with a persistent, feathery style, 3-5 cm long, at the tip.

C. hirsutissima Pursh, a herbaceous species with nodding flowers, occurs just south of the Park. Reference: Pringle (1997).

1. Flowers blue, nodding; leaflets 3 .... *C. occidentalis*
1. Flowers white, erect; leaflets 5-7 .... *C. ligusticifolia*

Clematis ligusticifolia Nutt. Stems 3-6 m long; plants sometimes covering shrubs and small trees; leaf blades divided into 5-7 ovate, lobed leaflets, 2-5 cm long; flowers unisexual, erect, stalked, several in a small inflorescence from leaf axils; sepals white, 6-10 mm long. July.

Rare in open montane forest or thickets, often along streams or at the edge of openings; West, expected East. B.C. to Man. south to CA, NM, and SD. Blackfeet used the bark and leaves to cure fever and colds. Other tribes used it as a horse medicine.

Clematis occidentalis (Hornem.) D. C. [*C. columbiana* (Nutt.) T. & G. misapplied]. Stems to 2 m long; leaf blades divided into 3 lance-shaped leaflets, 3-8 cm long with obscurely toothed margins; flowers long-stalked, nodding, solitary from leaf nodes; sepals blue, 2-5 cm long. May-July. Fig. 215.

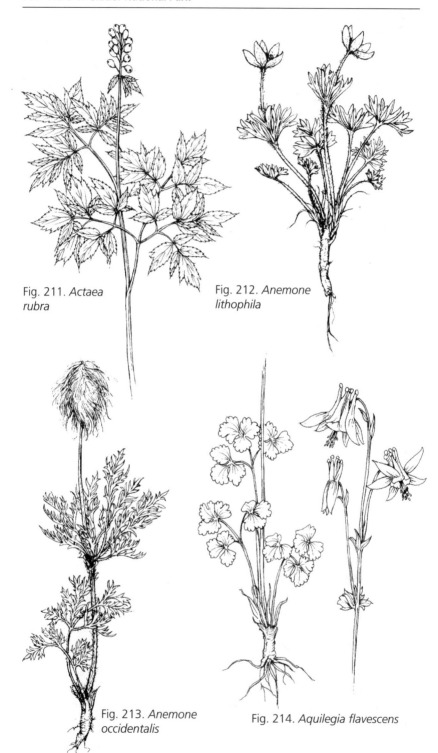

Fig. 211. *Actaea rubra*

Fig. 212. *Anemone lithophila*

Fig. 213. *Anemone occidentalis*

Fig. 214. *Aquilegia flavescens*

Common in open forest and thickets, montane or occasionally higher; East, West. Our plants are var. *grosseserrata* (Rydb.) Pringle. B.C. to N. B. south to OR, UT, CO, OH.

Delphinium L., Larkspur

Stems erect; basal leaves petiolate, palmately divided into narrowly oblong segments; stem leaves few, smaller; flowers stalked in a narrow inflorescence; sepals 5, petal-like, blue (ours), the upper elongated back into a tubular spur; petals 4, smaller than the sepals; stamens numerous; fruit with usually 3 cylindical chambers divergent above. Reference: Warnock (1997).

Plants are poisonous. *D. bicolor* and *D. nuttallianum* are sometimes difficult to distinguish.

1. Stems obviously hollow and usually ≥ 4 mm wide at mid-length **D. glaucum**
1. Stems usually < 4 mm wide and solid at mid-length 2
2. Lower petals lobed < 1/4 their length; upper petals white with blue lines .....
.... **D. bicolor**
2. Lower petals lobed 1/4-1/2 their length; upper petals usually bluish 3
3. Lower flowers much shorter than stalks; plants of habitats that dry by mid-summer; sepals flared; common .... **D. nuttallianum**
3. Lower flowers nearly as long as the stalks; plants of more permanently moist habitats; sepals cupped forward; rare .... **D. depauperatum**

Delphinium bicolor Nutt. Stems 10-40 cm high from fibrous roots; leaves mostly basal, the blades 3-6 cm wide; lower flowers shorter than the stalks; sepals 10-20 mm long, the lower ones longer; petals 5-7 mm long, the upper white with blue lines; fruits 15-18 mm long, hairy. May-July. Fig. 216.

Common in montane to alpine grasslands; East, expected West. B.C. to Sask. south to WA, WY, SD.

Delphinium depauperatum Nutt. Stems 25-50 cm high from a cluster of tuber-like roots; leaf blades 3-6 cm wide; inflorescence often branched; flowers about as long as the stalks; sepals 9-12 mm long; petals bluish; fruits 11-17 mm long, hairy, often glandular. July-Aug.

Rare in moist montane meadows; known only from the Camas Creek area. WA to MT south to CA, NV. Harvey's (1945) report of *D. burkei* Greene is referable here.

Delphinium nuttallianum Pritz ex Walpers. Stems 15-40 cm high from fleshy roots; leaves few, the blades 2-6 cm wide; lower flowers much shorter than the stalks; sepals 7-16 mm long, the lower ones longer; petals bluish; fruits 10-18 mm long, hairy. June-July.

Common in vernally moist, sometimes shallow soil of meadows, grasslands, rock outcrops and open forest, montane to alpine; East, West. B.C., Alta. south to CA, AZ, CO, NE. Based on his habitat description, Standley's (1921) report of *D. depauperatum* is probably referable here.

Delphinium glaucum Watson. Stems hollow, 60-100 cm high from deep fibrous roots; leaf blades 5-15 cm wide; inflorescence often branched with flowers longer than their stalks; sepals 9-15 mm long; lower petals blue, the upper paler; fruits 10-16 mm long, hairy. June.

Rare in montane grasslands; collected once north of Polebridge. AK south to CA, UT, and CO. Reports of *D. occidentale* Watson, a putative hybrid derivative of *D. glaucum*, are referable here.

Myosurus L., Mousetail

Myosurus minimus L. [*M. lepturus* (Gray) Howell]. Annual; stems 3-7 cm high; leaves basal, thread-like, 2-6 cm long; flower solitary, erect on stem tip; sepals 5, greenish, 2-3 mm long with tiny spurs at the base; petals absent; stamens 5-10; achenes 70-120 in a cylindrical head 15-50 mm long, short-beaked. Fig. 217.

Rare in vernally moist soil around shallow ponds; reported for the East Glacier area (Standley 1921). B.C. to Ont. south to CA, TX, NC. *M. apetalus* Gay [*M. aristatus* Benth. ex Hook.], with long-beaked achenes, occurs just east of the Park.

Ranunculus L., Buttercup, Crowfoot

Stems erect or prostrate and floating and rooting at the nodes; basal leaves long-petiolate; stem leaves alternate; flowers long-stalked, solitary or in open inflorescences; sepals 5, greenish; petals usually 5, yellow with a nectar-bearing spot at the base; stamens usually numerous; ovaries 5-many; fruit a cluster (head) of beaked, ovate, hard-coated seeds (achenes). Reference: Whittemore (1997).

Mature fruits as well as flowers are often necessary for positive identification. *R. sceleratus* L., a hollow-stemmed annual or short-lived perennial, occurs around montane shores in Waterton Park (Kuijt 1982).

1. Stems floating in water or prostrate and rooting at the nodes 2
1. Plants not aquatic; stems ascending or erect 5

2. Leaf blades lance-shaped, undivided with entire margins .... **R. flammula**
2. Leaf blades deeply lobed or divided 3

3. Flowers white; leaf blades divided into thread-like segments; aquatic plants . ... **R. aquatilis**
3. Flowers yellow; some ultimate leaf segments ≥ 1 mm wide; aquatic or terrestrial 4

4. Leaf blades divided into stalked leaflets; lawn and roadside weed .. **R. repens**
4. Leaf blades deeply lobed; plants of shallow water or mud .... **R. gmelinii**

5. Basal leaf blades divided into leaflets or lobed ≥ 1/2 their length 6
5. Margins of some basal leaves merely toothed or lobed < 1/2 their length . 16

6. Achenes hairy 7
6. Achenes glabrous or nearly so 9

7. Plant annual, < 10 cm high; achene with blade-like beak .... **R. testiculatus**
7. Perennial > 10 cm high 8

8. Basal leaves with 3 rounded lobes; achenes 2-3 mm long .... **R. uncinatus**
8. Basal leaves divided into > 3 linear lobes; achenes 2 mm long **R. pedatifidus**

9. Basal leaf blades divided to the midrib 10
9. Basal leaf blades lobed, not divided completely to the midrib 12

10. Achene with hooked beak ca. 0.5 mm long .... **R. acris**
10. Achene beak straight 1-5 mm long 11

11. Petals 4-6 mm long, the beak < 2 mm long .... **R. macounii**
11. Petals > 8 mm long, the beak > 2 mm long .... **R. orthorhynchus**

12. Stems ≤ 5 cm high; flowers 1 per stem .... **R. pygmaeus**
12. Stems > 5 cm long 13

13. Beak of achene mostly 1 mm long 14
13. Beak of achene ca. 0.5 mm long 15

14. Achene beak straight; petals 5-10 mm long .... **R. eschscholtzii**
14. Achene beak hooked; petals 2-3 mm long .... **R. uncinatus**

15. Basal leaf petioles hairy .... **R. acris**
15. Basal leaf petioles glabrous .... **R. verecundus**

16. Achene glabrous or nearly so 17
16. Achene densely hairy 19

17. Stems > 20 cm high .... **R. abortivus**
17. Stems < 20 cm high 18

18. Leaf just below the flowers (bract) deeply divided; petals 5-10 mm long .....
.... **R. glaberrimus**
18. Leaf just below flowers undivided; petals 3-5 mm long .... **R. cymbalaria**

19. Petals 3-8 mm long; achene beaked curved at the tip; leaf blades tapered
to the base .... **R. inamoenus**
19. Petals 8-15 mm long; achene beak straight; leaf blades indented at the
base .... **R. cardiophyllus**

Ranunculus abortivus L. Stems 10-50 cm high; basal leaf blades 1-4 cm long, fan-shaped with scalloped margins; stem leaves deeply lobed; inflorescence diffuse, branched, with narrow bracts; petals ca. 2 mm long, shorter than the sepals; seed head globose; achenes glabrous, 1-2 mm long with a tiny beak. June.

Rare in moist disturbed soil along montane roads, trails and streams; West. Most of temperate N. America. The more common *R. uncinatus* has deeply lobed basal leaves.

Ranunculus acris L., Tall Buttercup. Stems 25-80 cm high; basal leaf blades broadly triangular, sparsely long-hairy, 3-6 cm long and 3 times palmately divided and deeply lobed into oblong segments; stem leaves 3-lobed; inflorescence diffuse; petals 8-12 mm long; seed head globose; achenes glabrous, 2-3 mm long with a tiny beak. June-Aug.

Uncommon in moist montane and lower subalpine meadows and along roads and trails; East, West. Introduced from Europe. There are large infestations just southwest of Polebridge.

Ranunculus aquatilis L., Water Crowfoot [*Batrachium flaccidum* (Pers.) Rupr., *B. drouetii* (Schultz) Nyman]. Aquatic; stems weak, floating; leaf blades fan-shaped, 10-15 mm long, divided into numerous thread-like segments; flowers solitary at leaf nodes; petals white, 4-8 mm long; seed heads globose; achenes glabrous, 1-2 mm long with a tiny beak. July-Sept.

Common in shallow water of montane ponds, lakes and slow-moving streams; East, West. Our plants are var. *capillaceus* (Thuill.) D. C. Circumboreal south to most of N. America. Our other aquatic buttercups have yellow flowers.

Ranunculus cardiophyllus Hooker. Stems 10-30 cm high; basal leaves hairy, heart-shaped, 2-5 cm long with toothed margins; stem leaves divided into long, linear lobes; flowers few; petals 8-15 mm long; seed heads thimble-shaped, ca. 10 mm long; achenes hairy, 1-2 mm long with a straight beak. June.

Rare in montane grasslands; known only from the Belly River Valley. B.C. to Sask. south to WA, NM, SD. The hairy heart-shaped leaves are distinctive. Intermediates between *R. cardiophyllus* and the similar *R. pedatifidus* and *R. inamoenus* occur along the east edge of the Park.

Ranunculus cymbalaria Pursh [*Halerpestes cymbalaria* (Pursh) Greene]. Plants forming runners; stems 5-15 cm high; leaves all basal, the blades heart-shaped 1-2 cm long with scalloped margins; flowers few; petals 3-5 mm long, barely longer than the sepals; seed heads ovoid, 5-10 mm long; achenes ridged, 1-2 mm long with a straight beak.

Rare in mud around montane ponds and streams; reported for the East Glacier area (Standley 1921). Most of N. and S. America, Asia.

Ranunculus eschscholtzii Schlecht. [*R. helleri* Rydb., *R. saxicola* Rydb., *R. suksdorfii* Gray]. Stems ascending to erect, 8-25 cm high; basal leaf blades 1-3 cm long, 3- to 5-lobed with wedge-shaped segments; stem leaves 0-2, divided into linear lobes; flowers 1-3; petals 5-10 mm long; seed heads ovoid; achenes glabrous, 1-2 mm long with a straight beak. June-Aug. Fig. 218. Color plate 59.

Abundant in moist, subalpine and alpine meadows where there is ample snow cover; East, West. Our plants belong to the intergradent vars. *suksdorfii* (Gray) Benson and *eschscholtzii*. AK south to CA, NM; Asia. This is our most common high-elevation buttercup. The basal leaves show a great deal of variation. Standley's (1921) report of *R. alpeophilus* Nels. is probably also referable here.

Ranunculus flammula L. [*R. reptans* L.]. Stems slender, prostrate, rooting at the nodes; leaf blades narrowly lance-shaped, 1-4 cm long with entire margins; flowers 1 per leaf node; petals 3-5 mm long; seed head ovoid; achenes glabrous, 2-3 mm long with a short beak. July-Aug.

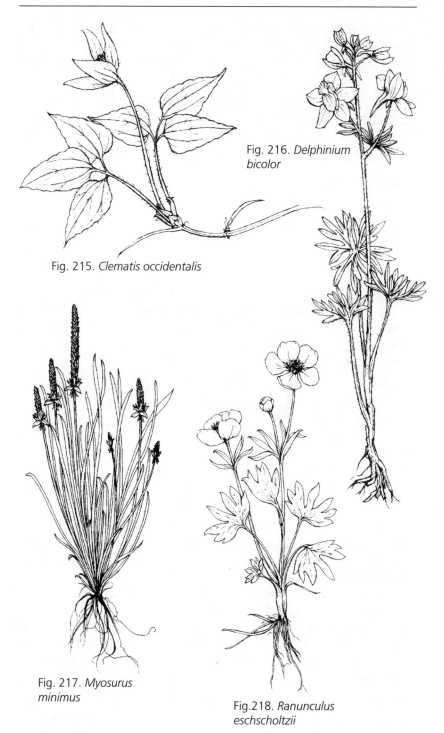

Fig. 216. *Delphinium bicolor*

Fig. 215. *Clematis occidentalis*

Fig. 217. *Myosurus minimus*

Fig.218. *Ranunculus eschscholtzii*

Common in mud around montane lakes and ponds; East, West. Our plants are var. *ovalis* (Bigelow) Benson. Circumboreal south to CA, NM, MN, NJ. The creeping stems with entire-margined leaves distinguish this plant.

Ranunculus glaberrimus Hooker. Glabrous; stems ascending, 3-10 cm high; basal leaf blades elliptic with entire margins, 1-3 cm long; stem leaves deeply lobed; flowers few; petals 5-10 mm long; seed heads hemispheric; achenes 1-2 mm long, hairy with a short, straight beak. April-May.

Uncommon in montane grasslands; East, expected West. Our plants are var. *ellipticus* Greene. B.C. to Sask. south to CA, NM, NE. One of our earliest wildflowers.

Ranunculus gmelinii D. C. [*R. purshii* Rich.]. Plants often aquatic; stems floating or prostrate and rooting at the nodes; leaf blades fan-shaped, 1-4 cm long, 1-2 times divided and lobed; submerged leaves more finely dissected; flowers 1-4 at branch tips; petals 4-10 mm long; seed heads ovoid, 4-5 mm long; achenes 1-2 mm long, glabrous with a small beak. Aug.

Uncommon in mud or shallow water of montane ponds, lakes, and slow streams; East, West. Both the hairy var. *limosus* (Nutt.) Hara and the glabrous var. *hookeri* (Don) Benson are reported for the Park. AK to Newf. south to most of N. America; Asia.

Ranunculus inamoenus Greene. Stems 10-25 cm high; basal leaf blades ovate to fan-shaped, lobed or scalloped on the upper half, 2-3 cm long; stem leaves divided to the base into linear lobes; flowers in axils of reduced upper leaves; petals 3-8 mm long; seed head cylindric, 5-10 mm long; achenes hairy, 1-2 mm long with a thin beak. May-June.

Rare in montane grasslands and meadows along the east margin of the Park. B.C. to Sask. south to WA, AZ, CO. Reported to be more common in Waterton Park. *R. abortivus* has smaller petals; see *R. cardiophyllus*.

Ranunculus macounii Britt. [*R. oreganus* (Gray) Howell]. Stems 20-50 cm high, erect to lax, sometimes rooting at the nodes; basal leaf blades hairy, 3-8 cm long, divided into 3 stalked, spade-shaped leaflets that are deeply 3-lobed with toothed margins; stem leaves smaller; flowers in axils of upper reduced leaves (bracts); petals 4-6 mm long; seed heads ovoid; achenes glabrous, 2-3 mm long with a straight beak. July.

Common in moist to wet, montane meadows, often around ponds or streams; East. Our plants are var. *macounii*. AK to Newf. south to CA, NM, MN. Standley (1921) reports the glabrous var. *oreganus* Gray from near Many Glacier.

Ranunculus orthorhynchus Hook. Stems hollow, 20-50 cm tall; foliage hairy; basal leaf blades 3-7 cm long, pinnately divided into 3-5 wedge-shaped, lobed leaflets; stem leaves more deeply divided; flowers long-stalked, in axils of upper reduced leaves (bracts); petals 9-14 mm long; seed heads globose; achenes glabrous, 3-4 mm long with a straight beak 2.5-3.5 mm long. May-June.

Rare in montane to lower subalpine wet meadows; known only from near Marias Pass. Our plants are var. *platyphyllus* Gray. AK south to CA, UT, WY.

Ranunculus pedatifidus Smith. Stems 8-20 cm high; basal leaf blades sparsely hairy, fan-shaped, 1-3 cm long, deeply divided into 5-7 narrow lobes; stem leaves similar, without petioles; flowers 1-8; petals 8-10 mm long; seed heads thimble-shaped, 8-10 mm long; achenes 2 mm long, finely hairy with a tiny hooked beak. June-July.

Uncommon in grasslands and turf, montane to alpine along the east front of the mountains. Our plants are var. *affinis* (R. Br.) Benson. Circumpolar south to NM, AZ. *R. verecundus* has smaller petals, while *R. eschscholtzii* has mostly glabrous leaves; see *R. cardiophyllus.*

Ranunculus pygmaeus Wahl. Stems 1-4 cm high often with 1 deeply divided leaf; basal leaf blades 5-9 mm long, broadly arrow-shaped with 3-5 lobes; flowers mostly 1; petals 2-4 mm long; seed head ovoid, 3-4 mm long; achenes 1 mm long, glabrous with a short, straight beak. July-Aug.

Rare in moist to wet soil of alpine turf, often on cliffs or cool slopes; East and along the Divide. Circumpolar south to CO.

Ranunculus repens L., Creeping Buttercup. Stems prostrate, rooting at the nodes; basal leaf blades hairy, 3-5 cm wide, divided into 3 ovate, lobed, and toothed leaflets; stem leaves becoming smaller and less divided; petals ca. 1 cm long; seed heads globose, ca. 4 mm long; achenes 2-3 mm long, glabrous with a curved beak. May-Aug.

Locally common in lawns around West Glacier and in the Lake McDonald area. Introduced from Europe.

Ranunculus testiculatus Crantz. Annual; stems 2-5 cm tall; leaves basal, 5-15 mm long, the blades divided into 3-7 linear lobes; flowers solitary; petals 5-8 mm long, quickly falling; seed head thimble-shaped, to 15 mm long, bristly; achenes short-hairy with a blade-like beak 3-4 mm long, twice the length of the body. May.

Rare in disturbed soil of grasslands and along roads; collected once along St. Mary Lake. Introduced from Eurasia.

Ranunculus uncinatus D. Don [*R. bongardi* Greene]. Stems 25-80 cm high; basal leaf blades 2-8 cm across, divided into 3 ovate, lobed leaflets; stem leaves with narrower lobes; petals 2-3 mm long, as long as the sepals; seed head globose; achenes 2-3 mm long, sparsely short-hairy with a curved beak. June-July. Fig. 219.

Common in moist, montane forest or thickets, often near streams or other openings; East, West. AK to Alta. south to CA, ID, CO. See *R. abortivus.*

Ranunculus verecundus Robins. Stems lax or ascending, 5-20 cm high; basal leaves fan-shaped, 1-3 cm wide, deeply and shallowly lobed; stem leaves divided into linear lobes; petals 3-5 mm long, barely longer than the sepals; seed head cylindical, 5-10 mm long; achenes 1-2 mm long, glabrous with a hooked beak. July-Aug.

Common in stony, often moist, sparsely vegetated soil of benches, moraine, and open slopes near or above treeline; East, West. AK to Alta. south to OR, ID, MT. See *R. pedatifidus.*

Thalictrum L., Meadow Rue

Rhizomatous; leaves 2-4 times divided into 3's, the ultimate leaflets ovate, stalked, lobed or toothed; flowers unisexual on separate plants (ours), stalked, in open, branched, leafy inflorescences; sepals 4-5, green or brown; petals absent; male flowers with numerous stamens; female flowers with 4-9 ovaries becoming a cluster of ellipsoid, beaked, hard-coated seeds (achenes).

1. Achenes 3-5 mm long; length of inflorescence branches not equal; leaflets prominently veined beneath .... *T. venulosum*
1. Achenes 5-8 mm long; branches of inflorescence nearly equal in length; veins of leaflets apparent, but not prominently protruding beneath .... *T. occidentale*

Thalictrum occidentale Gray. Stems 30-80 cm high; leaflets 10-30 mm long, 3-lobed above; branches of the inflorescence nearly equal in length; sepals 2-5 mm long, white to purplish; achenes 5-8 mm long, spreading to reflexed. June-Aug. Fig. 220.

Abundant in forest, common in meadows and avalanche slopes, montane and subalpine; East, West. AK to Sask. south to CA, UT, CO.

Thalictrum venulosum Trel. Stems 30-80 cm high; leaflets ca. 1 cm long, prominently veined, 2- to 3-lobed above; branches of the inflorescence differing in length; sepals greenish white, 2-4 mm long; achenes 3-5 mm long, spreading to erect. July.

Rare in montane grasslands, thickets, and open forest; East. Yuk. to Que. south to OR, NM, WI.

Trollius L., Globeflower

Trollius albiflorus (Gray) Rydb. [*T. laxus* Salisb. var. *albiflorus* Gray]. Perennial; stems 15-40 cm high, expanding with maturity; leaves petiolate, the blades broadly arrow-shaped, 4-8 cm wide, deeply palmately divided into ca. 5 toothed or lobed, ovate segments; flowers 1 per stem; sepals 5-7 petal-like, white, 10-20 mm long; petals inconspicuous; stamens many; ovaries 10-20, united; fruit a circular cluster of cylindrical capsules (follicles) ca. 1 cm high, opening at the top. July-Aug. Fig. 221.

Common in wet meadows and forest openings, often near melting snow, subalpine and lower alpine; East, West. B.C. to Alta. south to WA, ID, CO.

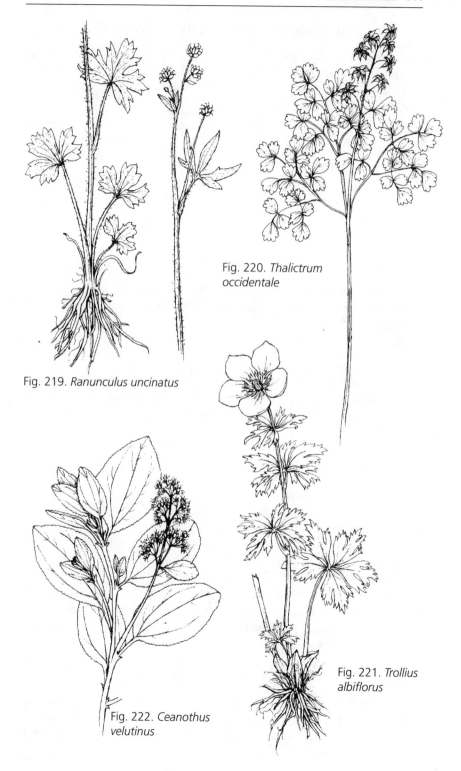

Fig. 219. *Ranunculus uncinatus*

Fig. 220. *Thalictrum occidentale*

Fig. 221. *Trollius albiflorus*

Fig. 222. *Ceanothus velutinus*

RHMNACEAE: BUCKTHORN FAMILY

Shrubs; leaves alternate, elliptic to ovate with finely toothed margins; flowers short-stalked, disk-shaped; calyx 5-lobed; petals usually 4-5, separate; stamens 4-5; ovary 1, on a disk at the base of the sepals; fruit a few-seeded capsule or firm-fleshed berry (drupe).

1. Flowers 2-5 in leaf axils; fruit a berry-like drupe .... **Rhamnus**
2. Flowers numerous in a branched inflorescence; fruit a capsule .... **Ceanothus**

Ceanothus L., Buckbrush, Wild Lilac

Leaves hairy below; flowers bisexual, white, numerous, in dense ovate clusters in a branched inflorescence; petals hood-shaped; stigmas 3; fruit a 3-chambered capsule.

1. Leaves shiny and sticky on the upper surface, persistent for > 1 year ....
.... **C. velutinus**
1. Leaves deciduous, not sticky or shiny above .... **C. sanguineus**

Ceanothus sanguineus Pursh. Stems erect, 1-2 m tall; twigs purplish; leaves deciduous, the blades 3-8 cm long; flowers 3-5 mm across; capsule 3-lobed, ca. 4 mm long. June-July.

Uncommon in dry, open montane forest, especially in recently burned areas; West. B.C. to MT south to CA.

Ceanothus velutinus Douglas ex Hooker. Stems 0.5-2 m high; twigs green; leaves fragrant, evergreen, 3-7 cm long, shiny; flowers ca. 4 mm across; capsule 3-5 mm long, 3-lobed with a low ridge on the back of each lobe. June-July. Fig. 222.

Common in open montane and lower subalpine forest, especially in recently burned areas; East, West. B.C., Alta. south to CA, CO, SD.

Rhamnus L., Buckthorn

Rhamnus alnifolia L'Her. Stems 0.5-2 m high; twigs short-hairy; leaves 3-12 cm long, nearly glabrous; flowers unisexual on separate plants (dioecious), ca. 4 mm across, green, stalked, 2-5 in leaf axils; petals lacking; stigmas 3; fruit a black, berry-like, 3-seeded drupe, 5-7 mm long, poisonous. May-June. Fig. 223.

Abundant in wet forest openings and along the margins of marshes, swamps, lakes and fens; montane; East, West. B.C. to Newf. south to CA, WY, NE, OH, NJ. *R. purshiana* D. C., with > 7 flowers per cluster, occurs just south and west of the Park.

ROSACEAE: ROSE FAMILY

Perennial herbs, shrubs and trees; leaves alternate with scale- or leaf-like stipules at the juncture with the stem; flowers bisexual, dish-shaped; sepals and petals mostly 5; stamens numerous; ovary superior or partly inferior with ≥ 10 cells, the lower portion of the calyx often fused to the fruit forming a hypanthium; fruit an achene (seed), berry, pome, or capsule. Reference: Hitchcock & Cronquist (1961), Holmgren (1997).

Genera are most easily distinguished on the basis of the fruit, but congeneric species are usually separated by flower or vegetative characters.

1. Plants woody, not dying back to the ground each winter (including plants with trailing woody stems) 2
1. Plants herbaceous, not woody above ground 14

2. Stems with thorns or prickles 3
2. Stems unarmed 5

3. Leaves lobed but not divided into leaflets; stems with thorns > 1 cm long .... **Crataegus**
3. Leaves divided into leaflets; stems with prickles < 1 cm long 4

4. Leaflets 3; fruit composed of many small segments .... **Rubus**
4. Leaflets 5-11; fruit not segmented .... **Rosa**

5. Leaves divided into leaflets 6
5. Leaves with entire or lobed margins but not divided into leaflets 8

6. Leaflets 7-15 .... **Sorbus**
6. Leaflets 3-5 7

7. Margins of leaflets entire .... **Pentaphylloides**
7. Margins of leaflets toothed .... **Rubus**

8. Plants low subshrubs with trailing stems .... **Dryas**
8. Shrubs with erect stems 9

9. Leaves with 3 or 5 major lobes, maple-like 10
9. Leaves with toothed or shallowly lobed margins 11

10. Largest leaves up to 7 cm long; fruit a dry capsule .... **Physocarpus**
10. Most leaves > 7 cm long; fruit a thimble-shaped raspberry .... **Rubus**

11. Flowers numerous in at least twice-branched, flat-topped to pyramidal inflorescences 12
11. Flowers few and/or in little-branched inflorescences 13

12. Leaves shallowly lobed the entire length .... **Holodiscus**
12. Leaves with toothed margins but entire near the base .... **Spiraea**

13. Leaf with finely toothed margins from tip to base .... **Prunus**
13. Basal third of leaf with entire margins .... **Amelanchier**

14. Leaves undivided .... **Dryas**
14. Leaves divided into leaflets 15

15. Leaflets 3-9, all attached at tip of petiole (palmate) 16
15. Leaflets ≥ 5, attached opposite each other on leaf axis (pinnate) 19

16. Petals white or red; fruit a juicy strawberry, blackberry, or raspberry 17
16. Petals yellow; fruit a dry cluster of seeds (achenes) 18

17. Calyx with small sepal-like bracts between the sepals; fruit a strawberry .... **Fragaria**
17. Calyx without small bracts; fruit a raspberry or blackberry .... **Rubus**

18. Leaflets 3, toothed only at the tip; petals smaller than sepals .... **Sibbaldia**
18. Leaflets 3-9, with toothed or lobed margins; petals often greater than sepals .... **Potentilla**

19. Leaflets divided almost to the midvein .... **Sanguisorba**
19. Leaflets toothed or shallowly lobed... 20

20. Styles elongate, persistent on the achene as a twisted beak or plume; fruit a spiny or cottony cluster of achenes .... **Geum**
20. Styles not elongate; cluster of achenes not appearing spiny or cottony ....
.... **Potentilla**

Amelanchier Medic., Serviceberry

Amelanchier alnifolia Nutt., Serviceberry, Saskatoon, Juneberry. Shrub with multiple stems to 4 m tall; twigs purplish; leaves broadly elliptic with toothed margins on the upper half, sparsely hairy beneath, petiolate, 2-5 cm long; flowers stalked in small, open clusters on twig tips; petals white, 8-12 mm long, spatula-shaped; styles usually 5; top of ovary hairy; hypanthium 1-2 mm long; fruit purplish, berry-like, juicy; seeds several. April-July. Fig. 224. Color plate 28.

Abundant in moist to dry forest, common in grasslands and avalanche slopes, montane and subalpine; East, West. AK to Que., south to CA, AZ, NM, MN. Fresh and dried fruits were an important food for Native Americans, often used in pemmican; they are also important food for bears and other wildlife.

Crataegus L., Hawthorn

Crataegus douglasii Lindl., Black Hawthorn. Shrub or small tree to 4 m high; twigs with nearly straight thorns 1-3 cm long; leaves petiolate, the blade 3-6 cm long, ovate with toothed margins especially above, sparsely hairy; flowers stalked, several in an open cluster at stem tips; petals white, circular 5-7 mm long; styles usually 5; stamens 10; hypanthium sparsely hairy; fruit a black, glabrous pome ca. 1 cm long with ca. 5 seeds. June-July. Fig. 225.

Common in montane meadows and open forest, often along streams or in other disturbed sites, uncommon on subalpine avalanche slopes; East, West. AK to Ont. south to CA, WY, SD. Native Americans ate the fruit and used the wood for tools. The fruits are prized by birds. *C. columbiana* Howell, with longer thorns and red fruit, occurs southwest of the Park.

Dryas L., Mountain Avens, Dryad

Mat-forming subshrubs with trailing, branched, woody stems; leaves petiolate, leathery and veiny above, white-wooly below; flowers solitary, long-stalked, saucer-shaped; petals and sepals 8-10; styles and stamens numerous, the former elongating and plume-like; fruit a cluster of seeds (achenes).

The feathery styles curl when moist and straighten when dry, helping to work the achene into the soil. The yellow-flowered species is uncommon at high elevations while the white-flowered is rare below subalpine.

1. Flowers with yellow petals and small bracts on the stalks ..... **D. drummondii**
1. Flowers white, without bracts on the stalks .... **D. octopetala**

Dryas drummondii Richardson ex Hooker, Yellow Mountain Avens. Leaf blades elliptic, 15-25 mm long; flowers stalks 5-25 cm high; calyx with stalked glands; petals yellow, elliptic, 8-12 mm long, nearly erect. June-Aug.

Common on gravel bars of montane rivers and streams, uncommon in stony moraine near or above treeline; East, West. AK to Que., south to OR, MT. The nodding flowers become erect in fruit with their cottony plumes.

Dryas octopetala L., Mountain Avens, Alpine Dryad. Leaf blades lance-shaped, 10-25 mm; flower stalks 3-8 cm high; calyx white- and black-hairy; petals white, elliptic, 10-15 mm long, spreading. June-Aug. Fig. 226. Color plate 54.

Abundant in stony, often calcareous soil of exposed slopes and ridges above treeline, less common in subalpine grasslands and moist turf; East, West. Our plants are ssp. *hookeriana* (Juz.) Hult. Circumpolar south to OR, ID, CO. Arguably our most common alpine plant. McGraw (1985) found that genetically distinct ecotypes occurred in dry fellfields and moist tundra.

Fragaria L., Strawberry

Herbaceous, stemless perennials with stolons (runners); leaves long-petiolate, with 3 toothed leaflets; flowers stalked, saucer-shaped, usually several in an open, stalked inflorescence; petals white, longer than the sepals; stamens ca. 20, styles numerous; fruit hemispheric, red, juicy, berry-like with scattered seeds imbedded.

1. Terminal tooth of central leaflet longer than the adjacent ones; upper surface prominently veiny .... **F. vesca**
1. Terminal tooth of middle leaflet shorter than adjacent ones; upper surface not prominently veiny .... **F. virginiana**

Fragaria vesca L., Woodland Strawberry [*F. bracteata* (Heller) Davis]. Leaflets 1-5 cm long, green, sparsely hairy on the upper surface and prominently veiny; inflorescence longer than the leaves; petals 8-11 mm long; fruit ca. 1 cm across. June-Aug.

Common in montane thickets and forest; East, West. Temperate Europe and N. America. Less common than *F. virginiana.*

Fragaria virginiana Duchesne, Wild Strawberry [*F. glauca* (Wats.) Rydb., *F. platypetala* Rydb.]. Similar to *F. vesca*; leaflets glabrous, blue-green,

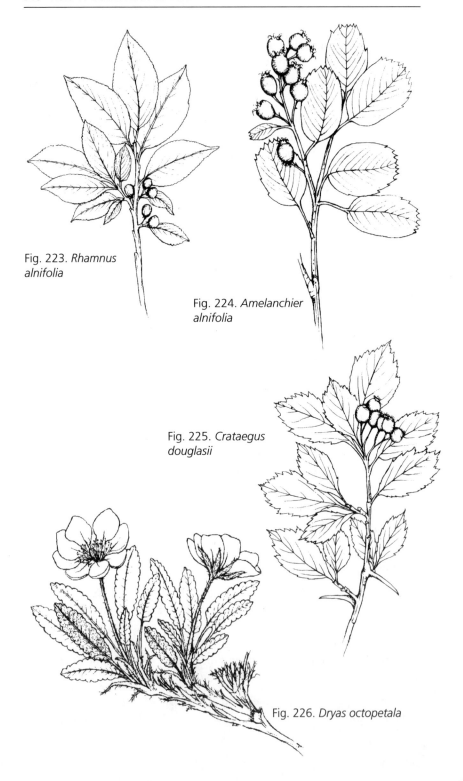

Fig. 223. *Rhamnus alnifolia*

Fig. 224. *Amelanchier alnifolia*

Fig. 225. *Crataegus douglasii*

Fig. 226. *Dryas octopetala*

inconspicuously veiny above; inflorescence as long or longer than the leaves; petals 5-12 mm long. May-Aug. Fig. 227.

Abundant in montane and subalpine meadows, grasslands, and forest; East, West. Native to most of temperate N. America. Native Americans ate the fruit and used the roots to treat diarrhea. Fruits are also eaten by birds and small mammals.

Geum L., Avens

Herbaceous, hairy perennials; leaves deeply pinnately divided with small segments among the main ones; flowers stalked, several in open, terminal inflorescences; stamens and styles numerous; fruit a hemispheric cluster of long-beaked seeds (achenes with persistent styles).

The fruit heads of *G. aleppicum*, *G. macrophyllum*, and *G. rivale* appear spiny with the achene beaks, while *G. triflorum* fruits are more cottony. Species of *Potentilla* have non-persistent styles.

1. Achene beaks > 2 cm long, plumose; leaves mainly basal; stems to 40 cm high with 1 reduced pair of leaves .... **G. triflorum**
1. Achene beaks < 2 cm long; stems mostly > 40 cm tall with leaves gradually reduced upward 2
2. Flowers cup-shaped with purple sepals and flesh-colored petals ..... **G. rivale**
2. Flowers saucer-shaped; petals yellow 3
3. Terminal leaflet nearly circular, much larger than adjacent lobes; achene beak with stalked glands .... **G. macrophyllum**
3. Terminal leaflet divided into 3 nearly equal lobes; achene beaks without glands .... **G. aleppicum**

Geum aleppicum Jacq. [*G. strictum* Soland]. Stems 30-80 cm high, spreading hairy above; lower leaf blades to 15 cm long, divided into 5-9 ovate, toothed segments, the terminal little wider than the adjacent ones; flowers saucer-shaped; sepals green, 5-8 mm long, reflexed; petals yellow, as long as sepals; achenes 3-4 mm long, long-hairy with a glabrous lower beak. June-July.

Uncommon in montane wet meadows and thickets, often along streams; East, West. Circumboreal south to CA, NM, MN, PA.

Geum macrophyllum Willd. Similar to *G. aleppicum*; stems to 1 m high; lower leaf blades to 20 cm long, the terminal leaflet orbicular, much larger than adjacent ones; sepals green, 4-5 mm long; petals yellow, longer than the sepals; lower beak of achenes with stalked glands. June-July. Fig. 228.

Common in wet meadows, thickets, and moist forest openings, often along streams and wetlands; montane and lower subalpine; East, West. Our plants are ssp. *perincisum* (Rydb.) Hult. AK to Newf. south to CA, NM, MN, NY; Asia.

Geum rivale L. Similar to *G. aleppicum;* flowers nodding, cup-shaped; petals flesh-colored, 7-10 mm long; sepals purple, longer than the petals; achene beak long-hairy and glandular. June-July.

Uncommon in wet soil of montane thickets, fens, marshes, and spruce forest; East, West. B.C. to Newf. south to WA, NM, IN, NJ; Europe, Asia.

Geum triflorum Pursh, Prairie Smoke, Old Man's Whiskers [*Sieversia ciliata* (Pursh) Don]. Stems 15-30 cm tall with thick, often branched rootstock; foliage soft-hairy; basal leaves narrowly oblong with numerous pinnate, lobed divisions; stem leaves 1 reduced pair; flowers mostly 3 per stem, nodding, cup-shaped, 8-15 mm long; calyx pink-purple with small, narrow bracts between the lobes; petals light yellow, barely longer than the sepals; achenes with a feathery beak 2-4 cm long. May-July. Fig. 229.

Common in mesic montane and subalpine grasslands; East, West. B.C. to Newf. south to CA, NM, IL, NY. Native Americans used an infusion of the roots to treat eye ailments, and the crushed seeds were used in perfume.

Holodiscus Maxim., Ocean-spray

Holodiscus discolor (Pursh) Maxim., Mountain Spray [*Sericotheca discolor* (Pursh) Rydb.]. Shrub with erect stems to 3 m tall; bark reddish gray; leaves petiolate, the blades 4-9 cm long, ovate, shallowly lobed and toothed, hairy; flowers small, white, saucer-shaped, numerous in a highly branched, pyramidal inflorescence 10-17 cm long at the branch tips; petals and sepals ca. 2 mm long; stamens 20; styles 5; seeds (achenes) hairy, 2 mm long. July-Aug. Fig. 230. Color plate 12.

Drier open forest and brushy slopes, montane; common West, uncommon East, and rare north of Many Glacier. B.C. to CA, ID, MT. Pacific slope Native Americans used the wood for making arrows.

Pentaphylloides Duhamel, Shrubby Cinquefoil

Pentaphylloides fruticosa (L.) Schw. [*P. floribunda* (Pursh) Löve, *Potentilla fruticosa* L.]. Highly branched shrub 10-100 cm high; leaves long-hairy, with 5 narrowly lance-shaped, pinnate, entire-margined leaflets 10-15 mm long; flowers 1-5 in open clusters in upper leaf axils; petals yellow, 8-13 mm long, longer than the sepals; achenes long-hairy. June-Aug. Fig. 231.

Abundant in moist grassland, meadows, fens, open forest, and exposed slopes and ridges at all elevations; East, West. Circumboreal south to CA, NM, MN, NJ. This plant has the greatest range of habitats of any in the Park, occurring in saturated peat of montane fens to dry, well-drained soil of alpine fellfields; it is absent only from deep low-elevation forest and high-elevation sites where snow lies late.

Physocarpus Maxim., Ninebark

Physocarpus malvaceus (Greene) Kuntze [*Opulaster malvaceus* (Greene) Kuntze]. Shrub with spreading to erect stems 1-2 m tall and striped gray

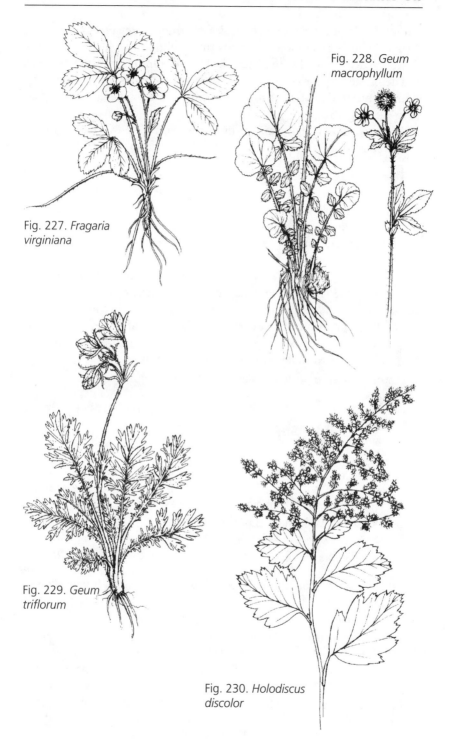

Fig. 228. *Geum macrophyllum*

Fig. 227. *Fragaria virginiana*

Fig. 229. *Geum triflorum*

Fig. 230. *Holodiscus discolor*

bark; foliage with star-shaped hairs; leaves petiolate, the blades ovate with 3 shallow lobes and toothed margins (resembling a maple), 2-7 cm long; flowers saucer-shaped with a cup-like hypanthium in the center, stalked, in a dense, hemispheric inflorescence on twig tips; petals white, 4 mm long; stamens numerous; styles 2-3; fruit a brittle 2- or 3-celled, hairy capsule. May-July. Fig. 232.

Locally common on warm, brushy slopes around West Glacier and along the Middle Fork Flathead River. B.C. to Alta. south to OR, UT, WY.

Potentilla L., Cinquefoil

Mostly herbaceous perennials; leaves petiolate, palmately (like fingers on a hand) or pinnately (paired on the axis) divided into leaflets with toothed or lobed margins; flowers saucer-shaped, stalked, in a terminal inflorescence; styles numerous; stamens usually ≥ 20; fruit a cluster of seeds (achenes) enveloped by the calyx.

A difficult genus due to frequent asexual reproduction; the taxonomy is far from stable (Barbara Ertter, personal communication). *P. nivea, P. quinquefolia,* and *P. uniflora* are very similar and occur in similar habitats. Careful examination of good material is required to separate them. Members of the closely related genus, *Geum,* have persistent styles that make the clusters of achenes appear spiny or cottony.

1. Leaves pinnately dissected; leaflets ≥ 5, attached opposite each other on the leaf axis 2
1. Leaves palmately dissected; leaflets 3 or more, all originating from the tip of the petiole 10
2. Leaflets with felt-like hair obscuring the lower surface 3
2. Leaflets glabrous or hairy, green of the lower surface visible 6
3. Plants stemless, spreading by runners; flowers solitary on long stalks ....
.... ***P. anserina***
3. Plants with leafy stems, without runners; flowers > 1 4
4. Leaflets densely white-hairy above and below .... ***P. hippiana***
4. Leaflets sparsely hairy to glabrous above 5
5. Petals > than the sepals; stipules at the base of the leaves not deeply lobed .
.... ***P. concinna***
5. Petals as long as sepals; stipules deeply lobed .... ***P. pensylvanica***
6. Rhizomatous plants of wetlands; stems red; petals purple .... ***P. palustris***
6. Non-rhizomatous upland plants; petals white or yellow 7
7. Stems lax to 15 cm high; leaflets divided .... ***P. ovina***
7. Stems erect, usually > 15 cm high, leaflets merely toothed 8
8. Leaflets with dense, felt-like hair along the lower midvein ... ***P. pensylvanica***
8. Leaflets sparsely hairy below 9
9. Petals as long as sepals; inflorescence branches stiffly erect; plants of deep grassland soil .... ***P. arguta***
9. Petals > sepals; inflorescence open; plants of stony habitats ... ***P. glandulosa***

10. Underside of leaflets densely white- or gray-hairy 11
10. Underside of leaflets not densely white- or gray-hairy 16

11. Lower leaves with 3 leaflets 12
11. Lower leaves with ≥ 5 leaflets 14

12. Stem with short, curly, appressed hair only .... *P. nivea*
12. Stem with some long, spreading hairs in addition to the appressed hair .. 13

13. Leaflets always 3 .... *P. uniflora*
13. Some leaves with 5 leaflets .... *P. quinquefolia*

14. Some lower leaves with 3 leaflets; alpine .... *P. quinquefolia*
14. Lower leaves all with ≥ 5 leaflets; lower elevations 15

15. Lower leaves with leaflets 1-2 cm long .... *P. argentea*
15. Lower leaves with leaflets > 2 cm long .... *P. gracilis*

16. Lower leaves with 3 leaflets .... *P. norvegica*
16. Lower leaves with ≥ 5 leaflets 17

17. Petals light sulphur-yellow, 9-12 mm long; hairs of petioles and stems stiffly
     spreading .... *P. recta*
17. Petals deeper yellow, 6-9 mm long; hairs more appressed 18

18. Larger leaflets > 3 cm long, toothed to the base, the veins conspicuously
     raised .... *P. gracilis*
18. Larger leaflets < 4 cm long, without teeth near the base, the lateral veins
     not conspicuous .... *P. diversifolia*

Potentilla anserina L., Silverweed. Plants stemless, spreading by runners;
leaf blades oblong, 5-20 cm long, pinnately divided into numerous lance-
shaped leaflets, silky-hairy beneath; flowers solitary on long stalks; petals
yellow, 6-10 mm long; stamens 10-25. achenes few-wrinkled. June-Aug.

Uncommon in vernally wet soil around ponds and depressions along
the east margin of the Park. Circumboreal south to CA, NE, IN, NY. Plants
have a growth form similar to strawberry. The roots were eaten by Native
Americans.

Potentilla argentea L. Stems 10-30 cm long, several from a branched
rootcrown; leaf blades with mostly 5 narrowly lance-shaped, palmate
leaflets, 1-2 cm long, cottony white-hairy below, deep green above; flowers
several in an open, branched inflorescence; petals yellow, ca. 2 mm long;
achenes wrinkled. July-Aug.

Common around roads, campgrounds and habitations; West. Introduced
from Europe.

Potentilla arguta Pursh [*Drymocalis arguta* (Pursh) Rydb.]. Stems 30-80
cm tall from a branched rootcrown; foliage stiff-hairy, glandular; leaf blades
ovate, 5-20 cm long, divided into 7-11 elliptic, deeply toothed leaflets,
reduced upward; flowers several, erect in a stiff, loosely congested
inflorescence; petals white to pale yellow, 7-9 mm long; achenes smooth.
June-July.

Common in deep soil of montane grasslands; East, West. AK to Que. south to OR, AZ, NM, MO, PA. P. glandulosa is usually smaller with a more spreading inflorescence, and it occurs in stonier soil.

Potentilla concinna Richardson. Stems lax, to 10 cm long, from a thick, branched rootcrown; foliage long-hairy; basal leaf blades ovate with 5-7 crowded, pinnate (sometimes appearing palmate), lance-shaped, deeply toothed leaflets, cottony white-hairy below; stem leaves with fewer leaflets; flowers few; petals yellow, ca. 6 mm long; achenes smooth. June.

Uncommon in sparsely vegetated soil of exposed sites in montane and lower subalpine grasslands; East. Our plants are var. *rubripes* (Rydb.) Hitchc. Alta to Man. south to NV, NM, SD. *P. ovina* is similar and occurs in similar habitats but the leaves are not cottony white-hairy beneath.

Potentilla diversifolia Lehm. [*P. glaucophylla* Lehm.]. Stems erect, 10-40 cm high from a branched rootcrown; foliage sparsely hairy; lower leaf blades spade-shaped, palmately (rarely pinnately) divided into 5-7 oblong, deeply toothed leaflets; stem leaves reduced; inflorescence branched, open; petals yellow, 6-9 mm long; achenes weakly net-ridged. July-Aug. Fig. 233.

Var. *diversifolia* is abundant in dry to moist turf and rocky slopes near or above treeline and common in subalpine grasslands; var. *perdissecta* (Rydb.) Hitchc. is uncommon in stony soil of exposed alpine ridges and slopes; East, West. Yuk. to Sask. south to CA, UT, NM. *P. gracilis* is hairier and usually occurs lower in elevation, but intermediate plants occur near Marias Pass; plants intermediate between *P. diversifolia* and *P. ovina* occur in the Two Medicine area.

Potentilla glandulosa Lindl. [*Drymocalis glandulosa* (Lindl.) Rydb., *D. pseudorupestris* Rydb.]. Stems 10-50 cm high; foliage glandular; lower leaves oblong, pinnately divided into 5-9 ovate leaflets 1-4 cm long; upper leaves reduced; flowers numerous in an open inflorescence; petals white to yellow, 6-10 mm long; achenes smooth with the style attached below the middle. June-July.

Var. *glandulosa*, mostly 30-50 cm high with petals as long as the sepals, is common in rocky soil of montane and lower subalpine open forest, grasslands, talus slopes, and brush fields; var. *pseudorupestris* (Rydb.) Breit., usually < 30 cm tall with petals longer than the sepals, is common in rocky soil of exposed ridges, slopes, and banks; subalpine and alpine, uncommon lower; East, West. B.C., Alta. south to CA, AZ, CO. See *P. arguta*.

Potentilla gracilis Dougl. ex Hook. [*P. nuttallii* Lehm., *P. blaschkeana* Turcz., *P. dichroa* Rydb., *P. viridescens* Rydb., *P. pulcherrima* Lehm., *P. filipes* Rydb., *P. g.* var. *rigida* (Nutt.) Wats.]. Stems 20-80 cm tall from a stout, branched rootcrown; foliage hairy; lower leaves palmately divided into 5-9 oblong, toothed or lobed leaflets 3-8 cm long; flowers numerous in an open inflorescence; petals yellow 5-8 mm long; achenes smooth. June-Aug.

Var. *brunnescens* (Rydb.) Hitchc., with glandular leaves and a glandular-hairy calyx, is uncommon in montane meadows and open forest; var. *flabelliformis* (Lehm.) Nutt. ex T. & G., with deeply lobed leaflets that are nearly glabrous above but densely white-hairy beneath and a mostly non-glandular calyx, is common in montane grasslands and meadows; var. *fastigiata* (Nutt.) Wats., with leaflets that are greenish on both surfaces and a mostly non-glandular calyx, is common in montane grasslands, meadows, and thickets; var. *pulcherrima* (Lehm.) Fern., similar to var. *flabelliformis* but with less deeply divided, gray-hairy leaflets, is uncommon in montane meadows and along roads; East, West. AK to Man. south to CA, NM, SD. Hairs on the stem and petioles of *P. recta* are stiffly spreading; see *P. diversifolia.*

Potentilla hippiana Lehm. Stems 15-30 cm tall from a thick, branched rootcrown; leaves densely white-hairy, narrowly ovate, pinnately divided into 5-11 oblong leaflets 1-3 cm long; flowers numerous in an open inflorescence; petals yellow, 5-7 mm long, barely longer than the sepals; achenes smooth with the style attached at the top. June-July.

Uncommon in sparsely vegetated montane grassland, occasionally to above treeline along the east edge of the Park. B.C. to Ont. south to AZ, NM, NE. *P. glandulosa* is glandular, and the leaflets of *P. pensylvanica* are more deeply dissected and not so densely hairy on the upper surface.

Potentilla nivea L. Mat-forming from a thick, branched rootcrown; stems 3-15 cm high; leaflets 3, elliptic, 5-15 mm long, white with dense, felt-like hair beneath, green above; flowers 10-15 mm across, 1-5 in an open inflorescence; petals yellow, sometimes with orange markings, 4-6 mm long, slightly longer than the sepals; achene smooth. June-July. Fig. 234.

Common in stony, sparsely vegetated soil of exposed ridges and slopes near or above treeline; East. Circumpolar south to B.C., NV, UT, CO. *P. uniflora* is usually 1-flowered and has some spreading hairs on the stem in addition to the appressed hairs.

Potentilla norvegica L. [*P. monspeliensis* L.]. Biennial with stems 15-40 cm high; foliage spreading-hairy; leaflets 3, 1-6 cm long, obovate; flowers in tight clusters in upper leaf axils; petals light yellow, 2-3 mm long, shorter than the sepals; achenes net-ridged. June-July.

Common on the margins of wetlands, and along roads in the montane zone; East, West. Circumboreal south to much of temperate N. America. Hitchcock and Cronquist (1961) believe this plant may be introduced in the Pacific Northwest.

Potentilla ovina Macoun. Stems lax to ascending, 5-20 cm high from a thick, branched rootcrown; foliage covered with long hair; leaves 2-7 cm long, pinnately divided into 9-21 leaflets that are further divided into 3-5 narrow lobes; flowers few in an open inflorescence; petals yellow, 3-6 mm

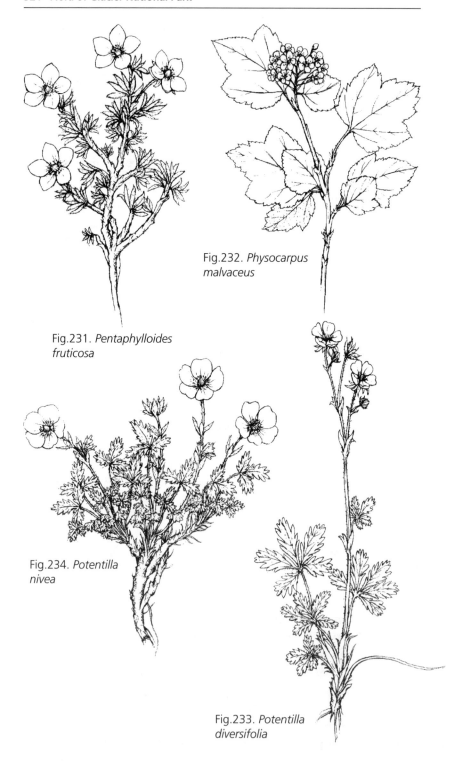

Fig.231. *Pentaphylloides fruticosa*

Fig.232. *Physocarpus malvaceus*

Fig.234. *Potentilla nivea*

Fig.233. *Potentilla diversifolia*

long, longer than the sepals; achene smooth with the style on top. June-July.

Uncommon in sparsely vegetated, exposed ridges and slopes of montane to alpine grasslands and turf along the east edge of the Park and near Marias Pass. B.C. to Sask. south to CA, UT, MN. *P. concinna* has fewer, less divided leaflets. Standley's (1921) report of *Geum turbinatum* Rydb. [=*Sieversia turbinata* (Rydb.) Greene] is referable here.

Potentilla palustris (L.) Scop. Stems prostrate to erect, reddish, 20-50 cm long, from shallow rhizomes; foliage sparsely hairy; lower leaves pinnately divided into 5-7 oblong, leaflets, 3-6 cm long; flowers long-stalked, solitary in axils of upper reduced leaves; petals dark red, 5-7 mm long, shorter than the purple sepals; achenes smooth with the style arising from the middle. July-Aug.

Common in wet or innundated, organic soil of montane fens and marshes; East, West. Circumboreal south to CA, WY, OH. The red petals and rhizomatous habit separate this from all other species of *Potentilla* and *Geum*.

Potentilla pensylvanica L. [*P. platyloba* Rydb.]. Stems ascending to erect, 15-30 cm tall, from a branched rootcrown; leaves oblong, 4-10 cm long, pinnately divided into 5-11 lance-shaped, deeply toothed leaflets, hairy, sparsely above, densely below; flowers in a compact inflorescence; petals yellow ca. 4 mm long, as long as the sepals; achenes smooth with bumps on top. June-July.

Reported for stony soil of grasslands around St. Mary (Standley 1921). AK to Que. south to CA, NM, NE, MN. See *P. hippiana.*

Potentilla quinquefolia Rydb. [*P. nivea* L. var. *pentaphylla* Breit.]. Stems 5-10 cm high from a small rootcrown; leaves ovate, palmately divided into mostly 5 lance-shaped, deeply toothed leaflets, 1-2 cm long, finely hairy above, gray-wooly below; inflorescence compact; petals 4-5 mm long, yellow, as long as the sepals; achenes smooth. June-July.

Locally common along the east margin of the Park on stony, sparsely vegetated, exposed slopes and ridges above treeline. B.C. to Sask. south to OR, UT, CO. *P. uniflora* and *P. nivea* have strictly 3 leaflets. *P. quinquefolia* is sometimes considered part of *P. hookeriana* Lehm. or *P. nivea.*

Potentilla recta L., Sulphur Cinquefoil. Stems 20-50 cm tall from a tough, fibrous root; foliage with stiff spreading hairs, leaves spade-shaped, palmately divided into 5-7 lance-shaped, sharply toothed leaflets, 2-7 cm long; inflorescence open, flat-topped; petals light yellow, 7-10 mm long, longer than the sepals; achenes with curved ridges. July-August.

Rare in montane grasslands, meadows, and disturbed residence areas; known from Big Prairie, West Glacier and Two Medicine. Introduced from

Eurasia. See *P. gracilis*. This plant is a serious invasive weed of low-elevation grassland south of the Park.

Potentilla uniflora Ledeb. [*P. ledebouriana* Porsild]. Similar to *P. nivea* but hairs of stems are spreading as well as felt-like and appressed; flowers usually 1-2, 15-25 mm across; petals longer than the sepals. June-July. Color plate 57.

Common in stony, sparsely vegetated soil of exposed slopes and ridges near or above treeline; East. AK to Que. south to B.C., OR, CO; Siberia. See *P. nivea, P. quinquefolia*.

Prunus L., Cherry, Plum

Shrubs or small trees; leaves undivided with toothed margins and small glands at the base of the petiole; stalked flowers cup-shaped, in stalked clusters near the stem tips; petals white; stamens 20-30; style 1; fruit a 1-seeded drupe.

P. mahaleb L., a commercial sweet cherry rootstock, has been observed along Lake McDonald, but does not appear to persist.

1. Leaves elliptic; inflorescence elongate with > 12 flowers .... ***P. virginiana***
1. Leaves oblong or lance-shaped; inflorescence hemispheric with < 12 flowers
.... 2

2. Leaf tips tapered to a point... ***P. pensylvanica***
2. Leaves rounded to the tip ....***P. emarginata***

Prunus emarginata (Dougl. ex Hook.) Walpers, Bitter Cherry [*P. corymbulosa* Rydb.]. Shrub or small tree 2-4 m tall; twigs purple; leaf blades oblong, rounded to the tip, 3-8 cm long; inflorescence hemispheric with ca. 5-8 flowers; petals 5-7 mm long; cherry red to black, 8-12 mm long. May-June. Color plate 24.

Common in montane and lower subalpine open forest, thickets, avalanche slopes, and along streams; West. Our plants are var. *emarginata*. B.C. to MT south to CA, UT, AZ.

Prunus pensylvanica L. Similar to *P. emarginata*; leaf blades 5-10 cm long, lance-shaped, long-tapered to the tip; flowers 4-6; cherry red. May-June.

Uncommon in montane open forest and thickets and lower subalpine avalanche slopes in the Two Medicine and Cut Bank drainages. B.C. to Newf. south to MT, CO, SD, LA, VA.

Prunus virginiana L., Chokecherry [*P. melanocarpa* (Nels.) Rydb.]. Usually a shrub to 5 m high; twigs purplish-gray; leaf blades elliptic, pointed at the tip, 4-10 cm long; inflorescence cylindric, many-flowered; petals 3-6 mm long; cherry red, becoming black, 8-11 mm long. May-July. Fig. 235.

Common in montane thickets, young forests, avalanche slopes, stream banks, and stony soil of grasslands; East, West. Our plants are var. *melanocarpa* (Nels.) Sarg. B.C. to Newf. south to CA, NM, TX, TN, NC.

Native Americans mixed the sweet but astringent cherries with meat to make pemmican; a decoction of the inner bark was used to treat colds. Birds eat the cherries, and they are an important source of bear food, especially in years with a poor huckleberry crop.

Rosa L., Rose

Shrubs with prickly stems; leaves pinnately divided into 7-11 leaflets with toothed margins; wing-like appendages (stipules) at the base of the petioles; stalked flowers solitary or in small clusters; calyx almost completely enclosing the ovary and attached to it (hypanthium); petals pink, shallowly lobed at the tip; stamens numerous; styles > 10; fruit a hip, the swollen, pulpy hypanthium enclosing the hairy seeds.

Native Americans ate the fruits and used a decoction of the roots to treat diarrhea. *R. acicularis* is the most common species of closed coniferous forest, while *R. woodsii* is the most common in more open or disturbed habitats.

1. Flowers usually in clusters of 2-5 2
1. Flowers solitary 3
2. Stems with stout, hooked prickles just below leaf attachments (fine straight prickles may also be present); common; usually > 40 cm tall .... **R. woodsii**
2. Stems bristly with fine straight prickles, but those below leaf nodes not noticeably different; plants < 40 cm tall, uncommon on east edge of Park .... **R. arkansana**
3. Mature hip lacking persistent sepals at the tip; petals 10-25 mm long; stems weak, lax .... **R. gymnocarpa**
3. Hips with sepals persistent on the top; petals ≥ 2 cm long; stems stout, erect 4
4. Stems with stout, hooked prickles just below leaf attachments (fine straight prickles may also be present); uncommon .... **R. nutkana**
4. Stems bristly with fine straight prickles, but those below leaf nodes not noticeably different; common .... **R. acicularis**

Rosa acicularis Lindl., Prickly Rose [*R. bourgeauiana* Crep., *R. sayi* Schwein.]. Stems 50-150 cm high, densely to sparsely covered with fine, straight prickles; leaflets 5-7, narrowly elliptic, sparsely hairy, 15-45 mm long with gland-tipped teeth; flowers usually solitary; petals 15-30 mm long; sepals persistent in fruit; hip globose to pear-shaped, 1-2 cm long, reddish-purple. June-July. Fig. 236.

Common in montane and lower subalpine forest and forest openings; East, West. Circumboreal south to B.C., CO, NE, MN, VT. Usually flowers earlier than *R. woodsii*. *R. nutkana* usually has larger, hooked prickles at leaf nodes and is less common.

Rosa arkansana Porter, Prairie Rose. Stems to 40 cm high, often partially dying back in winter, covered with fine, straight prickles; leaflets 9-11, narrowly elliptic, 1-4 cm long, sparsely hairy; flowers several in a cluster; petals 15-25 mm long; sepals persistent in fruit; hip globose, 8-15 mm long, purplish. June-Aug.

Rare in grasslands along the east edge of the Park. B.C. to Man. south to NM, TX, MO. This species becomes much more common just to the east. The more common *R. woodsii* may occur in the same habitats but is usually woodier.

Rosa gymnocarpa Nutt., Baldhip Rose. Stems lax, 30-100 cm high, usually sparsely and finely prickly; leaflets 5-9, elliptic, 5-30 mm long with gland-tipped teeth; flowers usually solitary; petals 8-15 mm long; sepals falling as the fruit develops; hip elliptic, ca. 1 cm long, red. June.

Uncommon in montane and lower subalpine forest; West. B.C. to MT south to CA, OR, ID. Our only species with hips lacking persistent sepals.

Rosa nutkana Presl. Stems 50-150 cm high, sparsely prickly, the largest prickles hooked, paired just below leaf nodes; leaflets 5-7, elliptic, 2-5 cm long; flowers usually solitary at tips of new twigs; petals 25-40 mm long; sepals persistent; hip globose, red to purple, ca. 1 cm long. June-July.

Uncommon in montane open forest and thickets; West, East. Our plants are var. *hispida* Fern. B.C. to MT south to CA, OR, UT, CO. See *R. acicularis*.

Rosa woodsii Lindl., Wood's Rose [*R. ultramontana* (Wats.) Heller, *R. pyrifera* Rydb., *R. fendleri* Crep.]. Stems 50-150 cm high, sparsely to densely prickly, larger hooked prickles paired just below leaf nodes; leaflets 5-9, ovate, 1-4 cm long, usually with gland-tipped teeth; flowers ≥ 2 in small clusters at branch tips; petals 15-25 mm long; sepals persistent in fruit; hip globose, 6-12 mm long, red. June-July. Color plate 26.

Var. *ultramontana* (Wats.) Jeps., with leaflets 2-5 cm long, is common in montane forest openings and thickets along streams and wetlands; East, West; var. *woodsii*, with leaflets 1-2 cm long, is common in moist sites in montane grasslands and open forest; East. Yuk to Que. south to CA, NM, TX. Our most common species of moist, open sites and disturbed areas.

Rubus L., Raspberry, Blackberry, Bramble

Shrubs or herbs with erect or trailing, sometimes prickly stems; leaves lobed or divided with toothed margins; flowers saucer-shaped, stalked, solitary or in few-flowered clusters; stamens and styles numerous; fruit a berry-like aggregate of small, juicy, 1-seeded drupes.

Berries that include the fibrous receptacle are called blackberries, while those that are thimble-like and come away from the receptacle when mature are raspberries. Standley (1921) reports a few stems of the cultivated evergreen blackberry, *R. laciniatus* Willd., escaped in the woods near East Glacier, but the plant has not been found in the Park since. The 3 trailing brambles are similar, but *R. acaulis* occurs in organic soil of fens; *R. pubescens* also occurs in fens but is more common on the margins or wet forest; *R. pedatus* is not associated with wetlands.

1. Stems with prickles 2
1. Stems unarmed 3
2. Main prickles hook-shaped; flowers usually > 4 per cluster ... **R. leucodermis**
2. Main prickles not hooked; flowers 1-4 per cluster .... **R. idaeus**
3. Stems woody, erect; leaves lobed but not divided into leaflets **R. parviflorus**
3. Stems trailing or not woody; leaves divided into leaflets.............................. 4
4. Erect shoots arising from subterranean stem (rhizome) .... **R. acaulis**
4. Erect shoots arising from stems trailing on the ground 5
5. Leaflets 3, the lateral ones not lobed or divided .... **R. pubescens**
5. Leaflets 5, or if 3, then lateral ones deeply divided .... **R. pedatus**

Rubus acaulis Michx. [*R. arcticus* L. ssp. *acaulis* (Michx.) Focke]. Stems herbaceous, unarmed, 5-20 cm high from long rhizomes; leaves divided into 3 ovate leaflets, 15-40 mm long; flowers solitary on stem tips; petals erect, pink to red, 10-15 mm long; raspberry globose, red, ca. 1 cm long. June.

Rare on hummocks in sphagnum fens, known from 3 sites in the north half of the Park; East, West. AK to Newf. south to B.C., CO, MN.

Rubus idaeus L., Red Raspberry [*R. strigosus* Michx.]. Shrubs with erect or lax, prickly stems 20-150 cm high; leaves divided into 3 or 5 lance-shaped leaflets 2-7 cm long, sometimes white-waxy beneath; flowers 1-4 in leaf axils; petals white, spreading, 4-6 mm long; raspberry hemispheric, red, ca. 1 cm across. June-July. Fig. 237.

Common along streams and on brushy, often burned-over slopes, especially rock slides, montane to lower subalpine; East, West. Our plants are var. *strigosus* (Michx.) Maxim. Circumboreal south to OR, NM, SD, MO.

Rubus leucodermis Dougl. ex T. & G., Black Raspberry. Stems erect to arching, 1-3 m long, armed with stout, curved prickles; leaves divided into 3 narrowly ovate leaflets, 15-60 mm long, densely, close white-hairy beneath; flowers several in a prickly, stalked, hemispheric cluster; petals white, ca. 8 mm long, shorter than the sepals; raspberry purple, hemispheric, 9-12 mm wide. June.

Locally common in open forest around Lake McDonald. B.C. to MT south to CA, NV, UT.

Rubus parviflorus Nutt., Thimbleberry. Stems unarmed, 50-150 cm tall with shredding bark; leaf blades spade-shaped, 5-lobed, maple-like, 6-15 cm long, sparsely hairy; flowers usually several in an open cluster on stem tips; petals white, 15-25 mm long; raspberry red, hemispheric, hairy, 10-20 mm wide. June-July. Color plate 34.

Abundant in montane and subalpine moist forest, thickets and avalanche slopes; East, West. AK to MN south to CA, NM, SD, Mex.

Rubus pedatus Smith. Stems unarmed, trailing, to 1 m long; erect branches herbaceous, very short with 2-4 leaves; leaves with long petioles and 3 leaflets; 1-3 cm long; the lateral ones divided almost to the base; flowers solitary, long-stalked; petals white, 6-12 mm long, spreading; blackberry red, composed of few druplets. June-July.

Rare in spruce forest on montane stream terraces, McDonald and Waterton valleys. AK to Alta. south to OR, ID, MT.

Rubus pubescens Raf., Dewberry. Unarmed stems trailing with short, erect shoots 3-20 cm high bearing 2-4 leaves; leaflets 3, ovate, 2-8 cm long; flowers 1-2 at shoot tips or directly from the main stem; petals white, 5-8 mm long; blackberry globose, red, 5-10 mm wide. June-July. Fig. 238.

Common in wet soil of montane spruce forests and shady margins of fens and marshes; West. B.C. to Lab. south to WA, CO, IN.

Sanguisorba L., Burnet
Sanguisorba occidentalis Nutt. [*S. annua* (Hook.) T. & G.]. Annual or biennial to 30 cm high; foliage glabrous; leaves 2-5 cm long, pinnately divided into 7-15 deeply divided leaflets, the ultimate segments linear; flowers numerous, sessile, subtended by a papery bract, clustered in a dense cylindrical head to 2 cm long; petals lacking; sepals 4, green, ca. 2 mm long; stamens 2; fruit a 1-seeded achene. June.

Rare in vernally moist, often compacted soil; collected once in a parking lot at the West Glacier maintenance area. B.C. to MT south to CA, ID.

Sibbaldia L.
Sibbaldia procumbens L. Low, mat-forming perennial herb with trailing stems; leaves petiolate, sparsely hairy, clover-like, divided into 3 oblong leaflets, 1-2 cm long with blunt, toothed tips; flowers saucer-shaped, 2-15 in stalked, compact clusters; petals yellow, ca. 1 mm long, shorter than the sepals; stamens 5; styles 5-15; fruit a cluster of small seeds. July-Aug. Fig. 239.

Abundant, often in sparsely vegetated soil of subalpine and alpine meadows and turf, usually where snow accumulates; East, West. Circumpolar south to CA, UT, CO, NH. Often found with *Juncus drummondii*, *Luzula hitchcockii*, or *Phyllodoce* spp. The plant often colonizes disturbed areas.

Sorbus L., Mountain Ash
Shrubs or small trees; leaves pinnately divided into numerous toothed leaflets; flowers short-stalked, numerous in densely branched, flat-topped to hemispheric inflorescences; petals white, stamens numerous; styles 2-5; fruit a small, globose, few-seeded pome.

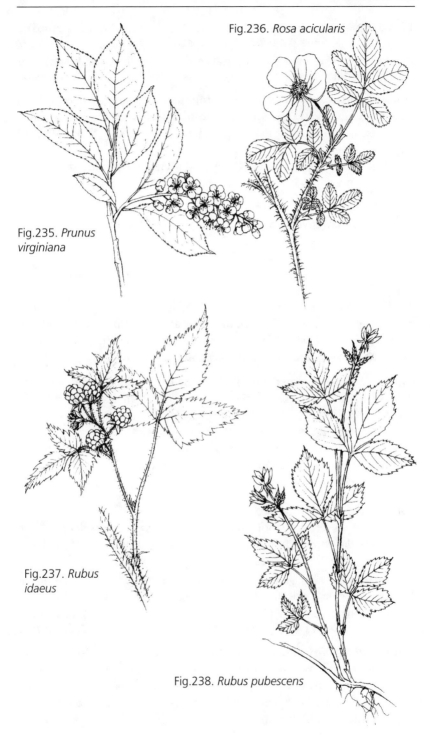

Fig.236. *Rosa acicularis*

Fig.235. *Prunus virginiana*

Fig.237. *Rubus idaeus*

Fig.238. *Rubus pubescens*

1. Introduced trees; leaflets ≥ 13 .... **S. aucuparia**
1. Native shrubs, some leaves with < 13 leaflets 2

2. Twigs and inflorescence red-hairy; leaflets rounded at the tip .... **S. sitchensis**
2. Twigs and inflorescence white-hairy; leaflets pointed .... **S. scopulina**

Sorbus aucuparia L., Rowan Tree, European Mountain Ash. Tree to 10 m tall; foliage glabrous to sparsely hairy; leaflets usually 13-15, lance-shaped, 2-6 cm long; inflorescence flat-topped; petals 2-4 mm long; fruit red, ca. 10 mm long. June.

A few trees along McDonald Creek west of Apgar. Introduced; birds eat the fruit and spread the seed of this commonly planted European ornamental.

Sorbus scopulina Greene [*S. sambucifolia* (Cham. & Schlect.) Roem.]. Shrub with stems 1-3 m high; foliage mostly glabrous; twigs white-hairy; leaflets 9-13, narrowly lance-shaped, pointed, 3-7 cm long; flowers 70-200 in a flat-topped inflorescence 6-15 cm across; petals 5-6 mm long; fruit orange to red, 5-8 mm long. June-July. Fig. 240.

Moist forest and avalanche slopes; abundant in the subalpine zone, less common montane; East, West. Our plants are var. *scopulina*. AK to Alta. south to CA, NM, SD. Our common mountain ash often occurs with *Xerophyllum tenax* and *Vaccinium membranaceum*.

Sorbus sitchensis Roemer. Shrub 1-3 m tall; twigs sparsely reddish-hairy; leaflets 7-11, oblong, rounded at the tip, 2-5 cm long, often sparsely reddish-hairy below; flowers 15-80 in a hemispheric inflorescence; petals 4-5 mm long; mature fruit waxy blue, ca. 10 mm long. July.

Uncommon in subalpine open forest, wet meadows, and heath, often near streams; East, West. Our plants are ssp. *sitchensis*. AK south to CA, OR, MT.

Spiraea L., Spiraea, Meadowsweet

Shrubs; leaves ovate with toothed margins; flowers cup-shaped, numerous in a dense, flat-topped inflorescence on stem tips; stamens ≥ 15; styles 5; fruit a cluster of 5 cylindric capsules.

1. Inflorescence conical, several times as long as broad .... **S. douglasii**
1. Inflorescence hemispheric, broader than long 2

2. Flowers white; leaves glabrous .... **S. betulifolia**
2. Flowers rose; leaves hairy on the margins .... **S. densiflora**

Spiraea betulifolia Pallas [*S. lucida* Dougl. ex Hook.]. Stems 20-80 cm high; leaf blades 2-9 cm long, glabrous; inflorescence 3-8 cm across; petals white, ca. 2 mm long; capsules 3 mm long. July-Aug. Color plate 45.

Abundant in montane and subalpine, moist to dry forest and avalanche slopes; East, West. Our plants are var. *lucida* (Dougl. ex Hook.) Hitchc. B.C. to Sask. south to OR, WY, SD.

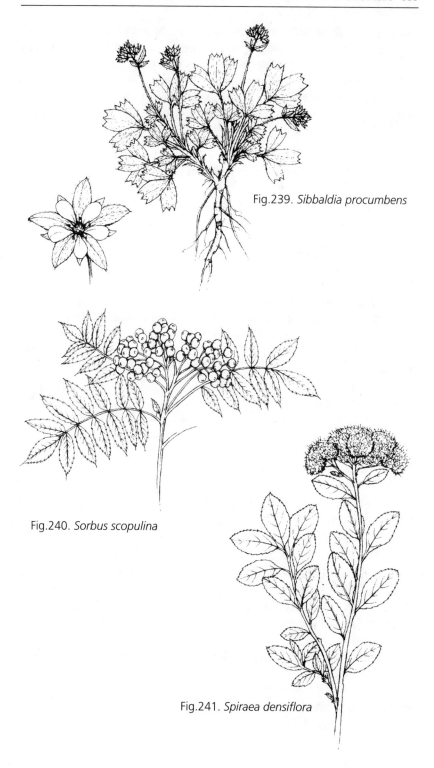

Fig.239. *Sibbaldia procumbens*

Fig.240. *Sorbus scopulina*

Fig.241. *Spiraea densiflora*

Spiraea densiflora Nutt. Stems 50-100 cm high; leaf blades 2-6 cm long, glabrous except for short hairs on the margins; inflorescence 2-5 cm across; petals rose, ca. 2 mm long; capsules ca. 3 mm long. July-Aug. Fig. 241. Color plate 51.

Common in stony soil of subalpine open forest, moist meadows, and avalanche slopes, uncommon lower; East, West. Our plants are var *densiflora*. B.C., Alta. south to OR, ID, MT.

Spiraea douglasii Hooker. Stems 1-2 m tall; leaf blades 3-10 cm long, glabrous; inflorescence conical, 6-20 cm long; petals pink or rose, ca. 2 mm long; capsule ca. 3 mm long. Aug.

Rare, collected once on the margins of a fen south of Polebridge. Our plants are var. *roseata* (Rydb.) Hitchc. AK south to MT, ID, CA.

RUBIACEAE: MADDER FAMILY

Galium L., Bedstraw

Herbaceous; stems 4-angled; leaves linear to narrowly lance-shaped, sessile, opposite or in whorls of 4-8; flowers bisexual, small, stalked, dish-shaped; sepals absent; corolla mostly 4-lobed, usually white; stamens 4; styles 2; ovary below the base of the corolla (inferior); fruit a 2-lobed, 2-seeded nutlet.

1. Leaves primarily opposite or in whorls of 4 2
1. Leaves primarily in whorls of 6 or more 4
2. Fruit and ovary glabrous; wetland plants; flowers 1-3 in leaf axils **G. trifidum**
2. Fruit and ovary hairy; plants of grasslands, meadows or forests 3
3. Plants annual; flowers solitary in leaf axils .... **G. bifolium**
3. Rhizomatous perennial; flowers numerous in a terminal, branched
   inflorescence .... **G. boreale**
4. Flowers yellow; stems erect; leaves with inrolled margins .... **G. verum**
4. Flowers white; stems lax; leaf margins not inrolled 5
5. Leaves 3-4 times as long as wide; rhizomatous perennial .... **G. triflorum**
5. Leaves > 4 times as long as wide; annual .... **G. aparine**

Galium aparine L., Cleavers, Goose Grass [*G. mexicanum* H. B. K.]. Annual; stems 10-100 cm long, prostrate or climbing with backward-pointing prickles; leaves narrowly oblong, 1-3 cm long, in whorls of 6-8; flowers few on short side-branches; corolla ca. 2 mm across; fruit 2-3 mm long with hooked bristles. June-July.

Uncommon in disturbed openings of montane and lower subalpine forest and thickets, often on moss-covered rocks; West, expected East. Circumboreal through most of N. America. Some authorities believe this species is not native in our area.

Galium bifolium Watson. Annual; stems erect, 5-20 cm high; leaves narrowly lance-shaped, 1-2 cm long, opposite or in whorls of 4; flowers 1 per leaf axil; corolla 3-lobed; fruit 2-4 mm long with short, hooked hairs. July.

Rare in moist, sparsely vegetated, shaded soil of montane forest and lower subalpine cliffs; East, expected West. B.C., MT south to CA, CO.

Galium boreale L., Northern Bedstraw. Rhizomatous perennial; stems erect, 10-50 cm high; leaves strap-shaped, 1-4 cm long in whorls of 4; flowers in branched, stalked clusters arising from upper leaf axils; corolla 3-5 mm wide; fruit ca. 2 mm long, short-hairy. June-Aug. Fig. 242. Color plate 3.

Montane and subalpine grasslands, meadows and open forest; abundant East, common West. Our plants are ssp. *septentrionale* (Roemer & Schultes) Iltis. Circumboreal south to CA, TX, MO, OH. A large proportion of stems are sterile. Native Americans used the roots to make a red or yellow dye.

Galium trifidum L. [*G. tinctorum* L.]. Rhizomatous perennial; stems weak, ascending, 10-30 cm long with rough-hairy edges; leaves strap-shaped, 6-14 mm long, in whorls of 4-6; flowers 1-3 in leaf axils; corolla 1-2 mm wide, 3-4 lobed; fruit glabrous, 1-2 mm long, the lobes becoming separate. July-Aug.

Common in montane and lower subalpine wet meadows, marshes, and fens; East, West. Circumboreal south to CA, AZ, TX, OK, GA. This is our only wetland bedstraw.

Galium triflorum Michaux, Sweet-scented Bedstraw. Rhizomatous perennial; stems lax, 20-50 cm long with prickles near the nodes; leaves narrowly elliptic, 1-4 cm long, in whorls of (4)6; inflorescence of open, mostly 3-flowered stalks arising from leaf axils; corolla 2-3 mm wide; fruit 1-2 mm long with hooked bristles. July-Aug.

Common in moist, montane and lower subalpine coniferous forest; East, West. Circumboreal south to CA, NM, OK, FL. Foliage has a faint vanilla smell upon drying. The annual *G. aparine* is similar.

Galium verum L., Yellow Bedstraw. Rhizomatous perennial; stems 30-80 cm high, erect; leaves linear with inrolled margins, 1-3 cm long, in whorls of 8-12; flowers densely clustered in a branched, terminal inflorescence; corolla yellow; fruit glabrous, ca. 1 mm long. Aug.

Disturbed montane meadows and open forest; known only from the Two Medicine campground area. Introduced from Europe.

SALICACEAE: WILLOW FAMILY

Deciduous, unisexual trees and shrubs; leaves simple, alternate on the stem with wing-like appendages (stipules) at the base of the petiole; flowers unisexual, subtended by a scale, numerous in unbranched cylindrical spikes (catkins, aments); male flowers of 1-many stamens; female flowers of 1 ovary with a style and 2-4 stigmas; fruit an ovoid capsule; seeds numerous, with a tuft of hair at the tip.

1. Trees with mostly single stems; bud scales >1, usually resinous .... **Populus**
1. Multi-stemmed shrubs (rarely small trees); bud scales 1, rarely resinous . **Salix**

Populus L., Poplar, Cottonwood

Relatively short-lived trees; buds resinous; leaves rounded at the base with toothed margins, turning yellow in autumn; catkins pendant.

Narrowleaf cottonwood (*P. angustifolia* James) occurs on gravel bars near Babb (Lynch 1955) and may occur in the Park. It can be distinguished by its narrower leaves and a petiole up to 1/3 the length of the blade; however, young plants of *P. balsamifera* may also have narrow leaves, but the petiole is longer than 1/3 the blade.

1. Petioles flat in cross section; leaves nearly round in outline .... *P. tremuloides*
1. Petioles round in cross section; leaves broadly lance-shaped ... *P. balsamifera*

Populus balsamifera L., Black Cottonwood, Balsam Poplar [*P. trichocarpa* T. & G. ex Hook., *P. hastata* Dode]. Tree to 40 m high with ascending branches; bark smooth and white at first, becoming gray and furrowed; leaf blades broadly lance-shaped, 5-15 cm long, pale below; female catkins 8-20 cm long; stigmas 3; capsules ovoid, 5-8 mm long. May.

Abundant in forest and gravel bars along montane rivers and streams; common on montane and lower subalpine avalanche slopes; East, West; common in aspen groves, East. Our plants are ssp. *trichocarpa* (T. & G. ex Hook.) Brayshaw. AK to Newf. south to CA, WY, WI. Occasionally found along roads or on disturbed slopes and banks. Native Americans used the buds for dye.

Populus tremuloides Michx., Quaking Aspen, Trembling Aspen. Tree 2-30 m high; bark smooth, pale, furrowed with age; leaf blades broadly spade-shaped to nearly orbicular, 2-9 cm long, pale beneath; female catkins 4-10 cm long; stigmas 2; capsule narrowly ovoid, 4-6 mm long. May. Fig. 243.

Often forming groves through vegetative propagation near streams and wetlands and in other places where the soil is somewhat moist, montane and lower subalpine; common West, abundant East where it can form large groves on cool slopes and in topographic depressions. AK to Newf. south to CA, NM, TN, NJ. The leaf petioles are flat rather than round in cross section, allowing the leaves to "tremble" with the slightest breeze. Trees are greatly dwarfed along the east edge of the Park where high winds and ice scouring are common in the winter.

Salix L., Willow

Shrubs (1 small tree); buds covered by a single scale, not resinous; male flowers of 1-7 stamens; stigmas 2 or 4. References: Dorn (1995), Dorn & Dorn (1997).

Mature catkins and mature leaves are useful for identification, but both are rarely available at the same time. Vegetative characters have a good deal of environmentally or developmentally induced variation. Understanding this variation is required for consistently correct determinations. Months when mature catkins are available are given. Characters in the key and descriptions are for mature leaves and catkins. New twigs of many willows contain a natural analog to aspirin. A decoction of these twigs was used by Native Americans to cure fever and treat pain. They also used willow stems to build sweat lodges, backrests, buckets, baskets, and boats.

1. Stems prostrate; mat-forming plants < 5 cm high 2
1. Erect shrubs or trees 3

2. Female catkins > 1 cm long; leaf blades tapered to the petiole ...... *S. arctica*
2. Female catkins ≤ 1 cm long; leaf blades rounded at the base .... *S. reticulata*

3. Lower surface of leaves with dense white or silver hair 4
3. Lower surface of leaves glabrous to moderately hairy at maturity............... 8

4. Leaf blades ≤ 4 cm long; female catkins < 2 cm long .... *S. brachycarpa*
4. Leaves and female catkins larger... 5

5. Twigs and upper leaf surfaces mostly obscured by dense white hair ....
.... *S. candida*
5. Upper leaf and twig surfaces hairy but still visible 6

6. Upper leaf surface with obvious impressed veins; plants usually < 1 m tall ....
.... *S. vestita*
6. Upper leaf surface not obviously veiny; plants > 1 m tall 7

7. 1- and 2-year-old twigs partly covered with pale wax (glaucous), especially
near the buds, not very hairy... *S. drummondiana*
7. Twigs not glaucous, often densely hairy .... *S. sitchensis*

8. Leaf blades ≥ 6 times as long as wide 9
8. Leaf blades < 6 times as long as wide 10

9. Mature leaf blades usually glabrate; female scales rounded ... *S. melanopsis*
9. Leaf blades usually hairy; female scales more pointed .... *S. exigua*

10. Leaf blades with toothed margins, sometimes only shallowly so.............. 11
10. Leaf blades with mostly entire margins 18

11. Petiole with bump-like glands where it meets the leaf blade 12
11. Petioles without glands... 13

12. Leaf blades lance-shaped, gradually tapered to a long point; mineral soil
along rivers and streams .... *S. lasiandra*
12. Leaf blades narrowly elliptic, short-pointed; wet organic soils of fens and
carrs .... *S. serissima*

13. Underside of leaf sometimes a little paler than the upper surface but
without a pale wax that can be rubbed off with the fingers (glaucous)...... 14
13. Underside of leaf glaucous 15

14. Mature leaves conspicuously hairy on 1 side at least .... *S. commutata*
14. Mature leaves mostly glabrous on both sides .... *S. boothii*

15. Leaf blades narrowly lance-shaped, long-pointed; female scales yellow, quickly falling; capsules glabrous .... *S. amygdaloides*
15. Leaf blades more elliptic, not long-pointed; female scales persistent, often dark 16

16. Young twigs red-purple and appressed-hairy; capsules hairy .... *S. bebbiana*
16. Twigs yellow or brown, not red-purple; capsules glabrous 17

17. Leaf blades narrowly lance-shaped, 3-6 times as long as wide ....
.... *S. eriocephala*
17. Leaf blades elliptic to ovate, 2-3 times as long as wide ....
.... *S. pseudomonticola*

18. Underside of leaf sometimes a little paler than the upper surface but without a pale wax that can be rubbed off with the fingers (glaucous) 19
18. Underside of leaf glaucous 20

19. Twigs oily; capsules hairy .... *S. barrattiana*
19. Twigs not oily; capsules glabrous .... *S. commutata*

20. Mature leaves without hair on the upper surface 21
20. Mature leaves hairy on the upper surface 23

21. Leaf blades wider above middle, often with some hair on upper midvein; plants of upland forest; capsules hairy .... *S. scouleriana*
21. Leaf blades widest at middle or below; plants of wet meadows or stream banks 22

22. Upper leaf surface and twigs shiny; lower leaf surface often sparsely hairy; capsules hairy .... *S. planifolia*
22. Leaves and twigs not shiny; leaves and capsules glabrous .... *S. farriae*

23. 1- and 2-year-old twigs partly covered with pale wax (glaucous), especially near the buds .... *S. geyeriana*
23. Twigs not glaucous 24

24. Leaf blades 2-3 cm long .... *S. brachycarpa*
24. Most leaf blades > 3 cm long 25

25. Young twigs red-purple and appressed-hairy .... *S. bebbiana*
25. Twigs dark, not red-purple 26

26. Leaf blades wider above middle; plants mostly of montane and lower subalpine forest .... *S. scouleriana*
26. Leaf blades widest at the middle or below; plants of open habitats near or above treeline .... *S. glauca*

Salix amygdaloides Anderss. Small tree or shrub 1-5 m tall; twigs yellow; leaf blades narrowly lance-shaped with finely toothed margins, 5-10 cm long, green above, waxy-pale beneath; female scales yellow; female catkins 3-10 cm long; capsules glabrous, 4-5 mm long with stalks 1-2 mm long; style < 0.5 mm long. June.

Montane river banks; reported by Standley (1921) for the Middle Fork Flathead River near West Glacier. B.C. to Que., south to WA, NM, AR, NJ.

Salix arctica Pallas [*S. petrophylla* Rydb., *S. anglorum* Cham.]. Mat-forming with trailing stems; leaf blades 1-4 cm long, waxy-pale beneath, often sparsely hairy, narrowly elliptic with entire margins; female scales brown or black, long-hairy; female catkins 1-3 cm long, with 25-50 flowers; capsules hairy, 4-7 mm long, sessile; style ca. 2 mm long. July-Aug. Fig. 244.

Abundant in moist to wet turf, often along streams or where snow accumulates, sometimes in talus, near or above treeline; East, West. Our plants are var. *petraea* (Anderss.) Bebb. Circumpolar south to CA, NM. Often occurs with *Dryas octopetala* and *Carex scirpoidea*. Stems may sometimes be erect in the shelter of rocks. *S. reticulata* has smaller catkins with fewer capsules and often forms denser mats. Standley's (1921) report of *S. cascadensis* Cock. is referable here (R. Dorn, personal communication).

Salix barrattiana Hook. Stems 30-100 cm tall; twigs reddish brown, hairy and oily; leaf blades 3-7 cm long, long-hairy on both sides, elliptic with entire margins; female scales black, long-hairy; female catkins 4-7 cm long, sessile; capsules hairy 3-5 mm long; styles ca. 1 mm long. July-Aug.

Rare along small streams on cool alpine slopes; known only from near Gunsight Pass. AK to B.C., WY. The long, sessile catkins and oily twigs are diagnostic.

Salix bebbiana Sarg., Bebb Willow [*S. b.* var *perrostrata* (Rydb.) Schneid]. Stems 2-4 m tall; twigs purplish, close-hairy; leaf blades 3-6 cm long, elliptic, often with shallowly toothed margins, green and sparsely hairy above, waxy-pale beneath; female scales tan, hairy; female catkins 2-5 cm long, stalked; capsules 6-9 mm long, hairy with stalks 2-5 mm long; style < 0.5 mm long. May-July.

Common in moist soil along montane streams and wetlands, generally not in very organic soil; East, West. AK to Lab. south to CA, NM, IN. The purplish, hairy twigs help distinguish this species.

Salix boothii Dorn [*S. pseudomyrsinites* Anderss., *S. myrtillifolia* Anderss. misapplied]. Stems 1-3 m tall; twigs brown, sparsely hairy; leaf blades 3-7 cm long, lance-shaped to elliptic, nearly glabrous, green above, slightly paler beneath, margins toothed; female scales dark brown, long-hairy; female catkins 2-5 cm long with short, leafy stalks; capsules glabrous 3-6 mm long with stalks 1-2 mm long; style ca. 0.5 mm long. May-July.

Common along montane and lower subalpine streams, lakes, and wetlands; East, West. B.C., Alta. south to CA, CO. Often with *S. planifolia* in fens and with *S. drummondiana* and *S. geyeriana* along streams and wetlands.

Salix brachycarpa Nutt. Stems 30-100 cm high; young twigs densely hairy; leaf blades 2-4 cm long, long-hairy especially beneath, elliptic with entire margins, waxy-pale beneath; female scales green to tan, hairy; female catkins 1-2 cm long with leafy stalks; capsules hairy, 3-5 mm long with stalks < 0.5 mm; style up to 1 mm long. July-Aug.

Common on rocky slopes and talus (especially limestone) near or above treeline and rare on hummocks in montane calcareous fens; East. AK south to OR, UT, NM. The two habitats are strikingly different. The small catkins help distinguish this species. See *S. candida*.

Salix candida Fluegge ex Willd. Stems 20-100 cm tall; young twigs densely white-hairy; leaf blades 3-9 cm long, narrowly oblong with entire margins, mostly densely white hairy above and beneath; female scales white-hairy, densely so below; female catkins 3-5 cm long, short-stalked; capsules white-hairy, 4-7 mm long; style ca. 1 mm long. June-July.

Uncommon in montane fens; East, West. AK to Newf. south to B.C., ID, CO, SD, IA, NJ. Occasional glabrate plants occur north of Lake McDonald. *S. brachycarpa* has shorter leaves.

Salix commutata Bebb [*S. c.* var. *denudata* Bebb]. Stems 20-150 cm tall; twigs dark and long-hairy; leaf blades 3-8 cm long, elliptic with entire to shallowly toothed margins, sparsely long-hairy above and beneath; female scales light brown, long-hairy; female catkins 2-5 cm long on leafy stalks; capsules glabrous, 4-5 mm long, short-stalked; style ca. 1 mm long. July-Aug.

Abundant along subalpine streams and ponds, often in areas of snow accumulation near treeline; East, West. AK to Alta. south to OR, ID, MT. *S. glauca* has narrower leaves that are waxy-pale beneath.

Salix drummondiana Barratt ex Hooker [*S. subcoerulea* Piper]. Stems 1-3 m high; twigs gray-waxy; leaf blades 3-7 cm long, narrowly elliptic to oblong with entire, inrolled margins, green above, densely silver-hairy below; female scales black, long-hairy; female catkins 2-7 cm long, sessile; capsules hairy, 3-6 mm long; styles ca. 1 mm long. May-July. Fig. 245.

Abundant along streams, lakes and wetlands, common on moist glacial moraine and talus and stream or river gravels, montane to near or above treeline; East, West. Yuk. to Sask. south to CA, NM. This is our most common erect willow, dominating in many typical early and late successional wetland and riparian habitats. The silvery felt on the leaf undersides with waxy twigs is diagnostic.

Salix eriocephala Michx. [*S. rigida* Muhl. var. *mackenzieana* (Hook.) Cronq., *S. mackenzieana* (Hook.) Anderss.]. Stems 2-4 m high; twigs red-brown; leaf blades thin, 4-10 cm long; lance-shaped to elliptic with toothed margins; green above waxy-pale beneath; female scales dark brown, glabrous; female catkins 3-6 cm long on leafy stalks; capsules glabrous, 3-7 mm long with a stalk 2-4 mm long; style ca. 0.5 mm long.

Uncommon on shores along forks of the Flathead River. Our plants are var. *mackenzieana* (Hook.) Dorn. Throughout most of Canada and U. S. *S. boothii* is very similar but has thicker leaves and shorter capsule stalks.

Salix exigua Nutt., Sandbar Willow [*S. interior* Rowlee misapplied]. Stems to 4 m tall; twigs brown, sparsely hairy; leaf blades 3-12 cm long, strap-shaped to narrowly lance-shaped with shallowly toothed margins, both surfaces gray-green or sometimes green and glabrous above; female scales yellow, hairy; female catkins 3-5 cm long, short-stalked; capsules 4-7 mm long, hairy or glabrous, short-stalked; style lacking. June-July.

Uncommon along lower montane rivers; our plants are var. *exigua*; East, West. AK to N.B. south to CA, Mex., LA, NJ. Leaves of the closely related *S. melanopsis* are usually dark green, glabrous and less narrow, but see *S. melanopsis*.

Salix farriae Ball. Stems up to 1.5 m high; twigs brown; leaf blades 2-5 cm long, lance-shaped to elliptic with entire margins, green above, waxy-pale beneath; female scales dark brown, glabrous to hairy; female catkins 1-4 cm long on short, leafy stalks; capsules glabrous, 4-6 mm long on stalks ca. 1 mm long; style ca. 0.5 mm long. June-Aug.

Uncommon in wet subalpine meadows, often near small streams or seeps; East, West. Yuk. south to ID, WY. *S. pseudomonticola* has leaves with toothed margins; *S. planifolia* has shiny twigs and upper leaf surfaces. Maguire's (1934) report of *S. monochroma* Ball for Gunsight Lake is probably referable here (R. Dorn, personal communication).

Salix geyeriana Anderss. Stems 1-3 m high; twigs gray-waxy with age; leaf blades 3-7 cm long, narrowly lance-shaped with entire margins, green above, paler beneath; female scales yellow to tan, hairy; female catkins 10-15 mm long on leafy stalks up to 1 cm long; capsules hairy, 3-6 mm long with stalks 1-2 mm long; style < 0.5 mm long. May-July.

Montane wet meadows, fens and alder or willow swamps; common West, uncommon East. B.C. to MT south to CA, NM. The waxy-pale twigs with sparsely hairy to nearly glabrous, waxy-pale leaf undersides is diagnostic.

Salix glauca L. Stems 50-100 cm tall; young twigs dark, hairy; leaf blades 2-5 cm long, narrowly elliptic with entire margins, sparsely long-hairy, waxy-pale beneath; female scales brown, hairy; female catkins 2-5 cm long on leafy stalks 5-15 mm long; capsules hairy, 4-6 mm long with a stalk 1-2 mm long; style 0.5-1 mm long. June-Aug.

Common in meadows, uncommon on limestone talus slopes near or above treeline, uncommon lower; East, expected West. Circumpolar south to UT, NM. Leaves of *S. commutata* are not waxy-pale beneath; those of *S. farriae* are not so hairy.

Salix lasiandra Benth. [*S. lucida* Muhl. ssp. *lasiandra* (Benth.) Murray, *S. caudata* (Nutt.) Heller]. Shrub 2-4 m high; leaf blades 3-15 cm long, lance-shaped with toothed margins and long-pointed tip, green above, paler beneath; female scales yellow, falling early; female catkins ca. 3 cm long on leafy branchlets; capsules glabrous, 4-7 mm long on stalks 1-2 mm long; style ca. 0.5 mm long.

Uncommon along lower montane rivers and streams; East, West. Our plants are var. *caudata* (Nutt.) Sudw. AK to CA, NM, SD. Small bump-like glands where the petiole meets the blade are found only in this species and *S. serissima* which grows in peat.

Salix melanopsis Nutt. [*S. exigua* Nutt. ssp. *melanopsis* (Nutt.) Cronq.]. Stems to 2 m high; twigs brown; leaf blades 4-7 cm long, mostly narrowly elliptic with shallowly toothed margins, green above, sometimes waxy-pale below, mostly glabrous; female scales yellow, hairy; female catkins 2-4 cm long, on tips of leafy branchlets; capsules 4-6 mm long, glabrous, on a stalk ca. 0.5 mm long; style short but evident. June-July.

Common along montane rivers, streams and lakes; East, West. B.C. to Alta. south to CA, CO. Uncommon forms of *S. melanopsis* with hairy and/ or narrower leaves can be confused with *S. exigua*.

Salix planifolia Pursh [*S. phylicifolia* L. ssp. *planifolia* (Pursh) Hiit.]. Stems 0.5-2 m tall; twigs red-brown, shiny; leaf blades 2-6 cm long, narrowly elliptic with entire margins; shiny green above, waxy-pale below; female scales black, long-hairy; female catkins 2-4 cm long, sessile; capsules 3-6 mm long, short-hairy, on stalks up to 1 mm long; style ca. 1 mm long. May-July.

Uncommon along streams and wetlands and in fens, montane and subalpine; East, West. Yuk to Lab. south to CA, NM, MN. The shiny twigs and upper leaf surfaces help distinguish this willow.

Salix pseudomonticola Ball [*S. monticola* Bebb misapplied]. Stems 1-3 m high; twigs dark; leaf blades 3-7 cm long, lance-shaped to elliptic, broadly rounded at the base with toothed margins, green above, waxy-pale below; female scales dark, long-hairy; female catkins 3-6 cm long, sessile; capsules glabrous, 4-7 mm long on stalks 1-2 mm long; style 0.5-1 mm long. June-July.

Uncommon in thickets, especially along montane streams and in wetlands; East. AK to Lab. south to B.C., ID, WY, SD. Leaf blades with red midveins near the base help distinguish this species.

Salix reticulata L., Snow Willow [*S. nivalis* Hook.]. Mat-forming subshrub with trailing stems; leaf blades 5-20 mm long, elliptic to orbicular with entire margins, green above, pale and veiny beneath; female scales yellow, nearly glabrous; female catkins 5-10 mm long, with few flowers; capsules hairy, 4-7 mm long, sessile; style 0.5 mm long. July-Aug.

Common in dry fellfields to moist turf of alpine slopes and ridges; East, West. Our plants are ssp. *nivalis* (Hook.) Löve et al. Circumpolar south to CA, NM. Often occurs with *S. arctica* and *Dryas octopetala*. See *S. arctica*.

Salix scouleriana Barratt ex Hook. Shrub (sometimes a small tree) 2-7 m tall, young twigs with short hairs; leaf blades 5-10 cm long, oblong-elliptic, widest above the middle with entire, inrolled margins, green above, pale-waxy beneath, often with a few brown hairs; female scales dark brown, hairy; female catkins 25-60 mm long, on stalks 3-10 mm long; capsules 5-8 mm long, short-hairy, with a stalk to 2 mm long; style ca. 0.5 mm long. May-June.

Common in montane and subalpine forest and avalanche slopes; East, West. AK to Man. south to CA, NM, SD. Our only species that is common in low- and mid-elevation, upland habitats; it often fails to flower in the shade. See *S. sitchensis*.

Salix serissima (Bailey) Fern., Autumn Willow. Stems 1-2 m high; twigs red or brown, shiny; leaf blades 4-8 cm long, narrowly elliptic with toothed margins, dark green above, pale beneath; female scales yellow, falling early; female catkins 15-30 mm long on leafy branchlets; capsules 6-9 mm long, reddish at maturity, on stalks 1-2 mm long; style < 0.5 mm long. June-Sept.

Known from a fen in the Belly River drainage; reported by Standley (1921) for a reach of Swiftcurrent Creek that is now under Lake Sherburne. Mack. to Lab. south to CO, SD, IN, NJ. Our only species with catkins that flower in spring but mature at the end of the growing season. See *S. lasiandra*.

Salix sitchensis Sanson ex Bong. Stems 1-3 m tall; twigs usually short-hairy; leaf blades 2-9 cm long, oblong-elliptic with entire margins, green and sparsely hairy above, satiny-hairy beneath; female scales brown, long-hairy; female catkins 3-8 cm long, short-stalked; capsules 3-5 mm long, short-hairy on stalks ca. 1 mm long; style 0.5 mm long. June-Aug.

Uncommon around montane and subalpine lakes, streams, and wet meadows; East, West. AK south to CA, ID, MT. *S. scouleriana* usually occurs in upland habitats and is pale-waxy on the leaf undersides; *S. drummondiana* usually has pale-waxy twigs.

Salix vestita Pursh [*S. v.* var *erecta* Anderss.]. Stems 20-100 cm tall; twigs brown; leaf blades 2-5 cm long, elliptic to ovate with entire, inrolled margins, rounded at the tip, green and veiny above, densely white-hairy and pale-waxy beneath; female scales brown, hairy; female catkins 2-4 cm long on tips of twigs; capsules 2-4 mm long, long-hairy, on stalks ca. 1 mm long; style < 0.5 mm long. July-Aug. Fig. 246.

Stony soil of wet meadows, open slopes, talus, and rock ledges, subalpine and alpine; abundant East, common West. B.C. to Que. south to WA, MT. The veiny leaves, densely white-hairy beneath, are distinctive.

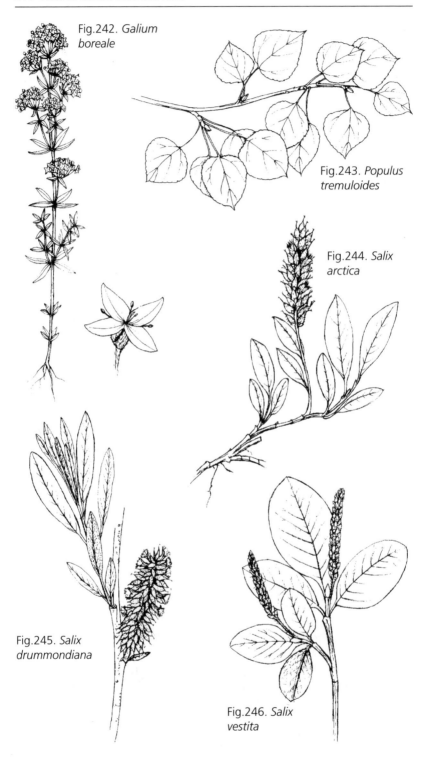

Fig.242. *Galium boreale*

Fig.243. *Populus tremuloides*

Fig.244. *Salix arctica*

Fig.245. *Salix drummondiana*

Fig.246. *Salix vestita*

SANTALACEAE: SANDALWOOD FAMILY

Glabrous, rhizomatous perennial herbs; leaves alternate; flowers bisexual; petals lacking; sepals 5, petal-like, united below and surrounding the ovary; stamens 5; style 1; fruit a berry-like drupe with the persistent calyx on top.

Members of this family are partially parasitic on the roots of neighboring plants.

1. Flowers several in a terminal inflorescence; fruits green to blue .... **Comandra**
1. Flowers usually 3, borne near the middle of the stem; fruits red .. **Geocaulon**

Comandra Nutt., Bastard Toad-flax

Comandra umbellata (L.) Nutt. [*C. pallida* D. C.]. Stems 10-30 cm high; leaves pale green, sessile, narrowly oblong, 1-3 cm long; flowers stalked in a branched, open, hemispheric inflorescence; sepals 2-4 mm long; berry green to blue, 4-8 mm long. July. Fig. 247.

Uncommon in grasslands along the east margin of the Park. Our plants are var. *pallida* (D. C.) Jones. B.C. to Newf. south to CA, AZ, TX, GA.

Geocaulon Fernald, Bastard Toad-flax

Geocaulon lividum (Rich.) Fern. [*Comandra livida* Rich.]. Stems 10-25 cm high; leaves short-petiolate, narrowly elliptic, 1-5 cm long, veiny; flowers in axils of middle leaves; sepals 1-2 mm long; berry red, 5-10 mm long. May.

Rare in moist, montane spruce forest; known only from the North Fork Flathead drainage. AK to Newf. south to WA, MT, MN, NY.

SAXIFRAGACEAE: SAXIFRAGE FAMILY

Perennial herbs; leaves alternate, often palmately lobed but not fully divided; flowers bisexual; calyx 5-lobed, the base partly surrounding the ovary; petals 5, separate; stamens 5 or 10; styles 2; ovary 2-chambered; fruit a many-seeded capsule.

Flowers are required for identification.

1. Stamens 5 2
1. Stamens 10 6
2. Basal leaves glabrous with entire margins .... **Parnassia**
2. Basal leaves with lobed or toothed margins 3
3. Stem with at least 1 leaf with a swollen petiole base .... **Suksdorfia**
3. Stems leafless 4
4. Lower inflorescence branched or with > 1 flower per node .... **Heuchera**
4. Lower nodes of inflorescence spike with only 1 flower 5
5. Petals lobed or divided... **Mitella**
5. Petals with entire margins... **Conimitella**
6. Stems leafless 7
6. Stem with at least 1 leaf well below the inflorescence 8

7. Petals pinnately divided into several linear lobes; capsule opening widely, appearing like a dish of seeds at maturity .... **Mitella**
7. Petals not deeply lobed; capsule opening at the tip(s) .... **Saxifraga**

8. Petals deeply lobed 9
8. Petals not lobed or absent 10

9. Plants < 25 cm high, of open habitats .... **Lithophragma**
9. Plants > 25 cm high, found in moist forest .... **Tellima**

10. Leaves leathery with a single clasping leaf at mid-stem .... **Leptarrhena**
10. Plants without a solitary clasping, leathery leaf at mid-stem 11

11. Petals linear; capsule with 2 unequal chambers; forest plants .... **Tiarella**
11. Petals oblong; capsule with equal chambers; plants of open, often stony and wet habitats .... **Saxifraga**

Conimitella Rydb.
Conimitella williamsii (D. C. Eat.) Rydb. Stems 20-40 cm high, leafless, glandular-hairy; basal leaves petiolate, 1-4 cm wide, broadly heart-shaped with shallowly lobed and toothed margins; flowers 5-10, short-stalked in a narrow, spike-like inflorescence; calyx cone-shaped; petals 5, longer than the sepals, white, spoon-shaped, 4-5 mm long; stamens 5; capsule 5-9 mm long. June.

Rare in montane grassland; known only from north of the Two Medicine Valley on the east edge of the Park. Alta. south to ID , WY. *Mitella* spp. are usually found in moister habitats, while species of *Heuchera* have more flowers.

Heuchera L., Alum-root
Stems glandular above; rootcrown branched, scaly; leaves basal, long-petiolate, spade-shaped with a rounded tip and shallowly lobed and toothed margins; flowers short-stalked in a narrow, spike-like inflorescence; stamens 5; capsule 2-beaked.

1. Calyx cup-shaped, 5-10 mm long .... **H. cylindrica**
1. Calyx dish-shaped, 2-3 mm long .... **H. parvifolia**

Heuchera cylindrica Douglas ex Hooker [*H. glabella* T. & G.]. Stem 20-60 cm high; leaf blades 15-40 mm wide with 5-7 lobes; flowers cup-shaped; calyx glandular, 5-10 mm long, the lobes 4-5, whitish; petals inconspicuous or absent. June-Aug. Fig. 248.

Common in stony soil of grasslands, rock outcrops, and talus slopes; montane, occasionally higher; East, West. Our plants are var. *glabella* (T. & G.) Wheelock. B.C., Alta. south to CA, NV, WY. Native Americans used a decoction of the roots to treat diarrhea; plants were also used to treat saddle sores on horses.

Heuchera parvifolia Nutt. ex T. & G. [*H. flabellifolia* Rydb.]. Stems 10-30 cm high; leaf blades 1-3 cm wide; flowers dish-shaped; calyx 2-3 mm long, green; petals white, longer than the sepals. June-July.

Rare in grasslands along the east margin of the Park; uncommon in stony soil of dry alpine turf in the Two Medicine area. Our plants are var. *dissecta* Jones. Alta. south to NV, AZ, NM. Native Americans used a decoction of the roots to treat stomach ailments.

Leptarrhena R. Br., Leather-leaf

Leptarrhena pyrolifolia (D. Don) R. Br. Stems 10-30 cm high with 1-2 sessile, ovate leaves in the middle; basal leaves oblong-ovate, 2-7 cm long, leathery, glabrous, green and shiny above with toothed margins; flowers short-stalked in a compact, branched, ovoid, terminal, glandular inflorescence; calyx cup-shaped; petals white, 2-3 mm long, longer than the sepals; stamens 10, longer than the petals; capsule 5-6 mm long, purplish. July-Aug. Fig. 249.

Common in moist to wet soil of meadows and cliffs, often along streams, near or above treeline. East, West. AK to WA, ID, MT.

Lithophragma Nutt., Prairie Star, Fringecup

Short-lived perennials; stems glandular; roots often with bulblets; leaves petiolate, the blades deeply lobed; flowers few, short-stalked in a terminal inflorescence; calyx glandular; petals longer than the sepals, deeply 3- to 5-lobed; stamens 10; styles 3. Reference: Taylor (1965).

1. Bulblets often replacing flowers; petals usually 5-lobed; basal leaf blades mostly glabrous .... *L. glabrum*
1. Above-ground bulblets absent; petals usually 3-lobed; basal leaves white-hairy .... *L. parviflorum*

Lithophragma glabrum Nutt. [*L. bulbifera* Rydb.]. Stems 5-25 cm high; leaf blades glabrous or sparsely hairy, 5-15 mm long; bulblets often replacing flowers; calyx cup-shaped; petals pinkish, 4-7 mm long, mostly deeply 5-lobed. May.

Uncommon in montane to subalpine grasslands; East, West. B.C. to Sask. south to CA, UT, CO, SD. Many of our plants lack flowers entirely; forms without bulblets occur south of the Park.

Lithophragma parviflorum (Hooker) Nutt. ex T. & G. Stems 10-30 cm high; basal leaf blades sparsely white-hairy, 1-3 cm wide; calyx conical; petals white, 5-10 mm long, usually 3-lobed. June-July. Fig. 250.

Common in montane and lower subalpine grasslands, meadows, and open forest; East, West. B.C., Alta. south to CA, CO, SD.

Mitella L., Mitrewort

Stems slender from a branched rootcrown, glandular at least above; leaves mostly basal, long-petiolate, mostly heart- or spade-shaped with shallowly palmately lobed and toothed margins; flowers short-stalked in a narrow, spike-like inflorescence; petals white or greenish, usually deeply divided into thread-like lobes; capsule opening widely to expose black seeds.

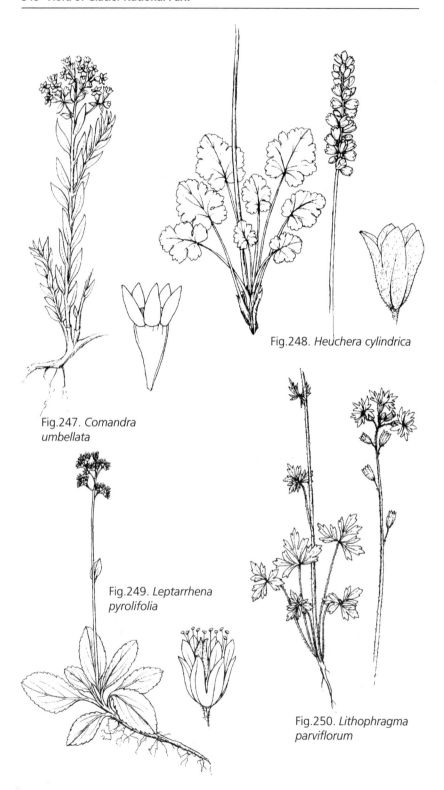

Fig.247. *Comandra umbellata*

Fig.248. *Heuchera cylindrica*

Fig.249. *Leptarrhena pyrolifolia*

Fig.250. *Lithophragma parviflorum*

There is broad ecological overlap among the species; however, *M. stauropetala* occurs in the driest habitats, *M. breweri* is often in areas with deep winter snow cover, and *M. nuda* is usually found in low swamp forest. *M. caulescens* Nutt., with leafy stems, occurs just south of the Park. See *Conimitella*.

1. Stamens 10; often in wet, montane spruce forest .... **M. nuda**
1. Stamens 5 2
2. Stamens between calyx lobes .... **M. pentandra**
2. Stamens in front of calyx lobes 3
3. Calyx saucer-shaped; petals with ≥ 5 lobes; leaf blades as wide as long ....
.... **M. breweri**
3. Calyx cone- or cup-shaped; petals 3-lobed; leaf blades somewhat longer
    than wide 4
4. Flowers all aligned on 1 side of the spike .... **M. stauropetala**
4. Flowers facing different directions on the spike............................ **M. trifida**

Mitella breweri Gray. Stems 8-30 cm tall; leaf blades nearly orbicular, 2-8 cm wide; calyx saucer-shaped, the lobes reflexed; petals 3-4 mm long with 5-9 pinnate lobes; stamens 5, opposite the sepals. June-Aug. Fig. 251.

Common in cool forests as well as moist cliffs and along streams, often where snow lies late; upper montane to lower alpine; East, West. B.C., Alta. south to CA, ID, MT.

Mitella nuda L. Stems 5-20 cm high; leaf blades 1-3 cm long; calyx saucer-shaped with spreading lobes; petals ca. 4 mm long with 7-9 pinnate lobes; stamens 10. June-July.

Common in moist to wet, montane spruce forest; East, West. AK to Newf. south to WA, MT, ND. Often growing with *Equisetum* spp.

Mitella pentandra Hooker. Stems 15-35 cm high; leaf blades 2-6 cm wide; calyx cup-shaped with spreading to reflexed lobes; petals 2-3 mm long with ca. 8 pinnate lobes; stamens 5, opposite the petals. June-Aug.

Common in moist to wet forest, cliffs and stony wet meadows, often along streams; montane to treeline; East, West. AK to Alta. south to CA, CO.

Mitella stauropetala Piper. Stems 20-40 cm high; leaf blades 2-6 cm wide, barely toothed; calyx conical with erect lobes and spreading tips; flowers aligned on 1 side of the spike; petals 2-4 mm long, 3-lobed; stamens 5, opposite the sepals. May-July.

Uncommon in dry to moist, montane forest, occasionally higher; East, West. WA to MT south to UT, CO.

Mitella trifida Graham [*M. violacea* Rydb.]. Stems 15-40 cm high; leaf blades 2-4 cm wide; calyx cup-shaped with erect lobes; petals often purplish, 1-3 mm long with 3 short lobes; stamens 5 opposite the sepals. June.

Uncommon in moist montane to subalpine forests and meadows; East, West. B.C., Alta. south to CA, MT.

Parnassia L., Grass-of-Parnassus

Foliage glabrous; leaves mostly basal, petiolate with entire margins; flower solitary; calyx broadly conical; petals white, ovate; stamens 5, alternating with 5 fringed scales (staminodia); stigmas usually 4, capsule 4-chambered, ca. 1 cm long.

This genus is sometimes treated as a separate family, the Parnassiaceae.

1. Leaves all basal .... *P. kotzebuei*
1. Stems with 1 sessile leaf near the middle 2

2. Petals fringed at the base; basal leaf blades wider than long ...... *P. fimbriata*
2. Petals ovate, longer than wide; petals not fringed 3

3. Petals 7-12 mm long; stem leaf ovate, clasping .... *P. palustris*
3. Petals 4-8 mm long; stem leaf lance-shaped, not clasping .... *P. parviflora*

Parnassia fimbriata Konig. Stems 10-40 cm high with 1 sessile, clasping leaf in the middle; basal leaf blades nearly orbicular, 1-5 cm broad with a wide sinus at the base; petals 8-12 mm long with thread-like fringes below the middle, longer than the sepals. July-Sept. Fig. 252.

Common in subalpine and lower alpine wet meadows and along streams and seeps at all elevations; East, West. Our plants are var. *fimbriata*. AK to Alta. south to CA, NM. Our most common and conspicuous grass-of-Parnassus.

Parnassia kotzebuei Cham. Stems 5-10 cm high, leafless; basal leaf blades broadly ovate, 5-15 mm long; petals 3-7 mm long, as long as the sepals. July-Aug.

Common but inconspicuous in moist to wet alpine turf and mossy rock ledges; East, West. AK to Greenl. south to WA, NV, CO; Asia.

Parnassia palustris L. [*P. montanensis* Fern. & Rydb.]. Stems 10-20 cm high with 1 sessile clasping leaf on the lower third; basal leaf blades ovate, 8-15 mm long; petals 8-12 mm long, longer than the sepals. July.

Rare in montane fens, wet meadows, and thickets; known only from near St. Mary and Waterton Lake. Our plants are var. *montanensis* (Fern. & Rydb.) Hitchc. Circumboreal south to CA, UT, CO, ND, MN.

Parnassia parviflora D. C. Stems 15-25 cm high with a lance-shaped leaf in the middle; basal leaf blades ovate, 1-3 cm long; petals 4-8 mm long, longer than the sepals. Aug.

Rare in fens, seeps and along streams; montane and alpine; East. B.C. to Que. south to CA, ID, MT, SD, MN. Hitchcock (Hitchcock et al. 1961) reports that this species intergrades with *P. palustris* in MT.

Saxifraga L.

Leaves basal and/or usually alternate; foliage often hairy and glandular above; flowers stalked; calyx saucer-shaped to conical; petals mostly white, oblong to ovate; stamens 10; styles 2; capsule 2-beaked, 2-chambered, many-seeded. Reference: Elvander (1984), Moss and Packer (1983).

S. occidentalis, S. nivalis, and S. subapetala can be difficult to distinguish and are part of an arctic-alpine complex in need of taxonomic revision from a circumpolar perspective. Species with all basal leaves may eventually be placed in a separate genus, Micranthes (Soltis et al. 1996).

1. Leaves all in basal rosettes; only greatly reduced bracts on the stem 2
1. At least 1-2 stem leaves present and nearly as large as the basal 11

2. Leaf blades orbicular, as wide as long 3
2. Leaf blades ovate to spatula-shaped, distinctly longer than wide 5

3. Stems < 10 cm high; flowers 1-3 per stem .... *S. rivularis*
3. Stems > 10 cm high; flowers numerous 4

4. Leave blades with toothed margins; petioles nearly glabrous; usually along streams .... *S. odontoloma*
4. Leaf blades shallowly lobed, the lobes 3-toothed; petioles long-hairy; usually on wet ledges .... *S. mertensiana*

5. Leaves 3-lobed at the tip 6
5. Leaf margins merely toothed 7

6. Leaves deeply 3-lobed > 1/2 of the way to the base .... *S. cespitosa*
6. Leaves shallowly 3-lobed < 1/3 of the way to the tip .... *S. adscendens*

7. Inflorescence open, lower branches or flowers well separated 8
7. Flowers in 1 or more head-like clusters 9

8. Flowers partly replaced by tiny plantlets; inflorescence densely glandular ......
.... *S. ferruginea*
8. Plantlets absent; inflorescence with few glands .... *S. lyallii*

9. Petals no longer than sepals, absent or falling early .... *S. subapetala*
9. Petals longer than sepals, more persistent 10

10. Sepals purplish; stems usually < 6 cm high; leaf blades < 2 cm long; flowers nearly sessile .... *S. nivalis*
10. Sepals mainly green; stems mostly > 8 cm high; some leaf blades > 2 cm long; flowers on stalks > 1 mm long .... *S. occidentalis*

11. Stems densely leafy, moss-like; leaves with entire margins 12
11. Stem leaves well separated; some basal leaves toothed or lobed 13

12. Leaves opposite; petals purple, 6-8 mm long .... *S. oppositifolia*
12. Leaves alternate; petals white with purple spots; 4-6 mm long ....
.... *S. bronchialis*

13. All but uppermost flower replaced by tiny red bulblets .... *S. cernua*
13. Flowers not replaced by bulblets 14

14. Basal leaf blades wider than long .... *S. rivularis*
14. Basal leaf blades oblong to spatula-shaped, longer than wide 15

15. Leaf blades 3-lobed at least 1/2 the length; plants forming small tufts ....
.... ***S. caespitosa***
15. Leaf blades shallowly lobed or toothed; most plants with only 1-2 stems ....
.... ***S. adscendens***

Saxifraga adscendens L. Stems 1-8 cm tall, often branched at the base; leaves 3-8 mm long, oblong, often with 3 shallow lobes at the tip, clustered at the base; foliage glandular; flowers long-stalked in an open bract-bearing inflorescence; calyx purplish, cup-shaped; petals 3-4 mm long. July-Aug.

Uncommon in vernally wet, gravelly soil of alpine turf and rock outcrops; East, West. AK to OR, UT, CO. *S. caespitosa* has more deeply lobed leaves and forms tufts; *S. debilis* and *S. cernua* have palmately lobed leaves with well-differentiated petioles.

Saxifraga bronchialis L. Mat-forming; sterile stems creeping, densely leafy; leaves needle-like, 5-10 mm long, spine-tipped; flowering stems 5-15 cm high, sparsely leafy; flowers in an open, branched inflorescence; calyx saucer-shaped; petals 4-6 mm long, purple-spotted. July-Aug. Fig. 253.

Abundant on talus slopes and cliffs and in gravelly soil of dry meadows and turf at all elevations. Our plants are ssp. *austromontana* (Wieg.) Piper. B.C., Alta. south to OR, ID, NM. The prickly leaves set this apart from our other saxifrages.

Saxifraga caespitosa L. Cushion-forming with a branched rootcrown and glandular stems 2-6 cm high; leaves mostly basal, narrowly oblong, 6-12 mm long, 3-lobed; flowers few in a terminal cluster; calyx conical; petals 3-6 mm long. July.

Uncommon in dry to moist alpine turf and rocky slopes; East, expected West. Circumpolar south to CA, NV, NM. See S. adscendens.

Saxifraga cernua L. Stems 5-20 cm high; roots often with bulblets; leaves petiolate, the blades 10-15 mm wide, broader than long with 3-7 shallow palmate lobes, reduced upward; flower solitary at the apex of a narrow inflorescence, the lower flowers replaced by purple bulblets; petals 6-10 mm long. July-Aug.

Uncommon in moist or wet gravelly soil of subalpine and alpine slopes and wet cliffs; East, West. Circumpolar south to WA, NV, NM, SD. The small red bulblets in the inflorescence are diagnostic.

Saxifraga ferruginea Graham. Stems 10-40 cm high; leaves basal, 2-10 cm long, spatula-shaped with toothed margins; flowers numerous in an open, branched inflorescence, some replaced by tiny plantlets; petals 4-6 mm long, the upper 3 with 2 yellow, basal spots. June-Aug.

Moist rock ledges and wet, gravelly soil along streams; common subalpine, uncommon lower and higher; East, West. AK to Alta. south to CA, ID, MT. The tiny plantlets can often be found taking root in moist moss beneath the parent plants.

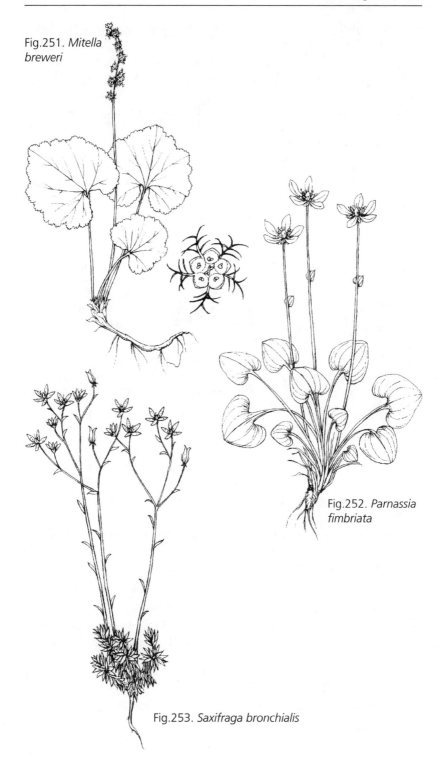

Fig.251. *Mitella breweri*

Fig.252. *Parnassia fimbriata*

Fig.253. *Saxifraga bronchialis*

Saxifraga lyallii Engler. Stems 7-25 cm high; foliage glabrous; leaves basal, short-petiolate, the blades 1-4 cm long, fan-shaped to ovate with toothed margins on the upper half; flowers in an open, few-branched, narrow inflorescence; calyx reddish, deeply lobed; petals 2-4 mm long. July-Aug. Fig. 254.

Abundant in wet soil of alpine gravelly meadows and rock outcrops, especially along streams; East, West. AK south to WA, ID, MT. See *S. odontoloma*; arguably our most common alpine saxifrage.

Saxifraga mertensiana Bong. Stems 10-25 cm high; roots often with bulblets; leaves basal, long-petiolate, the blades 2-5 cm wide, nearly orbicular with shallowly lobed margins; flowers in an open, branched inflorescence, some often replaced by plantlets; calyx deeply lobed; petals 2-4 mm long. July-Aug.

Common on wet cliffs or along small streams, subalpine and alpine; East, West. AK south to CA, ID, MT. *S. lyallii* has leaf blades longer than wide; *S. odontoloma* has glabrous petioles.

Saxifraga nivalis L. Stem 4-8 cm high with purple-tipped glandular hairs; leaves basal, short-petiolate, the blade 1-2 cm long, ovate with toothed margins, purple beneath; flowers 3-12 in a dense, globose cluster; calyx conical, purple; petals 2-4 mm long, white or purple. July.

Rare in moist turf and on wet rock ledges near or above treeline; known only from the Logan Pass area. Circumpolar south to UT, CO. Intergrades with small plants of *S. occidentalis*.

Saxifraga occidentalis Watson. Stems 10-25 cm high with purple-tipped glandular hairs; leaves basal, short-petiolate, the blade 1-4 cm long, elliptic with toothed margins, often red-hairy beneath; flowers many in a dense, erect-branched, narrow inflorescence, expanding in fruit; sepals nearly glabrous; petals 1-4 mm long; chambers of capsule nearly separate at maturity. May-Aug.

Common in vernally moist, usually stony soil of grasslands, meadows, and rock outcrops at all elevations; East, West. B.C., Alta. south to OR, NV, WY. See *S. nivalis*. Reports of *S. rhomboidea* Greene are referable here.

Saxifraga odontoloma Piper [*S. arguta* D. Don]. Stems 10-30 cm high from rhizomes; leaves basal, petiolate, the blades 2-8 cm wide, nearly orbicular with toothed margins; flowers in an open, branched, glandular inflorescence; calyx saucer-shaped, often purplish; petals 2-5 mm long with 2 basal spots. July.

Uncommon along streams and on wet, mossy rocks, montane and lower subalpine; East, expected West. B.C., Alta. south to CA, AZ, NM. Reported to hybridize with *S. lyallii* (Moss and Packer 1983), which has leaves much longer than wide. See *S. mertensiana*.

Saxifraga oppositifolia L., Purple Mountain Saxifrage. Cushion-forming; stems prostrate or ascending, densely leafy, 2-4 cm high; leaves opposite, 2-5 mm long, ovate with fringed margins; flowers solitary, short-stalked; calyx deeply lobed; petals light purple, 6-8 mm long, erect. May-June.

Uncommon in calcareous scree or stony soil of exposed slopes and moraine above treeline; East. Circumpolar south to OR, ID, WY. This distinctive plant flowers very early in its wind-blown habitat.

Saxifraga rivularis L. [*S. debilis* Engelm., *S. hyperborea* R. Br.]. Stems 2-8 cm high, forming small tufts; basal leaves petiolate, the blades 2-8 mm wide, fan-shaped with 3-5 shallow palmate lobes; stem leaves few, reduced; flowers 1-2; calyx conical, purplish; petals 3-4 mm long. July-Aug.

Uncommon in moist or wet gravelly soil of alpine slopes and cliffs; East, West. Our plants are ssp. *hyperborea* (R. Br.) Dorn. Circumpolar south to CA, AZ, CO. See *S. adscendens*.

Saxifraga subapetala Nelson [*S. oregana* Howell var. *subapetala* (Nels.) Hitchc.]. Stems 10-30 cm high; leaves basal, broadly petiolate, the blade 3-8 cm long, oblong elliptic with obscurely toothed margins; flowers nearly sessile in short-stalked clusters in a narrow inflorescence; calyx conical, purplish, the lobes longer than the petals; petals purplish, 1 mm long and quickly falling or absent entirely. June.

Rare in moist areas of subalpine grasslands; known only from the Marias Pass area. MT, WY. Similar plants in Canada are referred to *S. hieracifolia* Waldst. & Kit.

Suksdorfia Gray

Roots bearing bulblets; lower leaves petiolate, the blades fan-shaped, 2-4 cm wide palmately 3- to 7-lobed; upper leaves with expanded bases (stipules) clasping the stem; flowers stalked, bell-shaped in an open, glandular inflorescence; calyx conical; petals longer than sepals; stamens 5.

1. Petals white, spreading; stipules ovate .... **S. ranunculifolia**
1. Petals usually violet, nearly erect; some stipules lobed .... **S. violacea**

Suksdorfia ranunculifolia (Hooker) Engelm. [*Hemieva ranunculifolia* (Hook.) Raf.]. Stems 10-30 cm tall; lower leaves fleshy; upper leaves with an ovate-expanded base; inflorescence branched; petals white, 3-6 mm long, spreading. June-Aug. Fig. 255.

Uncommon in shallow soil of wet, montane to alpine cliffs and talus; East, West. B.C., Alta. south to CA, ID, MT.

Suksdorfia violacea Gray. Stems weak, 10-25 cm high; upper leaves with a lobed, expanded base; flowers few in an unbranched inflorescence; petals 5-8 mm long, violet or occasionally white, nearly erect. July.

Uncommon on wet montane cliffs; East, West. B.C., Alta. south to OR, ID, MT.

Tellima R. Br.

Tellima grandiflora (Pursh) Douglas ex Lindley. Stems glandular above, 40-70 cm high from a short rhizome; lower leaves petiolate, the blades nearly orbicular with a deep basal sinus, 5- to 7-lobed, 4-10 cm wide; upper leaves smaller, sessile; flowers many, short-stalked in a spike-like inflorescence; calyx cup-shaped with erect lobes; petals white to reddish, 5-8 mm long with 5-7 thread-like lobes at the tip; stamens 10. July. Fig. 256.

Uncommon in moist, montane forest, often near streams; West. AK south to CA, ID, MT.

Tiarella L., Foamflower

Tiarella trifoliata L. [*T. unifoliata* Hooker]. Stems 20-50 cm high, glandular above; leaves mostly basal, petiolate, the blade spade-shaped, 2-8 cm wide with 3-5 main lobes and toothed margins; flowers stalked, nodding in small groups in a narrow inflorescence; calyx cup-shaped with erect lobes; petals white, thread-like, 2-4 mm long; stamens 10, exserted; capsule with 2 unequal, ovoid chambers. June-Aug. Fig. 257.

Abundant in moist, montane to lower subalpine forest; East, West. Our plants are var. *unifoliata* (Hook.) Kurtz; however, plants with leaves fully divided into 3 leaflets are occasionally found mixed in. These are var. *trifoliata,* but the distinction is probably not worth recognizing in our area. AK south to CA, ID, MT.

SCROPHULARIACEAE: FIGWORT or SNAPDRAGON FAMILY

Annual or perennial herbs; flowers bisexual; sepals usually 4-5; corolla of mostly 4-5 united (at least at the base), unequal petals, often tubular and 2-lipped at the tip; stamens 2-5; ovary above the base of the calyx (superior); style 1; fruit a many-seeded, usually 2-chambered capsule.

Members of the tribe Rhinantheae (*Castilleja, Euphrasia, Melampyrum, Orthocarpus, Pedicularis, Rhinanthus*) are hemiparasites, gaining a portion of their nutrition by attaching their roots to those of their neighbors.

1. Leaves all basal; flower stalks arising from a basal rosette .... **Limosella**
1. Plants with leafy flowering stems 2

2. Leaves alternating around the stem (alternate) 3
2. Leaves paired on the stem (opposite) 10

3. Corolla saucer-shaped, bell-shaped or absent 4
3. Corolla tubular at the base and 2-lipped at the mouth; upper lip hood-like and enclosing the stamens 6

4. Corolla absent; anther stalks (filaments) purple; grasslands .... **Besseya**
4. Corolla saucer- or bell-shaped; disturbed sites 5

5. Corolla bell-shaped, 4-6 cm long, purplish ....**Digitalis**
5. Corolla saucer-shaped, 15-30 mm across, white to yellow .... **Verbascum**

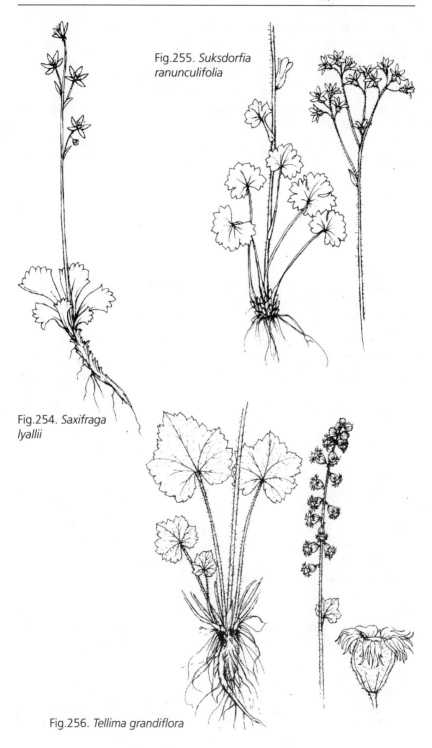

Fig.255. *Suksdorfia ranunculifolia*

Fig.254. *Saxifraga lyallii*

Fig.256. *Tellima grandiflora*

6. Corolla with a tubular spur projecting backward from the base of the calyx, resembling a snapdragon .... **Linaria**
6. Corolla without a spur 7

7. Flowers 5-6 mm long, much shorter than their stalks .... **Chaenorrhinum**
7. Flowers > 6 mm long, ca. as long or longer than their stalks 8

8. Flower bracts colored yellow or reddish .... **Castilleja**
8. Flower bracts green (sometimes tipped with pink) 9

9. Plants annual; leaves to 5 cm long, linear or with 3 linear lobes but otherwise the margins entire .... **Orthocarpus**
9. Plants perennial; some leaves > 5 cm long with toothed and often deeply lobed margins .... **Pedicularis**

10. Anther-bearing stamens 2 11
10. Anther-bearing stamens 4 12

11. Corolla saucer-shaped with 4 distinct lobes .... **Veronica**
11. Corolla tubular with 4-5 small lobes at the mouth .... **Gratiola**

12. Upper corolla lip 2-lobed, not hood-like or enclosing the anthers 13
12. Upper corolla lip not lobed but hood-like and enclosing the anthers 15

13. Stamens 5, 4 with anthers and 1 sterile .... **Penstemon**
13. Fertile stamens 4; sterile stamen absent 14

14. Corolla blue, 4-7 mm long, the lower lip boat-like and enclosing the anthers .... **Collinsia**
14. Corolla yellow or red, often > 7 mm long, tubular, without a boat-like lower lip .... **Mimulus**

15. Plants < 8 cm high; near or above treeline .... **Euphrasia**
15. Plants > 8 cm high; montane or lower subalpine 16

16. Calyx 4-lobed, inflated in fruit; corolla yellow .... **Rhinanthus**
16. Calyx tubular, 5-lobed; corolla white .... **Melampyrum**

Besseya Rydb., Kittentails

Besseya wyomingensis (Nelson) Rydb. [*Synthyris wyomingensis* (Nels.) Heller]. Perennial; foliage short-hairy; stems 5-25 cm high, expanding with maturity; basal leaves petiolate, the blades 2-5 cm long, ovate with toothed margins; stem leaves alternate, smaller, sessile; flowers sessile in an expanding spike; calyx 2-lobed; petals lacking; stamens 2, purple, long-exserted; capsule 5-6 mm long, oval, shallowly notched on top. May-June. Fig. 258.

Uncommon in montane grasslands and exposed alpine slopes and ridges along the east margin of the Park. B.C. to Sask. south to UT, CO, NE. The early flowering with spikes of purple stamens is diagnostic.

Castilleja Mutis ex L. f., Paintbrush

Perennial; stems usually several from a branched rootcrown; leaves alternate, sessile; flowers subtended by mostly shorter, colored, leaf-like bracts in dense terminal spikes; calyx 4-lobed, united below; corolla tubular with a long, arched upper lip (galea) and a shorter, 3-lobed, lower lip; stamens 4, enclosed in the galea; capsule ovoid, splitting in 2. Reference: Ownby (1959).

Closely related species are often difficult to distinguish due to interspecific gene flow via hybridization. For example, *C. miniata* intergrades with *C. rhexifolia*, which intergrades with *C. occidentalis*, which intergrades with *C. sulphurea*, etc. (Moss and Packer 1983).

1. Inflorescence red or purplish 2
1. Inflorescence whitish to yellow 4

2. Leaves of mid-stem lobed .... **C. hispida**
2. Mid-stem leaves with entire margins 3

3. Bracts orange to red; calyx lobes long-pointed; subalpine and lower ....
.... **C. miniata**
3. Bracts purplish; calyx lobes rounded or blunt; mostly alpine and subalpine ....
.... **C. rhexifolia**

4. Leaves of mid-stem lobed .... **C. cusickii**
4. Mid-stem leaves with entire margins 5

5. Bracts with 1-2 pairs of lateral lobes .... **C. lutescens**
5. Bracts with entire margins or 1 pair of short apical lobes 6

6. Galea almost as long as the tube .... **C. miniata**
6. Galea ≤ half as long as the tube 7

7. Plants usually < 20 cm high; corolla ca. as long as calyx .... **C. occidentalis**
7. Plants usually > 20 cm high; corolla longer than calyx .... **C. sulphurea**

Castilleja cusickii Greenm. [*C. lutea* Heller]. Stems 10-40 cm high; foliage soft long-hairy; lower leaves narrowly lance-shaped with entire margins; upper leaves and bracts with 1-3 pairs of pointed lobes; bracts yellow-tipped, longer than the flowers; calyx 20-30 mm long with rounded lobes; corolla yellow, ca. as long as the calyx, the galea much shorter than the tube. May-July.

Uncommon in montane grasslands; East, West. B.C., Alta. south to NV, ID, WY. See *C. lutescens.*

Castilleja hispida Benth. ex Hook. Stems 20-40 cm high; foliage long-hairy; leaves oblong, the lower entire; upper leaves and bracts 3- to 5-lobed at the tip; bracts red-tipped; calyx 15-30 mm long, the lobes blunt-tipped, corolla reddish green, 20-40 mm long, the galea as long as the tube. May-June. Color plate 2.

Common in montane grassland and dry open forest; East, West. B.C., Alta. south to OR, ID, MT. Standley's (1921) reports of *C. amplifolia* Rydb. and *C. bradburyi* (Nutt.) Don are probably referable here.

Castilleja lutescens (Greenm.)Rydb. Stems 15-60 cm tall; foliage short-hairy; leaves narrowly lance-shaped, mostly with entire margins; bracts yellow-tipped with 3-7 narrow, lateral lobes; calyx 15-25 mm long with long-pointed lobes; corolla yellow, just longer than the calyx, the galea much shorter than the tube. June-Aug.

Uncommon in montane grasslands; East. B.C., Alta. south to OR, ID, MT. *C. cusickii* has lobed leaves, and *C. sulphurea* has unlobed bracts.

Castilleja miniata Douglas ex Hooker [*C. vreelandii* Rydb., *C. lanceifolia* Rydb.]. Stems 20-60 cm high, sometimes branched; foliage glabrous to sparsely hairy; leaves lance-shaped with entire margins; bracts orange to red (sometimes white), entire or with 3-5 lobes toward the tip; calyx 15-25 mm long with long-pointed lobes; corolla red (rarely yellowish), longer than the calyx, the galea nearly as long as the tube. June-Aug.

Common in moist meadows, thickets and forest openings, often near streams, montane and subalpine; East, West. AK to Man. south to CA, NM. This species occurs in more habitats than any of our other paintbrushes. A white-flowered form occurs in aspen groves along the east margin of the Park.

Castilleja occidentalis Torrey. Stems 5-20 cm tall; foliage mostly sticky long-hairy; leaves narrowly lance-shaped with entire margins; bracts pale yellow to green with entire margins or nearly so; calyx 15-25 mm long with short, rounded lobes; corolla whitish, ca. as long as the calyx, the galea less than half as long as the tube. July-Aug.

Common in moist meadows and moraine near or above treeline; East, West. B.C., Alta. south to UT, NM. Often intergrades with *C. rhexifolia*, forming populations with flowers of every shade between white and purple.

Castilleja rhexifolia Rydb. [*C. lauta* Nelson]. Stems 10-40 cm tall; foliage glabrous to long-hairy; leaves lance-shaped with entire margins; bracts scarlet- to purple-tipped, entire or with 2-4 lateral lobes; calyx 15-25 mm long with bluntly rounded lobes; corolla purplish to red, longer than the calyx, the galea ca. half as long as the tube. July-Aug. Fig. 259. Color plate 48.

Moist meadows; abundant in the subalpine and alpine, uncommon lower; East, West. B.C., Alta. south to OR, UT, CO. Very similar to the yellow-flowered *C. sulphurea*; see *C. occidentalis*.

Castilleja sulphurea Rydb. Stems 15-30 cm high, often branched; foliage glabrous to sticky long-hairy; leaves narrowly lance-shaped with entire margins; bracts yellow-tipped, ovate with mostly entire margins or small apical lobes; calyx 15-25 mm long with short, pointed or rounded lobes; corolla yellow, just longer than the calyx, the galea less than half as long as the tube. July.

Uncommon in montane to alpine grasslands and dry meadows; East. B.C., Alta. south to UT, NM, SD. Moss and Packer (1983) report that *C. sulphurea* completely intergrades with *C. occidentalis* with increasing elevation in Alberta. See *C. lutescens*.

Chaenorrhinum Reich.
Chaenorrhinum minus (L.) Lange. Annual; stems 3-15 cm high; leaves mostly alternate, linear, 1-2 cm long; flowers long-stalked in leaf axils; calyx deeply 5-lobed; corolla blue, 5-6 mm long, 2-lipped, the upper lip 2-lobed, the lower yellow; stamens 4; capsule globose, ca. 5 mm long. July.

Rare in compacted soil around buildings and along roads; known from West Glacier and Polebridge. Introduced from Europe.

Collinsia Nutt.
Collinsia parviflora Lindl., Blue-eyed Mary. Annual; stem 5-30 cm high, often branched; leaves opposite, narrowly oblong, 1-3 cm long; flowers long-stalked, paired at leaf nodes or in a small terminal cluster; calyx 5-lobed; corolla 4-6 mm long, blue, 2-lipped with a broad spur at the base; the upper lip erect, 2-lobed, white; the lower lip 3-lobed; stamens 4; capsule ovate, 2-4 mm long with 2-4 seeds. June.

Common in sparsely vegetated soil of forest openings, meadows and rock outcrops, montane; East, West. AK to Ont. south to CA, CO.

Digitalis L., Foxglove
Digitalis purpurea L. Biennial; stems 50-100 cm high; leaves alternate, petiolate, the blade to 15 cm long, lance-shaped with toothed margins; flowers short-petiolate, horizontal in a crowded, narrow, 1-sided inflorescence; calyx of 5 separate sepals; corolla bell-shaped, pale purple-mottled, 3-5 cm long; stamens 4. June-July.

Rare in disturbed, moist forest; known only from along Lake McDonald. Introduced from Europe. Poisonous, the source of the heart stimulant, digitalis.

Euphrasia L., Eyebright
Euphrasia arctica Lange ex Rostrup. Annual; stems 1-8 cm tall; leaves to 1 cm long, opposite, ovate with toothed margins; flowers sessile in the axils of upper leaf-like bracts; calyx with 4 narrow lobes; corolla 2-lipped, 2-3 mm long, white and purple-spotted; upper lip hood-like (galea); lower lip 3-lobed; stamens 4; capsule hairy. July-Aug.

Rare and inconspicuous in moist turf near or above treeline; known only from northeast of Logan Pass. Our plants are var. *disjuncta* (Fern. & Wieg.) Cronq. AK to Newf. south to B.C., MT, MI, ME. One of our few alpine annuals.

Gratiola L., Hedge-hyssop
Gratiola ebracteata Benth. Annual: stems prostrate to ascending, 5-10 cm long, branched at the base; leaves opposite, narrowly lance-shaped, 1-4 cm long; flowers long-stalked, solitary in the axils of upper leaves; calyx deeply 5-lobed; corolla 8-10 mm long, whitish, tubular with 4-5 shallow lobes; stamens 2; capsule globose, 3-5 mm long. Aug.

Rare in drying mud at the margins of montane ponds near East Glacier and Marias Pass. B.C. to MT south to CA.

Limosella L., Mudwort

Limosella aquatica L. Stemless annual; leaves long-petiolate, the blade strap-shaped, 5-18 mm long; flowers long-stalked, from the center of the leaf rosette; calyx 5-lobed; corolla cup-shaped, 5-lobed, ca. 2 mm long, white; stamens 4; capsule ovate, longer than the calyx. Aug.

Uncommon in drying mud around montane ponds and lakes; known from Lake Sherburne and Two Medicine Lake. Circumboreal south to CA, NM, NE, MN.

Linaria Miller, Toadflax

Rhizomatous, glabrous perennials; leaves alternate with entire margins; flowers stalked, subtended by leaf-like bracts in a terminal inflorescence; calyx deeply 5-lobed; corolla yellow, 2-lipped with a conical spur at the base; upper lip erect, 2-lobed; lower lip 3-lobed, light orange; stamens 4; capsule ovoid.

Introduced from Europe; flowers of these pernicious weeds resemble those of the garden snapdragon.

1. Leaves ovate, < 8 times as long as wide, clasping the stem .... *L. dalmatica*
1. Leaves linear, > 8 times as long as wide .... *L. vulgaris*

Linaria dalmatica (L.) Miller, Dalmation Toadflax. Stems 50-80 cm high; leaves ovate, waxy, 2-5 cm long with a clasping base; corolla 25-40 mm long; capsules 7-8 mm long. July.

Rare in disturbed soil of montane grasslands, meadows and roadsides; West.

Linaria vulgaris Hill, Butter-and-Eggs. Stems 20-80 cm high; leaves linear, 2-4 cm long; corolla ca. 2 cm long; capsule 8-12 mm long. July-Aug. Fig. 260.

Locally common in disturbed areas of montane meadows and grasslands; East, West. A bad weed in the Many Glacier and St. Mary areas.

Melampyrum L.

Melampyrum lineare Desr., Cow-wheat. Annual; stems glandular, 10-25 cm high, branched above; leaves 2-5 cm long, opposite, narrowly lance-shaped, the upper often toothed; flowers sessile in axils of upper leaves and leaf-like bracts; calyx 5-lobed; corolla 2-lipped, 5-10 mm long, white; upper lip hood-like (galea); lower lip 3-lobed, yellow near the base; stamens 4; capsule ovate, flattened, beaked. July. Fig. 261.

Common in open, montane forest; West. B.C. to Newf. south to WA, MT, MN, GA.

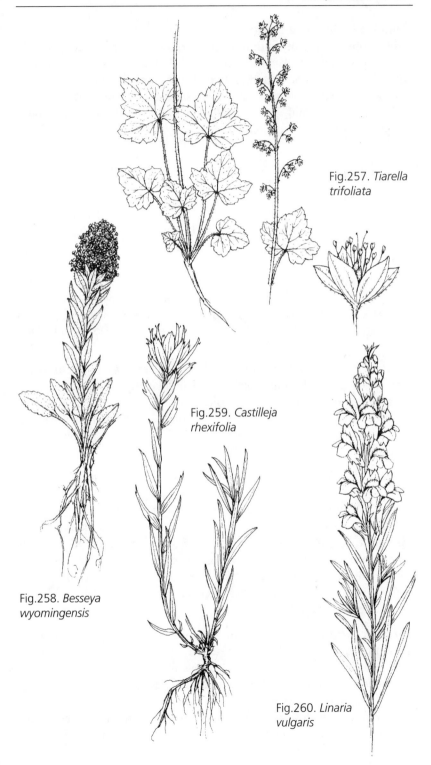

Fig.257. *Tiarella trifoliata*

Fig.259. *Castilleja rhexifolia*

Fig.258. *Besseya wyomingensis*

Fig.260. *Linaria vulgaris*

Mimulus L., Monkey-flower

Leaves opposite, usually with toothed margins; flowers stalked, paired in axils of leaves or leaf-like bracts; calyx with 5 short lobes; corolla tubular with 2 ridges along the bottom, 2-lipped at the mouth; upper lip 2-lobed, often erect; lower lip 3-lobed, spreading; stamens 4; capsule ellipsoidal.

M. breweri (Greene) Cov., a small annual with purple flowers, has been found just north of the Park in B.C.

1. Flowers rose-purple .... *M. lewisii*
1. Flowers yellow 2
2. Plant diminutive annual; corolla ≤ 15 mm long 3
2. Plant perennial; corolla > 15 mm long 5
3. Leaf blades narrowly elliptic, tapered to the petiole .... *M. breviflorus*
3. Leaf blades ovate, abruptly narrowed to the petiole 4
4. Uppermost calyx lobe the longest .... *M. guttatus*
4. Calyx lobes nearly equal .... *M. floribundus*
5. Calyx lobes nearly equal .... *M. moschatus*
5. Uppermost calyx lobe the longest 6
6. Stems ≤ 20 cm long; flowers 1-5; upper subalpine and alpine .... *M. tilingii*
6. Stems usually > 20 cm long; flowers often > 5; montane to lower subalpine
.... *M. guttatus*

Mimulus breviflorus Piper. Annual 3-8 cm high; leaves petiolate, the blade narrowly elliptic, 5-10 mm long; foliage glandular-hairy; corolla yellow, 6-8 mm long, only slightly 2-lipped. May-June.

Rare in vernally moist, shallow soil of montane cliffs in the Middle Fork Flathead drainage. WA to MT south to CA, NV. Occurs with *M. floribundus*, which has ovate leaf blades.

Mimulus floribundus Lindl. Annual; stems weak, 2-15 cm high; leaves petiolate, the blades ovate, 1-4 cm long; foliage glandular-hairy; corolla yellow 7-15 mm long, yellow, red-spotted, only slightly 2-lipped. June-Aug.

Uncommon in vernally moist, shallow, often mossy soil of montane cliffs; East, West. B.C., Alta. south to CA, NM. See *M. breviflorus*. A specimen from Missoula Co. similar to those from Glacier Park was annotated to *M. floribundus* var. *membranaceus* (Nels.) Grant (a relatively glabrous form) by C. L. Hitchcock; however, Robert Meinke believes these plants are better referred to *M. patulus* Pennell.

Mimulus guttatus D. C. [*M. hallii* Greene]. Annual to perennial, often spreading by runners; stems erect to lax, 10-50 cm long; leaves short-petiolate below, sessile above, ovate, 1-5 cm long; foliage glabrous to sparsely glandular-hairy; uppermost calyx lobe the largest; corolla 1-4 cm long, yellow, red-spotted, strongly 2-lipped. July-Aug.

Common in shallow water of streams or on wet cliffs, montane to lower subalpine; East, West. Found throughout most of N. America. Plants of permanently moist sites are large and perennial, while those of vernally moist sites may be small and annual. Standley's report (1921) of *M. glabratus* H. B. K. is referable here. See *M. tilingii.*

Mimulus lewisii Pursh. Rhizomatous perennial; stems 20-80 cm high; leaves sessile, ovate, 2-6 cm long; foliage sticky long-hairy; corolla pinkish-red, 3-5 cm long, strongly 2-lipped; anthers hairy. July-Aug. Fig. 262. Color plate 49.

Wet meadows and along streams; abundant subalpine, less common higher and lower; East, West. AK south to CA, UT, WY. Our only red monkeyflower and one of our most showy wildflowers.

Mimulus moschatus Douglas ex Lindley. Slenderly rhizomatous perennial; stems lax, 5-30 cm long; leaves petiolate, ovate, 1-8 cm long; foliage slimy long-hairy; corolla yellow, 15-20 mm long, somewhat 2-lipped. July.

Uncommon in montane wet forest openings or along streams and ponds; West. B.C. to MT south to CA, UT, CO.

Mimulus tilingii Regel [*M. caespitosus* Greene misapplied]. Rhizomatous perennial; stems 5-20 cm high; leaves petiolate, the blades ovate, 1-2 cm long; foliage glandular-hairy; uppermost calyx lobe the largest; corolla yellow, 2-3 cm long, strongly 2-lipped. July-Aug.

Common in wet gravel or shallow water along streams near or above treeline; East, West. Our plants are var. *tilingii.* AK south to CA, NM. Similar to *M. guttatus,* but the flowers are noticeably larger compared to the leaves.

Orthocarpus Nutt., Owl Clover

Annual; stems mostly unbranched; leaves alternate; flowers sessile, subtended by leaf-like bracts in dense terminal spikes; calyx 4-lobed; corolla tubular with an arched upper lip (galea) and a 3-toothed, sac-like, lower lip nearly as long as the galea; stamens 4, enclosed in the galea; capsule many-seeded.

1. Bracts of upper flowers pink; leaves of mid-stem divided .... *O. tenuifolius*
1. Flower bracts green; mid-stem leaves undivided .... *O. luteus*

Orthocarpus luteus Nutt. Stems 10-20 cm high; leaves numerous, undivided, linear, 1-3 cm long; foliage stiff-hairy; bracts green, broader than the leaves with a pair of lateral lobes; corolla 9-12 mm long, yellow. June-Aug.

Uncommon in grasslands along the east margin of the Park. B.C. to Ont. south to CA, NM, NE, MN. Blackfeet used the plant to dye skins red.

Orthocarpus tenuifolius (Pursh) Benth. Stems 10-20 cm high; leaves 1-3 cm long, those of mid-stem divided into 3-5 linear lobes, short-hairy; bracts ovate, with pink tips and sometimes a pair of lateral lobes; corolla yellow, 14-20 mm long; galea just longer than the lower lip. July-Aug. Fig. 263.

Rare in montane grasslands; known only from along St. Mary Lake. B.C. to MT south to OR, ID.

Pedicularis L., Lousewort

Mostly glabrous perennials; stems clustered; leaves basal and alternate on the stem, pinnately lobed or toothed, petiolate below; flowers mostly sessile, subtended by green bracts in a dense, terminal spike; calyx mostly 5-lobed; corolla tubular at the base, 2-lipped with a beak-like upper lip (galea) and a 3-lobed lower lip; stamens 4, included in the galea; capsule flattened.

1. Leaves with toothed margins but otherwise undivided .... *P. racemosa*
1. Leaves deeply pinnately lobed 2
2. Corolla pink-purple; galea with a long, upturned tubular beak ....
.... *P. groenlandica*
2. Corolla primarily yellow or white with a downturned beak 3
3. Corolla yellow, often tinged purple; flower bracts unlobed .... *P. bracteosa*
3. Corolla white with a semicircular beak; flower bracts mostly divided ....
.... *P. contorta*

Pedicularis bracteosa Benth. Stems 25-70 cm high; basal leaves few or sometimes absent, deeply lobed into narrowly oblong, toothed segments 1-5 cm long; stem leaves similar but sessile; spikes dense; bracts narrowly lance-shaped; corolla yellow or purplish, 10-20 mm long; galea arched with a short beak. June-Aug. Fig. 264.

Open forest and meadows; abundant subalpine, common lower and higher; East, West. Our plants are var. *bracteosa*. B.C., Alta. south to CA, CO. Often found with beargrass.

Pedicularis contorta Benth., Parrot's-beak. Stems 15-30 cm high; leaves mostly basal, the blades 2-15 cm long, deeply lobed into narrowly oblong, toothed segments; flowers well separated in the spike; bracts deeply lobed; corolla white, 8-10 mm long; galea with a long beak curved into a semicircle; lower lip wide and nearly vertical. June-Aug.

Common in subalpine and alpine grassland and meadows, uncommon in open forest and montane grasslands; East, West. Our plants are var. *contorta*. B.C., Alta. south to CA, ID, WY.

Pedicularis groenlandica Retz., Elephant's-head. Stems 10-60 cm high; leaves 3-20 cm long, deeply lobed into lance-shaped, toothed segments; spikes dense; bracts small, deeply lobed; corolla magenta, 10-15 mm long; galea prolonged into an upturned, tubular beak ca. 1 cm long. July-Aug.

Common in wet meadows and fens, montane to lower alpine; East, West. B.C. to Lab. south to CA, NM. The elephant trunk-like galea is distinctive.

Pedicularis racemosa Douglas ex Hooker, Sickletop. Stems 20-60 cm tall; leaves undivided, 4-8 cm long, narrowly lance-shaped with toothed margins, mostly on the stem; flowers stalked, well-sparated in a leafy-bracted

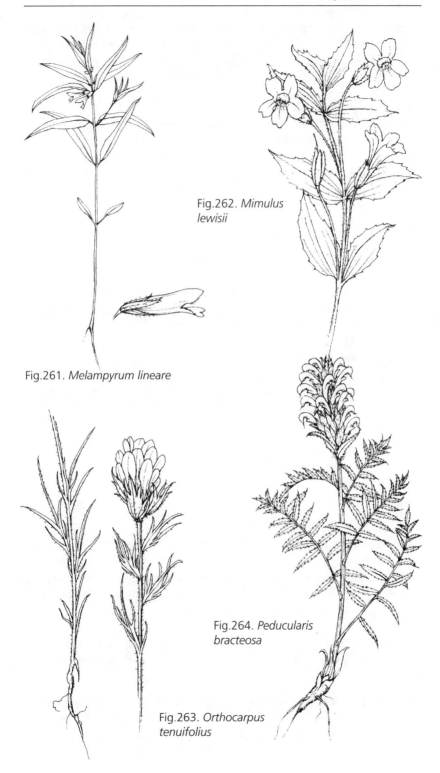

Fig.262. *Mimulus lewisii*

Fig.261. *Melampyrum lineare*

Fig.264. *Peducularis bracteosa*

Fig.263. *Orthocarpus tenuifolius*

inflorescence; calyx divided in 2; corolla white, 12-15 mm long; galea with a long beak curved into a semicircle; lower lip broad and horizontal. July.

Uncommon in montane and lower subalpine forest; West, expected East. Our plants are ssp. *alba* Pennell. B.C., Alta. south to CA, NM.

Penstemon Mitch., Beardtongue

Perennial, mostly herbaceous with stems clustered on a branched, woody root crown; leaves undivided, the basal petiolate; stem leaves opposite on the stem, sessile; flowers stalked or sessile, clustered in axils of upper reduced leaves in a narrow or branched inflorescence; calyx deeply 5-lobed; corolla tubular with 2 spreading lips at the mouth; upper lip 2-lobed; lower lip 3-lobed; stamens 5, 4 fertile, the other without an anther but hairy at the tip; capsule ovoid.

Three of our species are endemic to the Northern Rockies of eastern B.C., adjacent Alberta south to ID and MT.

1. Anthers wooly; corolla ≥ 3 cm long 2
1. Anthers glabrous or sparsely hairy; corolla mostly < 3 cm long 3
2. Leaf blades elliptic; tufts of basal leaves present......................... *P. ellipticus*
2. Leaves narrowly lance-shaped, all on the stem .... *P. lyallii*
3. Mid and upper stem leaf blades with toothed margins 4
3. Stem leaf blades with entire margins 5
4. Sterile stamen with long, yellow hair most of its length; corolla > 20 mm long ....*P. eriantherus*
4. Sterile stamen hairy just on the tip; corolla mostly < 20 mm long ....
.... *P. albertinus*
5. Inflorescence sparsely glandular-hairy ....*P. attenuatus*
5. Inflorescence glabrous 6
6. Flowers white to yellow .... *P. confertus*
6. Flowers blue 7
7. Corolla < 12 mm long; moist habitats... *P. procerus*
7. Corolla > 14 mm long; dry habitats .... *P. nitidus*

Penstemon albertinus Greene [*P. virens* Pennell]. Stems 10-40 cm high; basal leaf blades narrowly ovate, often with toothed margins, 15-40 mm long; stem leaves narrowly lance-shaped with obscurely toothed margins; foliage glabrous; inflorescence sparsely glandular, composed of long-stalked flower clusters; corolla blue, 10-16 mm long; capsule ca. 5 mm long. June-July. Fig. 265.

Common in grasslands, rock outcrops, and dry, open forest; montane or occasionally higher; East, West. B.C., Alta., ID, MT. In some plants only the upper stem leaves have toothed margins.

Penstemon attenuatus Douglas ex Lindley. Stems 20-40 cm high; leaves petiolate, the blade 4-10 cm long, narrowly oblong; stem leaves narrowly lance-shaped with entire margins, sessile; foliage glabrous; inflorescence sparsely glandular, composed of long-stalked flower clusters; corolla blue, 14-20 mm long; capsule 6-8 mm long. July.

Rare in open montane and subalpine forest; collected once above Lake McDonald. Our plants are var. *attenuatus*. WA to MT south to OR, ID, WY. Plants of *P. albertinus* with obscurely toothed leaves could key here.

Penstemon confertus Douglas ex Lindley. Stems 10-50 cm tall; basal leaves 1-10 cm long, narrowly oblong lance-shaped with entire margins, glabrous; stem leaves sessile; inflorescence of dense flower clusters; corolla cream-colored, 8-10 mm long; capsule 4-5 mm long. June-July.

Common in usually stony soil of grassland and dry meadows; montane, occasionally higher; East, West. B.C., Alta., OR, ID, MT. Our only yellowish-flowered penstemon.

Penstemon ellipticus Coulter & Fisher. Plants with prostrate woody stems and erect herbaceous stems 5-15 cm high; leaves 10-25 mm long, elliptic, often with shallowly toothed margins, glabrous; inflorescence glandular-hairy, small, composed of paired, stalked flowers; corolla light lavender, 30-35 mm long; anthers wooly; capsule 8-11 mm long. July-Aug. Fig. 266.

Rock outcrops, moraine, and stony, exposed slopes; abundant upper subalpine, common lower and higher; East, West. B.C., Alta., ID, MT. Our only shrubby penstemon, sometimes mistaken for *P. fruticosus* (Pursh) Greene.

Penstemon eriantherus Pursh. Stems 10-30 cm high; basal leaf blades 3-8 cm long, spatula-shaped with toothed margins; stem leaves narrowly oblong; foliage short-hairy; inflorescence glandular, composed of short-stalked flower clusters; corolla light purple, 2-3 cm long; sterile stamen with long, yellow hairs; capsule 7-12 mm long.

Rare in sparsely vegetated, stony soil of montane grasslands; reported for the East Glacier area (Standley 1921). Our plants are var. *eriantherus*. B.C., Alta. south to OR, CO, NE.

Penstemon lyallii Gray [*P. linearifolius* Coult. & Fish.]. Stems 30-70 cm tall; basal leaves absent; stem leaves 3-12 cm long, narrowly lance-shaped with toothed margins; inflorescence open, glandular, composed of long-stalked flower clusters; corolla pale purple, 3-4 cm long; anthers wooly; capsule 10-14 mm long. July-Aug.

Common in stony, sparsely vegetated soil of montane and subalpine rock outcrops and open slopes; East, West. B.C., Alta., WA, ID, MT. The tall stems without basal leaves are distinctive.

Penstemon nitidus Douglas ex Benth. Stems 10-25 cm high; basal leaf blades 2-7 cm long, ovate with entire margins, glabrous and waxy; stem leaves similar; inflorescence of dense flower clusters; corolla 12-16 mm long, bright blue; capsules 9-12 mm long. May-June.

Rare in sparsely vegetated soil of montane grasslands and rock outcrops; East and near Marias Pass. Our plants are var. *nitidus*. B.C. to Man. south to WA, CO, ND.

Penstemon procerus Douglas ex Graham. Stems 15-40 cm tall, curved at the base; basal leaf blades 2-8 cm long, oblong with entire margins, glabrous; stem leaves lance-shaped; inflorescence of dense, stalked flower clusters; corolla 6-10 mm long, dark blue; capsule 4-5 mm long. June-July.

Uncommon in moist montane meadows; East, West. Our plants are var. *procerus*. AK to Man. south to CA, CO.

Rhinanthus L.

Rhinanthus crista-galli L., Yellow Rattle [*R. minor* L., *R. borealis* (Sterneck) Chab.]. Annual; stems 15-50 cm high, mostly unbranched; leaves opposite, 2-4 cm long, lance-shaped with toothed margins; flowers nearly sessile, paired in axils of upper leaves or leaf-like bracts; calyx orbicular with 4 pointed lobes, inflated in fruit; corolla barely exserted, yellow, 1-2 cm long, 2-lipped with a hood-like upper lip (galea) and a shorter, 3-lobed lower lip; stamens 4; capsule coin-shaped, 10-15 mm wide. July-Aug. Fig. 267.

Common in montane to lower subalpine grasslands, meadows and dry open forest, often where disturbed; East, West. Circumboreal south to OR, CO, NY. Plants may be common one year and rare the next.

Verbascum L.

Biennial; stems mostly unbranched; basal leaves petiolate, in a rosette; stem leaves sessile, alternate; flowers saucer-shaped, short-stalked in a long, narrow inflorescence, lengthening in fruit; calyx of 5 sepals; corolla with 5 nearly orbicular lobes; stamens 5; capsule broadly ellipsoid.

Introduced from Eurasia.

1. Lower leaves glabrous; inflorescence glandular-hairy; flowers stalked .... ***V. blattaria***
1. Foliage densely hairy; plants not glandular; flowers nearly sessile . ***V. thapsus***

Verbascum blattaria L., Moth Mullein. Stems 30-80 high; leaves oblong, shallowly lobed, 3-10 cm long; foliage glabrous below; inflorescence glandular-hairy; corolla white, 2-3 cm across; anthers purple-hairy.

Rare in disturbed, often stony soil of montane grasslands and meadows; reported by Standley for the St. Mary Valley but not collected since.

Verbascum thapsus L., Mullein. Stems 40-150 cm high; leaves oblong-ovate, 10-40 cm long; foliage densely yellow-, short-hairy; corolla yellow, 15-20 mm across; upper 3 stamen stalks yellow-hairy. Aug.

Rare in disturbed soil of montane meadows and along roads and trails; East, West.

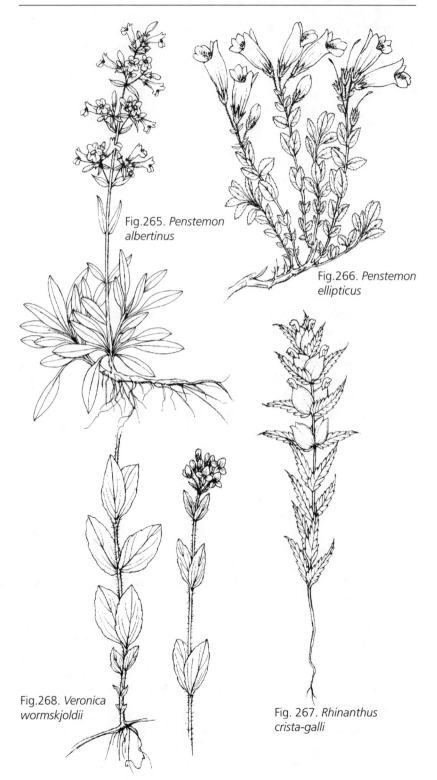

Fig.265. *Penstemon albertinus*

Fig.266. *Penstemon ellipticus*

Fig.268. *Veronica wormskjoldii*

Fig. 267. *Rhinanthus crista-galli*

Veronica L., Speedwell, Betony

Herbaceous annuals or perennials; leaves mostly opposite; flowers saucer-shaped, usually stalked in terminal or axillary, spike-like inflorescences; calyx deeply 4-lobed; corolla 4-lobed, the lowest usually smaller; stamens 2, exserted; style exserted; capsule flattened, ovate to heart-shaped.

The shape of the mature capsules is often helpful in distinguishing species.

1. Flowers in spikes arising from upper leaf axils 2
1. Flowers in a spike-like inflorescence terminating the stem 6

2. Lower leaves and stem hairy; plants of lawns and disturbed uplands 3
2. Plants glabrous, occurring around streams and ponds 4

3. Leaves sessile with 10-22 marginal teeth .... *V. chamaedrys*
3. Leaves short-petiolate with ≥ 24 marginal teeth .... *V. officinalis*

4. Leaves petiolate at mid-stem, 2-4 times as long as wide .... *V. americana*
4. Leaves sessile, > 4 times as long as wide 5

5. Capsule ovoid, only obscurely notched on top .... *V. catenata*
5. Capsule wider than high, flattened, deeply notched on top ...... *V. scutellata*

6. Plants annual with a slender taproot; flowers 2-3 mm across 7
6. Plants perennial with slender rhizomes; flowers 3-10 mm across 9

7. Upper stem leaves with 1-many lobes .... *V. verna*
7. Stem leaves sometimes toothed but not lobed 8

8. Leaves strap-shaped; flowers white .... *V. peregrina*
8. Leaves ovate; flowers blue .... *V. arvensis*

9. Stems creeping at the base, short-hairy; flowers 3-4 mm across ....
.... *V. serpyllifolia*
9. Stems with long, spreading hairs, not creeping; flowers 6-10 mm across ......
.... *V. wormskjoldii*

Veronica americana Schwein. ex Benth., American Brooklime. Glabrous perennial; stems prostrate, rooting at the nodes; leaves 1-6 cm long, short-petiolate, broadly lance-shaped with toothed margins; flower stalks 5-12 mm long; spikes paired at leaf nodes; corolla blue, 5-10 mm across; capsule ovate, unlobed. June-July.

Common in slow, shallow water or wet banks of montane streams, fens, and wet meadows; East, West. Most of N. America, Asia. Our common streamside speedwell.

Veronica arvensis L. Annual; stems erect or lax, often branched at the base, 5-10 cm high; leaves hairy, nearly sessile, 5-10 mm long, ovate with toothed margins; flowers sessile, subtended by linear bracts in a terminal spike; corolla blue, 2-3 mm wide; capsule heart-shaped, 2-lobed above. May-June.

Rare on gravelly roadsides; East, West. Introduced from Eurasia.

Veronica catenata Pennell. Perennial; stems lax or ascending; leaves 1-5 cm long, narrowly lance-shaped, sessile with a clasping base and mostly entire margins; fruit stalks 3-8 mm long; spikes paired at leaf nodes; corolla white, ca. 5 mm across; capsule ovoid, obscurely notched on top. July.

Rare in slow water of montane streams; East, West. B.C. to Que. south to CA, NM, NE, OK, PA; Eurasia.

Veronica chamaedrys L. Rhizomatous perennial; stems lax 10-30 cm long; leaves sessile, sparsely hairy, 15-40 mm long, ovate with toothed margins; flower stalks 5-10 mm long; spikes 1-2 at leaf nodes; corolla dark blue, 8-12 mm across; capsule slightly 2-lobed on top. June-July.

Rare in lawns at West Glacier. Introduced from Europe. The more common *V. officinalis* has short-petiolate, more finely toothed leaves and smaller flowers.

Veronica officinalis L. Perennial; stems lax, rooting at the nodes; leaves short-petiolate, 15-50 mm long, ovate with toothed margins, hairy; flowers nearly sessile in spikes in upper leaf nodes; corolla light blue, 4-8 mm wide; capsule triangular. June-Sept.

Locally common in lawns around buildings and campgrounds; uncommon along trails in moist montane forest; West. Introduced from Europe. See *V. chamaedrys*. Perhaps our only exotic that invades under a closed forest canopy.

Veronica peregrina L. Annual; stems 2-10 cm high, glandular-hairy; leaves 5-20 mm long, strap-shaped with entire or toothed margins; flowers sessile, subtended by alternate, linear bracts in a terminal spike; corolla white, 2-3 mm wide; capsule heart-shaped, shallowly-lobed. May-June.

Uncommon and inconspicuous in disturbed, sparsely vegetated soil of montane grasslands and roadsides; East, West. Our plants are ssp. *xalapensis* (H. B. K.) Pennell. Throughout much of N. and S. America, Asia.

Veronica scutellata L. Rhizomatous, mostly glabrous perennial; stems often lax and rooting at the nodes; leaves sessile, 2-5 cm long, narrowly lance-shaped with entire or obscurely toothed margins; flower stalks 6-10 mm long; inflorescences in the leaf axils; corolla blue, 6-10 mm wide; capsule flattened, distinctly 2-lobed.

Uncommon in mud along montane streams and ponds; East. Circumboreal south to CA, CO, ND, VA. *V. americana* has wider, short-petiolate leaves.

Veronica serpyllifolia L. Rhizomatous perennial; stems prostrate at the base, 5-15 cm high; leaves 5-15 mm long, elliptic with obscurely toothed margins, the lower ones petiolate; flowers nearly sessile in a terminal spike; corolla blue, 3-4 mm wide; capsule nearly orbicular, shallowly lobed. June-July.

Common in moist, sparsely vegetated soil of montane forest or thickets, especially along streams, lakes and trails; East, West. Our plants are mainly ssp. *humifusa* with glandular flower stalks. Circumboreal south to much of

N. America. Ssp. *serpyllifolia*, with non-glandular flower stalks, has been collected near Logging Lake; it is thought to be introduced from Europe.

Veronica verna L. Annual; stems 5-10 cm high; lower leaves lance-shaped, 5-10 mm long, with toothed margins; upper leaves oblong with 1-many lobes at the base; flowers sessile, subtended by lobed to entire bracts in a terminal spike; corolla blue, 2-3 mm wide. June.

Rare along montane roads and other disturbed areas; collected along St. Mary Lake. Introduced from Europe.

Veronica wormskjoldii Roem. & Schult. [*V. alpina* L. var. *alterniflora* Fern.]. Rhizomatous perennial; stems 5-25 cm high, hairy; leaves 1-3 cm long, narrowly ovate with obscurely toothed margins; flowers short-stalked in a short, glandular, terminal spike; corolla deep blue, 6-10 mm across; capsule ovate, glandular-hairy, shallowly lobed. July-Aug. Fig. 268.

Abundant in moist meadows and turf, often near streams near or above treeline, uncommon lower; East, West. AK to Greenl. south to CA, NM.

SOLANACEAE: POTATO FAMILY

Solanum L., Nightshade

Leaves alternate, petiolate; flowers bisexual, stalked in open inflorescences from leaf axils; calyx 5-lobed; corolla star-shaped with 5 reflexed lobes; stamens 5, the anthers forming a cone around the single style; ovary above the base of the calyx (superior); fruit a many-seeded berry.

Our species are introduced from Eurasia.

1. Perennial vine; berries red; some leaves with basal lobes .... **S. dulcamara**
1. Annual; berries black; leaves unlobed .... **S. sarrachoides**

Solanum dulcamara L., Bittersweet Nightshade. Rhizomatous perennial vine, woody toward the base, climbing on shrubs; leaf blades 2-7 cm long, arrow-shaped, often with a pair of ovate lobes at the base; foliage glabrous or sparsely hairy; inflorescence branched with 10-25 flowers; corolla blue, the lobes 4-8 mm long; berry red, 8-11 mm long. July-Aug. Fig. 269.

Rare in disturbed, montane forest openings; collected once near Lake McDonald.

Solanum sarrachoides Sendt. Annual; stems 15-30 cm high; leaves petiolate, the blade 2-6 cm long, ovate with entire margins; foliage sticky, spreading-hairy; flowers few in each cluster; corolla white, 5-10 mm across; berry black, ca. 8 mm long.

Rare around roads and buildings; collected once at West Glacier.

URTICACEAE: NETTLE FAMILY

Urtica L., Nettle

Urtica dioica L. [*U. lyallii* Wats.]. Rhizomatous perennial; stems 4-angled, 50-120 cm tall; leaves opposite, petiolate, the blade 7-15 cm long, lance-shaped with toothed margins; foliage prickly; flowers small, unisexual, sessile, in interrupted clusters on drooping, branched inflorescences at upper leaf nodes; calyx 4-lobed, 1-2 mm long; petals absent; stamens 4; style 1; ovary above the base of the calyx (superior); fruit a hard-coated seed (achene) as long as the sepals. June-July. Fig. 270.

Common in moist, often highly organic soil of montane and subalpine meadows and open forest, often where there has been some disturbance; East, West. Our plants are ssp. *gracilis* (Ait.) Selander. Circumboreal through much of N. America. Ssp. *holosericea* (Nutt.) Thorne was reported for East and West Glacier by Standley (1921).

VALERIANACEAE: VALERIAN FAMILY

Valeriana L.

Glabrous perennial herbs; basal leaves petiolate; stem leaves opposite; flowers unisexual or bisexual, in a branched, hemispheric, terminal inflorescence; calyx of numerous feather-like bristles that expand in fruit; corolla tubular with 5 equal lobes at the mouth; stamens 3, exserted; ovary below the base of the calyx and corolla (inferior); stigma 3-lobed; fruit a hard-coated seed (achene).

Native Americans used the roots of *V. dioica, V. occidentalis,* and *V. sitchensis* to treat stomach ailments. *V. occidentalis* and *V. dioica* intergrade in our area, and the distinction between the two is not recognized by all authorities.

1. Basal leaves narrowly oblong, gradually tapered to the base .... ***V. edulis***
1. Basal leaf blades elliptic or divided, more abruptly narrowed to the petiole . 2
2. Corolla 5-8 mm long; lateral lobes of some stem leaves ≥ 2 cm wide ....
.... ***V. sitchensis***
2. Corolla 1-4 mm long; lateral lobes of stem leaves < 2 cm wide 3
3. Lateral lobes of stem leaves strap-shaped < 1 cm wide .... ***V. dioica***
3. Lateral lobes of stem leaves ovate, some 1-2 cm wide .... ***V. occidentalis***

Valeriana dioica L. [*V. septentrionalis* Rydb.]. Stems 15-40 cm high; basal leaf blades 2-5 cm long, elliptic with entire margins; stem leaves with 1-3 pairs of strap-shaped lateral lobes and a larger, oblong terminal lobe; flowers both bisexual and female; corolla white, 2-4 mm long; achenes 3-5 mm long, glabrous. June.

Common in dry to moist montane forest and grasslands; East, West. Our plants are ssp. *sylvatica* (Rich.) Mey. Yuk. to Newf. south to WA, WY. Stem leaves of *V. occidentalis* have wider lateral lobes.

Valeriana edulis Nutt. ex T. & G. Stems 15-30 cm tall; basal leaves 7-15 cm long, the blades narrowly oblong; stem leaves pinnately divided into 3-9 strap-shaped lobes; flowers bisexual and unisexual; corolla white 1-4 mm long; achenes 2-5 mm long, glabrous or hairy.

Rare in montane and subalpine moist meadows and dry, calcareous ridges along the east margin of the Park. B.C. to MT, south to Mex., also MN, IA, Ont. Becomes much more common just south of the Park. Roots were cooked and eaten by some Native American tribes.

Valeriana occidentalis Heller. Stems 15-80 cm high; basal leaf blades 3-7 cm long, elliptic, entire or with basal lobes; stem leaves with 1-3 pairs of ovate lateral lobes and a larger terminal lobe; flowers both bisexual and female; corolla white, 3-4 mm long; achenes 3-5 mm long, mostly short-hairy. June-July.

Uncommon in moist montane forest and thickets; East, West. ID and MT south to CA, AZ, CO, SD.

Valeriana sitchensis Bong. Stems 30-70 cm high; basal leaves often absent; lower stem leaves petiolate, the blades divided into 3-9 ovate, toothed or entire segments 1-4 cm wide; flowers bisexual; corolla white or pinkish, 5-8 mm long; achenes glabrous, 3-6 mm long. July-Aug. Fig. 271.

Abundant in subalpine meadows, open forest and avalanche slopes, occasionally higher or lower, usually with abundant snow cover; East, West. AK to Que. south to CA, ID, MT.

VERBENACEAE: VERVAIN FAMILY

Verbena L., Vervain

Verbena bracteata Lag. & Rodr. Mostly perennial; stems prostrate, 10-50 cm long, branched; leaves opposite, 1-5 cm long, short-petiolate, ovate with deeply lobed and sharp-toothed margins; foliage stiff-hairy; flowers bisexual, sessile in the axils of long, narrow bracts in a terminal spike; calyx 5-lobed, stiff-hairy; corolla tubular, 5-lobed, the upper 2 slightly larger, blue, ca. 4 mm long; stamens 4; fruit 4-lobed and separating into 4 hard seeds (nutlets) 2 mm long. Aug. Fig. 272.

Rare in disturbed soil along roads and buildings; collected at West Glacier. Native to much of the U. S. but probably introduced in the Park.

VIOLACEAE: VIOLET FAMILY

Viola L., Violet

Herbaceous perennials, often with runners (stolons); leaves basal and alternate on the stem, the blade lance- to heart-shaped, mostly with toothed margins and a pair of appendages (stipules) at the base of the petiole; flowers long-stalked, bisexual, usually nodding; sepals 5, separate; petals 5, separate,

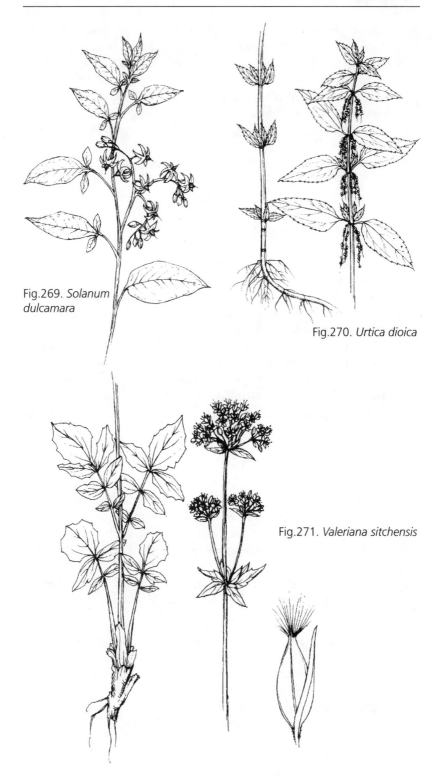

Fig.269. *Solanum dulcamara*

Fig.270. *Urtica dioica*

Fig.271. *Valeriana sitchensis*

pansy-like, the lowest with a sac-like spur at the base, the upper 2 erect, the lower 3 spreading, often hairy; stamens 5; style 1, longer than the stamens; ovary above the base of the calyx (superior); fruit a many-seeded, 3-chambered capsule.

Plants often produce inconspicuous, non-opening flowers (cleistogamous) near ground level.

1. Petals primarily yellow, sometimes tinged with purple 2
1. Petals primarily white or blue 5

2. Leaf blades ovate to lance-shaped, longer than wide 3
2. Leaf blades heart-shaped to nearly orbicular, as wide or wider than long 4

3. Leaf blades thick, deeply scalloped on the margins, purplish at least on the veins .... *V. purpurea*
3. Leaf blades with nearly entire margins, not thick or purplish .... *V. nuttallii*

4. Leaf blades rounded, mostly flat to the ground .... *V. orbiculata*
4. Leaf blades with a pointed tip, many erect or ascending .... *V. glabella*

5. Petals primarily blue to purple 6
5. Petals primarily white, sometimes tinged with blue 8

6. Flowers light blue; leaves wider than long .... *V. palustris*
6. Flowers deep blue or purple; leaves longer than wide 7

7. Style tip glabrous; erect leafy stems lacking; wet habitats ... *V. nephrophylla*
7. Style tip hairy; plants with leafy stems; forest or meadows .... *V. adunca*

8. Plants with erect, leafy stems; style tip sparsely hairy .... *V. canadensis*
8. Plants lacking significant leafy stems; style glabrous 9

9. Petals tinged with blue on the back .... *V. palustris*
9. Petals with blue lines but not blue tinged 10

10. Plants with stolons; leaves < 3 cm wide, no wider than long . *V. macloskeyi*
10. Plants without stolons; leaves wider than long, some > 3 cm wide ....
.... *V. renifolia*

Viola adunca Smith [*V. montanensis* Rydb.]. Rhizomatous; stems often lax, 1-5 cm long; leaf blades lance-shaped to ovate, 5-30 mm wide; flowers blue, 8-20 mm long; style with short hairs at the tip. April-Aug.

Var. *adunca*, > 5 cm tall with hairy leaves, occurs in moist montane forest openings or thickets; common West, rare East; var. *bellidifolia* (Greene) Harrington, shorter with glabrous leaves, is known from moist meadows at Logan Pass. The similar *V. nephrophylla* usually occurs in wet meadows and fens. AK to Greenl. south to CA, CO, SD, MN.

Viola canadensis L. Stems 10-40 cm high, often with runners from the base; leaf blades 5-10 cm long, broadly heart-shaped with a lobed base and pointed tip; petals white, 10-12 mm long, yellow at the base with purple lines; style with sparse long hairs. April-July.

Common in moist, open forest, often with aspen; montane, occasionally higher; East, West. AK to N. S. south to AZ, NM, TN, SC. A few plants often flower late into summer. Leaves are usually larger than our other white-flowered violets, which lack leafy stems.

Viola glabella Nutt. Rhizomatous; stems 5-20 cm tall; leaf blades 3-8 cm wide, broadly heart-shaped with a pointed tip; petals 8-14 mm long, yellow (fading to cream) with purple lines; style short-hairy at the tip. May-July. Fig. 273.

Abundant in moist montane and subalpine forest, often along streams or in other openings; East, West. AK to Alta. south to CA, MT. *V. orbiculata* will usually have a few leaves from the previous year.

Viola macloskeyi Lloyd. Plants with rhizomes and usually stolons, stemless; leaf blades 1-2 cm long, ovate and shallowly lobed at the base; flower stalks 3-6 cm high; petals 5-10 mm long, white with purple lines; style glabrous at the tip. May-July.

Common in fens, wet meadows, or along streams, montane and subalpine; East, West. B.C. to Newf. south to most of U. S. Standley's (1921) reports of *V. palustris* are mostly referable here. Petals of *V. palustris* are larger and bluish.

Viola nephrophylla Greene [*V. sororia* Willd. in part]. Plants without stolons, stemless; leaf blades 2-5 cm wide, heart-shaped or ovate with a shallowly lobed base; flower stalks 5-15 cm high; petals 10-20 mm long, blue, the lower 3 white at the base; style glabrous at the tip. May-June.

Common in montane wet meadows, marshes and fens; East, West. Our plants are var. *cognata* (Greene) Hitchc. B.C. to Newf. south to CA, NM, MN, NY. *V. adunca* has a longer spur and is usually found in drier habitats.

Viola nuttallii Pursh [*V. linguaefolia* Nutt.]. Stems 2-5 cm high; leaf blades 2-6 cm long, lance-shaped with nearly entire margins; petals 8-12 mm long, yellow with purple lines and often reddish brown on back; style tip hairy. June-July.

Uncommon in montane grasslands and subalpine meadows; East, expected West. Our plants are var. *major* Hook. B.C. to Man. south to CA, AZ, NM, KS. Leaves of *V. purpurea* are deeply toothed and purplish beneath.

Viola orbiculata Geyer ex Hooker. Plants stemless with stolons; leaf blades 2-4 cm wide, nearly orbicular, notched at the base; flower stalks 2-10 cm long, erect or lax; petals yellow with purple lines, 5-15 mm long; style tip hairy. April-July. Fig. 274. Color plate 15.

Moist montane and lower subalpine coniferous forest; abundant West, common East. B.C., Alta. south to OR, ID, MT. Leaves lay flat on the ground and remain green through the winter.

Viola palustris L. Plants stemless with stolons; leaf blades 2-5 cm long, nearly orbicular with a notched base; flower stalks 5-15 cm long; petals 10-13 mm long, white or bluish with purple lines; style tip glabrous. May-June.

Rare in wet soil of montane seeps, streams, wet meadows, and marshes; collected once north of Apgar. B.C. to Newf. south to CA, UT, CO, NH. See *V. macloskeyi*.

Viola purpurea Kellogg. Rhizomatous; stems 2-5 cm high; leaves 5-10 mm wide, ovate with large, rounded marginal teeth, thick and purple-tinged; petals 5-12 mm long, yellow with brown lines; style tip hairy. June.

Rare in vernally moist soil of montane grasslands; known only from southwest of Marias Pass. Our plants are var. *venosa* (Wats.) Brain. WA to MT south to CA, AZ, CO. See *V. nuttallii*.

Viola renifolia Gray. Stemless; leaf blades 2-6 cm long, nearly orbicular, broader than long with a notched base; flower stalks 2-7 cm high, shorter than the leaves; petals 10-15 mm long, white with purple lines; style tip glabrous.

Rare in wet alder thickets and spruce forest, montane; East, West. B.C. to ME south to WA, CO. Both *V. macloskeyi* and *V. palustris* usually have stolons; their leaves are not wider than long.

VISCACEAE: MISTLETOE FAMILY

Arceuthobium Bieb., Dwarf Mistletoe

Parastic perennials with slender, branched stems; leaves opposite, minute; flowers unisexual in small clusters along upper stem; sepals and stamens 3 on male flowers; sepals 2 on female flowers; corolla absent; ovary inferior; fruit a berry with 1 sticky seed. Reference: Hawksworth and Wiens (1972).

1. Plants parasitic on Douglas fir (*Pseudotsuga*) .... **A. douglasii**
1. Plants paratitic on pines (*Pinus*) .... **A. americanum**

Arceuthobium americanum Nutt. [*Razoumofskyia americana* (Nutt.) Kuntze]. Stems yellowish, 2-10 cm long; fruit blue, 2-3 mm long. Fig. 275.

Uncommon on lodgepole pine; montane and subalpine; East (Standley 1921). B.C. to Ont, south to CA, NM. The presence of this plant can cause the dense, short branching of "witches' broom." *A. laricis* (Piper) St. John is found on larch; it occurs on Flathead National Forest and probably occurs in the Park as well.

Arceuthobium douglasii Engelm. Similar to *A. americanum*; stems green, 1-3 cm long; fruit green, 3-5 mm long. Uncommon on Douglas fir, montane; West, expected East. B.C. to CA, NM.

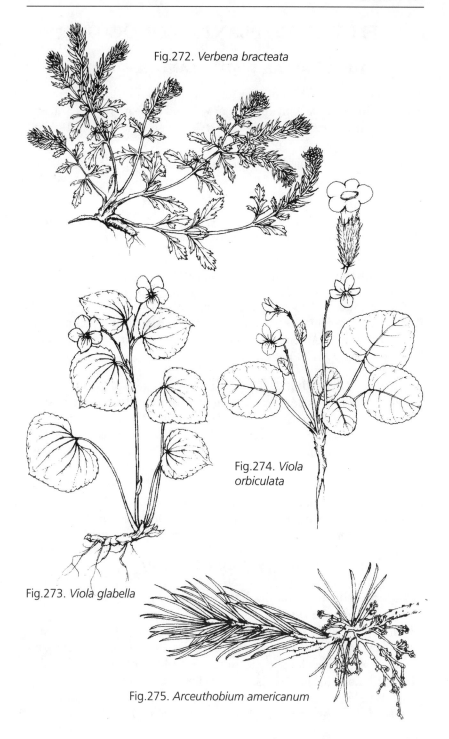

Fig.272. *Verbena bracteata*

Fig.274. *Viola orbiculata*

Fig.273. *Viola glabella*

Fig.275. *Arceuthobium americanum*

FLOWERING PLANTS: MONOCOTS

ALISMATACEAE: WATER-PLANTAIN FAMILY

Aquatic or emergent perennial herbs; leaves basal, long-petiolate, erect or floating; inflorescence with opposite or whorled branches; sepals 3, green; petals 3, white; stamens ≥ 6; ovaries and styles numerous; fruits hard-coated seeds (achenes).

1. Leaf blades elliptic; fruits in a wheel-like cluster .... **Alisma**
1. Leaves arrow-shaped; fruits in a globose cluster .... **Sagittaria**

Alisma L., Water-plantain
Alisma plantago-aquatica L. [*A. brevipes* Greene, *A. trivale* Pursh]. Stems to 60 cm high; leaves petiolate, the blade elliptic, 5-20 cm long; inflorescence twice-branched, the branches whorled; flowers bisexual; petals 4-6 mm long; achenes 2-3 mm long in a segmented, wheel-like cluster. July. Fig. 276.

Rare in mud of shallow montane marshes and stream or lake shores; East, West. Our plants are var. *americanum* Schul. & Schul. Circumboreal south to most of N. America. *A. gramineum* Lej., with narrower leaf blades, is reported for just north of the Park (Kuijt 1982).

Sagittaria L., Arrowhead
Sagittaria cuneata Sheld., Wapato. Stems to 50 cm long from tuber-bearing rhizomes; floating or emergent leaves with blades 2-8 cm long, narrowly arrow-shaped; submerged leaves grass-like; flowers unisexual, stalked, 2-3 per inflorescence node, subtended by narrow, leaf-like bracts; lower flowers female; petals 7-10 mm long; achenes 2-3 mm long in a globose cluster.

Rare in mud of shallow water of ponds; reported by Standley (1921) for the East Glacier area. AK to Lab. south to CA, NM, TX. The tubers are prized by waterfowl and were eaten by Native Americans. Plants in deeper water usually remain vegetative, with only strap-shaped submergent leaves.

ARACEAE: ARUM FAMILY

Lysichiton Schott, Skunk Cabbage
Lysichiton americanus Hult. & St. John [*L. kamtschatcensis* Schott]. Perennial herb 30-50 cm high with a skunky odor; leaves basal, petiolate, 40-100 cm long when fully expanded, the blades elliptic with entire margins; flowers bisexual, small, tightly clustered in a cylindrical spike enclosed at first by a yellowish, lance-shaped, hood-like bract (spathe) to 20 cm long; sepals 4, yellowish, minute; petals lacking; stamens 4; ovary and stigma solitary; fruit a 2-seeded berry. May. Fig. 277. Color plate 11.

Rare in montane alder or spruce swamps; known only along the North Fork Flathead River. AK to B.C. south to CA, ID, MT. After flowering, the leaves resemble those of tobacco, but they don't enlarge in dry years.

CYPERACEAE: SEDGE FAMILY

Annual or perennial herbs; stems often 3-angled; leaves alternate, grass-like, usually with a lower portion that sheaths the stem and a spreading upper portion (blade); flowers often unisexual, small, simple, each borne in the axil of a scale and arranged in spike-like clusters; petals and sepals lacking; flower parts reduced to a papery sac or bristles; stamens 3, ovary superior with 1-3 stigmas; seed (achene) 1.

Carex and *Kobresia* are distinctive by the presence of a sac (perigynium) surrounding the seed, while the other genera have bristles. Species of *Schoenoplectus* and *Trichophorum* are placed in *Scirpus* by some authorities.

1. Achene surrounded in a membranous wrapper or sac (perigynium); bristles lacking; flowers unisexual 2
1. Perigynium lacking; achene with bristles arising at the base; flowers bisexual 3

2. Perigynium a closed sac fully enclosing the achene .... **Carex**
2. Perigynium partly open, exposing the achene .... **Kobresia**

3. Leaf sheaths on lower 1/3 of stem without blades, but upper leaf blades well developed .... **Dulichium**
3. Leaf blades lacking entirely or leaves well developed at the base of the plant 4

4. Leaves appear to be lacking, all reduced to bladeless sheaths 5
4. At least 1 leaf with a distinct blade present... 6

5. Spikelets solitary .... **Eleocharis**
5. Spikelets numerous .... **Schoenoplectus**

6. Achene bristles > 6, elongate, forming a cottony tuft .... **Eriophorum**
6. Achene bristles 0-6, elongate in only 1 species 7

7. Spikelets numerous in a diffusely branched inflorescence .... **Scirpus**
7. Spikelets solitary 8

8. Bract subtending spikelet stem-like; plants aquatic .... **Schoenoplectus**
8. Bract subtending spikelet scale-like; plants forming tufts or tussocks .... **Trichophorum**

Carex L., Sedge

Perennial herbs; leaves grass-like, alternate and/or basal, entire-margined; stems often 3-angled; herbage mostly glabrous; flowers unisexual, arranged in unisexual or bisexual spikes, unisexual spikes with male flowers segregated into either the top or bottom portion, at least the lowest subtended by a leaf-like bract; each flower subtended by a papery, green to black scale; male flowers of 3 stamens; female flower an ovary enclosed in a papery, sac-like perigynium; style 1; fruit an achene (seed) that is 2-sided when there are 2 stigmas and 3-sided when there are 3 stigmas. References: Hermann (1970), Standley (1985), Murray (1969).

This is the largest genus in the Park. Mature achenes are often needed for positive identification; these are obtainable in June or later for low-elevation upland species and in July or later for wetland or high-elevation plants.

1. Plants unisexual; flowers all male or all female .... Group A
1. Both male and female flowers on the same plant 2

2. Stems with a solitary, terminal spike, male flowers above (plants with a tightly clustered head of several spikes may appear to have a solitary spike but will not key here) .... Group B
2. Spikes > 1 per stem 3

3. Terminal spike all male (sometimes a few perigynia at the base) and narrower than the female spikes below.. 4
3. Terminal spike bisexual or all female, mostly similar to the lower ones......... 5

4. Achene (seed) 3-sided; stigmas 3 .... Group C
4. Achene 2-sided, coin- or lens-shaped; stigmas 2 .... Group D

5. Achene (seed) 3-sided; stigmas 3 .... Group E
5. Achene 2-sided, coin-shaped; stigmas 2 6

6. Bisexual spikes with male flowers above the female (or all spikes unisexual).. .... Group F
6. Bisexual spikes with male flowers at the base 7

7. Inflorescence head-like; spikes tightly clustered, barely distinguishable as separate (sometimes the lowest one separate) .... Group G
7. At least the lower spikes somewhat separate and easily distinguishable (many plants will key in both Groups G and H)................................. Group H

Group A : Plants Unisexual (Dioecious)

1. Spike solitary 2
1. Spikes >1, sometimes tightly clustered and appearing solitary 3

2. Leaves < 1 mm wide; perigynia glabrous; organic soil of fens . **C. gynocrates**
2. Some leaves > 1 mm wide; perigynia hairy; meadows and turf .... **C. scirpoidea**

3. Inflorescence 8-17 mm long, tightly clustered .... **C. stenophylla**
3. Inflorescence longer, head-like or not 4

4. Scales light brown; rhizome not scaly; grasslands..................... **C. douglasii**
4. Scales dark brown; rhizome scaly; wet meadows to fens 5

5. Perigynia < 3 mm long; plants of peaty soil in fens .... **C. simulata**
5. Perigynia > 3 mm long; plants of mineral soil in wet meadows .... **C. praegracilis**

Group B: Stems with a Solitary, Terminal Spike

1. Perigynia 5-6 mm long, 1-3 per spike, separate ....**C. geyeri**
1. Perigynia < 5 mm long, usually > 3 per spike, clustered 2

2. Plants of dry habitats; stems tightly clustered among old brown stem bases 3
2. Plants rhizomatous; stems not clustered among old stem bases 6

3. Perigynia short-hairy; montane .... ***C. filifolia***
3. Perigynia glabrous or nearly so; alpine or subalpine 4

4. Male flowers comprising ca. the top half of the spike .... ***C. elynoides***
4. Male flowers few, inconspicuous at the tip of the spike 5

5. Perigynia narrower toward the tip .... ***C. pyrenaica***
5. Perigynia broader at the top .... ***C. nardina***

6. Perigynia hairy .... ***C. scirpoidea***
6. Perigynia glabrous 7

7. Scales subtending, perigynia long and leaf-like .... ***C. backii***
7. Female scales about as long as the perigynia 8

8. Achene 2-sided, coin-shaped .... ***C. gynocrates***
8. Achene 3-sided 9

9. Female scales falling as the perigynia mature and become horizontal or
   nearly so 10
9. Female scales persistent in mature spikes; perigynia erect or nearly so 11

10. Stems clumped; rhizomes lacking; leaves rolled, ca. 1 mm wide ....
    .... ***C. pyrenaica***
10. Plants rhizomatous, sod-forming; some leaves > 1 mm wide ... ***C. nigricans***

11. Perigynia widest at the blunt tip; swamps, thickets .... ***C. leptalea***
11. Perigynia tapered to the tip; grassland, tundra 12

12. Leaves < 1 mm wide; perigynia beak 2-lobed .... ***C. obtusata***
12. Some leaves > 1 mm wide; perigynia beak not lobed .... ***C. rupestris***

Group C: Plants with a Narrow Terminal Male Spike; Achene 3-Sided

1. Perigynia hairy, at least on the beak 2
1. Perigynia glabrous 8

2. Stems usually ≤ 25 cm tall; terminal male spike < 15 mm long 3
2. Stems mostly > 25 cm high; terminal spike > 15 mm long 7

3. Perigynia with a 2-lobed beak ca. 1 mm long 4
3. Perigynia with an inconspicuous beak 6

4. Only the perigynium beak hairy .... ***C. petricosa***
4. Body of perigynia hairy, montane 5

5. Terminal male spike 5-12 mm long; rhizomes lacking .... ***C. rossii***
5. Terminal spike > 10 mm long; plants rhizomatous .... ***C. heliophila***

6. Terminal male spike 3-6 mm long; wet forest .... ***C. concinna***
6. Terminal spike 7-15 mm long; dry forest .... ***C. concinnoides***

7. Leaves 1-2 mm wide; wet organic soil of fens .... ***C. lasiocarpa***
7. Leaves 2-5 mm wide; wet meadows .... ***C. pellita***

8. Perigynia with a distinct beak ≥ 0.5 mm long 9
8. Beak of perigynia not distinct, < 0.5 mm long 14

9. Terminal male spike of most plants > 20 mm long 10
9. Terminal male spike of most plants < 20 mm long 13

10. Upper leaf sheaths hairy; female scales long-pointed .... *C. atherodes*
10. Leaf sheaths glabrous; scales not long-pointed 11

11. Perigynia pointed sharply upward, ± erect .... *C. vesicaria*
11. Perigynia spreading at about right angles to the spike axis 12

12. Leaves rolled, whitish-waxy on the upper surface .... *C. rostrata*
12. Leaves ± flat, upper surface bright green .... *C. utriculata*

13. Perigynia 2-3.5 mm long, all spreading .... *C. viridula*
13. Perigynia 3.5-7 mm long, the lower reflexed .... *C. flava*

14. Plants of wet, organic soils of montane fens 15
14. Plants of grasslands, meadows and turf 17

15. Leaf-like bract subtending the lowest spike sheathing the stem for > 5 mm
.... *C. capillaris*
15. Sheaths of bracts < 4 mm long 16

16. Terminal male spike 13-27 mm long .... *C. limosa*
16. Terminal male spike 4-12 mm long .... *C. paupercula*

17. Female scales green to light brown 18
17. Female scales black or dark brown 19

18. Lowest female spikes on spreading stalks .... *C. capillaris*
18. Female spikes on erect or ascending stalks .... *C. livida*

19. Perigynia plump; montane and lower subalpine grasslands ..... *C. raynoldsii*
19. Perigynia flattened; near or above treeline 20

20. Female scale with a short awn at the tip .... *C. spectabilis*
20. Female scales pointed or rounded but without an awn-tip 21

21. Perigynia ovate; leaves 2-4, 2-4 mm wide; lowest spike drooping ....
.... *C. podocarpa*
21. Perigynia nearly orbicular; leaves usually > 4, 2-8 mm wide; lowest spike
often ascending .... *C. paysonis*

Group D: Plants with a Narrow Terminal Male Spike; Achene 2-Sided

1. Perigynia 3.5-5 mm long, lustrous .... *C. saxatilis*
1. Perigynia < 4 mm long, not lustrous 2

2. Leaf-like bract subtending the lowest spike sheathing the stem for 3-12 mm;
mature perigynia yellowish .... *C. aurea*
2. Bracts sheathless 3

3. Perigynia with raised veins apparent on the ventral (closest to the spike axis)
face 4
3. Perigynia without apparent veins or with obscure impressed veins 5

4. Plants forming distinct tussocks; spreading rhizomes lacking; leaves 2-4 mm
wide .... *C. lenticularis*
4. Plants rhizomatous; some leaves > 4 mm wide .... *C. nebrascensis*

5. Mature perigynia inflated, pillow-like; rare .... *C. aperta*
5. Perigynia flattened; common .... *C. aquatilis*

Group E: Achene 3-Sided; Male Flowers below Female in Bisexual Spikes

1. Some female scales leaf-like and as long as the spike .... **C. backii**
1. Female scales not leaf-like and long 2

2. At least the lower spikes nodding on long stalks .... **C. mertensii**
2. Spikes sessile or on erect or ascending stalks 3

3. Spikes < 10 mm long .... **C. media**
3. Some spikes > 10 mm long 4

4. Rhizomatous plants of montane fens .... **C. buxbaumii**
4. High elevation plants forming tufts or tussocks 5

5. Perigynia loosely 3-sided, ovate with sloping sharply from the tip, well filled
   by the achene .... **C. atrosquama**
5. Perigynia flattened with a small achene in the center, nearly orbicular,
   broadly rounded to the tip 6

6. Female scales smaller than the perigynia; perigynia with a deeply cleft beak .
   .... **C. epapillosa**
6. Female scales as large as the perigynia; perigynium beak shallowly cleft ....
   .... **C. albonigra**

Group F: Achene 2-Sided; Male Flowers above in Bisexual Spikes

1. Spikes well-separated with 1-3 perigynia each .... **C. disperma**
1. Spikes with > 3 perigynia per spike 2

2. Inflorescence branched below; i.e., some spikes attached to side branches . 3
2. All spikes attached directly to the main inflorescence axis (continuation of
   the stem) 4

3. Perigynia 4-5 mm long, triangular, broadest near the base .... **C. stipata**
3. Perigynia 2-4 mm long, ovate, broadest above the base .... **C. cusickii**

4. Stems forming tussocks; creeping rhizomes lacking 5
4. Stems arising singly or in small groups from rhizomes 6

5. Spikes clustered in an ovoid head; grasslands .... **C. hoodii**
5. Inflorescence elongate; marshy areas .... **C. diandra**

6. Inflorescence 5-15 mm long 7
6. Inflorescence > 15 mm long 9

7. Stems lax and rooting at the nodes; rare in montane fens .... **C. chordorrhiza**
7. Stems erect; not in fens 8

8. Plants of wet alpine turf .... **C. maritima**
8. Plants of montane grasslands .... **C. stenophylla**

9. Leaf sheaths green and white on the side opposite where the blade meets
   the stem .... **C. sartwellii**
9. Upper leaf sheaths white or brown, not green and white striped 10

10. Perigynia ca. 2 mm long; plants of fens .... **C. simulata**
10. Perigynia > 3 mm long; grasslands to wet meadows 11

11. Perigynia 4-6 mm long .... **C. siccata**
11. Perigynia 3-4 mm long 12

12. Stem mostly > 20 cm high, sharply 3-angled and roughened near the top .. .. *C. praegracilis*
12. Stem < 20 cm high, rounded and smooth .... *C. douglasii*

Group G: Achenes 2-Sided; Male Flowers Below; Spikes Tightly Clustered

1. Leaf-like bract at the base of inflorescence longer than the inflorescence .... *C. athrostachya*
1. Inflorescence bract smaller than the inflorescence 2

2. Perigynia, including the beak, > 5 mm long 3
2. Perigynia < 5 mm long 6

3. Perigynia > 6 mm long .... *C. petasata*
3. Perigynia 5-6 mm long 4

4. Female scales dark brown .... *C. haydeniana*
4. Female scales tan 5

5. Beak of perigynia saw-edged to the tip .... *C. tahoensis*
5. Very tip (0.5 mm) of perigynium beak smooth .... *C. phaeocephala*

6. Perigynia lens-shaped, filled by the seed, without a wing-like margin 7
6. Seed filling just the central portion of the perigynium, leaving a thin wing-margin 10

7. Spikes 2-4 8
7. Spikes 5-15 9

8. Stems < 20 cm high; female scales dark brown; alpine turf .... *C. lachenalii*
8. Stems > 20 cm; female scales pale; montane fens .... *C. tenuiflora*

9. Spikes 7-15, 5-10 mm long .... *C. arcta*
9. Spikes mostly 5-8, 4-6 mm long .... *C. laeviculmis*

10. Perigynia beak flattened and saw-edged to the tip 11
10. Very tip (0.5 mm) of perigynium beak smooth 13

11. Perigynia 3-4 times as long as wide .... *C. crawfordii*
11. Perigynia 1-2 times as long as wide 12

12. Body of perigynia nearly orbicular with a well-defined beak .... *C. brevior*
12. Perigynia more ovate, tapered gradually to the ill-defined beak .... *C. bebbii*

13. Female scales as long and wide as the perigynia and nearly concealing them 14
13. Female scales smaller than perigynia, leaving the beaks and upper edges exposed 15

14. Leaves 1-2 mm wide, firm .... *C. phaeocephala*
14. Some leaves > 2 mm wide .... *C. platylepis*

15. Achene filling only the center of the perigynium, leaving a broad wing-margin .... *C. microptera*
15. Achene nearly filling the perigynium, the wing-margin narrow 16

16. Perigynia green, contrasting sharply with the red-brown scales ..... *C. preslii*
16. Perigynia and female scales both copper-brown with little contrast between them .... *C. pachystachya*

Group H: Achenes 2-Sided; Male Flowers Below; Lowest Spikes Well Separated

1. Lowest spike with a stalk ≥ 5 mm long; stems lax or ascending .... *C. lenticularis*
1. Spikes sessile or nearly so 2

2. Perigynia, including the beak, 2-3 mm long 3
2. Perigynia ≥ 3 mm long 6

3. Perigynia widely spreading at maturity; spikes appearing star-like . *C. interior*
3. Perigynia erect or ascending in the spikes 4

4. Spikes with mostly 15-30 perigynia.. *C. canescens*
4. Spikes with mostly 5-10 perigynia.. 5

5. Perigynia with a small beak ≤ 1/3 length of the body .... *C. brunnescens*
5. Beak of perigynia distinct, ca. 1/2 length of body.................. *C. laeviculmis*

6. Perigynia 6-8 mm long .... *C. petasata*
6. Perigynia 3-6 mm long 7

7. Perigynia lens-shaped, filled by the seed, without a wing-like margin.......... 8
7. Seed filling just the central portion of the perigynium, leaving a thin wing-margin 9

8. Beak of perigynium 1-2 mm long, margins saw-edged .... *C. deweyana*
8. Perigynium beak ≤ 1 mm long, margins barely saw-edged ...... *C. laeviculmis*

9. Perigynia beak flattened and saw-edged to the tip 10
9. Very tip (0.5 mm) of perigynium beak smooth 12

10. Perigynia 5-6 mm long .... *C. tahoensis*
10. Perigynia < 5 mm long 11

11. Inflorescence stiff with the spikes close enough together to be touching; perigynia 3-4 times as long as wide... *C. crawfordii*
11. Inflorescence somewhat bent and flexuous; spikes well-separated; perigynia twice as long as wide .... *C. foenea*

12. Female scales as long and wide as the perigynia and nearly concealing them 13
12. Female scales smaller than perigynia, exposing the beak and upper edges 14

13. Leaves mostly basal; perigynia > 4.5 mm long .... *C. praticola*
13. Lower third of stem leafy; perigynia 4-4.5 mm long .... *C. platylepis*

14. Perigynia green contrasting sharply with the red-brown scales ...... *C. preslii*
14. Perigynia and female scales both copper-brown with little contrast between them .... *C. pachystachya*

Carex albonigra Mack. Plants clumped on short rhizomes; stems 10-25 cm tall; leaves 2-7 mm wide, clustered near the stem base; spikes 5-20 mm long, bisexual with males below, short-stalked with 2-4 clustered in a loose head; perigynia dark, broadly elliptic, short-beaked, flattened, 3-4 mm long; female scales black-purple with light margins above, as long as perigynia; stigmas 3.

Common in dry to moist turf near or above treeline; East. B.C., Alta. south to AZ, CA. Most common in the Two Medicine area. Reports of *C. nova* Bailey (Bamberg and Major 1968) are probably referable here. See *C. atrosquama.*

Carex aperta Boott. Stems 30-60 cm tall, loosely clustered on short rhizomes; leaves 2-6 mm wide; spikes 3-4, the uppermost male, the lower 2-3 female, stalked, erect, widely separated, 1-4 cm long; perigynia copper-colored, elliptic, inflated, 2-3 mm long; female scales dark with a pale midvein, longer than the perigynia; stigmas 2.

Rare on montane lake shores and stream banks; known from 1 location near St. Mary, expected West. B.C. to Alta south to OR, ID. MT. The inflated, coppery perigynia separate this from *C. aquatilis, C. nebraskensis* and *C. lenticularis.*

Carex aquatilis Wahl. [*C. substricta* (Kuken.) Mack.]. Stems 30-100 cm tall, somewhat clustered on long rhizomes; leaves 2-5 mm wide; spikes well separated, subtended by leaf-like bracts, 15-50 mm long on erect or spreading stalks, mostly unisexual, the terminal spike male; perigynia green to tan with dark speckles, elliptic, flattened, 2-4 mm long with a minute beak; female scales brown with a pale midvein, narrower than the perigynia; stigmas 2.

Common in fens and wet soil along streams and ponds in the montane zone; East, West. Circumboreal south to CA and AZ. *C. nebraskensis* usually has wider leaves, and *C. lenticularis* forms large tussocks without rhizomes.

Carex arcta Boott. Plants forming tussocks with stems 30-50 cm tall; leaves on lower half of stem, 1-4 mm wide; spikes 7-15, 5-10 mm long, sessile and loosely clustered, bisexual with male flowers below; perigynia spreading, green, 2-4 mm long, ovate and tapered to a broad, saw-edged beak; female scales pale with transparent margins and a green midvein, shorter than the perigynia; stigmas 2.

Locally common in wet (sometimes drying later) soil of thickets, small streams and around wetlands, montane to rarely subalpine; East, more common West. Yuk. to Que. south to CA, MT. May respond positively to disturbance. *C. canescens* has fewer spikes, and the perigynia have shorter, less distinct beaks.

Carex atherodes Spreng. Stems 40-80 cm tall from deep rhizomes; leaves 3-8 mm wide, upper sheaths hairy; spikes 2-7 cm long, unisexual with males above; perigynia green or tan, 7-10 mm long, lance-shaped tapering to a long beak with 2 spreading tips; female scales narrower than perigynia, pale and papery, with a long awn; stigmas 3.

Locally common around montane lakes and ponds, often in meadows that are innundated early but dry later; East, West. Circumboreal south to CA, UT, CO. The tall stems with hairy leaf sheaths are distinctive.

Carex athrostachya Olney. Plants forming tussocks; stems to 80 cm tall without leaves at the base; leaves only on lower half of stem, 1-4 mm wide; spikes 4-10, bisexual with male flowers below, 5-10 mm long, sessile and crowded into a head subtended by at least 1 leaf-like bract longer than the inflorescence; perigynia narrowly lance-shaped, tapered to an indistinct beak, green or tan, 3-5 mm long; female scales tan to brown with papery margins and a pointed tip, smaller than perigynia; stigmas 2.

Locally common in drying mud around montane streams, ponds, and wet meadows; East, West. AK to CA, CO. The narrow perigynia and long bract of the inflorescence help distinguish this sedge.

Carex atrosquama Mack. [*C. atrata* L. var. *atrosquama* (Mack.) Cronq.]. Stems 10-30 cm high, forming small tussocks with old leaves at the base; leaves only on lower stem, 1-5 mm wide; spikes 3-5, 1-2 cm long with short stalks, loosely clustered, erect, the uppermost bisexual, the lower all female; perigynia ca. 3 mm long, ovate with a tiny beak, brown, loosely 3-sided; female scales dark, shorter than the perigynia; stigmas 3.

Uncommon in moist to rather dry, subalpine and lower alpine meadows; East, West. B.C., Alta. south to OR and CO. The perigynia of *C. epapillosa* and *C. albonigra* are more flattened, broadly elliptic, rounded to the tip.

Carex aurea Nutt. Stems 3-30 cm tall from rhizomes; leaves 1-4 mm wide; spikes 2-5, well separated, stalked, 5-20 mm long, subtended by leaf-like bracts, unisexual, the uppermost male; perigynia ovate, inflated, rounded on top, 2-3 mm long, light green to golden when mature, loosely aggregated in the spikes; female scales shorter than perigynia, light brown with a pale center and white margins; stigmas 2. Fig. 278.

Common in moist soil around montane to lower alpine streams, ponds, and wetlands; East, West. AK to Newf. south to CA, AZ, PA. Plants are often inconspicuous in higher vegetation, but the loosely aggregated, golden perigynia are distinctive. Long-stalked female spikes may arise from the base.

Carex backii Boott. Stems 5-30 cm high, forming small tussocks; leaves 2-6 mm wide; spikes 2-7 mm long not including the elongate scales, few-flowered, widely separated, the lower long-stalked among the leaves, the terminal sessile, bisexual, but male flowers inconspicuous; perigynia green, ca. 5 mm long, elliptic with a short beak at one end and a short stalk at the other; female scales elongate, leaf- or bract-like; stigmas 3.

Rare in montane grasslands and thickets; West. B.C. to Que. south to OR, UT, CO. The leaf-like female scales are distinctive.

Carex bebbii Olney ex Fern. Plants forming dense tussocks; stems 20-60 cm high without leaves at the very base; leaves long, 1-4 mm wide; spikes 4-12, 4-9 mm long, sessile, densely aggregated into an elongate head, bisexual with male flowers below; perigynia erect, green to tan, lance-shaped and tapered to the tip, flattened, 3-4 mm long; female scales papery, tan with a green midvein, smaller than perigynia; stigmas 2.

Uncommon in montane wet meadows and streambanks; West, expected East. B.C. to Newf. south to WA, ID, CO. The closely related *C. crawfordii* Fer. has narrower perigynia.

Carex brevior (Dewey) Mack. ex Lunnel. Stems 30-80 cm high, forming tussocks, leaves lacking from the very base; leaves flat or V-shaped, 2-4 mm wide; spikes 3-6, sessile, clustered at the stem tip, 6-10 mm long, bisexual with male flowers below; perigynia greenish, orbicular with a short, tapered beak, nearly vertical, flattened, 3-5 mm long; female scales light tan, narrower and usually shorter than the perigynia; stigmas 2.

Collected once in wet soil near Apgar. B.C. to Que. south to OR, AZ. The nearly orbicular perigynia help separate this species from *C. bebbii.*

Carex brunnescens (Pers.) Poir. Stems 20-60 cm high, forming dense tussocks; leaves 1-2 mm wide, on lower half of stem; spikes sessile, 3-6 mm long, well separated, bisexual, the male flowers below; perigynia green or pale brown, ovate, ca. 2 mm long, with a short, broad beak; female scales whitish-papery with a dark midvein, shorter than the perigynia; stigmas 2.

Rare in montane wet meadows and thickets; collected near Lake McDonald and Lincoln Lake. Circumboreal south to OR, UT. *C. canescens* has more perigynia in each larger spike, and *C. laeviculmis* has perigynia with more distinct beaks.

Carex buxbaumii Wahl. [*C. polygama* Schkuhr.]. Stems 25-80 cm high from long rhizomes, without leaves at the very base; leaves 1-4 mm wide, on lower half of the stem; spikes 2-5, 1-3 cm long, the uppermost bisexual with males below, the lower female and short-stalked, subtended by leaf-like bracts, not strongly clustered; perigynia elliptic, nearly beakless, green, 3-4 mm long; female scales larger than the perigynia, long-tapered, dark with a pale midvein; stigmas 3.

Montane fens and sometimes wet meadows; common West, uncommon East. Circumboreal south to CA, CO. The green perigynia with dark, long-pointed scales are distinctive.

Carex canescens L. [*C. curta* Good]. Stems 20-60 cm high, forming loose tussocks on short rhizomes; leaves 1-3 mm wide, clustered on the lower stem; spikes 4-8, sessile, well separated, 5-10 mm long, bisexual, male flowers below; perigynia light green, ca. 2 mm long, ovate with an indistinct beak; female scales pale, papery with a green midvein, shorter than the perigynia; stigmas 2.

Common in wet meadows, marshes, and fens, usually in organic soils in the montane and lower subalpine zones; East, West. Circumboreal south to CA, AZ. See *C. brunnescens.*

Carex capillaris L. Stems to 30 cm high, forming small tussocks; leaves 1-3 mm wide, mainly basal; spikes 2-5, 4-17 mm long, unisexual, the terminal male, the others female and long-stalked, subtended by narrow, leaf-like bracts, well-separated and often nodding; perigynia ca. 3 mm long, narrowly elliptic with an indistinct beak, somewhat inflated, green to brown; female scales green or tan with pale margins, shorter than the perigynia; stigmas 3.

Wet, usually organic soil of stream banks, wet meadows, fens, and turf, uncommon montane to common in the alpine; East, West. Circumboreal south to OR, UT, CO. Hermann (1970) recognizes the alpine var. *capillaris* and the low-elevation var. *major* Drej.

Carex chordorrhiza L. Stems trailing, clothed in old leaves, giving rise to fertile stems 10-30 cm high; leaves V-shaped in cross section, 1-2 mm wide; spikes 3-5, sessile, clustered in a small, dense head that appears to be 1 spike, bisexual with the male flowers above; perigynia ca. 3 mm long, 1-5 per spike, ovoid with a short beak, brown; female scales brown, as long as the perigynia; stigmas 2.

Rare in montane sphagnum fens; known from 2 locations in the North Fork Flathead drainage. Circumboreal south to MT, IA, NY. The lax stems lying on the wet peat are diagnostic.

Carex concinna R. Br. Stems lax, 5-20 cm tall, sometimes tufted, from rhizomes; leaves basal, 1-3 mm wide; spikes 3-4, 3-10 mm long, unisexual, the male terminal, ca. 5 mm long, inconspicuous, female spikes below, short-stalked, loosely clustered; perigynia hairy, few per spike, 2-3 mm long, lightbulb-shaped with a minute beak; female scales brown with pale centers, shorter than the perigynia; stigmas 3.

Uncommon in wet, often calcareous soil of montane (often spruce) forest; West. Yuk. to Newf. south to OR, CO, SD. The more common *C. concinnoides* has a larger male spike.

Carex concinnoides Mack. Stems 10-40 cm high from long rhizomes; leaves 3-5 mm wide; spikes 2-4, unisexual; the male terminal, 8-12 mm long; the female below, sessile, 5- to 12-flowered, loosely clustered, ca. 5 mm long; perigynia hairy, ca. 3 mm long, elliptic with a short beak and stalk; female scales brown with pale margins, as long as the perigynia; stigmas 4.

Common in montane, often dry, coniferous forest; West. B.C., Alta. south to CA, MT. See *C. concinna*.

Carex crawfordii Fern. Schk. Stems 15-60 cm high, forming tussocks; leaves 1-4 mm wide; spikes 5-12, sessile, 5-10 mm long, the upper clustered, the lower sometimes separated, bisexual with the male flowers below; perigynia 4-5 mm long, narrowly lance-shaped with flattened margins; female scales light brown with a green center, shorter than the perigynia; stigmas 2.

Uncommon in moist montane meadows and forest openings; West. B.C. to Newf., south to CA, NM, FL. Perigynia of *C. petasata* are larger, and those of *C. bebbii* are more ovate.

Carex cusickii Mack. Plants forming large tussocks; stems 30-80 cm high, without leaves at the very base; leaves 3-5 mm wide with red-dotted sheaths copper-colored above and opposite the blades; spikes sessile, 5-6 mm long, numerous in a branched inflorescence, bisexual with male flowers above; perigynia light brown, ca. 3 mm long, lance-shaped, tapered to an indistinct, greenish beak; female scales papery, light brown, mostly as large as the perigynia; stigmas 2.

Uncommon in wet, organic soil of fens and marshes; collected in the McDonald Valley. B.C. south to CA, ID, WY. *C. diandra* has narrower leaves and inflorescences.

Carex deweyana Schwein. [*C. leptopoda* Mack]. Stems 20-60 cm high, forming tussocks; leaves 2-4 mm wide, on lower half of stem; spikes 5-15 mm long, well separated, bisexual with male flowers below, the lower subtended by long, narrow bracts; perigynia 3-5 mm long, strictly erect, green to tan, sessile, narrowly elliptic with a beak 1-2 mm long; female scales papery, white to brown with a green midvein, nearly as long as the perigynia; stigmas 2.

Moist montane to lower subalpine forest; common West, uncommon East. B.C. to Lab. south to CA, AZ, PA; Asia. *C. brunnescens* and *C. canescens* occur in wetter habitats and have perigynia with smaller beaks.

Carex diandra Schrank. Stems 30-70 cm tall forming large tussocks; leaves 1-3 mm wide with red-dotted sheaths, not copper-colored, opposite the blades; spikes sessile, 3-7 mm long, numerous in a short-branched inflorescence, bisexual with male flowers above; perigynia dark brown, 2-3 mm long, lance-shaped, tapered to an indistinct beak; female scales papery, light brown with a light margin, not as large as the perigynia; stigmas 2.

Uncommon in montane fens and swamps; West. Circumboreal south to WA, CO, PA. See *C. cusickii*.

Carex disperma Dewey. Stems 10-40 cm tall, lax, sometimes forming loose clumps from long, slender rhizomes; leaves long, all on lower half of the stem, 1-2 mm wide; spikes sessile, well separated, few-flowered, to 4 mm long, bisexual with male flowers above; perigynia 2-3 mm long, green to light tan, broadly elliptic with a short, broad beak; female scales pale with a green midvein, papery, shorter than the perigynia; stigmas 2.

Common in wet organic soil, often beneath shrubs, in montane swamps, fens, and spruce or cedar forest; East, West. Circumboreal south to CA, AZ, NJ. The few-flowered, sessile spikes with the moist, shady habitat distinguish this sedge. *C. leptalea* has a similar habit but only 1 spike.

Carex douglasii Boott. Stems 5-15 cm high from slender rhizomes; leaves 1-3 mm wide, all on lower stem but not at the very base; spikes several, 8-12 mm long, sessile, clustered at stem tip, unisexual, all usually either male or female; perigynia 4-5 mm long, pale brown, lance-shaped and narrowed to a slender beak; female scales light brown with a pale midvein and margins, longer than the perigynia; stigmas 2.

Uncommon in montane grasslands, often in compacted soil, along the east margin of the Park. B.C. to Man. south to CA, NM. Our only dioecious dryland sedge.

Carex elynoides Holm. Similar to *C. filifolia*; stems 5-10 cm high; spike solitary 8-15 mm long, bisexual with male flowers forming the upper half; perigynia 2-4 mm long, with little or no hair; female scales brown and larger than the perigynia.

Rare in dry alpine turf; known only from north of Two Medicine Lake. MT south to NV, UT, CO. *C. nardina* has few male flowers and is more common; *Kobresia myosuroides* is also similar, but the perigynium is split down 1 side. Glacier Park is the northernmost location for this sedge; it becomes very common in southwest Montana.

Carex epapillosa Mack. [*C. atrata* L. var. *erecta* Boott]. Similar to *C. atrosquama*; forming tussocks; leaves only on lower stem, 3-6 mm wide; spikes 4-6, 1-2 cm long with short stalks, loosely clustered, erect, the uppermost bisexual, the lower all female; perigynia 3-4 mm long, broadly elliptic with a tiny beak, brown, flattened; female scales dark, narrower than the perigynia; stigmas 3.

Uncommon in moist, subalpine meadows; East, West. B.C., Alta. south to CA, UT, CO. See *C. atrosquama*.

Carex filifolia Nutt. Stems 5-20 cm high with old leaf sheaths at the base, forming dense tufts or mats; leaves wiry, ca. 0.5 mm wide, mainly basal; spike solitary, narrow, 1-2 cm long, bisexual with male flowers above; perigynia 3-5 mm long, short-hairy, narrowly ovoid with a short beak; female scales brown with wide, white margins, larger than the perigynia; stigmas 3.

Uncommon in dry soil of montane grasslands, occasionally higher, along the east margin of the Park. Yuk to Man. south to CA, NM, TX. The similar *C. elynoides* occurs near or above treeline.

Carex flava L. Stems 10-50 cm high, forming tussocks; leaves 2-5 mm wide; spikes unisexual, the terminal male, the 2-5 female below nearly sessile, 8-15 mm long, well separated, subtended by leaf-like bracts; perigynia spreading, yellowish, 4-6 mm long, ovate with a distinct recurved beak; female scales copper-colored with a green midvein; stigmas 3.

Common in wet, often gravelly and calcareous soil of meadows, fens, and shores of lakes and streams in the montane and subalpine zones; East, more common West. Circumboreal south to B.C., ID, MT. *C. viridula* has smaller perigynia with straight beaks.

Carex foenea Willd. [*C. aenea* Fern.]. Plants bunch-forming; stems 30-80 cm high; leaves 2-5 mm wide, on lower half of stem but not at the base; spikes 3-8, sessile, 6-25 mm long, female and crowded above but bisexual and separate below; perigynia green to pale brown, thin margined, tapered to finely toothed beak, 3-5 mm long, 2-3 times as long as wide; female scales light brown with lighter margins, as long and wide as perigynia; stigmas 2.

Moist to dry soil at mid-elevations; collected along Swiftcurrent and Waterton lakes. Yuk to Lab. south to B.C., ID MT.

Carex geyeri Boott, Elk Sedge. Stems 15-50 cm high without leaves at the base, loosely clustered on branched rhizomes; leaves long, 1-3 mm wide; spike solitary with male flowers above and only 1-3 well-separated perigynia below; perigynia green, 5-6 mm long, thumb-shaped, rounded on top; female scales sharp-pointed, brown with a pale midrib and margins, longer than perigynia; stigmas 3.

Common in montane and subalpine open forests, grasslands, and meadows; more common in forests West, and in grasslands East. Alta. and B.C. to Ca, CO. Easily recognized by the few, distinctly shaped perigynia. Leaves remain green during winter; stems flower near the ground in early spring, then elongate in fruit.

Carex gynocrates Wormsk. ex Drejer [*C. dioica* L. var *gynocrates* (Drejer) Ostenf.]. Stems 5-20 cm high from long rhizomes; leaves to 1 mm wide, mostly basal; spike solitary, 10-15 mm long, bisexual (sometimes unisexual) with male flowers above; perigynia horizontal at maturity, dark brown, 3-4 mm long, narrowly ovoid with a distinct, downcurved beak; female scales light brown with pale margins, shorter than the perigynia; stigmas 2.

Uncommon in wet, calcareous, organic soil of montane to alpine fens and seep areas; East, West. AK to Newf. south to OR, CO, MI, PA. The female scales of *C. nigricans* and *C. pyrenaica* fall before maturity.

Carex haydeniana Olney [*C. nubicola* Mack.]. Stems 10-25 cm tall, forming small tussocks; leaves all on lower stem, 2-4 mm wide; spikes 4-7, sessile, densely clustered on stem tips, 7-10 mm long, bisexual with males below; perigynia brown and green-edged, 5-6 mm long, ovate with flattened edges, tapering to a prominent beak; female scales brown, narrower than perigynia; stigmas 2. Fig. 279.

Common in moist to wet, often stony soil of subalpine and alpine meadows and turf; East. B.C., Alta. south to CA, CO. See *C. microptera*.

Carex heliophila Mack. [*C. pensylvanica* Lam. var. *digyna* Boeck., *C. inops* Bailey ssp. *heliophila* (Mack.) Crins]. Stems 5-20 cm high, forming small

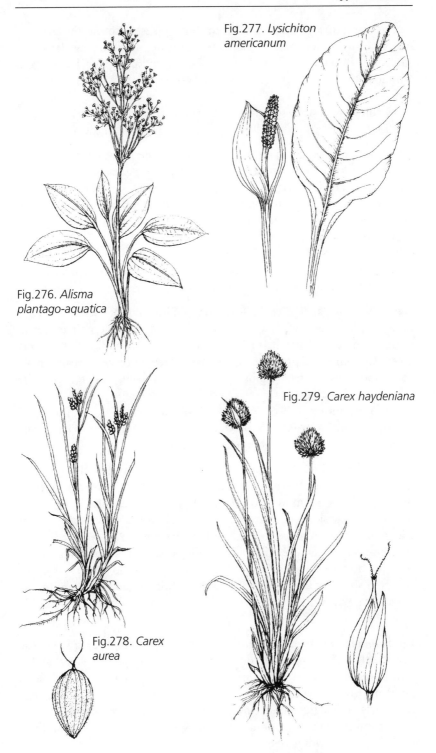

Fig.277. *Lysichiton americanum*

Fig.276. *Alisma plantago-aquatica*

Fig.279. *Carex haydeniana*

Fig.278. *Carex aurea*

clumps from rhizomes; leaves 1-2 mm wide; spikes unisexual, the terminal male, the lower 1-2 female, sessile, usually separate, 4-6 mm long, the lowest subtended by a leaf-like bract; perigynia 3-4 mm long, green, hairy, ovoid with a distinct beak; female scales tan to brown with a green middle; as long as perigynia; stigmas 3.

Uncommon in stony soil of montane grasslands; East. Alta. to Man. south to NM, IN. Plants may have long-stalked female spikes at the base.

Carex hoodii Boott. Stems 30-60 cm tall with leafless bases, forming large, dense tussocks; leaves 2-4 mm wide on basal half of stem; spikes 4-8, sessile, bisexual with male flowers above, densely clustered into an ovoid head 1-2 cm long; perigynia 3-5 mm long, brown with green margins, flattened, elliptic, tapered to an indistinct beak; female scales as long as perigynia, brown with a green midvein; stigmas 2.

Grasslands and meadows, common in the montane zone, less common higher; East, West. B.C. to Sask. south to CA, CO, SD. *C. microptera* and *C. pachystachya* have shorter heads with female flowers above the males.

Carex interior Bailey. Stems 20-50 cm high, forming tussocks; leaves 1-2 mm wide, all on lower stem; spikes 3-6, ca. 5 mm long, sessile, well separated, the uppermost bisexual with male flowers below, others all female; perigynia green, 2-3 mm long, spreading, ovate, tapered to an indistinct beak; female scales tan with white margins and a green midvein, shorter than the perigynia; stigmas 2.

Locally common in wet, organic soil of montane fens, meadows and swamps; East, West. B.C. to Newf. south to CA, Mex., PA. The widely spreading perigynia help distinguish this species.

Carex lachenalii Schkuhr. [*C. bipartita* All.]. Stems in small clusters, 5-20 cm high; leaves mostly basal, 1-3 mm wide; spikes 2-4, sessile, clustered at the stem tip, 5-10 mm long, bisexual with male flowers below; perigynia brown, ca. 3 mm long, elliptic with a short beak, erect; female scales brown with a pale middle, concealing the perigynia; stigmas 2.

Common in wet, organic soil on cool slopes, rock ledges, and along streams above treeline; East, West. Circumpolar south to CO, UT. The few, clustered spikes and perigynia that are not thin-edged help identify this sedge. Graff's (1922) report of *C. heleonastes* Ehrh. is probably referable here.

Carex laeviculmis Meinsh. Stems slender, 30-70 cm high, forming dense tussocks; leaves 1-2 mm wide, on lower half of stem; spikes 3-8, sessile, 4-6 mm long, the lower well separated, bisexual, the male flowers below; perigynia green, ovate, 2-4 mm long, tapered to a distinct beak; female scales pale with a green midvein, shorter than the perigynia; stigmas 2.

Uncommon in moist, montane thickets and forest; West. AK south to CA, MT. See *C. brunnescens*.

Carex lasiocarpa Ehrh. Stems 30-80 cm high from stout rhizomes; leaves long, 1-2 mm wide, lacking from the stem bases; spikes narrow, bisexual, the upper male, the lower female, nearly sessile, 1-5 cm long, well separated and subtended by leaf-like bracts; perigynia hairy, purplish, 3-4 mm long, narrowly ovoid with a short, divided beak; female scales brown with a pale middle, mostly longer than the perigynia; stigmas 3.

Locally abundant in saturated organic soil of montane and lower subalpine fens; West, less common East. Our plants are var. *americana* Fern. Circumboreal south to WA, ID, MT. This sedge often forms monocultures in fens; see *C. pellita*.

Carex lenticularis Michx. [*C. kellogii* Boott, *C. goodenovii* Gay, *C. plectocarpa* Hermann, *C. eleusinoides* Turcz. misapplied]. Stems erect or lax, 15-80 cm high, forming tussocks; leaves 2-4 mm wide; spikes 3-6, cylindric, short-stalked, well separated, erect, 1-3 cm long, mostly unisexual, the lower female; perigynia 2-3 mm long, green, ovate to broadly lance-shaped with an indistinct beak; female scales dark with pale midveins, shorter than the perigynia; stigmas 2. Fig. 280.

Var. *lipocarpa* (Holm) Standley, with a terminal male spike and erect stems, is abundant in wet, often stony and sparsely vegetated habitats along streams and lake shores and occasionally in fens at all elevations; var *dolia* (Jones) Standley, with a bisexual terminal spike and lax stems, is local and uncommon in wet, stony soil of shallow meltwater streams and around ponds at or above treeline; West, East. AK to Lab. south to CA, CO, MA; Asia. Prostrate stems lying in shallow water of high-elevation habitats distinquish var. *dolia*, while var. *lipocarpa* can be told from *C. aquatilis* by the tussock growth form. The two varieties intergrade near Logan Pass.

Carex leptalea Wahl. Stems weak, 10-40 cm high, forming loose clumps from finely branched rhizomes; leaves ca. 1 mm wide; spike ca. 5 mm long, solitary with male flowers above and few erect perigynia; perigynia ca. 2 mm long, green, narrowly elliptic with a thick stalk at the base and flattened on top; female scales green to tan, pointed, shorter than the perigynia; stigmas 3.

Common, often on hummocks in wet, montane fens, thickets, and spruce or cedar forests; East, West. AK to Lab. south to CA, CO, TX. Often found with *C. disperma*, which has several small spikes.

Carex limosa L. Stems 20-50 cm tall from long rhizomes and yellow roots; leaves few, 1-2 mm wide, absent at the stem base; spikes 2-4, unisexual, the uppermost male, the lower female (few male flowers at the tip), 10-25 mm long, long-stalked, well separated, nodding, subtended by leaf-like bracts; male spikes 10-30 mm long; perigynia green to tan, 2-4 mm long, elliptic with a short beak; female scales brown, mostly as long as perigynia; stigmas 3.

Common in montane fens; West, expected East. Circumboreal south to CA, UT, WY. The nodding spikes with fen habitat distinguish this and the similar *C. paupercula,* which has a smaller male spike.

Carex livida (Wahl.) Willd. Stems 10-30 cm high from long rhizomes; leaves mostly basal, 1-3 mm wide; spikes unisexual, the uppermost male, the lower 1-3 female, short-stalked, loosely clustered, 12-40 mm long; perigynia pale green, ca. 4 mm long, narrowly elliptic, tapered at both ends; female scales brown with a pale center, mostly as long as the perigynia; stigmas 3.

Rare in calcareous, montane fens; known only from north of Apgar. Circumboreal south to CA, CO. Plants arise from the rhizome, remain vegetative for one or more years, flower once, and then perish. The pale color of the plants is distinctive.

Carex maritima Gunn. [*C. incurviformis* Mack. var. *danaensis* (Stacey) Hermann]. Stems 2-15 cm high, loosely clustered on stout rhizomes; leaves ca. 1 mm wide, mostly basal; spikes 3-5, bisexual with inconspicuous male flowers above, sessile, clustered in a dense head 5-10 mm long; perigynia glossy brown, ca. 3 mm long, narrowly ovoid, tapering to an indistinct beak; female scales brown with pale margins, shorter than the perigynia; stigmas 2.

Rare in wet turf along small streams above treeline; known only from the Continental Divide south of St. Mary Lake. Our plants are var. *incurviformis* (Mack.) Boivin. Circumpolar south to CA, CO.

Carex media R. Br. [*C. norvegica* Retz. sensu lato, *C. halleri* Gunn.]. Stems 10-40 cm tall, loosely aggregated on rhizomes; leaves on basal half of stem, 1-3 mm wide; spikes 2-5, 3-10 mm long, the uppermost bisexual with male flowers below, the lower female, short-stalked, aggregated; perigynia green to tan, 2-3 mm long, elliptic with a short beak; female scales black, shorter than the perigynia; stigmas 3.

Rare along subalpine streams, and lakes; East. AK to Lab. south to WA, NM.

Carex mertensii Prescott ex Bong. Stems 40-100 cm tall, loosely clustered on short rhizomes; leaves absent at the base, 3-8 mm wide; spikes cylindric, 15-35 mm long, long-stalked, loosely aggregated, bisexual with male flowers below; perigynia numerous, vertical, 4-5 mm long, green to tan, flattened, elliptic with a minute beak; female scales dark with a pale midvein, shorter than the perigynia, falling at maturity; stigmas 3.

Widespread but never abundant in thickets and moist open forest and along streams and roads in the montane and lower subalpine zones; East, West. AK to CA, ID, MT. The large, nodding spikes make this one of our most attractive sedges.

Carex microptera Mack. [*C. festivella* Mack.]. Stems 20-80 cm tall, forming tussocks; leaves 2-4 mm wide, all on basal half of the stem; spikes sessile, clustered into a globose head 1-2 cm long, bisexual with male flowers below; perigynia green to light brown, 3-5 mm long, flattened, ovate, gradually tapered to the beak; female scales brown, shorter than perigynia; stigmas 2.

Common in moist meadows, grasslands and shores of streams and lakes, montane to treeline; East, West. B.C. to Sask. south to CA, AZ, SD. *C. microptera, C. haydeniana, C. pachystachya, C. preslii, C. praticola,* and *C. platylepis* are closely related and difficult to distinguish based on size of the perigynia and the female scales. At one time they were all treated as a single species. *C. limnophila* Hermann, with small, narrow perigynia, occurs just south of the Park.

Carex nardina Fries [*C. hepburnii* Boott]. Stems 5-15 cm high, forming dense tufts or mats with old leaf sheaths at the base; leaves mostly basal, 0.5 mm wide; spike solitary, to 15 mm long, bisexual with a few male flowers above; perigynia greenish, nearly erect, ca. 4 mm long, flattened, narrowly elliptic with a small beak; female scales brown with a pale middle, as long as perigynia; stigmas 2 or 3. Fig. 281.

Abundant in low turf to stony soil of exposed ridges and slopes near or above treeline; East, West. Circumpolar south to WA, UT, CO. See *C. elynoides.*

Carex nebrascensis Dewey, Nebraska Sedge. Stems 30-90 cm high from long rhizomes; leaves 4-10 mm wide, V-shaped basally, bluish-green; spikes cylindric, unisexual, the upper 1-2 male, the lower female, short-stalked, well separated, 15-60 mm long, the lowest subtended by a long, leaf-like bract; perigynia green to light tan, 3-4 mm long, ovate with a short, divided beak; female scales dark with a pale midvein, long-pointed, narrower than the perigynia; stigmas 2.

Uncommon in wet meadows and along streams, montane; East, West. B.C. to Alta. south to CA, NM, KS. *C. nebrascensis* has nerves on the perigynia, while *C. aquatilis* does not. The wide, blue-green leaves help distinguish this species.

Carex nigricans C.A. Meyer. Stems 10-30 cm high, leafless at the very base, loosely aggregated on rhizomes; leaves 1-3 mm wide, on the lower half of the stem; spike solitary, 1-2 cm long, unisexual with male flowers above; perigynia green below and dark above, 3-5 mm long, spreading at maturity, narrowly lance-shaped with an indistinct beak; female scales dark, short, falling at maturity; stigmas 3.

Abundant in moist meadows and along streams, especially where snow lies late near or above treeline; East, West. AK south to CA, UT, CO. *C. nigricans* forms solid colonies in depressions where snow is deeper. *C.*

pyrenaica forms small, dense tussocks and usually occurs in drier or better drained soils.

Carex obtusata Lilj. Stems 5-15 cm high from slender rhizomes; leaves mostly basal, 1-2 mm wide, spike solitary, 5-10 mm long, bisexual with male flowers above; perigynia few, shiny brown, 3-4 mm long, elliptic with a short, divided beak; female scales pale brown, as long as the perigynia; stigmas 3.

Common but inconspicuous in montane and subalpine grasslands, less common in alpine turf; East. Yuk. to Man. south to NM, CO, NE; Eurasia.

Carex pachystachya Cha. ex Steud. Stems 30-60 cm high, forming tussocks; leaves 2-4 mm wide; spikes 7-12 mm long, sessile, aggregated into an oblong head, bisexual with male flowers below; perigynia copper-colored, 3-5 mm long, flattened along the edges, ovate, gradually tapered to the beak; female scales brown with pale margins, shorter than perigynia; stigmas 2.

Common in dry to moist, montane and subalpine grasslands and meadows; West, more common East. AK to CA, CO. See *C. microptera*. *C. preslii* has green perigynia that contrast sharply with the dark scales.

Carex paupercula Michx. [*C. magellanica* Lam. in part]. Similar to *C. limosa*; stems clustered, 20-50 cm tall; female spikes 7-12 mm long, often with a few male flowers at the base; male spikes 4-12 mm long; perigynia green to tan, 2-4 mm long, elliptic without a distinct beak; female scales falling at maturity.

Rare in montane fens; known only from near Camas Creek. Circumboreal south to WA, UT, CO. See *C. limosa*.

Carex paysonis Clokey [*C. tolmiei* Boott]. Stems 15-40 cm with dried leaves at the base, loosely aggregated on short, branched rhizomes; leaves numerous, on basal half of the stem, 2-8 mm wide; spikes unisexual, 1-2 male above, 2-6 female below, short-stalked, well-separated, mostly erect, 10-20 mm long; perigynia 2-4 mm long, green and purplish, flattened, broadly ovate with a minute beak; female scales blackish with a blunt tip and pale midvein, narrower but as long as the perigynia; stigmas 3. Fig. 282.

Common in moist to wet meadows and turf near or above treeline; West, East. AK to OR, ID, WY. *C. raynoldsii* occurs in the montane and lower subalpine zones; *C. podocarpa* has more nodding female spikes and fewer leaves; *C. spectabilis* has narrower perigynia and pointed scales.

Carex pellita Willd. [*C. lanuginosa* Michx.]. Similar to *C. lasiocarpa*; leaves 2-5 mm wide; perigynia 3-5 mm long; female scales purplish-brown with a pale middle, narrower than the perigynia; stigmas 3.

Uncommon in wet mineral soil of montane stream banks, wet meadows and marshes; East, West. B.C. to Que. south to CA, AZ, TX. *C. lasiocarpa* is usually found in organic soils.

Carex petasata Dewey. Stems 30-60 cm tall, forming large tussocks; leaves 2-3 mm wide, absent at the very base of the stem; spikes 3-6, sessile, 8-15 mm long, somewhat aggregated, bisexual, the male flowers below; perigynia green to tan, 6-8 mm long, erect, flattened, lance-shaped and tapered gradually to the beak with a cylindric tip; female scales light brown, as large as the perigynia; stigmas 2.

Uncommon in montane and lower subalpine grasslands and meadows; East, West. B.C. to Sask. south to CA, AZ. The similar *C. tahoensis* has perigynia with flattened, deeply divided tips and slightly smaller perigynia.

Carex petricosa Dewey. Stems 10-25 cm high from rhizomes; leaves 1-2 mm wide with rolled margins; spikes mostly unisexual, the upper 1-2 male, the lower 2-3 female, stalked, erect, mostly separate, 8-12 mm long; perigynia brown, 4-6 mm long, lance-shaped with an indistinct beak; female scales brown with pale margins, shorter than the perigynia; stigmas 3.

Rare in stony, calcareous turf near or above treeline; East. AK south to MT.

Carex phaeocephala Piper. Stems 5-30 cm high, forming dense tussocks; leaves 1-2 mm wide, mainly on lower half of stem; spikes 2-5, sessile, 7-12 mm long, aggregated at stem tips, bisexual with male flowers below; perigynia tan with green margins, 4-6 mm long, erect, flattened, narrowly elliptic, tapered to an indistinct beak; female scales as long as the perigynia, brown with paler margins and midvein; stigmas 2.

Common in usually dry, stony soil of meadows and turf near or above treeline; East, West. B.C., Alta. south to CA, CO. The low stature with light brown spikes and narrow perigynia help distinguish this sedge. Reports of *C. leporinella* Mack. (Bamberg and Major 1968) may be referable here.

Carex platylepis Mack. Stems 25-60 cm tall, forming small tussocks; leaves 3-5 mm wide, wrinkled, absent from the very base of stems; spikes 5-8, sessile, loosely aggregated in a head 15-35 mm long, bisexual with male flowers below; perigynia green to tan, ca. 4 mm long, erect, flattened on the margins, narrowly elliptic, gradually tapered to the indistinct beak; female scales red-brown with a green midvein and pale margins, as long as the perigynia; stigmas 2.

Rare in montane grasslands and open forest; West. B.C., Alta. south to ID, WY. *C. petasata* has larger perigynia, and scales of *C. pachystachya* are shorter than the perigynia.

Carex podocarpa R. Br. Stems 15-50 cm, loosely aggregated on short, branched rhizomes; leaves 2-4, 2-4 mm wide, absent from the very base; spikes unisexual, the solitary male above, 2-4 female below, stalked, often nodding, well-separate, 1-2 cm long; perigynia ca. 4 mm long, dark purplish brown, flattened, narrowly ovate with a minute beak; female scales dark, rounded or broadly pointed at the tip but narrower than the perigynia; stigmas 3.

Abundant in moist or wet alpine and subalpine meadows, especially near streams; East, West. AK south to ID, MT. See *C. paysonis*. Cronquist (Hitchcock et al. 1969) mistakenly called this *C. spectabilis*.

Carex praegracilis Boott. Stems 30-50 cm tall from long rhizomes; leaves 1-3 mm wide, absent from the very base of stems; spikes 5-15, sessile, loosely aggregated, up to 1 cm long, bisexual with male flowers above (some plants may be entirely male or female); perigynia dark brown, 3-4 mm long, ovate, gradually tapering into the beak; female scales light brown with a pale midvein and margins, as long as the perigynia; stigmas 2.

Rare in moist soil of montane meadows, often around ponds or streams; East, West. B.C. to Man. south to CA, Mex. The similar *C. simulata* has perigynia with smaller beaks and is usually confined to fens.

Carex praticola Rydb. Stems 20-60 cm tall, forming small tussocks; leaves 1-4 mm wide, absent from the very base of stems; spikes 2-7, sessile, loosely aggregated, 7-10 mm long, bisexual with male flowers below; perigynia green to tan, 4-6 mm long, erect, flattened on the margins, narrowly elliptic, gradually tapered to the indistinct beak; female scales as long as the perigynia, light brown with paler margins; stigmas 2.

Common in montane meadows, moist forest and along lake shores; East, West. AK to Greenl. south to CA, CO, SD. See *C. microptera*. Reports of *C. tenera* Dewey and *C. tincta* Fern. are referable here.

Carex preslii Steud. Stems 20-40 cm high, forming tussocks; leaves mainly near the base, 1-4 mm wide; spikes 3-8, sessile, 5-10 mm long, clustered at the stem tip, bisexual with the male flowers below; perigynia 3-4 mm long, green, flattened on the margins, ovate, tapered to an indistinct beak; female scales brown, shorter than the perigynia; stigmas 2.

Common in moist subalpine meadows, often near streams, uncommon lower; East, West. B.C. to Alta. south to CA, MT. See *C. microptera, C. pachystachya*. The green perigynia contrasting with the brown scales helps distinguish this sedge.

Carex pyrenaica Wahl. Stems 5-20 cm high, forming dense tussocks; leaves ca. 1 mm wide, crowded near the base; spike solitary, 8-15 mm long, bisexual with male flowers above; perigynia light to dark brown, 3-5 mm long, spreading at maturity, narrowly lance-shaped with an indistinct beak; female scales dark, short, falling at maturity; stigmas 3.

Common in moist to dry, well-drained soil of meadows and rocky slopes near or above treeline, often where snow lies late; East, West. Circumpolar south to OR, UT, CO. See *C. nigricans*.

Carex raynoldsii Dewey. Stems 20-80 cm tall, loosely aggregated on short rhizomes with old leaves at the base; leaves 2-7 mm wide; spikes unisexual, the terminal male, the lower 2-5 female, 1-2 cm long, stalked and loosely aggregated; perigynia green or yellowish, ca. 4 mm long, inflated, ovate with a short, distinct beak; female scales black; shorter than the perigynia; stigmas 3.

Common in montane and lower subalpine, moist grasslands and meadows; East, West. B.C. to Alta. south to CA, UT, CO. The green, pillow-like, inflated perigynia help distinguish this sedge.

Carex rossii Boott. Stems 5-40 cm high, forming small, loose tussocks; leaves 1-2 mm wide; spikes unisexual, the terminal male, the lower 2-5 female, short-stalked, few-flowered, loosely aggregated, the lowest long-stalked, arising from near the stem base; perigynia green, ca. 4 mm long, short-hairy, elliptic with a distinct beak and a stalk at the opposite end; female scales green to brown, shorter than the perigynia; stigmas 3.

Common in montane and subalpine forest, less common in grasslands and above treeline; East, West; Yuk. south to CA, AZ, MI. The long-stalked spikes arising from the base help distinguish this inconspicuous sedge; it lacks the dark female scales of *C. concinna* and *C. concinnoides*. High-elevation plants with smaller, short-beaked perigynia have been separated out as *C. deflexa* Hornem. or *C. brevipes* Boott, but the variation appears continuous rather than discrete.

Carex rostrata Stokes. Similar to *C. utriculata* but stem cross section nearly round; leaves mostly 2-4 mm wide, with inrolled edges, pale-waxy on the upper surface.

Rare in montane fens; known only from southeast of Polebridge. Circumboreal south to ID, MT. See *C. utriculata*.

Carex rupestris Allioni, Curly Sedge. Stems 4-15 cm high, loosely aggregated on shallow rhizomes; leaves mostly basal, 1-3 mm wide, curled; spike solitary, 5-12 mm long, bisexual with male flowers above; perigynia few, ca. 4 mm long, green below, brown above, narrowly elliptic with a short, indistinct beak; female scales brown with pale margins, as long as the perigynia; stigmas 3.

Uncommon in gravelly, often calcareous soil of grasslands or low turf, often on exposed slopes or ridges, montane to alpine; East. Circumboreal south to UT, CO. The curled, relatively broad leaves help distinguish this sedge from other small, single-spike species.

Carex sartwellii Dewey. Stems 30-100 cm high from long rhizomes; leaves 2-5 mm wide, lacking at the base of the stem; spikes 5-10 mm long, sessile, numerous in a congested inflorescence 3-5 cm long, bisexual with male flowers below; perigynia 2-4 mm long, green to light brown, ovate, tapered to a short, indistinct beak; female scales light brown with pale margins, as long as the perigynia; stigmas 2.

Marshes and along streams in the montane zone; rare East, uncommon West. Alta. to Ont., NY, south to CO, MO. *C. cusickii* and *C. diandra* have similar inflorescences, but form tussocks without rhizomes.

Carex saxatilis L. [*C. physocarpa* Presl.]. Stems 20-50 cm tall from long rhizomes; leaves 2-4 mm wide; spikes unisexual, the uppermost male, the lower 1-3 female, 10-25 mm long, short-stalked, well separated, the lowest one often drooping, subtended by leaf-like bracts; perigynia tan to brown, 3-5 mm long, erect, elliptic with a short beak; female scales dark with a light midvein, shorter than the perigynia; stigmas 2.

Uncommon in wet, gravelly soil on lake shores and in seep areas, montane to alpine; East, West. Circumboreal south to WA, UT, CO. Our plants are var. *major* Olney. Standley's (1921) report of *C. miliaris* Michx. may be referable here (Hermann 1956).

Carex scirpoidea Michx. Stems with only leaf sheaths at the base, 10-30 cm high from short rhizomes; leaves 1-3 mm wide; spike solitary, narrowly cylindric, 10-25 mm long, unisexual; plants either male or female; perigynia erect, 2-3 mm long, hairy, flattened, ovate with a short beak and short stalk; female scales hairy, dark brown with pale margins, longer than the perigynia; stigmas 3. Fig. 283.

Common in moist subalpine and alpine grassland and turf, less common lower; East, West. Our plants are ssp. *scirpoidea*. B.C., Alta. south to CA, UT, CO. Reports of ssp. *stenochlaena* (Holm) Löve & Löve were not confirmed by Dunlop and Crow (1999).

Carex siccata Dewey. Stems 15-40 cm high from long rhizomes; leaves 1-3 mm wide, lacking from the very base of the stem; spikes 6-12, sessile, up to 1 cm long, loosely aggregated, unisexual or bisexual with the male flowers below; perigynia 4-6 mm long, ovate, tapered to a long beak with flattened margins; female scales brown with a pale midvein and silvery margins, shorter than the perigynia; stigmas 2.

Rare in montane grasslands in the North Fork Flathead drainage. Alta. to ME south to WA, AZ.

Carex simulata Mack. Similar to *C. praegracilis*, but stems 20-60 cm tall; spikes 8-25, unisexual or bisexual with male flowers above; perigynia dark brown, shiny, ca. 2 mm long, ovate, with a short beak; female scales longer than the perigynia.

Rare in montane fens; known from the Belly River drainage. WA to Sask. south to CA, UT.

Carex spectabilis Dewey. Stems 30-60 cm tall, loosely clumped; leaves 2-5 mm wide, lower ones lacking blades; spikes unisexual, the uppermost male,

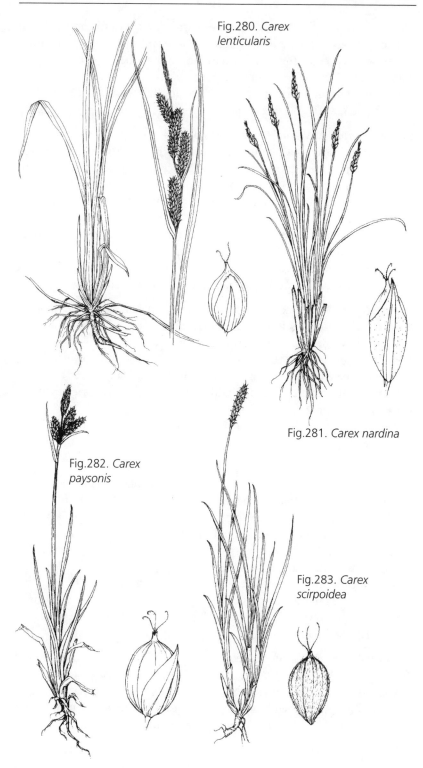

Fig.280. *Carex lenticularis*

Fig.281. *Carex nardina*

Fig.282. *Carex paysonis*

Fig.283. *Carex scirpoidea*

2-4 female below, short-stalked, well-separated, erect to spreading, 12-30 mm long; perigynia 4-5 mm long, green and purplish, flattened, narrowly elliptic with a pointed tip; female scales shorter than the perigynia, blackish with a pointed tip and pale midvein; stigmas 3.

Common in subalpine and alpine moist meadows and turf; East, West. AK south to CA, ID, WY. See *C. paysonis.*

Carex stenophylla Wahl. [*C. eleocharis* Bailey]. Stems 5-20 cm high from long rhizomes; leaves ca. 1 mm wide, V-shaped in cross section, all near the stem base; spikes sessile, ca. 5 mm long, closely aggregated, bisexual with male flowers above; perigynia brown, ca. 3 mm long, ovoid with a short, indistinct beak; female scales as long as the perigynia, brown with a pointed tip,; stigmas 2.

Rare in dry, sparsely vegetated areas of montane grasslands; known only from near St. Mary. Our plants are ssp. *eleocharis* (Bailey) Hult. Circumboreal south to CA, AZ, IA. *C. praegracilis* is usually taller and found in moister habitats; *C. heliophila* has more widely spaced spikes.

Carex stipata Muhl. ex Willd. Stems 30-60 cm tall, strongly 3-angled, forming dense tussocks; leaves 5-11 mm wide with sheaths cross-corrugated opposite the blade; spikes small, sessile, very numerous, densely aggregated in a branched head 3-10 cm long, bisexual with male flowers above; perigynia green to tan, 4-5 mm long, narrowly triangular, tapered to a long, flattened beak; female scales tan with a green midvein, shorter than the perigynia; stigmas 2.

Uncommon in moist soil along montane lakes, ponds and streams; West. AK to Newf. south to CA, NM, FL; Japan. *C. cusickii* has smaller perigynia and lacks cross-corrugated leaf sheaths.

Carex tahoensis Smiley [*C. xerantica* Bailey misapplied]. Stems hollow, to 30 cm high, forming tussocks; leaves 1-2 mm wide; spikes 6-12 mm long, erect, well separated, bisexual with male flowers at the base; perigynia brown, 4-6 mm long, narrowly lance-shaped with thin wing-margins, tapered to a flattened beak ca. 1 mm long; female scales white-margined, covering the perigynium; stigmas 2.

Uncommon in montane grasslands at the southeast edge of the Park; Alta. to CA. Material from Glacier Park and elsewhere in the Rocky Mountains that keys to *C. xerantica* in Hermann (1970) was determined to be *C. tahoensis* by A. A. Reznicek.

Carex tenuiflora Wahl. Stems weak, 20-60 cm high, forming loose tussocks from slender rhizomes; leaves ca. 1 mm wide; spikes 2-4, aggregated into a head 5-10 mm long, bisexual with male flowers below; perigynia pale green, 3-4 mm long, elliptic with an indistinct beak; female scales pale with a green midvein, shorter than the perigynia; stigmas 2.

Rare on hummocks in sphagnum fens; known only from northeast of Polebridge. Circumboreal south to B.C., MT, Sask. *C. disperma* and *C. leptalea* also have lax stems and beakless perigynia and occur in fens, but the former has widely separated spikes, while the latter has only 1 spike.

Carex utriculata Boott, Beaked Sedge. Stems 50-100 cm tall from long rhizomes, cross section 3-angled; leaves mostly 5-12 mm wide, green, not waxy on the upper surface; spikes unisexual, 2-4 male above, 2-5 female below, cylindric, 3-8 cm long, lower ones stalked, well separated, subtended by leaf-like bracts; perigynia 3-8 mm long, tan, shiny, ovoid, tapering to a stout beak; female scales light brown, narrower and shorter than the perigynia or sometimes with long point; stigmas 3. Fig. 284.

Abundant in wet soil to standing water of marshes, fens, swamps, and around streams and ponds, montane to near treeline; East, West. Circumboreal south to CA, NM, NE, IN. This is our most common coarse sedge. It has usually been called *C. rostrata*, but this name correctly applies to a similar but much less common species confined to northern fens and bogs.

Carex vesicaria L. [*C. exsiccata* Bailey]. Stems 30-90 cm high forming large tussocks; leaves 3-7 mm wide; spikes unisexual, 2-4 males above, the 1-3 females below are nearly sessile, 2-5 cm long, separate, erect, subtended by leaf-like bracts; perigynia 6-8 mm long, nearly erect, reddish to tan, ovoid, inflated, gradually tapered to a long beak; female scales smaller than the perigynia, brown with a paler center; stigmas 3. Color plate 17.

Var. *major* Boott [=*C. exsiccata*], with perigynia that taper gradually into the indistinct beak, is common in montane and lower subalpine marshes and wet meadows, often on lake shores and other areas that become dry in late summer; East, West. Var. *vesicaria*, with perigynia distinctly contracted to the beak, is uncommon in montane fens; West. Circumboreal south to CA, NM, IN. Perigynia of *C. utriculata* have a more distinct beak and are more nearly horizontal. The similar *C. lacustris* Willd., with a membranous ligule 12-40 mm long where the leaf blade joins the sheath, occurs just south of the Park.

Carex viridula Michx. [*C. oederi* Retz. var. *viridula* (Michx.) Kuekenth.]. Stems 10-30 cm tall, forming small tussocks; leaves 1-2 mm wide; spikes unisexual, the terminal male, the 2-4 females below mostly sessile, 5-12 mm long, somewhat aggregated, subtended by leaf-like bracts; perigynia spreading, green to tan, 2-3 mm long, narrowly ovate with a distinct beak; female scales light brown with a green midvein; stigmas 3.

Uncommon in montane fens and wet, gravelly, calcareous soil of shores and stream banks; East, West. AK to Newf, south to CA, CO, NJ. See *C. flava* with which it often grows.

Dulichium Pers.

Dulichium arundinaceum (L.) Britt. Strongly rhizomatous; stems 20-70 cm high; leaves without blades on lower stem, those of upper stem with short blades 2-8 mm wide; flowers bisexual, each subtended by 6-9 brown bristles and a scale that encloses them; spikelets cylindric, 10-25 mm long, ca. 6 flowers each; spikes with opposing rows of 7-10 spikelets, borne in axils of upper leaves; stigmas 2.

Locally common in montane fens of the McDonald Valley. B.C. to Newf. south to CA, ID, MT, MN, FL. Easily recognized by the bladeless leaves below and short, spreading leaves above.

Eleocharis R. Br., Spike Rush

Rhizomatous perennials; stems nearly round in cross section; leaves reduced to bladeless sheaths; spikelets solitary on stem tips, composed of spirally arranged flowers; flowers bisexual, composed of an ovary with 2-3 stigmas, 1-3 stamens and 0-6 bristles, subtended by a scale; achene (seed) with a bulge or crown (tubercle) on top.

Mature achenes are needed for positive identification. *E. rostellata* (Torr.) Torr., with leaves that arch over and root at the tip, occurs in spring-fed fens just south of the Park. See *Trichophorum caespitosum*.

1. Achene ca. 1 mm long; scales ca. 1.5 mm long; stems thread-like .... *E. acicularis*
1. Achene and scales larger 2
2. Spikelets with ca. 5 flowers .... *E. pauciflora*
2. Spikelets with at least 10 flowers 3
3. Stigmas 2; achene 2-sided with a triangular tubercle .... *E. palustris*
3. Stigmas 3, achene 3-sided with a dish-like tubercle .... *E. tenuis*

Eleocharis acicularis (L.) R. & S. Stems slender, 3-8 cm high, forming small tufts; spikelets 2-4 mm long; scales brown with a green middle, ca. 1.5 mm long; bristles 3-4 or lacking; achenes pale, ca. 1 mm long, ovoid with a pointed-disk tubercle.

Common but inconspicuous in shallow water or drying mud at the edge of montane ponds and lakes; East, West. Circumboreal south to CA, Mex., FL.

Eleocharis palustris (L.) R. & S. Stems 10-70 cm tall, little tufted, purple at the base; spikelets 5-20 mm long; scales brown, 2-5 mm long; bristles barbed, 4-6; achenes yellow to brown, 2-sided, ca. 2 mm long with a triangular tubercle. Fig. 285.

Common in soil that is wet early in the season to shallow water of montane lakes, ponds, marshes, fens, and wet meadows; East, West. Circumboreal south through U. S. This is the common conspicuous, low-elevation spike rush.

Eleocharis pauciflora (Lightf.) Link [*E. quinqueflora* (Hartm.) Schwarz]. Stems 10-25 cm tall, forming small tufts; spikelets 4-8 mm long with ca. 5 flowers; scales brown, 2-6 mm long; achene brown, 3-sided, 2-3 mm long with a beak-like tubercle.

Moist to wet, often peaty soil of fens, seeps, and along meltwater streams; rare montane, uncommon alpine; East, West. Circumboreal south to CA, NM, IL. Standley's (1921) report of *E. tenuis* is probably referable here.

Eleocharis tenuis (Willd.) Schultes. Stems 5-30 cm high, little tufted; spikelets 2-7 mm long, 10- to 30-flowered; scales 2-3 mm long, brown with pale margins; achene yellow, 3-sided, 1-2 mm long; tubercle dish-like with a central beak.

Uncommon in more sparsely vegetated areas of montane fens; West. Our plants are var. *borealis* (Svenson) Gleason. B.C. to Que. south to MT.

Eriophorum L., Cotton Grass

Rhizomatous (ours) perennials; leaves grass-like; spikelets 1-several, composed of spirally arranged flowers, appearing like a cottony tuft 2-4 cm high in fruit; flowers bisexual, with an ovary with 3 stigmas, 1-3 stamens and 20-30 long bristles, subtended by a scale; achenes 3-sided often with a short beak.

All of our species occur in montane to lower subalpine fens; the presence of one rather than another in any particular fen is enigmatic. See *Trichophorum alpinum.*

1. Spikelet solitary at stem tip .... *E. chamissonis*
1. Spikelets > 1 2
2. Inflorescence subtended by 1 leaf-like bract.................................... *E. gracile*
2. Inflorescence subtended by > 1 leaf-like bract 3
3. Midvein of scales disappearing well before the tip .... *E. polystachion*
3. Midvein prominent, extending to tip of scale .... *E. viridicarinatum*

Eriophorum chamissonis C. A. Meyer. Stems 25-50 cm high; upper leaves bladeless; lower leaf blades 1-2 mm wide; spikelet solitary, scales dark; bristles white, tan toward the base. Fig. 286.

Locally common; East, West. Circumboreal south to OR, WY, MN. This is our only cotton grass with solitary spikelets.

Eriophorum gracile Koch. Stems slender, 25-60 cm tall; leaves up to 2 mm wide, V-shaped in cross section; spikelets 2-5, stalked, subtended by 1 short erect, leaf-like bract; scales greenish-black with a midrib not reaching the tip; bristles white.

Rare; West. Circumboreal south to CA, CO, PA.

Eriophorum polystachion L. [*E. angustifolium* Roth]. Stems 15-50 cm tall; leaves flat, 2-5 mm wide; spikelets 2-8, nodding on stalks, subtended by at

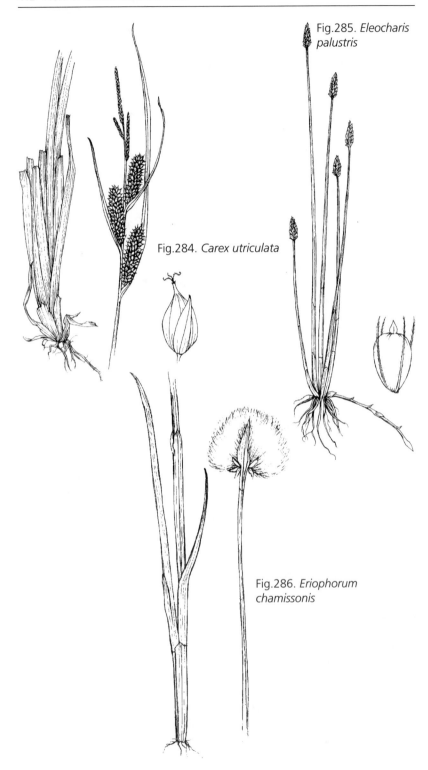

Fig.285. *Eleocharis palustris*

Fig.284. *Carex utriculata*

Fig.286. *Eriophorum chamissonis*

least 2 leaf-like bracts; scales brown to green-black with a midrib not reaching the pale tip; bristles white.

Uncommon; East, West. Circumboreal south to OR, UT, NM.

Eriophorum viridicarinatum (Engelm.) Fern. Stems 25-60 cm high; leaves 2-5 mm wide; spikelets 2-10, nodding on stalks, subtended by several leaf-like bracts; scales black-green with a pale midvein that reaches the tip; bristles white or pale brown.

Rare; East, West. AK to Newf. south to B.C., ID, CO.

Kobresia Willd.

Plants forming tussocks or mats; leaves grass-like; flowers unisexual; male flowers of 3 anthers; female flowers of an ovary with 3 stigmas partly wrapped in a membranous scale (perigynium); 1 female and ≥ 1 male flower clustered behind a scale, these spikelets spirally arranged in spikes; fruit a 3-sided seed (achene).

Very similar in general appearance to *Carex* spp.; the open perigynium is diagnostic.

1. Inflorescence a solitary, linear spike .... *K. myosuroides*
1. Inflorescence of several aggregated spikes .... *K. simpliciuscula*

Kobresia myosuroides (Vill.) Fiori & Paol. [*K. bellardi* (All.) Degl.]. Stems 5-15 cm tall with old brown leaf sheaths at the base; leaves ca. 0.5 mm wide, erect; spikelets in a solitary linear spike 8-15 mm long, the upper male, the lower bisexual; scales brown, 2-4 mm long; perigynia light brown, 3-4 mm long; achenes 2-3 mm long, brown, narrowly ovoid with a small beak.

Uncommon in moist to sometimes dry, often exposed, alpine turf; East. Circumpolar south to CA, NM. Very similar in habitat and general appearance to *Carex nardina* and *C. elynoides*.

Kobresia simpliciuscula (Wahl.) Mack. Stems 5-20 cm high with old leaves at the base; leaves ca. 1 mm wide; spikelets in many sessile, aggregated spikes ca. 5 mm long, terminal one male, lower ones bisexual; scales brown; perigynia ca. 2.5 mm long; achenes ca. 3 mm long, brown, narrowly lance-shaped with a short beak.

Rare in permanently moist, organic soil of alpine turf on cool, gentle slopes; East. Circumboreal south to CO.

Schoenoplectus (Rchb.) Palla

Perennial; leaves grass-like or lacking; flowers bisexual with 2-3 stamens and 1-6 bristles at the base of the ovary, each subtended by a scale and spirally arranged in spikelets; inflorescence terminal, appearing lateral but actually subtended by a stem-like bract; fruit a beaked seed (achene) that is 2-sided when there are 2 stigmas and 3-sided when there are 3 stigmas. Reference: Strong (1994).

1. Emergent; stems > 3 mm wide; inflorescence branched .... *S. acutus*
1. Plants aquatic; stems < 3 mm wide; spikelet solitary .... *S. subterminalis*

Schoenoplectus acutus (Muhl ex Bigelow) A. & D. Löve, Hardstem Bulrush, Tule. [*Scirpus acutus* Muhl. ex Bigelow, *S. occidentalis* (Wats.) Chase]. Stems 50-200 cm tall, round in cross section, from stout rhizomes; leaves all basal, blades reduced to membranous scales; spikelets many, 7-10 mm long, stalked clusters in a branched inflorescence; bract short, erect and stem-like; scales ca. 4 mm long, short-hairy, brown with dark red spots, pointed at the tip; achene 2-sided, dark brown, shiny, 2-3 mm long. Fig. 287.

Uncommon around montane ponds and lakes; East, West. Throughout temperate N. America. This plant is abundant in more arid regions of western N. America. The roots were eaten by Native Americans.

Schoenoplectus subterminalis (Torr.) Sojak [*Scirpus subterminalis* Torr.]. Plants rhizomatous, aquatic; stems lax, round in cross section, 20-80 cm long; leaves narrow, floating; spikelet solitary, 7-12 mm long, subtended by an erect bract longer than the spikelet; scales membranous, light brown, 4-6 mm long; achene 3-sided, 2-4 mm long; bristles white, minutely barbed.

Rare in shallow water of lakes and ponds; Maguire (1939) reported sterile plants in Lake McDonald. AK to Newf. south to OR, ID, MT. Stems float on the water rather than emerging.

Scirpus L., Bulrush, Clubrush

Scirpus microcarpus Presl. Perennial with 3-angled stems 30-80 cm high from rhizomes; leaves grass-like, flat, 6-10 mm wide; flowers bisexual with 2-3 stamens and 1-6 white bristles at the base of the ovary, each subtended by a scale and spirally arranged in spikelets; spikelets 3-6 mm long, sessile in numerous, small, stalked clusters in an open, hemispheric inflorescence subtended by 2-4 leaf-like bracts; scales dark, marked with green; achenes 2-sided, ca. 1 mm long; bristles white. Fig. 288.

Uncommon in montane marshes and along streams; East, West. AK to Newf. south to CA, AZ, NM, WV. The open inflorescence with numerous spikelets is diagnostic. Many species often placed in *Scirpus* are herein placed in *Schoenoplectus* and *Trichophorum*.

Trichophorum Persoon

Plants perennial, tussock-forming; leaves grass-like; flowers bisexual with 3 stamens and 1-6 bristles at the base of the ovary, each subtended by a scale and spirally arranged in a solitary, terminal spikelet subtended by an inconspicuous scale-like bract; fruit a beaked, 3-sided seed (achene) with 3 stigmas. Reference: Hultén (1968).

T. pumilum (Vahl) Schinz & Thell., similar to *T. caespitosum* but much smaller, occurs just south and east of the Park.

Trichophorum alpinum (L.) Pers. [*Eriophorum alpinum* L., *Scirpus hudsonianus* (Michx.) Fern.]. Stems 3-angled, 10-40 cm high, loosely clustered on short rhizomes; leaves ca. 1 cm long, mainly basal; spikelet solitary on the stem tip, 5-7 mm long; bract beaked, shorter than the spikelet; scales brown; achene 3-sided, ca. 1.5 mm long; bristles white, long, exserted in fruit.

Rare in wet organic soil; known from near Apgar and the Many Glacier area. Circumboreal south to B.C., MT. True cotton grasses (*Eriophorum*) have > 6 bristles per flower.

Trichophorum caespitosum (L.) Hartm. [*Scirpus cespitosus* L.]. Stems round in cross section, 15-40 cm high with old stems at the base, forming large, dense tussocks; leaves solitary, basal, small; spikelet solitary on the stem tip, 2-5 mm long with 2-4 flowers; bract beaked, as long as spikelet; scales brown; achene 3-sided, ca. 1.5 mm long. Fig. 289.

Rare in wet organic soil; known from north of Apgar and and alpine slopes near Logan Pass. Circumboreal south to OR, ID, UT, MT. This plant may appear to be an *Eleocharis*, but for the small bract subtending the spikelet.

HYDROCHARITACEAE: WATERWEED FAMILY

Elodea Michx., Waterweed
Elodea canadensis Richardson. Submerged, aquatic perennial; leaves opposite below but in whorls of 3 above, strap-shaped, 6-15 mm long, sessile with finely toothed margins; flowers unisexual on separate plants, floating, on long stalks from leaf axils; petals 3, white, 4-5 mm long; sepals 3, 2-4 mm long, united below into a slender tube to 15 cm long; stamens 9; stigmas 3; fruit a capsule ca. 6 mm long. July. Fig. 290.

Rare in shallow to deep water of montane lakes; known only from St. Mary Lake. B.C. to Que. south to CA, UT, CO, OK, AL.

IRIDACEAE: IRIS FAMILY

Perennial herbs; leaves folded in half lengthwise and enfolding next higher leaf at the base (equitant); flowers bisexual, stalked, subtended by cylindrical, leaf-like bracts; sepals 3, petal-like; petals 3; ovary below base of sepals (inferior), stamens 3; style 1; fruit a 3-chambered capsule.

1. Leaves ≥ 5 mm wide; flowers ca. 6 cm long; sepals and petals unlike .... **Iris**
1. Leaves < 5 mm wide; flowers ca. 1 cm long; sepals and petals alike ....
.... **Sisyrinchium**

Iris L., Iris, Flag
Iris missouriensis Nutt., Blue Flag. Stems 20-50 cm high from rhizomes; leaves basal, 20-40 cm long, sword-shaped; flowers 2-4; sepals spreading,

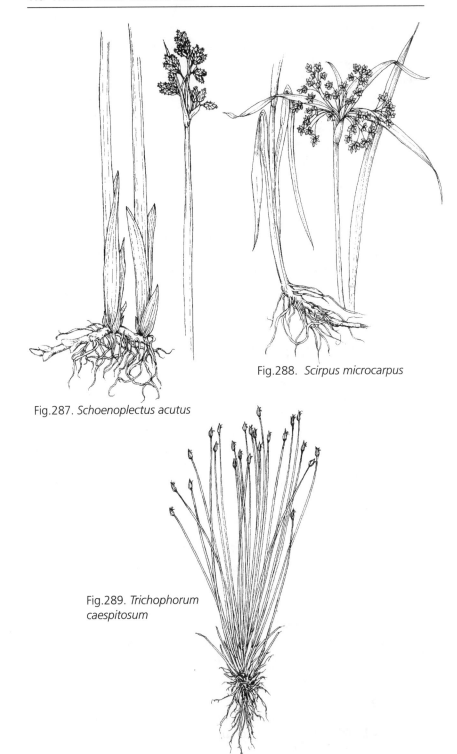

Fig.287. *Schoenoplectus acutus*

Fig.288. *Scirpus microcarpus*

Fig.289. *Trichophorum caespitosum*

5-6 cm long, oblong, whitish with a yellow spot and purple lines; petals blue, erect, shorter than the sepals; styles 3, petal-like, 2-4 cm long; capsule ellipsoid, 2-4 cm long. June. Fig. 291.

Rare in moist, montane meadows and open aspen forest along the east margin of the Park. B.C., Alta. south to CA, NM, SD. Blue flag increases under grazing because it is poisonous.

Sisyrinchium L., Blue-eyed Grass

Sisyrinchium montanum Greene [*S. angustifolium* Mill., *S. mucronatum* Michx., *S. sarmentosum* Suskd. ex Greene misapplied]. Stems 10-50 cm high, flattened and thin-edged, forming small clumps; leaves linear, 1-4 mm wide; flowers 2-5, blue, erect or spreading; sepals and petals alike, 6-12 mm long, capsule globose 3-6 mm wide. June-July. Fig. 292.

Moist grassland and meadows; common montane, uncommon higher; East, West. Yuk. to Newf. south to AZ, CO, IN, NY.

JUNCACEAE: RUSH FAMILY

Mostly perennial grass-like herbs; flowers lily-like, bisexual; petals 3, sepals 3, lance-shaped, undifferentiated (tepals); 1 or 2 small bracts sometimes present at the base of the tepals; stamens mostly 6; ovary superior with 3 stigmas; fruit a globose to lance-shaped capsule. Reference: Brooks and Clemants (2000).

1. Leaves glabrous; capsule with many seeds .... **Juncus**
1. Leaves with at least some long hairs; capsule 3-seeded .... **Luzula**

Juncus L., Rush

Leaf sheaths open with overlapping margins; leaves sometimes tubular with membranous crosswalls; flowers in a terminal inflorescence; inner tepals often shorter; capsules with numerous seeds sometimes with tail-like appendages. Reference: Hermann (1975).

Months given below are those when mature capsules are present; these are needed for positive identification.

1. Plants fibrous-rooted annual with flowers in the axils of leaves as well as in terminal clusters .... ***J. bufonius***
1. Plants perennial without flowers in leaf axils 2
2. Plants alpine, to 8 cm high; leaves all basal, tubular 3
2. Plants not as above 4
3. Flowers 1-2, exceeded by subtending, leaf-like bract; capsules indented on top ....***J. biglumis***
3. Flowers 2-4, as long as subtending, papery bracts; capsules rounded on top ....***J. triglumis***
4. Stems without leaves or leaves reduced to papery sheaths 5
4. Stem leaves present and conspicuous 7

5. Stems with 1-4 flowers .... *J. drummondii*
5. Stems with > 4 flowers 6

6. Inflorescence apparently attached near middle of stem (stem-like portion above inflorescence is actually considered a bract) .... *J. filiformis*
6. Inflorescence apparently attached on upper third of the stem .... *J. balticus*

7. Leaves iris-like, each folded in half and enfolding those above at the base (equitant) .... *J. ensifolius*
7. Leaves not equitant 8

8. Leaf blades round in cross section and divided crosswise into sections (visible when held up to light) 9
8. Leaf blades flat or v-shaped in cross section 13

9. Stems with a single head-like cluster of flowers; subalpine to alpine .... *J. mertensianus*
9. Most stems with > 1 head; habitat usually lower 10

10. Flowers in clusters of 3-12 11
10. Most flower heads with > 12 flowers each 12

11. Petal-like tepals mostly > 3 mm long .... *J. nevadensis*
11. Tepals 2-3 mm long .... *J. alpinoarticulatus*

12. Capsules rounded to the small pointed tip; anthers as long or longer than anther stalks .... *J. nevadensis*
12. Capsules tapered to a long point; anthers shorter than anther stalks .... *J. nodosus*

13. Each flower with a pair of small bracts at the base in addition to a single bract at the base of the flower stalk 14
13. Each flower with only 1 bract at the base of the flower or flower stalk ... 16

14. Flowers 2-7 per stem .... *J. parryi*
14. Flowers usually > 7 per stem 15

15. Inflorescence compact, flower clusters short-branched .... *J. confusus*
15. Inflorescence open, some flower clusters well separated from the others .... *J. tenuis*

16. Seeds without tails .... *J. longistylis*
16. Seeds with tails as long as the body 17

17. Petal-like tepals uniformly brown .... *J. castaneus*
17. Tepals with a broad green mid-stripe .... *J. regelii*

Juncus alpinoarticulatus Chaix [*J. alpinus* Vill.]. Rhizomatous; stems 10-35 cm high, forming tufts; stem leaves 1-3, tubular; inflorescence open with clusters of 3-6 flowers at tips and forks of erect branches; tepals brown, 2-3 mm long with bluntly pointed tips; capsules longer than tepals, rounded on top with a short nipple; seeds with minute appendages. July-Sept.

Common in wet, gravelly soil of montane and lower subalpine stream banks and around ponds and fens; East, West. AK to Newf. south to WA, UT, CO, MN. Reports of *J. articulatus* L. are referable here.

Juncus balticus Willd., Baltic Rush, Wiregrass [*J. arcticus* Willd. var. *balticus* (Willd.) Traut.]. Rhizomatous; stems 15-80 cm tall, arising singly; stem leaf blades lacking; inflorescence branched, appearing to arise from the side of the upper stem, subtended by an erect stem-like bract; tepals brown, lance-shaped, 4-5 mm long, subtended by 2 small bracts; capsules ovoid with a short nipple, smaller than the tepals; seeds with minute appendages. July-Sept.

Common in montane and lower subalpine wet meadows, fens, low areas in grasslands, and along streams; East, West. Our plants are var. *montanus* Engelm. Circumboreal south to much of N. America and S. America.

Juncus biglumis L. Stems 2-10 cm high, forming small tufts; leaves basal, tubular; flowers mostly 2, sessile, in a terminal cluster, subtended by a leaf-like bract slightly higher than the flowers; tepals brown, 3-4 mm long; capsules larger than the tepals, indented on top; seeds with a minute appendage. July-Aug.

Local and uncommon in wet, cold, shallow soil of alpine seeps, rock ledges, and along streams; East, and along the Continental Divide. Circumpolar south to CO.

Juncus bufonius L., Toad Rush. Annual with stems 3-20 cm high, branched at the base; leaves flat or inrolled, to 1 mm wide; flowers sessile, 1-3 per cluster along ascending branches of the inflorescence; tepals narrow, tapered, 4-6 mm long, green with white margins, subtended by 2 small bracts; capsules shorter than the tepals, rounded on top. July-Aug.

Local and uncommon in drying mud of montane meadows and along streams and ponds; East, West. Circumboreal south to most of N. America. Our only annual rush.

Juncus castaneus J. E. Smith. Stems 5-30 cm high, arising singly from rhizomes; leaves with inrolled margins; 4-9 flowers clustered in 1-3 stalked heads subtended by a short leaf-like bract; tepals brown, narrow, tapered, 5-6 mm long; capsule long-pointed, longer than the tepals; seeds with long appendages. July-Aug.

Uncommon in alpine seeps, wet rock ledges, and along streams; East. Circumpolar south to NM.

Juncus confusus Coville. Stems 15-40 cm high, forming tussocks; leaves flat, < 1 mm wide; flowers in small, short-stalked clusters in a compact inflorescence greatly exceeded by the leaf-like bract; tepals 3-4 mm long, lance-shaped with green middle and clear margins, subtended by 2 small bracts; capsule shorter than the tepals, indented on top. July-Sept.

Common in moist, often disturbed soil of montane and lower subalpine meadows, open forest, rock outcrops, and along trails; East, West. B.C. to Sask. south to CA, CO. Many reports of the similar *J. tenuis* Willd. are referable here.

Juncus drummondii Mey. Stems 10-30 cm high, forming tussocks; stem leaf blades lacking; flowers 1-5, short-stalked in a small, terminal inflorescence, subtended by an erect stem-like bract; tepals narrow, 4-7 mm long, long-pointed, green with brown margins, subtended by 2 small bracts; capsule equal to or longer than the tepals; indented on top; seeds with long appendages. July-Sept.

Var. *drummondii*, with mature capsules ca. as long as the tepals, is abundant in moist subalpine and alpine meadows, open forest, and turf, often along streams or where snow accumulates; var. *subtriflorus* (Mey.) Hitchc., with capsules > 1 mm longer than the tepals, occurs near or above treeline around Logan Pass and Granite Park; East, West. AK to Alta. south to CA, NM. Often found with *Luzula hitchcockii* and *Sibbaldia procumbens*. See *J. parryi*.

Juncus ensifolius Wikst. [*J. saximontanus* A. Nels., *J. tracyi* Rydb.]. Rhizomatous, stems 15-60 cm high, arising singly; leaves folded flat in half, iris-like, 2-6 mm wide; 5-many sessile flowers clustered in several globose, long-stalked heads; tepals brown, long-pointed, 2-4 mm long, subtended by 1 small bract; stamens 3 or 6; capsule ca. as long as the tepals, rounded on top. July-Aug.

Abundant in moist to wet soil around seeps, ponds, lakes, and along streams and ditches, montane to subalpine; East, West. AK to Que. south to CA, NM, TX. Var. *ensifolius* has 3 stamens and dark brown heads; var. *montanus* (Engelm.) Hitchc. is more common and has 6 stamens and lighter heads. *J. tracyi* is now considered a form of var. *montanus* having seeds with long appendages (Brooks and Clemants 2000).

Juncus filiformis L. Rhizomatous, stems 20-60 cm high, loosely clustered; stem leaf blades lacking; 7-15 short-stalked flowers in a compact inflorescence subtended by an erect stem-like bract, appearing to arise from the middle of the stem; tepals green, 3-5 mm long, long-pointed, subtended by 2 small bracts; capsule rounded on top, shorter than the tepals. July-Aug.

Uncommon around montane lakes and streams, common in fens; East, West. Circumboreal south to OR, UT, WY, MI, PA. The inflorescence appearing to arise from the middle of the stem is diagnostic. The plant is most common in or near the McDonald Valley.

Juncus longistylis Torrey. Stems 20-60 cm tall, clustered on rhizomes; leaves flat, 1-3 mm wide; 3-8 flowers in 1-5 short-stalked, hemispheric clusters subtended by a short, papery bract; tepals 5-6 mm long, brown with a green middle and silvery margins; capsule shorter than the tepals, indented on top. July.

Uncommon on the margins of montane ponds, streams, and wetlands; East, West. B.C. to Ont. south to CA, NM, NE. The silvery-margined tepals help distinguish this rush.

Juncus mertensianus Bong. Stems 10-30 cm high, forming small tussocks; stem leaves tubular, 1-2 mm wide; flowers in a solitary, terminal, globose cluster exceeded by the subtending leaf-like bract; tepals brown, 3-5 mm long, narrow; capsule just shorter than the tepals, rounded on top. July-Sept. Fig. 293.

Abundant in permanently wet soil along subalpine and alpine streams, lakes and seeps, less common lower; East, West. AK to Alta. south to CA, NM; Asia. A faithful inhabitant of rocky soil along high-elevation rivulets where the single dark flower head is diagnostic.

Juncus nevadensis Wats. Rhizomatous, stems 10-40 cm high; stem leaves tubular; flowers in several small, hemispheric clusters well separated on erect branches of the inflorescence; tepals brown, long-pointed, 3-4 mm long; capsule as long as the tepals, narrowly rounded on top. July-Aug.

Rare along gravelly shores, collected once near Many Glacier. Our plants are var. *badius* (Suksd.) Hitchc. B.C., Alta. south to CA, NM.

Juncus nodosus L. Rhizomatous; stems 15-40 cm high, arising singly; stem leaves tubular; flowers numerous in several stalked, hemispheric clusters exceeded by the subtending leaf-like bract; tepals 3-4 mm long, light brown, subtended by 1 papery bract; capsule long-tapered, longer than the tepals. July-Aug.

Uncommon in wet meadows and wet, gravelly soil along montane streams, ponds and lakes; East, West. B.C. to Newf. south to CA, NM, NE, PA.

Juncus parryi Engelm. Stems 10-30 cm tall, forming tussocks; only 1 stem leaf with a tubular blade; inflorescence of 1-3 stalked flowers, exceeded by the subtending, erect stem-like bract; tepals 5-6 mm long, long-pointed with clear margins, subtended by 2 small bracts; capsule narrow, as long as the tepals. Aug.

Uncommon in stony soil of subalpine and alpine meadows, often near streams or shores; East, West. B.C. to Alta. south to CA, CO. The more common *J. drummondii* lacks any stem leaf blades.

Juncus regelii Buch. Rhizomatous; stems 20-40 cm high, loosely clustered; leaves flat, 1-3 mm wide; inflorescence of 1-3 loose, many-flowered clusters, subtended by a short leaf-like bract; tepals 4-6 mm long, brown with a green middle; capsule rounded on top, ca. as long as the tepals; seeds with long appendages. July-Sept.

Uncommon in montane and lower subalpine seeps, marshes and wet meadows; East. B.C. to MT south to CA, UT, WY.

Juncus tenuis Willd. Similar to *J. confusus*; leaves mostly basal, flat, ca. 1 mm wide; flowers in a branched inflorescence 4-6 cm long; tepals 3-5 mm long.

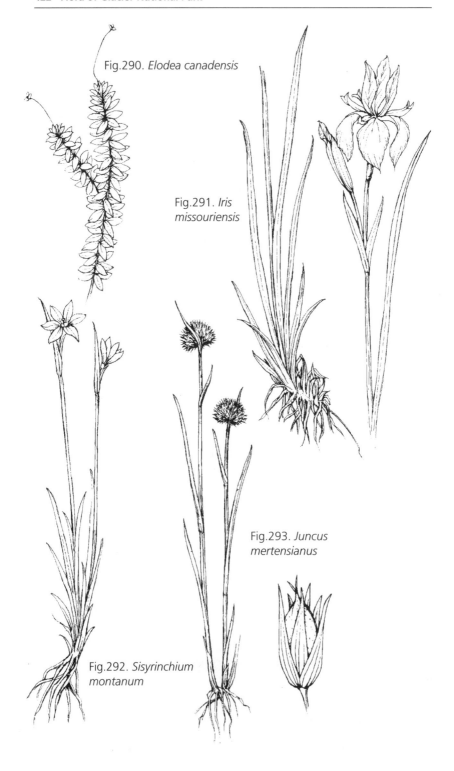

Fig.290. *Elodea canadensis*

Fig.291. *Iris missouriensis*

Fig.293. *Juncus mertensianus*

Fig.292. *Sisyrinchium montanum*

Sparsely vegetated soil of montane grassland and open forest, especially along roads and trails; collected once at West Glacier. Throughout much of N. America, Eurasia. More common in the Great Plains and east, perhaps introduced in the Park.

Juncus triglumis L. [*J. albescens* (Lange) Fern.]. Stems 5-20 cm high, forming small tufts; leaves basal, tubular; flowers usually 2-4 in a small terminal head subtended by 2 papery bracts; tepals 3-5 mm long, tan or whitish; capsule rounded on top, as long as the tepals; seeds with appendages ca. as long as the body. Aug.

Local and uncommon in permanently wet, shallow, gravelly soil along alpine meltwater streams and seeps along the Continental Divide. Our plants are var. *albescens* Lange. Circumpolar south to UT, CO.

Luzula D. C., Wood Rush

Leaves with a closed sheath and flat blades often fringed with long hairs; flowers solitary or clustered, subtended by 2 small, papery bracts; inflorescence open or spike-like; tepals brown; stamens 6; capsule 3-seeded. Reference: Swab (2000).

1. Flowers ≥ 3 in cylindical or globose clusters 2
1. Flowers 1-2 on tips of branches in an open inflorescence 4

2. Inflorescence a single cylindrical drooping spike .... *L. spicata*
2. Inflorescence of ≥ 2 globose or cylindrical clusters 3

3. Inflorescence with clusters of 3-6 flowers; alpine .... *L. arcuata*
3. Clusters with 8-15 flowers; montane .... *L. multiflora*

4. Capsules > 2.5 mm long with a beak nearly 1 mm long .... *L. hitchcockii*
4. Capsules < 2.5 mm long with a minute beak 5

5. Plants < 20 cm tall; stem leaves 2-3, rarely > 5 mm wide .... *L. piperi*
5. Plants > 20 cm tall; some plants with > 3 stem leaves, some > 5 mm wide ...
.... *L. parviflora*

Luzula arcuata (Wahl.) Sw. Stems 5-15 cm high, forming small tufts; leaves 2-3 mm wide, the basal often purplish; inflorescence of few arching and stalked clusters of 3-6 flowers; tepals 2-3 mm long; capsule as long as the tepals. Aug.

Rare in moist alpine turf; collected twice along the Continental Divide. Our plants are ssp. *unalaschkensis* (Buchan.) Hult. Circumpolar south to WA, MT.

Luzula hitchcockii Hamet-Ahti [*L. glabrata* (Hoppe) Desv. misapplied, *Juncoides glabratum* (Hoppe) Sheld.]. Stems 10-30 cm high from rhizomes; leaves 5-6 mm wide, often turning golden brown; flowers 1-2 on the spreading branch ends of an open inflorescence; tepals 2-3 mm long; capsule barely longer than the tepals. June-Aug. Fig. 294.

Abundant in open subalpine forest and meadows, often where snow lies late; East, West. B.C., Alta. south to OR, ID, MT. Often with *Xerophyllum tenax, Juncus drummondii* and *Vaccinium* spp. The honey-colored leaves help distinguish this plant later in the summer. See *L. piperi*.

Luzula multiflora (Ehrh.) LeJ. [*Juncoides campestre* (L.) Kuntze, *L. campestris* (L.) D. C., *L. comosa* Mey.]. Stems 10-40 cm high, forming tussocks; leaves 2-6 mm wide; inflorescence of several stalked, short-cylindric, many-flowered clusters; tepals ca. 3 mm long; capsule ca. as long as the tepals. June-July.

Common in dry, montane forest openings; West. Our plants are ssp. *multiflora*. AK to CA and much of northern U. S.

Luzula parviflora (Ehrh.) Desv. [*Juncoides parviflorum* Ehrh.]. Stems 20-50 cm high, forming tussocks; stem leaves 3-10 mm wide, mostly glabrous; flowers 1-2 on the ends of arching branches in an open inflorescence; tepals ca. 2 mm long; capsule slightly longer than the tepals. June-Aug.

Common in subalpine meadows, uncommon in montane and subalpine forest; East, West. Ssp. *fastigiata* (Mey.) Hamet-Ahti occurs at higher elevations than does ssp. parviflora. Circumboreal south to CA, NM.

Luzula piperi (Cov.) Jones [*Juncoides piperi* Cov., *L. wahlenbergii* Rupr. misapplied]. Stems 10-20 cm high, forming tussocks; stem leaves 2-3, 2-3 mm wide; flowers single on the ends of arching branches in an open inflorescence; tepals ca. 2 mm long; capsule slightly longer than the tepals. July-Aug.

Common in subalpine and low alpine meadows; East, West. AK and Que. south to WA, MT. *L. hitchcockii* has larger flowers.

Luzula spicata (L.) D. C. [*Juncoides spicatum* (L.) Kuntze]. Stems 10-30 cm high, forming small tufts; leaves 1-3 mm wide; flowers short-stalked, numerous in a terminal, nodding, cylindical cluster 1-3 cm long; tepals 2-3 mm long; capsule shorter than the tepals. July-Aug.

Common in alpine and high subalpine turf and grasslands; East, West. Circumpolar south to CA, CO. The arching, solitary spike inflorescence is diagnostic.

JUNCAGINACEAE: ARROW-GRASS FAMILY

Triglochin L., Arrow-grass
Triglochin palustris L. Perennial herbs with short rhizomes; Stems 15-40 cm high; leaves basal, linear, succulent, shorter than the stem; flowers bisexual, short-stalked in a narrow spike-like inflorescence; petals and sepals (tepals) 3 each, alike, greenish, 1-2 mm long, quickly falling; stamens 6; ovary 3-chambered with 3 stigmas; fruit erect, narrowly club-shaped, 5-7 mm long. July. Fig. 295.

Rare in wet, alkaline, often organic soils of montane fens and shallow wetlands; known only from near Lake McDonald and the Belly River. Circumboreal south to CA, NM, IA, NY; S. America.

LEMNACEAE: DUCKWEED FAMILY

Lemna L., Duckweed

Minute, frond-like, floating aquatic plants with 1 root beneath (ours); flowers unisexual, minute and rarely visible. Plants reproduce mainly by vegetative division. Circumboreal south through most of N. America.

1. Fronds oval, 2-5 mm long .... *L. minor*
1. Fronds shaped like a canoe paddle, 6-10 mm long .... *L. trisulca*

Lemna minor L. [*L. turionifera* Landolt]. Frond oval, 2-5 mm long; solitary or in small clusters; root 1-12 mm long. Fig. 296.

Rare in permanent water of montane ponds and lakes; collected near Polebridge and East Glacier. Waterfowl may start populations that are later extinguished by a hard winter. Our plants have been referred to *L. turionifera* Landolt based on microscopic characters.

Lemna trisulca L. Fronds shaped like canoe paddles, 6-10 mm long, connected together; root solitary or absent.

Rare in permanent water of montane ponds and lakes; collected near Polebridge and East Glacier.

LILIACEAE: LILY FAMILY

Herbaceous perennials; leaves alternate with entire margins; flowers bisexual, radially symmetrical; sepals and petals 3 each, separate, often identical and called tepals; stamens usually 6; ovary above the base of the sepals (superior) with 1 style and 3 stigmas; fruit a berry or 3-chambered capsule.

A report of *Lloydia serotina* (L.) Sw. (Bamberg and Major 1968) has not been verified, though it occurs nearby to the south. *Lilium philadelphicum* L., with large orange flowers, occurs just north of the Park and should be sought in the Belly River country.

1. Leaves lance-shaped, elliptic or ovate, < 8 times as long as wide 2
1. Leaves linear, grass-like, > 8 times as long as wide 8

2. Flowers mostly 1-3 per stem; tepals > 15 mm long 3
2. Flowers usually > 3 per stem; tepals < 15 mm long 5

3. Leaves 3 in a whorl at the top of the stem .... **Trillium**
3. Leaves basal or nearly so 4

4. Tepals hairy, ca. 20 mm long; fruit a blue berry .... **Clintonia**
4. Tepals glabrous, > 25 mm long; fruit a capsule .... **Erythronium**

5. Flowers stalked, 1-3 in the axils of leaves or on branch tips 6
5. Flowers ≥ 3 in a terminal inflorescence 7

6. Flowers at the tips of branches .... **Prosartes**
6. Flowers in leaf axils .... **Streptopus**

7. Plants > 80 cm high; tepals green, ≥ 8 mm long; fruit a capsule .... **Veratrum**
7. Plants < 60 cm high; tepals 2-5 mm long; fruit a berry .... **Smilacina**

8. Flowers 1-4 9
8. Flowers > 4 10

9. Flower yellow, turning brick-red; fruit not lobed .... **Fritillaria**
9. Flower white to purple; fruit 3-lobed .... **Calochortus**

10. Flowers stalked, all arising from the stem tip, forming a flat-topped or
    hemispheric inflorescence (umbel); foliage onion-scented .... **Allium**
10. Inflorescence elongate 11

11. Tepals 2-4 mm long; leaves folded lengthwise .... **Tofieldia**
11. Tepals ≥ 6 mm long; leaves not strongly folded 12

12. Basal leaves wiry, numerous, forming tussocks .... **Xerophyllum**
12. Basal leaves few, not tussock-forming 13

13. Tepals white with a basal spot; capsule with a pointed beak ...... **Zigadenus**
13. Tepals red-green to blue; capsule rounded on top 14

14. Tepals blue, 2-3 cm long; moist montane meadows .... **Camassia**
14. Tepals reddish-green, 10-15 mm long; forests, cliffs, turf .... **Stenanthium**

Allium L., Onion

Foliage and underground bulbs onion-scented; bulbs often clustered; leaves linear, flat or tubular, arising near stem base; flowers cup-shaped, stalked from the stem tip in a flat-topped or hemispheric cluster (umbel), subtended by 1-3 sheathing, membranous bracts; tepals in 2 whorls of 3; fruit an ovoid, 3- or 6-seeded capsule.

Wild onions were used by Native Americans in cooking.

1. Flowers rose-purple, ≥ 8 mm long; wet areas .... *A. schoenoprasum*
1. Flowers white or pink, ≤ 8 mm long; grassland to open forest 2

2. Umbel nodding in flower .... *A. cernuum*
2. Umbel erect 3

3. Umbel with small bulbs at the base of the flower stalks .... *A. geyeri*
3. Umbel without bulbs 4

4. Outer bulb coat reticulate-fibrous; east of the Divide .... *A. textile*
4. Outer bulb coat membranous; southwest of Marias Pass .... *A. fibrillum*

Allium cernuum Roth, Nodding Onion. Bulbs slender with a membranous outer coat; stem 10-40 cm high; leaves several, 2-4 mm wide, u-shaped in cross section, shorter than the stem; umbel nodding; tepals 4-6 mm long, pink; stamens exserted; capsule with 6 horns on top. June-Aug. Fig. 297. Color plate 25.

Meadows, grasslands and open forest (especially aspen); common montane, less common subalpine; East, West. B.C. to Ont. south to OR, AZ, Mex., GA. The distinctive nodding umbel becomes more erect in fruit.

Allium fibrillum Jones ex Abrams. Bulbs ovoid with a membranous outer coat; stem 3-15 cm high; leaves 2, 1-3 mm wide, u-shaped in cross section, as long as the stem; tepals 5-6 mm long, white with green midveins; stamens shorter than tepals; capsule nearly round on top. May-June.

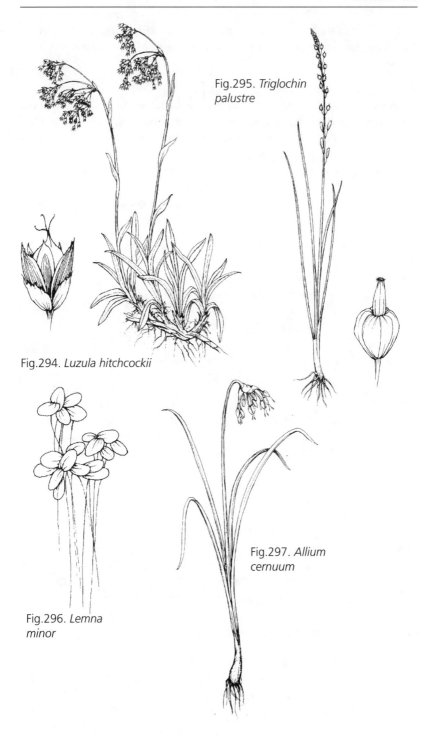

Fig.295. *Triglochin palustre*

Fig.294. *Luzula hitchcockii*

Fig.297. *Allium cernuum*

Fig.296. *Lemna minor*

Locally common in shallow, vernally moist soil of montane to lower subalpine grasslands and rock outcrops southwest of Marias Pass. WA, OR, ID, MT.

Allium geyeri Wats. [*A. fibrosum* Rydb.]. Bulbs narrowly ovoid with a fibrous outer coat; stem 20-40 cm high; leaves 3-5, u-shaped in cross section, 1-3 mm wide, shorter than the stem; umbels with numerous small bulbs at the base; tepals 6-8 mm long, pink; stamens shorter than the tepals; capsule nearly round on top. June.

Uncommon in montane grasslands, meadows, and aspen groves; East. Our plants are var. *tenerum* Jones. B.C., Alta. south to AZ, NM, TX. Montana plants have 5 sets of chromosomes and are sexually sterile, reproducing only by the vegetatively produced bulbs in the inflorescence.

Allium schoenoprasum L., Chives [*A. sibiricum* L.]. Bulbs narrowly ovoid with a membranous outer coat; stems 20-60 cm tall; leaves 1-7 mm wide, tubular, hollow, shorter than the stem; tepals 8-14 mm long, bright purple; stamens shorter than the tepals; capsule rounded on top. July-Aug. Fig. 298.

Wet meadows, especially along streams and lakes; abundant subalpine, common montane; East, West. Our plants are var. *sibiricum* (L.) Hartm. Circumboreal south to OR, CO, MN, NY. This is the wild progenitor of the chives of commerce.

Allium textile Nels. & Macbr. [*A. nuttallii* Wats.]. Bulbs ovoid with a fibrous outer coat; stem 5-25 cm high; leaves 2, u-shaped in cross section, 1-3 mm wide, as long as the stem; tepals 5-7 mm long, white with a brown midvein; stamens shorter than the tepals; capsules with 6 rounded knobs on top. June-July.

Uncommon in montane grasslands along the east margin of the Park. Alta. to Man. south to NV, NM, MN.

Asparagus L., Asparagus

Asparagus officinalis L. Stems branched, 1-2 m tall from rhizomes; leaves scale-like, subtending 1-several small branchlets, 8-15 mm long, that function as leaves; flowers bell-shaped, pendant, 1-2 in leaf axils; tepals 3-5 mm long, greenish-white; fruit a red berry 6-8 mm wide. June.

Rare in disturbed meadows and along roads; collected once near West Glacier. A garden escape introduced from Europe.

Calochortus Pursh, Mariposa, Sego Lily

Stems unbranched from bulbs; foliage glabrous, leaves few, shorter than the stem (ours); flowers large, bowl-shaped, few, stalked, subtended by leaf-like bracts; sepals narrow, green; petals wedge-shaped with a narrow base, colorful, hairy on the inside; stamens shorter than the petals; ovary and capsule 3-angled.

The size of petals varies greatly among years, possibly depending on soil moisture.

1. Leaf solitary, basal; flowers white .... *C. apiculatus*
1. Stem with several leaves; flowers purplish .... *C. macrocarpus*

Calochortus apiculatus Baker. Stem 10-30 cm high; leaf solitary, basal, flat, 5-18 mm wide; petals white with a dark, hairy, circular spot near the base; stamens long-pointed; capsule nodding. June-Aug. Fig. 299.

Common in montane and lower subalpine grasslands, drier meadows and forest openings; East, West. B.C., Alta. south to WA, ID, MT. The bulbs were eaten by Native Americans.

Calochortus macrocarpus Dougl. Stem 20-50 cm tall; leaves linear, u-shaped in cross section; petals light purple, 3-4 cm long, hairy and light-colored around the dark triangular spot near the base; stamens narrow. July.

Rare in montane grasslands; known only from the lower slopes of the Apgar Mountains. B.C. to MT south to CA, NV. The large purple flowers are striking.

Camassia Lindl., Camas

Camassia quamash (Pursh) Greene [*Quamasia quamash* (Pursh) Cov.]. Stems 30-60 cm tall from bulbs; foliage glabrous; leaves linear, 5-15 mm wide, basal; flowers star-shaped, light to dark blue, stalked in a narrow terminal inflorescence that elongates with maturity; tepals narrow, 1-2 cm long; fruit an ellipsoidal, 3-angled capsule, 10-15 mm long. June-July. Fig. 300.

Common in deep soil of moist montane meadows and grasslands; East, West. Our plants are var. *quamash*. B.C., Alta. south to CA, UT, WY. Native Americans gathered the bulbs after flowering and roasted them for a long time before eating.

Clintonia Raf., Bead Lily

Clintonia uniflora (Schult.) Kunth. Stems 6-15 cm tall from extensive rhizomes; leaves few, short-petiolate, the blade 8-15 cm long, narrowly elliptic, hairy beneath; flower white, bell-shaped, solitary at the stem tip; tepals lance-shaped, hairy, 1-2 cm long; fruit a several-seeded, lustrous blue berry, ca. 1 cm long. June-July. Fig. 301.

Moist montane and lower subalpine coniferous forest; common East, abundant West. AK to Alta. south to CA, ID, MT. The solitary blue berry is distinctive.

Erythronium L., Glacier Lily

Erythronium grandiflorum Pursh. Foliage glabrous; stems 10-40 cm high from a fleshy, elongate root; leaves 2, basal, short-petiolate, narrowly elliptic,

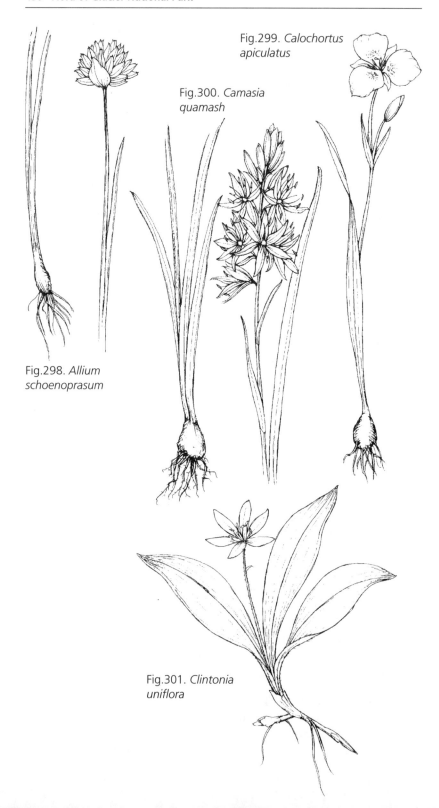

Fig.299. *Calochortus apiculatus*

Fig.300. *Camasia quamash*

Fig.298. *Allium schoenoprasum*

Fig.301. *Clintonia uniflora*

fleshy, 10-20 cm long; flowers solitary or few, nodding; tepals narrowly lance-shaped, 25-35 mm long; anthers red or yellow; fruit an erect, oblong ovoid, 3-lobed, many-seeded capsule 3-6 cm long. May-Aug. Fig. 302. Color plate 50.

Var. *grandiflorum*, with yellow flowers, is abundant in montane and subalpine open forest and moist subalpine meadows; East, West; var. *candidum* (Piper) Abrams, with white flowers, occurs in open montane forest between Polebridge and Apgar. Fritz-Sheridan (1988) presented evidence that the 2 taxa are separate species. B.C., Alta. south to CA, UT, and CO. Flowers soon after snow release. Tepals extend forward over the anthers and stigmas at night and on rainy days, but reflex backward when it is sunny. Native Americans occasionally ate the roots, and they are an important staple for many grizzly bears that turn over a great deal of sod harvesting them.

Fritillaria L., Fritillary

Fritillaria pudica (Pursh) Spreng., Yellow Bell. Stems 10-25 cm high from a bulb; leaves few, strap-shaped, 3-10 cm long, clustered near mid-stem; flower yellow, bell-shaped, usually solitary, nodding on the stem tip; tepals oblong, overlapping, 12-20 mm long; capsule oblong-ovoid, ca. 2 cm long, erect, not lobed. May. Fig. 303.

Rare in montane grasslands; East, West. B.C. to Alta. south to CA, UT, WY. Native Americans ate the bulbs. Tepals turn reddish-brown with age.

Prosartes D. Don, Fairy-bell

Rhizomatous, stems leafy, branched, 30-80 cm high; leaves sessile, ovate with pointed tips; flowers stalked, few, drooping on branch tips; tepals narrow, white or yellowish, 8-18 mm long; stamens exserted; fruit a red berry. Our species formerly placed in *Disporum*. Reference: Shinwari et al. (1994).

1. Fruit rounded on top with a bumpy surface; leaves glabrous above ....
.... *P. trachycarpa*
1. Fruit with a point on top and a smooth surface; leaves sparsely hairy above ..
.... *P. hookeri*

Prosartes hookeri Torr. [*Disporum reganum* (Wats.) Mill., *D. hookeri* (Torr.) Britt.]. Stems short-hairy above; leaves 5-15 cm long, hairy below, above and on the margins; ovary smooth, sometimes hairy; berry 7-9 mm long with a small point on top. June-July.

Uncommon in moist, montane coniferous forest; West and the Waterton Valley. B.C., Alta. south to OR, ID, MT.

Prosartes trachycarpa Wats. [*Disporum trachycarpum* (Wats.) Benth. & Hook.]. Stems short-hairy; leaves 4-12 cm long, hairy below and on margins but glabrous above; ovary minutely bumpy, glabrous; berry 7-10 mm long, rounded on top. May-June. Fig. 303.

Abundant in moist montane forest and thickets; East, West. B.C. to Ont. south to OR, AZ, CO, NE. Our common fairy-bell.

Smilacina Desf., False Solomon's-seal, Solomon's Plume

Rhizomatous; leaves sessile; flowers small, white; tepals narrow, spreading; ovary globose; fruit a few-seeded berry.

1. Flowers short-stalked in a branched inflorescence; ripe berries red; tepals 1-3 mm long .... **S. racemosa**
1. Flowers stalked in a narrow inflorescence; berries green and dark striped; tepals 3-5 mm long .... **S. stellata**

Smilacina racemosa (L.) Desf. [*Vagnera racemosa* (L.) Morong, *Maianthemum racemosum* (L.) Link]. Stem 20-70 cm high; foliage short-hairy; leaves narrowly elliptic, 7-15 cm long; flowers congested in a branched, pyramidal inflorescence, 5-10 cm long; tepals 1-3 mm long; stamens longer than the tepals; berry red, 5-7 mm long. May-July. Color plate 27.

Common in montane open forest, thickets and avalanche slopes; East, West. AK to N. S. south to CA, AZ, CO, MO, GA. Our more common forest species.

Smilacina stellata (L.) Desf. [*Vagnera amplexicaulis* (Nutt.) Morong, *V. stellata* (L.) Morong, *V. sessiliflora* (Nutt.) Greene, *Maianthemum stellatum* (L.) Link]. Stems 10-40 cm tall, often zigzag above; leaves lance-shaped, 3-10 cm long, often folded, hairy below; flowers stalked in a narrow, open inflorescence, 1-5 cm long; tepals 3-5 mm long; stamens shorter than the tepals; berry green with dark stripes, 7-10 mm long. June. Fig. 305.

Common in montane grasslands, moist meadows and open forest, especially aspen; East, West. AK to Newf. south to most of U. S. The green and dark striped berries are distinctive. Native Americans used powdered root to staunch bleeding.

Stenanthium Gray

Stenanthium occidentale Gray, Bronze Bells. Stem 10-50 cm tall from a long bulb; foliage glabrous; leaves linear, 16-20 cm long, mostly basal; flowers green to reddish-brown, bell-shaped, nodding, stalked in a narrow, open inflorescence; tepals 8-10 mm long, narrow with spreading tips; fruit a 3-beaked capsule, 15-20 mm long. July-Aug. Fig. 306.

Common in moist, shallow soil of coniferous forest and cool cliffs; montane to alpine; East, West. B.C. to Alta. south to CA, ID, MT. Inconspicuous due to the dull flowers and few narrow leaves.

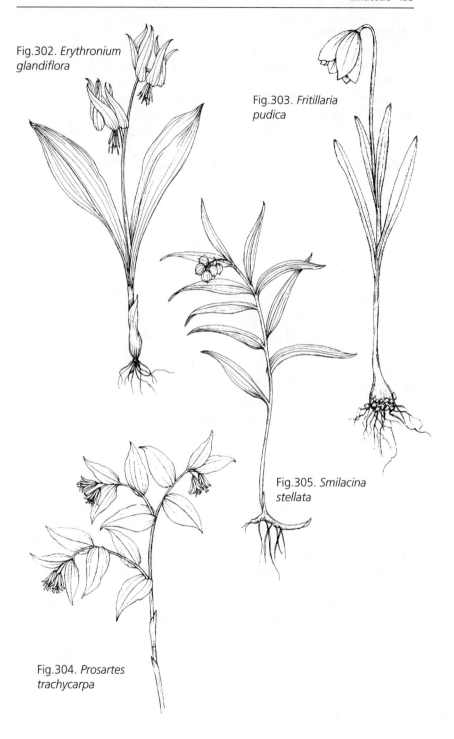

Fig.302. *Erythronium glandiflora*

Fig.303. *Fritillaria pudica*

Fig.305. *Smilacina stellata*

Fig.304. *Prosartes trachycarpa*

Streptopus Michx.

Streptopus amplexifolius (L.) D. C., Twisted-Stalk. Stems branched, 30-100 cm high from rhizomes; leaves 4-10 cm long, ovate with pointed tips, clasping the stem; flowers bell-shaped, yellowish, 1-2, pendant from leaf axils on bent stalks; tepals narrow, ca. 1 cm long, spreading at the tip; fruit a yellow or red berry, 10-12 mm long. July. Fig. 307.

Common in moist montane and lower subalpine forest and thickets, especially along streams; East, West. Our plants are var. *chalazatus* Fassett. AK to Newf. south to CA, NM, MN, VT. The bent flower stalks are distinctive.

Tofieldia Huds., False Asphodel

Short-rhizomatous; leaves narrow, sheathing the stem, mostly basal; flowers small, white, short-stalked in a dense, terminal cluster; tepals narrow, ca. as long as the stamens; fruit a globose capsule.

1. Stems glandular, > 10 cm tall; capsules 5-6 mm long .... ***T. occidentalis***
1. Stems glabrous, < 10 cm long; capsules 2-3 mm long .... ***T. pusilla***

Tofieldia occidentalis Wats. [*Tofieldia intermedia* Rydb., *T. glutinosa* (Michx.) Pers. in part, *Triantha occidentalis* (Wats.) Gates ssp. *montana* (Hitchc.) Packer]. Stems 10-40 cm high, glandular-hairy above; leaves trough-shaped, 3-25 cm long; tepals ca. 4 mm long; capsule 5-6 mm long. July-Aug. Fig. 308.

Common in wet meadows and fens, especially around lakes and streams; subalpine and alpine, uncommon lower; East, West. AK to CA, WY.

Tofieldia pusilla (Michx.) Pers. Stem 3-10 cm high; leaves 1-3 cm long, iris-like, folded along the midvein and enfolding each other at the base; tepals ca. 2 mm long; capsule 2-3 mm long. July-Aug.

Very local in permanently moist, often shallow soil of alpine turf; East. Circumpolar south to B.C., MT. Usually obscured by taller grasses and sedges, often with *Kobresia simpliciuscula*.

Trillium L., Trillium, Wake-robin

Trillium ovatum Pursh. Stems 10-40 cm tall from a short rhizome; leaves 3, 4-12 cm long, broadly ovate with pointed tips, whorled at stem tip just below the solitary, stalked flower; sepals 3, narrow; petals white (fading pink), 25-40 mm long; fruit a broadly ovoid capsule, 3-ridged. May. Fig. 309.

Common in moist montane forest; West. B.C., Alta. south to CA, CO. Most common at the lowest elevations.

Veratrum L., False Hellebore, Corn Lily

Veratrum viride Ait. [*V. eschscholtzii* (R.& S.) Gray]. Stems leafy, 1-2 m tall from thick rhizomes; leaves 10-30 cm long, sessile, elliptic, pleated, twisted;

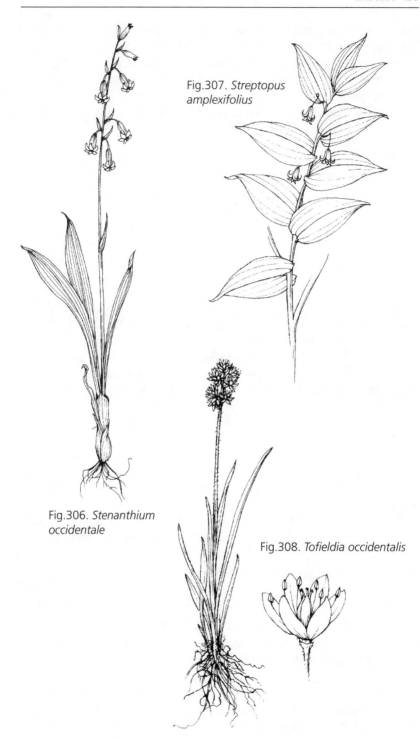

Fig.307. *Streptopus amplexifolius*

Fig.306. *Stenanthium occidentale*

Fig.308. *Tofieldia occidentalis*

flowers green, short-stalked, numerous, crowded in a long, narrow, inflorescence with many ascending branches; tepals 6-9 mm long, lance-shaped, spreading; fruit an ovoid capsule 1-2 cm long. July-Aug. Fig. 310.

Common in moist, open forest, meadows and avalanche slopes, montane and subalpine; East, West. AK to Alta. south to CA, ID, MT. Plants resemble corn stalks. Native Americans used the poisonous root to treat headache and to commit suicide.

Xerophyllum Michx.

Xerophyllum tenax (Pursh) Nutt., Beargrass. Stems leafy, 50-120 cm high from a short, thick, often branched rhizome; basal leaves linear, wiry, 15-60 cm long, forming large tussocks; stem leaves reduced upward; flowers white on long, ascending stalks in an inflorescence that is hemispheric at first but elongates to as much as 50 cm as more flowers come into bloom; tepals ca. 6 mm long, oblong, spreading, shorter than the stamens; fruit a globose capsule, 5-7 mm long. June-Aug. Fig. 311. Color plate 39 and cover photograph.

Coniferous forest, avalanche slopes and meadows; abundant subalpine, common montane; East, West. B.C., Alta. south to CA, ID, MT. Particulary common on subalpine slopes where fire has reduced tree cover. Well known for periodic mass flowering and one of the Park's most spectacular wildflower displays. Plants often have several leaf rosettes, each of which dies back after flowering. Native Americans used the root to treat wounds and other injuries.

Zigadenus Michx., Death Camas

Stems erect from bulbs; foliage glabrous; leaves linear, mainly basal, reduced on the stem; flowers white, stalked in a narrow, unbranched inflorescence; tepals spoon-shaped with a spot near the base; fruit a 3-lobed, pointed-ovoid capsule.

The bulbs of these plants are poisonous and may be mistaken for true camas (*Camassia*) when they occur together and the above-ground parts have withered.

1. Flowers not overlapping; tepals 7-10 mm long; capsule 15-20 mm long .... *Z. elegans*
1. Flowers crowded; tepals 4-5 mm long; capsule 8-15 mm long *Z. venenosus*

Zigadenus elegans Pursh. Stems 10-60 cm high; flowers widely spaced on spreading stalks in the inflorescence; tepals 7-10 mm long with a green spot near the base; stamens as long as the tepals; capsule 1-2 cm long. July-Aug. Fig. 312.

Moist meadows and open forest as well as dry, stony, calcareous soil of exposed slopes and ridges, montane to alpine; common East, less common West. AK to Man. south to OR, AZ, Mex. The ability to thrive in limestone fellfields as well as wet meadows is unusual.

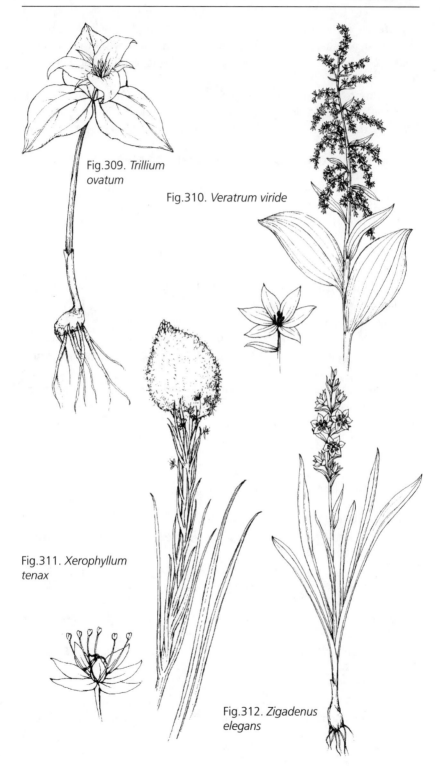

Fig.309. *Trillium ovatum*

Fig.310. *Veratrum viride*

Fig.311. *Xerophyllum tenax*

Fig.312. *Zigadenus elegans*

Zigadenus venenosus Wats. Stems 20-40 cm high; flowers crowded, overlapping on ascending stalks; tepals 4-5 mm long with a yellow spot near the base; stamens a little longer than the tepals; capsule 8-15 mm long. June-July.

Uncommon in montane grassland; East, West. Our plants are var. *gramineus* (Rydb.) Walsh. B.C. to Sask. south to CA, CO, NE. The mashed bulbs were used as a poultice for sprains and bruises by Native Americans.

ORCHIDACEAE: ORCHID FAMILY

Herbaceous perennials; roots often fleshy; leaves with entire margins, often sheathing the stem, alternate or all basal; flowers bisexual, solitary or short-stalked; sepals 3; petals 3, unlike, the lower (lip) often enlarged and modified to form a sac or spur; fertile stamen usually solitary and united with the style; ovary below the base of the sepals (inferior); fruit a capsule with numerous minute seeds. Reference: Luer (1975).

Species of *Coeloglossum, Piperia* and *Platanthera* were formerly placed in *Habenaria.*

1. Stems leafless; leaves all basal or absent entirely 2
1. Leaves on the lower 1/3 of the stem or higher 8

2. Flowers solitary .... **Calypso**
2. Flowers > 1 3

3. Green leaves completely absent; stems yellow to purple .... **Corallorhiza**
3. Green basal leaves present 4

4. Basal leaf solitary (rarely 2)... 5
4. Basal leaves > 1 6

5. Lip lobed and spotted .... **Amerorchis**
5. Lip with entire margins, not spotted .... **Platanthera**

6. Upper stem glandular-hairy; flowers without a down-turned, nectar-holding
   spur .... **Goodyera**
6. Stems glabrous; flowers with a spur 7

7. Leaves opposite, nearly orbicular, appressed to the ground .... **Platanthera**
7. Leaves not opposite not strictly appressed to the ground, narrowly elliptic ....
   .... **Piperia**

8. Leaves 2, opposite each other at mid-stem .... **Listera**
8. Leaves usually > 2, alternate 9

9. Flowers 1-3; lip petal > 12 mm long, pouch-shaped .... **Cypripedium**
9. Flowers mostly > 3; lip petal usually < 12 mm long 10

10. Flowers purplish .... **Epipactis**
10. Flowers white or greenish-white 11

11. Upper stem glandular-hairy... **Goodyera**
11. Stems glabrous 12

12. Flowers without a down-turned, nectar-holding spur, spirally arranged in a
    spike .... **Spiranthes**
12. Flowers with a spur, not spirally arranged 13

13. Lip petal 3-lobed at the tip .... **Coeloglossum**
13. Lip rounded on the tip 14
14. Leaves on lower third of stem only .... **Piperia**
14. Leaves present above mid-stem .... **Platanthera**

Amerorchis Hultén

Amerorchis rotundifolia (Banks ex Pursh) Hult. [*Orchis rotundifolia* Pursh]. Stem leafless, 10-25 cm high; basal leaf solitary, 3-7 cm long, glabrous, broadly elliptic; flowers 2-8, sessile, ca. 15 mm long, subtended by small, narrow, leaf-like bracts; sepals narrowly ovate; lower sepals spreading; upper sepal and narrow upper petals white to rose, forming a hood over the lip; lip petal 6-9 mm long, purple-spotted, oblong, 3-lobed, the middle lobe largest and shallowly indented at the tip, with a curved tubular spur at the base. July.

Rare in montane spruce forest, often near calcareous seeps or fens; known only from the Belly River drainage. AK to Greenl. south to WY, MI, NY.

Calypso Salisb., Fairy Slipper

Calypso bulbosa (L.) Oakes [*Cythera bulbosa* (L.) House]. Leafy stems lacking; basal leaf solitary, petiolate, the blade 3-6 cm long, elliptic; flower solitary on an erect stalk 5-20 cm high; sepals and upper petals alike, 1-2 cm long, lance-shaped, purple; lip petal forming a white and purple spotted pouch, ca. 2 cm long, open at the top, yellow within; capsule club-shaped, ca. 1 cm long. May-June. Fig. 313. Color plate 35.

Locally common in montane coniferous forest; East, West. Our plants are var. *americana* (R. Br.) Luer. Circumboreal south to CA, AZ, CO, MI, NY. Pollinated by bumblebees who search the nectarless flowers in vain. Amount of spotting on the lip is variable.

Coeloglossum Hartm.

Coeloglossum viride (L.) Hartm. [*Habenaria viridis* (L.) R. Br.]. Stems leafy, 15-40 cm high; leaves 3-10 cm long, oblong-elliptic with sheathing bases; inflorescence 2-15 cm long; flowers leafy-bracted, green; lip petal 7-11 mm long, broadly strap-shaped with 3 small lobes at the tip; spur short, sac-like. June-July. Fig. 314.

Rare in open montane forest, especially aspen; East, West. Our plants are var. *virescens* Muhl. Luer. Circumboreal south to WA, NM, IL, NC.

Corallorhiza Chat., Coral-root

Plants purple or yellow, lacking chlorophyll; stems clustered on a coral-like rhizome; leaves reduced to scales; flowers short-stalked, several in a spike-like inflorescence; sepals nearly equal, cupped forward; upper petals shorter but similar to the sepals; lip petal wider, oblong, often with a small lobe on either side at the base; capsules ellipsoid, pendant.

The sole source of nutrition for these orchids is soil-inhabiting fungi that invade the coral-like rhizome and are digested. All species are more common west of the Divide.

1. Stems yellow 2
1. Stems red to purple 3
2. Lip petal 3-5 mm long; capsule ca. 10 mm long............................. **C. trifida**
2. Lip petal 6-8 mm long; capsule 15-20 mm long .... **C. maculata** (albino form)
3. Petals striped; lip petal without small lobes at the base .... **C. striata**
3. Petals spotted or clear; lip with tiny basal lobes 4
4. Lip petal clear or with 1-2 spots .... **C. mertensiana**
4. Lip with several spots .... **C. maculata**

Corallorhiza maculata Raf. Spotted Coral-root [*C. multiflora* Nutt.]. Stem purple, 20-60 cm high; flowers 10-30; sepals and upper petals 6-10 mm long, purple; lip petal white with purple spots and small basal lobes; capsule 15-20 mm long. June-July. Fig. 315.

Locally common in montane coniferous forest; East, West. B.C. to Newf. south to CA, NM, IN, NC. This species parasitizes root rot fungi. A yellow form of this species, sometimes mistaken for *C. trifida,* is common just to the south and may occur in the Park as well.

Corallorhiza mertensiana Bong. Stems reddish, 15-50 cm tall; flowers 10-30; sepals and upper petals pink, spreading, 8-10 mm long; lip petal deep pink with small, narrow, basal lobes; capsule 15-25 mm long. June-July.

Rare in moist, montane, coniferous forest; East, West. AK to CA, ID, WY.

Corallorhiza striata Lindl., Striped Coral-root. Stems purplish, 15-40 cm tall; flowers 7-25; sepals and upper petals cupped forward, 10-16 mm long with 3-4 longitudinal stripes; lip petal 8-12 mm long, white with purple stripes, unlobed, a rounded spur at the base; capsule 12-25 mm long. June. Color plate 32.

Uncommon in montane coniferous forest; East, West. B.C. to Newf. south to CA, UT, Mex. Vegetatively similar to *C. maculata,* but the striped petals are distinctive.

Corallorhiza trifida Chat., Yellow Coral-root. [*C. innata* R. Br.]. Stems pale yellow, 10-25 cm high; sepals and petals 4-6 mm long, yellow to white, cupped forward; lip petal white, 3-5 mm long with indistinct basal lobes; capsule ca. 1 cm long. June-July.

Uncommon in moist montane forest, thickets and under shrubs in fens; East, West. Circumboreal south to CA, ID, CO, IN, NJ. This inconspicuous plant may be more common than records indicate. See *C. maculata.*

Cypripedium L., Lady's-slipper

Stems hairy, leaves several (ours), broadly lance-shaped, pleated, sheathing the stem at the base; flowers 1-3, stalked, large, each subtended by a leaf-like bract; sepals and upper petals similar, spreading; lip petal pouch-like, style column arched over the opening at the top.

In addition to the following species, *C. calceolus* L., the yellow lady's-slipper, occurs south and east of the Park in calcareous fens and wet meadows.

1. Sepals twisted, > 20 mm long .... **C. montanum**
1. Sepals not twisted, 10-15 mm long .... **C. passerinum**

Cypripedium montanum Dougl. ex Lindl., White Lady's-slipper. Stems 20-60 cm high; foliage glandular-hairy; leaves 5-16 cm long; sepals and upper petals brown-purple, narrow, 3-6 cm long, twisted; lip petal white, 2-3 cm long. June-July. Fig. 316.

Rare in moist to dry, montane forest; East, West. AK to CA, WY. Usually in deep, moist soil but sometimes in shallow, limestone-derived soil.

Cypripedium passerinum Richards. Similar to *C. montanum*; sepals 10-15 mm long, green, the uppermost wider; upper petals white ca. 15 mm long; lip petal 12-16 mm long, white with purple mottling. June-July.

Rare in wet, organic soil of wet, montane forests and seeps; East, West. AK to Que. south to MT. The forest-fen ecotone is a common habitat.

Epipactis Sw., Helleborine

Epipactis gigantea Dougl. ex Hook., Giant Helleborine [*Serapias gigantea* (Hook.) Eat.]. Stems 30-80 cm high from rhizomes; foliage hairy, leaves 5-15 cm long, narrowly lance-shaped with sheathing bases; flowers 3-15 in a 1-sided, arching inflorescence; sepals ovate, ca. 15 mm long, similar, reddish green; upper 2 petals similar to sepals, brownish red; lip petal ca. 15 mm long, projecting forward, sac-like at the base, 3-lobed, the middle lobe narrowed in the middle; capsule pendant, 20-35 mm long. July.

Rare around spring-fed fens; known from near Lake McDonald. B.C. to CA, AZ, Mex.

Goodyera R. Br., Rattlesnake Orchid

Plants rhizomatous; leaves all basal, petiolate, mottled with white; foliage glandular-hairy; flowers greenish-white, subtended by a scale-like bract, sessile in a long, narrow, congested spike; lower sepals spreading; upper sepal and upper 2 petals united into a hood over the pouch-like lip petal.

The variegated leaves are diagnostic.

1. Inflorescence ≥ 6 cm long; hood ≥ 5 mm long; leaf midribs white ....
.... **G. oblongifolia**
1. Inflorescence 3-6 cm long; hood < 4 mm long; midrib not white .. **G. repens**

Fig.313. *Calypso bulbosa*

Fig.314. *Coenoglossum viride*

Fig.315. *Corallorhiza maculata*

Fig.316. *Cypripedium montanum*

Goodyera oblongifolia Raf. [*Peramium decipiens* (Hook.) Piper]. Stems 15-40 cm high; leaf blades 3-8 cm long, elliptic; inflorescence 6-12 cm long, spiraled or 1-sided; sepals and petals 6-9 mm long. July-Aug. Fig. 317.

Abundant in montane and lower subalpine coniferous forest; East, West. AK to N. S. south to CA, AZ, NM. Often with *Clintonia*, *Chimaphila* and *Vaccinium* spp. Only a fraction of the plants bloom in any given year.

Goodyera repens (L.) R. Br. Stems 10-20 cm high; leaf blades ovate, 15-30 mm long; inflorescence 3-6 cm long, 1-sided; sepals and petals 3-5 mm long. Aug.

Rare in open coniferous forest; known only from near Upper Kintla Lake. Circumboreal south to AZ, NM, SD, TN, NC.

Listera R. Br., Twayblade

Leaves 2, sessile, opposite near mid-stem; flowers greenish, stalked, in a narrow, open inflorescence; sepals and upper petals similar, spreading, narrowly lance-shaped; lip petal often 2-notched or lobed at the broad tip.

More common west of the Continental Divide.

1. Lip petal divided nearly halfway into pointed, divergent lobes .... **L. cordata**
1. Lip shallowly lobed or indented 2
2. Lip petal blunt at the tip with barely a shallow indentation; common ....
.... **L. caurina**
2. Lip with distinct lobes ca. 1/5 length of the petal 3
3. Lip short-hairy, at least marginally, narrowed at the base . **L. convallarioides**
3. Lip not hairy, base nearly as wide as the tip .... **L. borealis**

Listera borealis Morong. Stems 5-20 cm high; leaves ovate, 2-5 cm long, sparsely hairy; lip petal 7-12 mm long, broadly oblong with 2 rounded lobes, slightly narrowed in the middle. June.

Rare in spruce forest, often at the margin of wetlands; known only from near Apgar. AK to Que. south to WA, UT, CO.

Listera caurina Piper [*Ophrys caurina* (Piper) Rydb.]. Stems 10-30 cm high, glandular-hairy; leaves broadly elliptic, 3-7 cm long; lip petal 4-7 mm long, pear-shaped, the broad tip shallowly indented. July.

Common in moist, montane coniferous forest; East, West. AK to CA, WY. *L. cordata* has a smaller, deeply divided lip.

Listera convallarioides (Sw.) Nutt. [*Ophrys convallarioides* (Sw.) Wight]. Stems 8-25 cm high, glandular-hairy; leaves broadly elliptic, 3-7 cm long; lip petal 8-12 mm long, nearly horizontal, oblong-triangular with rounded lobes at the tip. July.

Moist montane forest or thickets, often on the margins of wetlands; rare East, uncommon West. AK to Newf. south to CA, UT, CO, MI, NY; Asia.

Listera cordata (L.) R. Br. [*Ophrys cordata* L.]. Stems 6-25 cm high; leaves 1-3 cm long, broadly ovate, shallowly lobed and surrounding the stem at the base; lip petal 3-4 mm long, divided into long, pointed, divergent lobes. June-July. Fig. 318.

Common in moist montane and lower subalpine forest; East, West. Circumboreal south to CA, UT, NM, MI, NC. Our most common twayblade. See *L. caurina.*

Piperia Rydb., Rein Orchid

Leaves glabrous, all basal or nearly so; flowers small, greenish or white, sessile, subtended by a small, leaf-like bract, numerous in a narrow spike inflorescence; sepals and petals spreading; lip petal little differentiated from the others, the base prolonged into a downward-pointing, nectar-holding spur longer than the lip.

Found in drier habitats than species of *Platanthera.*

1. Spur 3-5 mm long .... *P. unalascensis*
1. Spur > 7 mm long .... *P. elegans*

Piperia elegans (Lindl.) Rydb. [*Platanthera elegans* Lindl., *Habenaria elegans* (Lindl.) Boland]. Similar to *P. unalascensis*; stems 20-50 cm high; flowers overlapping in the inflorescence; lip petal 4-5 mm long, lance-shaped; spur linear, much longer than the lip. June.

Rare in drier montane forest; known only from the Belton Hills; B.C. to MT south to CA.

Piperia unalascensis (Sprengel) Rydb. [*Habenaria unalascensis* (Sprengel) Wats.]. Stems 20-60 cm high; leaves 6-15 cm long, oblong with sheathing bases; inflorescence 10-20 cm long; flowers well separated below, yellowish green; lip petal 2-5 mm long, ovate; spur club-shaped, barely longer than the lip. July-Aug. Fig. 319.

Moist to often dry montane forest, thickets or stream banks; East, West. AK to Que. south to CA, CO, SD.

Platanthera, Bog Orchid

Foliage glabrous; flowers small, greenish or white, sessile, subtended by a small, leaf-like bract, few to many in a narrow spike inflorescence; lower sepals spreading; upper sepal and upper petals partly grouped together into a hood; lip petal longer, lance-shaped with a rounded or blunt tip, the base prolonged into a downward-pointing, nectar-holding spur.

1. Leaves all basal, usually 1-2; stem leaves absent.. 2
1. Stems leafy 3
2. Leaves 2, opposite, appressed to the ground, nearly round in outline............
.... *P. orbiculata*
2. Leaves 1-3, alternate, oblong to narrowly elliptic .... *P. obtusata*
3. Flowers bright white; lip distinctly broadened at the base .... *P. dilatata*
3. Flowers yellowish or greenish; lip narrow, gradually tapered 4

4. Spur ca. 1/2 as long as lip petal .... *P. stricta*
4. Spur ca. as long as the lip .... *P. hyperborea*

Platanthera dilatata (Pursh) Lindl. ex Beck [*Habenaria dilatata* (Pursh) Hook.]. Stems 20-80 cm high, leafy; leaves 4-16 cm long, broadly lance-shaped with sheathing bases; inflorescence 4-20 cm long; flowers white or greenish; lip petal 5-8 mm long, broad-based with a tongue-shaped tip; spur slender, curved. June-Aug. Fig. 320.

Both var. *dilatata*, with the spur at least as long as the lip, and var. *albiflora* (Cham.) Ledeb., with the spur much shorter than the lip, are common in wet soil of montane and lower subalpine wet meadows and fens and along streams and ditches; East, West. AK to Greenl. south to CA, NM, SD, MI, NY. Our most common bog orchid can often be seen growing in moist roadside ditches along Lake McDonald.

Platanthera hyperborea (L.) Lindl. [*Habenaria hyperborea* (L.) R. Br.] Stems 10-50 cm high, leafy; leaves 6-15 cm long, lance-shaped; inflorescence 2-10 cm long; flowers yellowish-white; lip 4-7 mm long, lance-shaped, tapered to the narrow tip; spur cylindrical, as long as the lip. June-July. Color plate 18.

Uncommon in montane fens and wet meadows; East, West. Our plants are var. *hyperborea*. AK to Greenl. south to CA, NM, NE, IN, NY. *P. stricta* has more elliptic leaves and a shorter spur. Standley's (1921) report of *Habenaria sparsiflora* Wats. is probably referable here.

Platanthera obtusata (Banks ex Pursh) Lindl. [*Habenaria obtusata* (Pursh) Richards]. Stems leafless, 10-20 cm high; basal leaf usually 1, the blade 3-10 cm long, narrowly elliptic; inflorescence 2-8 cm long; flowers few, greenish white; lip 5-8 mm long, strap-shaped with a broad base, reflexed; spur slender, as long as the lip. June.

Rare on the forested margins of montane fens and wet meadows, often beneath spruce; East, West. Circumboreal south to OR, UT, CO, WI, NY.

Platanthera orbiculata (Pursh) Lindl. [*Habenaria orbiculata* (Pursh) Torrey]. Stem leafless; 30-60 cm high; leaves 2, opposite, appressed to the ground, broadly elliptic, 6-16 cm long; inflorescence 5-20 cm long; flowers well separated, greenish white; lip petal white, 10-15 mm long, strap-shaped; spur slender, longer than the lip, reflexed. July-Aug.

Uncommon in moist montane forest; West. AK to Newf. south to OR, MT, IN, GA. Our plants are var. *orbiculata*. The large, fleshy, paired leaves on the ground are diagnostic. Widespread in the right habitat but there are rarely more than a few plants in any one place.

Platanthera stricta Lindl. [*Habenaria saccata* Greene, *H. stricta* (Lindl.) Wats., *P. saccata* (Greene) Hult.]. Stems leafy, 20-80 cm tall; leaves 4-12 cm long, elliptic with sheathing bases; inflorescence 6-25 cm long; flowers

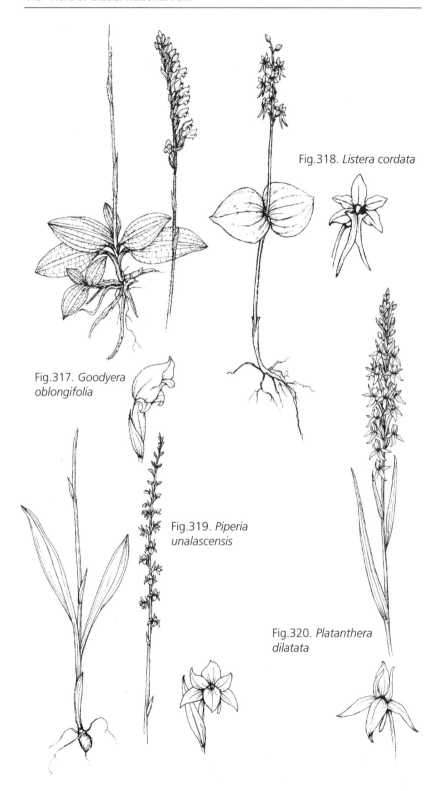

Fig.318. *Listera cordata*

Fig.317. *Goodyera oblongifolia*

Fig.319. *Piperia unalascensis*

Fig.320. *Platanthera dilatata*

overlapping, greenish; lip petal 5-7 mm long, narrowly oblong; spur sac-like, shorter than the lip. June-Aug.

Common in wet meadows, fens, thickets and moist forest openings, montane and lower subalpine; East, West. AK to Alta. south to CA, NM. See *P. hyperborea*.

Spiranthes L. C. Rich., Ladies' Tresses

Spiranthes romanzoffiana Cham. [*Ibidium romanzoffianum* (Cham.) House]. Stem 10-30 cm tall; foliage glabrous; leaves narrowly lance-shaped, sheathing the stem at the base, 5-12 cm long, reduced to mere scales above; flowers white, spirally arranged in the crowded spike inflorescence, 3-8 cm long; sepals and upper petals lance-shaped, cupped forward, forming a hood; lip petal 9-12 mm long, tongue-shaped, constricted in the middle, the wavy-margined tip reflexed down. Aug. Fig. 321. Color plate 20.

Uncommon in montane and subalpine wet meadows and fens; East, West. AK to Newf. south to CA, NM, NE, IA, NY.

POACEAE (GRAMINEAE): GRASS FAMILY

Annual or perennial, herbaceous; stems hollow, usually round in cross section; leaves alternate, the blade strap-shaped with parallel veins, the basal portion (sheath) sheathing the stem and forming a small, hairy or membranous extension (ligule) where it joins the blade; flowers usually bisexual, composed of a canoe-shaped bract (lemma) enfolding at the base another, usually smaller bract (palea), 3 stamens, and an ovary with 2 styles and 2 stigmas; flowers arranged in 2-sided spikelets; spikelets with 2 sterile canoe-like bracts (glumes) at the base, stalked or not; lemmas, paleas and glumes may have a needle-like awn arising from the back or tip; fruit a seed- like grain, often enclosed in the lemma and palea. Reference: Hitchcock (1950).

Grass plants increase vegetatively by forming new tillers at the base. Each tiller is a stem with 1 or more leaves. In species with erect tillers, the plant will form tussocks or bunches of side-by-side stems (bunchgrasses) In other species, some or all of the tillers grow horizontally. These horizontal tillers are rhizomes (under the ground) or stolons (above ground). These rhizomes or stolonsgive rise to erect tillers from their nodes. In rhizomatous or stoloniferous grasses, the plants will be single or small tufts of tillers spaced throughout the surface of the ground (sod) rather than distinct large tussocks as in bunchgrasses. This is the second largest plant family in the Park. Months during which grasses are flowering or have intact spikelets are given in the descriptions.

1. Flowers converted to bulb-like plantlets 2
1. Flowers not converted to bulblets 3
2. Plants of montane disturbed sites .... ***Poa bulbosa***
2. Plants from near or above treeline .... ***Festuca vivipara***

3. Spikelets sessile in 1 or more narrow spikes .... Group A
3. Spikelets stalked in an open to congested inflorescence 4

4. Spikelets of 2 glumes (rarely absent) and 1 fertile floret .... Group B
4. Spikelets with 2 or more fertile florets 5

5. Spikelets with at least 1 glume longer than the lowest lemma .... Group C
5. Lowest lemma longer than both glumes .... Group D

Group A: (Spikelets Sessile)

1. Inflorescence of > 1 spike 2
1. Inflorescence of a solitary, terminal spike 3

2. Spikelets with an awn from the sterile lemma .... **Echinochloa**
2. Glumes and lemmas without awns .... **Beckmannia**

3. Inflorescence dense, separate spikelets not readily discernable, not a true
   spike; short stalks present at the nodes upon close examination 4
3. Spikelets truly sessile, often widely spaced and discernable as separate 6

4. Stem nodes hairy .... **Muhlenbergia**
4. Stem nodes glabrous 5

5. Lemmas awned; glumes unawned .... **Alopecurus**
5. Lemmas unawned; glumes awned .... **Phleum**

6. Annual plants along roads and railroads; leaf blades 1-2 cm wide .... **Triticum**
6. Perennial plants; most leaves < 1 cm wide 7

7. Spikelets aligned with 1 edge adjacent to the axis of the inflorescence ....
   .... **Lolium**
7. Spikelets with the broad face adjacent to the inflorescence axis 8

8. Each inflorescence node with ca. 9 awn-like lemmas and glumes but only 1-
   3 fertile florets 9
8. Nodes with 1-2 spikelets, > 3 fertile florets and fewer sterile awn-like
   lemmas and glumes 10

9. Alpine and subalpine slopes .... **Elymus**
9. Montane moist or disturbed areas .... **Hordeum**

10. Spikelets densely spaced, at least 4 times longer than the distance between
    nodes of the spike, ascending to nearly horizontal .... **Agropyron**
10. Spikelets not so dense, 1-3 times the distance between nodes of the spike,
    ascending to nearly erect .... **Elymus**

Group B: (Spikelets Stalked, 1-Flowered)

1. Inflorescence dense, cylindrical, spike-like, branches not apparent 2
1. Inflorescence narrow or open but not cigar- or cigarette-shaped 4

2. Stem nodes hairy .... **Muhlenbergia**
2. Stem nodes glabrous 3

3. Lemmas awned; glumes unawned .... **Alopecurus**
3. Lemmas unawned; glumes awned .... **Phleum**

4. Mature lemmas bony, harder than the glumes 5
4. Mature lemmas not hardened, as soft as the glumes 7

5. Lemmas without awns .... **Panicum**
5. Lemmas awned 6

6. Awns < 15 mm long .... **Oryzopsis**
6. Awns > 20 mm long .... **Stipa**

7. Both glumes (exclusive of awns) shorter or equal to the lemma ....
.... **Muhlenbergia**
7. At least 1 glume longer than the lemma 8

8. Stiff hairs at the lemma base (callus) at least as long as the lemma ....
.... **Calamagrostis**
8. Callus hairs lacking or very short 9

9. Leaves ≤ 8 mm wide; plants usually < 60 cm high 10
9. Larger leaves > 10 mm wide; plants > 60 cm high 11

10. Spikelets 5-6 mm long .... **Hierochloe**
10. Spikelets < 4 mm long .... **Agrostis**

11. Inflorescence ≥ 10 cm wide, open with spreading branches .... **Cinna**
11. Inflorescence narrow, dense, < 10 cm wide .... **Phalaris**

Group C: Spikelets Stalked, > 1-Flowered, 1 or Both Glumes > Lowest Lemma.

1. Lemmas unawned or merely awn-tipped (≤ 1 mm long) .... **Koeleria**
1. Lemmas with an awn > 1 mm long 2

2. Glumes ≤ 8 mm long 3
2. Glumes ≥ 10 mm long 4

3. Awn arising from at or below the middle of the lemma .... **Deschampsia**
3. Awn absent or arising from between 2 teeth at tip of the lemma ... **Trisetum**

4. Plants annual, growing in human-disturbed habitats .... **Avena**
4. Plants perennial of grasslands, meadows and open forest 5

5. Lemmas hairy on the back, at least along the margins .... **Danthonia**
5. Lemmas glabrous on back with a few hairs at the very base .. **Helictotrichon**

Group D: Spikelets Stalked, > 1-Flowered; Both Glumes < Lowest Lemma.

1. Lemmas with awns 2
1. Lemmas unawned 6

2. Leaf sheaths open at least 1/3 their length, the edges overlapping, not
   forming a closed tube 3
2. Leaf sheaths forming a closed tube like a soda straw to near the top 4

3. Lemma awned from the very tip .... **Festuca**
3. Awn arising from base of two long teeth at the tip of the lemma ... **Trisetum**

4. Awn ca. 1 mm long; inflorescence of congested clusters of spikelets **Dactylis**
4. Awn of lemma > 1 mm long; inflorescence not of clustered spikelets 5

5. Branches of inflorescence spreading, solitary at the nodes of the axis **Melica**
5. Branches of inflorescence spreading to erect, > 1 per node (whorled) ....
.... **Bromus**

6. Spikelets with 2 flowers 7
6. Spikelets with > 2 flowers 8

7. Glumes and lemmas with ragged blunt tips ....**Catabrosa**
7. Glumes and lemmas pointed .... **Koeleria**

8. Lemmas rounded at the tip 9
8. Lemmas sharp-pointed at the tip 10

9. Largest glume with 3 veins; leaf sheaths open, the edges overlapping, not forming a closed tube like a soda straw .... **Puccinellia**
9. Largest glume with 1 vein; leaf sheaths forming a closed tube to near the top .... **Glyceria**

10. At least 1 glume as long or nearly as long as the lowest lemma 11
10. Both glumes definitely shorter than the lowest lemma 12

11. Plants of dry meadows or grasslands .... **Koeleria**
11. Plants of forest or moist to wet meadows .... **Trisetum**

12. Leaf sheaths forming a closed tube like a soda straw to near the top **Melica**
12. Leaf sheaths open at least 1/3 their length, the edges overlapping, not forming a closed tube 13

13. Lemmas 7-8 mm long .... **Festuca**
13. Lemmas ≤ 6 mm long .... **Poa**

Agropyron Gaertn., Wheatgrass

Agropyron cristatum (L.) Gaertn., Crested Wheatgrass [*A. desertorum* (Fisch. ex Link) Schult]. Perennial; stems 40-60 cm tall, tussock-forming; leaves 2-4 mm wide with a short ligule; inflorescence 4-8 cm long; spikelets sessile, 8-12 mm long with 5-7 flowers, closely spaced but widely spreading in the spike inflorescence; glumes short-awned, nearly equal; lemmas ca. 6 mm long with awns 2-4 mm long; palea nearly as long as the lemma. July-Aug.

Uncommon along unshaded roads and trails, montane; East, West; introduced from Russia. Many species placed in *Agropyron* by some authorities are herein placed in *Elymus*.

Agrostis L., Bentgrass

Perennial, ligules membranous, spikelets numerous, 1-flowered, stalked in a branched inflorescence; glumes nearly equal, 1-nerved, pointed but mostly awnless; lemma shorter or equal to the glumes; palea smaller than the lemma, minute or absent in some species.

1. Plants rhizomatous; stems mostly solitary, not tufted; anthers > 1 mm long .. .. ***A. stolonifera***
1. Plants forming tufts or tussocks; anthers ca. 0.5 mm long 2

2. Inflorescence diffuse, branches spreading .... ***A. scabra***
2. Inflorescence narrow, branches erect or ascending 3

3. Habitat montane; largest ligule usually > 3 mm long .... ***A. exarata***
3. Alpine and subalpine plants; ligules < 3 mm long 4

4. Plants mostly < 10 cm high; palea present; ligules < 1 mm long ...**A. humilis**
4. Plants mostly >10 cm high; palea absent; ligules ≥ 1 mm long 5

5. Palea minute; leaves to 1 mm wide ....**A. variabilis**
5. Palea 2/3 length of lemma; some leaves > 1 mm wide ....**A. thurberiana**

Agrostis exarata Trin. Stems 20-60 cm high, forming small tussocks; leaves flat, 2-6 mm wide; ligule 3-8 mm long, blunt, ragged; inflorescence narrow, 5-15 cm long, branches ascending; glumes 2-3 mm long; lemma shorter than glumes, unawned or short-awned; palea minute; anthers ca. 0.5 mm long. July-Sept. Fig. 322.

Moist meadows and open forest, especially around streams and ponds; common montane, uncommon subalpine; East, West. Our plants are ssp. *minor* (Hook.) Hitchc. AK south to CA, NM, Mex., NE. This is the common narrow- inflorescence bentgrass of low elevations.

Agrostis humilis Vasey. Stems 3-10 cm high, forming small tufts; leaves to 1 mm wide; ligule ≤ 1 mm long, blunt; inflorescence narrow, 1-3 cm long, purplish, branches ascending; glumes 1-2 mm long; lemma as long as glumes, awnless; palea nearly as long as lemma; anthers ca. 0.5 mm long. Aug.

Uncommon in moist meadows and heath, usually along small streams or in snow accumulation areas near or above treeline; East, West. B.C., Alta. south to OR, UT, CO. See *A. variabilis*.

Agrostis scabra Willd., Tickle Grass [*A. hiemalis* (Walt.) B. S. P.]. Stems 10-60 cm tall, forming small tussocks; ligule 2-3 mm long, rounded, ragged; leaves 1-3 mm wide; inflorescence 10-25 cm long, diffuse, branches spreading; glumes 2-3 mm long, unequal; lemma unawned; palea lacking; anthers < 0.5 mm long. July-Aug.

Widespread and common in montane to alpine grasslands, meadows, wetlands and open forest; East, West. AK to Newf. south to most of U. S., Asia. The plant's ability to occur in fens as well as stony grasslands and alpine turf is exceptional. Size of the inflorescence varies immensely.

Agrostis stolonifera L., Redtop [*A. alba* L., *A. palustris* Huds.]. Stems 20-80 cm high from extensive rhizomes; leaves flat 2-5 mm wide; ligule blunt, ragged, 3-5 mm long; inflorescence 5-30 cm long, reddish, pyramidal, the branches whorled and widely ascending in flower; glumes 2-3 mm long; lemma 2/3 as long as glumes, unawned; palea 2/3 as long as the lemma; anthers > 1 mm long. July-Aug.

Common in montane and lower subalpine meadows, moist grassland, fens, stream banks, and gravel bars, uncommon along roads and in lawns; East, more common West. Introduced from Eurasia. A. S. Hitchcock (1950) believes that smaller plants with narrow leaves and a purplish inflorescence are native to N. America. *A. tenuis* Sibeth. is similar to *A. stolonifera*, but the branches of the inflorescence are without spikelets near the base; it is commonly introduced in western MT and could be found in the Park.

Agrostis thurberiana A. S. Hitchc. Stems 10-30 cm tall, forming tufts; ligules 1-3 mm long, blunt; leaves 1-3 mm wide; inflorescence narrow, 4-8 cm long, purplish, branches ascending; glumes 2-3 mm long, pointed; lemma as long as the glumes, unawned; palea 2/3 as long as the lemma; anthers ca. 0.5 mm long. July-Sept.

Common in moist soil of turf and rock ledges near or above treeline, occasionally lower; East, West. AK to CA, UT, CO. See *A. variabilis.*

Agrostis variabilis Rydb. Stems 8-25 cm tall, forming small tufts; leaves ca. 1 mm wide; ligules ca. 1-2 mm long, blunt, ragged; inflorescence narrow, 2-6 cm long, purplish with short, erect branches; glumes ca. 2 mm long; lemma shorter than the glumes, unawned; palea minute; anthers ca. 0.5 mm long. Aug-Sept.

Uncommon in moist, subalpine and alpine meadows; East, West. B.C., Alta. south to CA, CO. *A. humilis* is mostly smaller, and *A. thurberiana* has an inflorescence with longer and less erect branches.

Alopecurus L., Foxtail

Perennial; leaves flat; ligules membranous; spikelets flattened, 1-flowered, very short-stalked in a dense, cylindric inflorescence; glumes long-hairy, equal; lemma as long as the glumes, awned from the back; palea lacking.

Often mistaken for timothy (*Phleum* spp.), which has awned glumes much longer than the lemma.

1. Inflorescence ca. 5 mm wide; wet habitats .... **A. aequalis**
1. Inflorescence ca. 1 cm wide; grasslands .... **A. alpinus**

Alopecurus aequalis Sobol. [*A. aristulatus* Michx.]. Stems 20-50 cm high, often spreading or ascending from small tufts; leaves 2-4 mm wide; ligule 4-8 mm long, inflorescence 1-5 cm long; glumes 2-3 mm long. July-Aug. Fig. 323.

Common in mud or shallow water at the margin of ponds, wetlands or slow streams, montane; East, West. Circumboreal south to Mex. All but the inflorescence may be submerged in shallow water.

Alopecurus alpinus Smith [*A. occidentalis* Scribn. & Tweedy]. Stems 30-80 cm tall, forming clumps from short rhizomes; ligule 1-3 mm long; leaves 3-5 mm wide; inflorescence 1-5 cm long, ca. 1 cm wide; glumes 3-4 mm long. June-Aug.

Uncommon in montane grasslands and meadows; East, West. AK to Newf. south to ID, UT, CO.

Fig.321. *Spiranthes romanzoffiana*

Fig.322. *Agrostis exarata*

Fig.323. *Alopecurus aequalis*

Avena L., Oats

Avena sativa L. Anuual; stems 30-80 cm high; leaves 3-10 mm wide; ligules membranous, 3-6 mm long; spikelets 2- or 3-flowered, drooping, several in an open, branched inflorescence; glumes 2-3 cm long, equal; lemmas glabrous, shorter than the glumes, the lowest one with an awn from the back; palea shorter than the lemma. Aug.

Rare in open, disturbed areas, often where horses frequent; collected once near Lake McDonald. Introduced from Eurasia. Does not persist in the wild.

Beckmannia Host, Slough Grass

Beckmannia syzigachne (Steud.) Fern. [*B. erucaeformis* (L.) Host misapplied]. Annual; stems 30-80 cm high; ligules membranous, 6-11 mm long; leaves flat, 3-8 mm wide; spikelets 1-flowered, arranged in 2-column spikes, 1-2 cm long, these arranged on erect branches in a narrow inflorescence; glumes flattened, orbicular, 2-3 mm long; lemma as long as the glumes but narrower; palea as long as the lemma. July-Aug. Fig. 324.

Common in mud or shallow water on the margins of montane ponds and slow streams; East, West. AK to N. S. south to CA, NM, MO. Spikelets resemble rattlesnake rattles.

Bromus L., Brome Grass

Annual or perennial; leaves flat; sheaths closed to near the top; ligules membranous; spikelets several-flowered, stalked in a branched inflorescence; glumes unequal; lemmas longer than the glumes, often awned; palea shorter than the lemma.

All the annual bromes are introduced, but it appears that only *B. tectorum* is truly persistent in the Park.

1. Plants annual; remains of previous year's growth not apparent at the base of the plant, root system lacking any perennial tissue 2
1. Plants perennial 5

2. Inflorescence with erect or ascending branches 3
2. Inflorescence open, diffuse, the branches spreading or drooping 4

3. Branches of inflorescence shorter than the spikelets .... ***B. mollis***
3. Inflorescence branches as long or longer than the spikelets . ***B. commutatus***

4. Awns of lemmas 12-14 mm long .... ***B. tectorum***
4. Awns < 10 mm long .... ***B. secalinus***

5. Spikelets flattened; lemmas V-shaped in cross section .... ***B. carinatus***
5. Spikelets more cylindric; lemmas rounded on back 6

6. Awn of lemma 1-2 mm long or lacking 7
6. Awn of lemma 3-8 mm long 8

7. Lemmas glabrous .... ***B. inermis***
7. Lemmas hairy .... ***B. pumpellianus***

8. Lemmas hairy over the entire back .... ***B. anomalus***
8. Lemmas hairy on the margins but not in the center 9

9. Awn of lemma 3-5 mm long; ligule ca. 1 mm long .... ***B. ciliatus***
9. Awn of lemma 6-8 mm long; ligule 3-5 mm long .... ***B. vulgaris***

Bromus anomalus Rupr. ex Fourn. Perennial; stems 30-80 cm high forming tussocks; ligules 1-2 mm long; leaf blades 5-15 mm wide; sheaths glabrous; inflorescence open with drooping spikelets 15-30 mm long; lemmas 9-14 mm long, hairy, the awn 3-7 mm long. July-Aug.

Uncommon in aspen groves along the east margin of the Park. B.C. to Sask. south to CA, AZ, NM, TX. The similar *B. ciliatus* and *B. vulgaris* have lemmas that are hairy on the margins only.

Bromus carinatus Hook. & Arn., Mountain Brome [*B. marginatus* Nees, *B. polyanthus* Scribn.]. Perennial; stems 60-100 cm tall, forming tussocks; ligules 1-3 mm long; leaf blades 3-12 mm wide; sheaths hairy; inflorescence narrow with short, erect branches; spikelets 2-4 cm long, flattened; lemmas 10-15 mm long, hairy, the awn 2-5 mm long. July-Aug.

Montane and lower subalpine grassland, meadows, avalanche slopes, and open forest, especially aspen; abundant East, common West. Our plants are var. *carinatus*. AK to Alta. south to Mex., NM, SD. Often used in wildlands revegetation projects. An apparent hybrid with *B. ciliatus* or *B. anomalus* was once collected near Lake Sherburne.

Bromus ciliatus L. [*B. richardsonii* Link.]. Perennial; stems 50-80 cm high, forming tussocks; ligules ca. 1 mm long; leaf blades 4-8 mm wide; sheaths hairy or glabrous; inflorescence open with drooping branches; spikelets 15-23 mm long; lemmas rounded on back, 10-15 mm long, long-hairy on the margins, the awn 3-5 mm long. July-Aug. Fig. 325.

Uncommon in montane grasslands, meadows, and open forest, especially aspen; East, West. AK to Newf. south to Mex. *B. vulgaris* has longer ligules and is more common in coniferous forest. See *B. anomalus*

Bromus commutatus Schrad. Annual; stems 20-50 cm high; ligules ca. 1 mm long; leaf blades 1-3 mm wide; sheaths short-hairy; inflorescence open with ascending branches; spikelets 15-30 mm long; lemmas rounded on back, 8-10 mm long with an awn of equal length. July.

Rare in stony soil of open montane slopes and banks; collected twice near West Glacier. Introduced from Europe.

Bromus inermis Leys, Smooth Brome, Hungarian Brome. Perennial; stems 50-100 cm high from long rhizomes; ligule 1-3 mm long; leaf blades 5-10 mm wide; foliage mostly glabrous; inflorescence somewhat crowded and narrow; branches ascending or spreading; spikelets cylindric, 15-30 mm long, often purple-banded; lemmas 9-12 mm long, glabrous, the awns short or absent. July-Aug.

Common along roads, near habitations, and in disturbed montane meadows and grasslands; East, West. Introduced from Europe. Smooth brome was apparently introduced in some parts of the Park by horse packers attempting to increase forage for their stock.

Bromus mollis L. Annual; stems 20-90 cm high; ligules ca. 1 mm long; leaves 1-4 mm wide; foliage soft-hairy; inflorescence narrow with short, erect branches; spikelets 1-2 cm long; lemmas 6-9 mm long, V-shaped in cross section; awn 6-10 mm long. July.

Rare in disturbed soil; collected once on a roadside near Lake McDonald. Introduced from Europe.

Bromus pumpellianus Scribn. [*B. inermis* Leys. ssp. pumpellianus (Scribn.) Wagn.]. Rhizomatous perennial similar to *B. inermis*; foliage at least somewhat hairy; branches of inflorescence ascending; lemmas hairy. July-Aug.

Montane to alpine grasslands and meadows; common East, uncommon West. AK south to ID, CO, SD, MI. Often with *Pentaphylloides fruticosa* and *Festuca scabrella*.

Bromus secalinus L. Annual; stems 20-90 cm tall; ligules 1-3 mm long; leaves 2-8 mm wide, hairy; inflorescence nodding, open with ascending branches; spikelets 1-2 cm long; lemmas 7-9 mm long, rounded on back, the awn shorter than the lemma.

Rare in disturbed soil; reported for West Glacier by Standley (1921) but not collected since. Introduced from Europe.

Bromus tectorum L., Cheatgrass. Annual with stems 20-50 cm tall; ligules 1-3 mm long; leaves 2-3 mm wide; foliage soft-hairy; inflorescence open with spreading branches; spikelets 1-2 cm long, often drooping; lemmas 10-12 mm long, V-shaped in cross section; awn 12-14 mm long. June-July.

Uncommon in montane grasslands and along roads and around campgrounds and buildings; East, West. Introduced from Europe. Our most common annual brome.

Bromus vulgaris (Hook.) Shear. Tussock-forming perennial with stems to 100 cm high; ligules 3-5 mm long; leaves 5-10 mm wide; foliage glabrous to short-hairy; inflorescence diffuse with spreading branches; spikelets 15-25 mm long, drooping; lemmas 8-10 mm long, long-hairy, V-shaped in cross section; awn 5-8 mm long. July-Aug.

Common in montane and lower subalpine, moist coniferous forest or sometimes meadows; East, West. B.C. to Alta. south to CA, WY. See *B. ciliatus, B. anomalus.*

Calamagrostis Adans., Reedgrass

Perennial, often rhizomatous; ligules membranous; leaf sheaths open; spikelets 1-flowered, stalked in a branched inflorescence; glumes pointed,

V-shaped in cross section, nearly equal; lemma shorter or equal to the glumes, short-awned from the back with long, stiff hairs at the base (callus); palea shorter than lemma. Reference: Greene (1993).

1. Inflorescence open, the branches spreading .... *C. canadensis*
1. Inflorescence narrow, congested, the branches erect 2

2. Callus hairs ca. 2/3 as long as the lemma .... *C. stricta*
2. Callus hairs < 1/2 as long as the lemma 3

3. Awn as long as the glumes; glumes 4-5 mm long; top of most leaf sheaths (collar) with hairs .... *C. rubescens*
3. Awn well exserted from the glumes; glumes 5-8 mm long; without hairs at the collar .... *C. purpurascens*

Calamagrostis canadensis (Michx.) Beauv., Bluejoint. Stems 60-100 cm high, clustered on rhizomes; ligules 3-8 mm long, ragged; leaves 3-10 mm wide; inflorescence pyramidal with spreading branches at flowering; glumes 3-5 mm long, often purplish; lemma with a straight awn; callus hairs as long as the lemma. July-Aug. Fig. 326.

Abundant in montane and lower subalpine wet meadows and fens and around ephemeral ponds, lakes, and streams, often in forest openings; East, West. AK to Que. south to most of U. S. The variation within this species has been described by a dizzying array of different varieties and subspecies.

Calamagrostis purpurascens R. Br. [*C. vaseyi* Beal]. Stems 30-70 cm high, forming tussocks; ligules 2-4 mm long; leaves 1-4 mm wide, usually rolled inward; inflorescence narrow with erect branches; glumes 5-8 mm long, often purplish; lemma 4-7 mm long, the awn longer than the glumes; callus hairs short. July-Aug.

Common in stony soil of montane and subalpine meadows, grasslands, turf, and talus; East, West. AK to Greenl. south to CA, CO, MN, Asia. The veiny, leathery leaves help identify this grass.

Calamagrostis rubescens Buckl., Pinegrass. Stems 50-80 cm high, forming tufts from rhizomes; ligules 1-5 mm long; leaves flat, 2-4 mm wide, short-hairy at the top of the sheath; inflorescence narrow with erect branches; glumes 4-5 mm long; lemma with a bent awn just longer than the glumes; callus hairs short. Aug.

Abundant in montane and subalpine coniferous forest, occasionally in meadows; East, West. B.C. to Man. south to CA, CO. Plants rarely flower except following fire. Hair at the top of the sheath is a good vegetative character to identify this common plant.

Calamagrostis stricta (Timm) Koeler [*C. inexpansa* Gray, *C. neglecta* (Ehrh.) Gaertn.]. Stems 40-80 cm high from rhizomes; ligules 1-8 mm long, ragged; leaves 1-3 mm wide, often rolled inward; inflorescence congested, narrow with erect branches; glumes 3-5 mm long, often purplish; lemma with a straight awn as long as the glumes; callus hairs just shorter than the lemma. July-Aug.

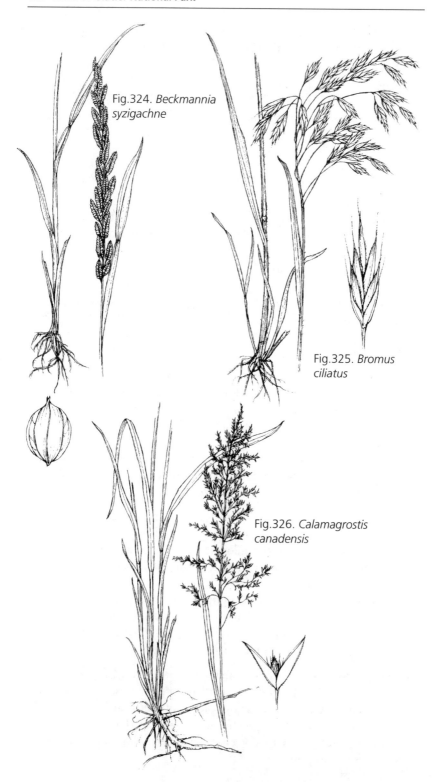

Fig.324. *Beckmannia syzigachne*

Fig.325. *Bromus ciliatus*

Fig.326. *Calamagrostis canadensis*

Common in montane marshes and wet meadows; East, West. Both ssp. *inexpansa* (Gray) Greene with ligules 4-8 mm long and ssp. *stricta* with ligules < 4 mm long are reported for the Park (Standley 1921). Circumboreal south to CA, NM, MO, VA.

Catabrosa Beauv., Brook Grass

Catabrosa aquatica (L.) Beauv. Rhizomatous perennial; stems 10-50 cm long, ascending to prostrate; ligules membranous, 2-8 mm long; leaves flat, 2-5 mm wide; spikelets 2-flowered, stalked on spreading, whorled branches of an open inflorescence; glumes blunt, ca. 1-2 mm long, unequal, shorter than lemmas; lemmas 2-3 mm long with ragged, blunt tips, rounded on back, tan, unawned; palea as long as the lemma.

Rare in mud or shallow water around montane ponds and slow streams; reported for East Glacier by Standley (1921). AK to Greenl. south to OR, AZ, IA, Eurasia.

Cinna L., Wood Reed

Cinna latifolia (Trev.) Griseb. Rhizomatous perennial; stems 50-100 cm tall; ligules membranous, 3-8 mm long; leaves flat, 5-15 mm wide; spikelets 1-flowered, stalked in a diffuse, pyramidal inflorescence; glumes sharp-pointed, 3-5 mm long, nearly equal; lemma as long as glumes with a short awn, V-shaped in cross section; palea as long as the lemma. July-Sept. Fig. 327.

Common in moist, montane and lower subalpine forest, often along streams; East, West. Circumboreal south to CA, NM, TN, NC. Leaves are wider than other forest-growing grasses with 1-flowered spikelets.

Dactylis L.

Dactylis glomerata L., Orchard Grass. Perennial; stems 60-100 cm high, forming tussocks; ligules membranous, 3-9 mm long; leaves flat, 2-8 mm wide; spikelets 3- to 6-flowered, flattened, short-stalked in clusters in a narrow inflorescence with mostly ascending branches; glumes 4-6 mm long, nearly equal, short awn-tipped; lemmas 5-8 mm long, awn-tipped, longer than the glumes; palea nearly as long as lemma. July.

Rare in moist, disturbed soil; collected along roads and trails near the head of Lake McDonald. Introduced from Eurasia. Commonly seeded for hay or revegetation and sometimes spread by horses.

Danthonia Lam. & D. C., Oatgrass

Tuft- or tussock-forming perennials; leaf sheaths open; ligules composed of hairs; spikelets 3- to 8-flowered, long-stalked, few in an open inflorescence; glumes sharp-pointed; lemmas shorter than the glumes, rounded on back, forked at the tip with a twisted awn arising between the teeth; palea shorter than the lemma.

Some plants produce a few asexual (cleistogamous) flowers in the sheaths of lower leaves.

1. Lemmas hairy over the entire back (sometimes sparse) 2
1. Lemmas hairy on the margins only 3
2. Glumes 10-12 mm long; awns 5-8 mm long .... *D. spicata*
2. Glumes 20-22 mm long; awns 10-15 mm long .... *D. parryi*
3. Spikelets usually solitary .... *D. unispicata*
3. Spikelets > 1 4
4. Spikelets on short, erect or ascending stalks .... *D. intermedia*
4. Spikelets on long, spreading stalks .... *D. californica*

Danthonia californica Boland [*D. americanum* Scribn.]. Stems 40-80 cm high; leaves 1-3 mm wide, sometimes inrolled; foliage often long-hairy; spikelets 2-5 on spreading stalks; glumes 9-15 mm long; lemmas 7-12 mm long, short-hairy on the margins; awns 5-10 mm long. July.

Uncommon in montane grasslands and open forest; East, West. B.C., Alta. south to CA, NM; Chile.

Danthonia intermedia Vasey, Timber Oatgrass. Stems 10-40 cm high; leaves 1-3 mm wide, sometimes inrolled; spikelets usually 5-10 on ascending stalks; glumes 10-15 mm long; lemmas 7-10 mm long with long hair on the margins; awns to 10 mm long. July-Aug. Fig. 328.

Abundant in montane grasslands, common on dry rock outcrops and subalpine to alpine meadows; East, West. AK to Newf. south to CA, NM, MI.

Danthonia parryi Scribn. Stems 30-50 cm high; leaves 1-4 mm wide, inrolled; spikelets 3-8 on ascending stalks; glumes 15-20 mm long; lemmas 10-13 mm long, hairy over most of the back; awns 10-15 mm long. July.

Uncommon in montane grasslands along the east margin of the Park. B.C., Alta. south to NM. This grass is common along the Front Range of the Rockies south and north of the Park.

Danthonia spicata (L.) Beauv., Poverty Oatgrass [*D. pinetorum* Piper]. Stems 20-70 cm high; leaves 1-2 mm wide, inrolled; spikelets several on short, erect stalks; glumes 10-12 mm long; lemmas 3-5 mm long, hairy over the entire back; awns 5-8 mm long. July-Aug.

Rare in subalpine meadows and open forest, collected once in the McDonald drainage. Our plants are var. *pinetorum* Piper. AK to Newf. south to OR, NM, eastern U. S.

Danthonia unispicata (Thurb.) Munro. Stems 10-25 cm tall; leaves 1-2 mm wide, inrolled; foliage long-hairy; spikelet usually solitary, erect; glumes 10-23 mm long; lemmas 9-12 mm long, long-hairy on the margins; awns 6-8 mm long. June-July.

Uncommon in stony soil of exposed outcrops and montane grassland ridges; East, West. B.C., Alta. south to CA, CO.

Deschampsia Beauv., Hair Grass

Tuft- or tussock-forming perennials; ligules membranous, hairy; leaf sheaths open; spikelets shiny or purplish, 2- or 3-flowered, short-stalked in a branched inflorescence; glumes sharp-pointed; lemmas shorter than the glumes, hairy at the base (callus), blunt-tipped with a straight or bent awn from the back; palea shorter than the lemma.

1. Inflorescence narrow with erect branches .... *D. elongata*
1. Inflorescence open with spreading branches 2
2. Leaf blades mostly 4-6 mm wide; awn of lemma 2-3 mm long ....
.... *D. atropurpurea*
2. Leaf blades 1-3 mm wide; awn of lemma 3-4 mm long .... *D. caespitosa*

Deschampsia atropurpurea (Wahl.) Scheele [*Aira atropurpurea* Wahl., *Vahlodea atropurpurea* Wahl.]. Stems 20-40 cm tall; ligules blunt, 1-4 mm long; leaves flat, 3-5 mm wide; spikelets in a diffuse inflorescence; glumes equal, 3-5 mm long; lemmas 2-3 mm long with callus hairs half as long; awn 2-3 mm long. July-Aug. Fig. 329.

Common in moist meadows, open forest, and rock ledges, often where snow persists, subalpine to alpine; East, West. Circumpolar south to CA, CO, NH. Most common in high subalpine forests.

Deschampsia caespitosa (L.) Beauv., Tufted Hairgrass [*Aira cespitosa* L.]. Stems 20-100 cm high; ligules ca. 4 mm long; leaves 1-3 mm wide, inrolled; spikelets in a diffuse inflorescence; glumes unequal, 2-6 mm long; lemmas 2-4 mm long with callus hairs 1 mm long; awn 3-4 mm long. July-Aug.

Common in montane to alpine wet meadows and shores of rivers, streams, and lakes; East, West. Our plants are ssp. caespitosa. Circumpolar south to Mex., WI, GA. Tussocks are often elevated well above ground level.

Deschampsia elongata (Hook.) Munro [*Aira elongata* Hook.]. Stems 10-60 cm high; ligules pointed, 3-9 mm long; leaves ca. 1 mm wide; inflorescence narrow with erect branches; glumes 3-6 mm long; lemmas 2-3 mm long with callus hairs half as long; awn 3-4 mm long. June-Aug.

Common in open montane and lower subalpine forest; East, West. AK south to Mex. Often found along trails and old roads.

Echinochloa Beauv.

Echinochloa crus-galli (L.) Beauv., Barnyard Grass. Annual; stems 30-70 cm high, flattened, erect or ascending; ligules lacking; leaves 5-15 mm wide; spikelets arranged in sessile, cylindric clusters in an open, spike-like inflorescence, each one 3-4 mm long and composed of 1 fertile and 1 sterile floret; glumes unequal; second glume and sterile lemma as long as the floret; sterile lemma long-awned; fertile lemma shiny, hardened, unawned. July.

Fig.328. *Danthonia intermedia*

Fig.327. *Cinna latifolia*

Fig.329. *Deschampsia atropurpurea*

Rare along roads and other disturbed habitats, often where wet at least part of the year, collected once near Polebridge. Introduced from Eurasia.

Elymus L., Wild Rye, Wheatgrass

Perennial; ligules membranous; leaf sheaths open; spikelets several-flowered, sessile, 1-4 per node in a spike inflorescence; glumes equal, sometimes awned; lemmas longer than the glumes rounded on back, sometimes awned; palea as long as the lemma.

Jones' (1910) report of tHe low elevation species, *E. triticoides* Buckl., for Blackfoot Glacier is probably based on *E. trachycaulus*, although I haven't seen the specimen. *E. macounii* Vasey, a hybrid between *E. trachycaulus* and *Hordeum jubatum* with a slender spike < 1 cm wide, has been collected in grasslands around St. Mary. There is a great deal of hybridization among species both within and between *Elymus* and *Agropyron* as traditionally conceived. Barkworth and Dewey (1985) proposed a classification based on differences in chromosomal complements. Their system, which divides the group into numerous genera, seems untenable because some genera are composed entirely of species that are hybrids between species of other genera. Furthermore, species that are barely distinguishable in the field, such as *E. lanceolatus* and *E. smithii*, are placed in different genera. It seems more sensible to combine members of such a widely hybridizing complex into a single genus as proposed by Gould (1947).

1. Spikelets > 1 per inflorescence node 2
1. Spikelets 1 per node 6
2. Awns of lemmas ≥ 2 cm long; spike often nodding or curved 3
2. Spike erect or nearly so; awns ≤ 2 cm long 4
3. Glumes awl-shaped with 1 nerve in the middle; subalpine and alpine ....
.... ***E. elymoides***
3. Glumes narrowly lance-shaped, 2-3 nerved; montane .... ***E. canadensis***
4. Awns of the lemma 1-2 cm long .... ***E. glaucus***
4. Awns of lemma < 1 cm long 5
5. Ligule < 1 mm long; glumes and lemmas densely hairy .... ***E. innovatus***
5. Ligule ≥ 3 mm long; glumes and lemmas glabrous or sparsely hairy ....
.... ***E. cinereus***
6. Plants forming tussocks; rhizomes short or lacking 7
6. Rhizomatous, stems arising singly or nearly so 8
7. Anthers 4-6 mm long; spikelets barely touching each other in the spike ....
.... ***E. spicatus***
7. Anthers 1-3 mm long; spikelets strongly overlapping .... ***E. trachycaulus***
8. Leaf blades flat, 5-10 mm wide .... ***E. repens***
8. Leaf blades usually rolled, < 5 mm wide 9
9. Spikelets barely touching each other in the spike .... ***E. spicatus***
9. Spikelets strongly overlapping in the inflorescence 10

10. Glumes broadest above mid-length; lemmas hairy .... *E. lanceolatus*
10. Glumes widest below middle; lemmas usually glabrous .... *E. smithii*

Elymus canadensis L., Canada Wildrye. Stems 20-80 cm high; leaves 3-12 mm wide; ligules ca. 1 mm long; spike crowded, nodding, 8-15 cm long, appearing bristly; spikelets 2 per node; glumes awned; lemmas 8-14 mm long, the awn 2-4 cm long.

Rare in open montane forest; reported by Standley (1921) for West Glacier. AK to Que. south to CA, NM, TX, MO, SC.

Elymus cinereus Scribn. & Merr., Basin Wildrye [*E. condensatus* Presl, *E. piperi* Bowden, *Leymus cinereus* (Scribn. & Merr.) Löve]. Stems 60-150 cm tall, forming large tussocks; leaves 8-15 mm wide; ligules 3-7 mm long; spike narrow, stiff, 7-20 cm long; spikelets 2 per node; glumes sharp-pointed 10-20 mm long; lemmas pointed or short-awned. July.

Rare in vernally wet grassland habitats along the east margin of the Park and collected once along the road near the head of Lake McDonald. B.C. to Sask. south to CA, AZ, CO.

Elymus elymoides (Raf.) Swezey, Squirreltail [*Sitanion hystrix* (Nutt.) Smith]. Tussock-forming perennial; stems 10-50 cm high; ligules membranous, short; leaves 1-4 mm wide; spikelets few-flowered, sessile, 2 per node in a spike inflorescence 2-15 cm long, only the central lemma fertile; glumes usually 4 per node, awl-shaped, awns 5-9 cm long; lemmas awned, some glume-like; palea with 2 ridges on back. July.

Common in dry meadows, talus, and rocky slopes, subalpine and lower alpine; East, West. Our plants are var. *hystrix*. B.C. to Sask. south to CA, TX. Reports of *Agropyron scribneri* Vasey are referable here. *Hordeum* spp. are similar, but they occur at lower elevations; mature awns are more widely spreading in *E. elymoides*.

Elymus glaucus Buckl. Stems 30-100 cm high; leaves 5-10 mm wide; ligules 1 mm long; spike stiff, narrow, crowded, 5-15 cm long; spikelets 2 per node; glumes pointed or awn-tipped; lemmas 10-12 mm long, with an awn 1-2 cm long. June-Sept. Fig. 330.

Abundant in moist, montane and subalpine, coniferous, aspen, or riparian forest and avalanche slopes, less common in grasslands and meadows; East, West. Our plants are mainly var. *glaucus*, but var. *jepsonii* Davy has been collected in lodgepole pine forest, West. AK to Ont. south to CA, NM, IA, MI. One of our most ubiquitous forest grasses.

Elymus innovatus Beal. Stems 40-100 cm tall, forming small clumps from rhizomes; ligules < 1 mm long, leaves 2-4 mm wide, inrolled; spike 4-10 cm long, narrow; spikelets 2 per node; glumes hairy, 5-12 mm long, awn-tipped; lemmas 7-9 mm long, hairy, short-awned. July.

Rare in open montane forest of river valleys; collected twice in the northeast part of the Park (Hermann 1956). AK to Ont. south to MT, WY, SD.

Elymus lanceolatus (Scribn. & Smith) Gould, Thick-spike Wheatgrass [*Agropyron dasystachyum* (Hook.) Scribn.]. Stems 40-80 cm tall from rhizomes, sometimes forming small clumps; leaves 1-3 mm wide; inflorescence 8-12 cm long; spikelets overlapping, 12-18 mm long with 4-8 flowers, 1 per node; glumes sharp-pointed; lemmas 8-10 mm long, sharp-pointed, usually hairy. July-Aug.

Rare in montane grasslands; East, West. B.C. to Man. south to CA, AZ, NE.

Elymus repens (L.) Gould., Quackgrass [*Agropyron repens* (L.) Beauv., *Elytrigia repens* (L.) Nevski]. Stems 40-80 cm high from long rhizomes; leaves flat, 5-10 mm wide; inflorescence 5-15 cm long; spikelets with 4-8 flowers, overlapping, 1 per node; glumes 8-12 mm long, awn-tipped; lemmas awn-tipped.

Uncommon in campgrounds and residence areas; West. Introduced from Eurasia.

Elymus smithii (Rydb.) Gould, Western Wheatgrass [*Agropyron smithii* Rydb., *Pascopyrum smithii* (Rydb.) Löve]. Stems 30-60 cm high from long rhizomes; leaves 2-4 mm wide, rolled when dry; inflorescence 7-12 cm long; spikelets 1-2 cm long, strongly overlapping with 6-10 flowers, 1 per node; glumes sharp-pointed; lemmas 10-12 mm long, sharp-pointed or short-awned, glabrous or gray waxy. July.

Uncommon in montane grasslands along the east edge of the Park and collected once along Lake McDonald, where perhaps it was introduced. B.C. to Ont. south to OR, AZ, TX, TN. *E. lanceolatus* has glumes widest at the middle with 5-7 nerves, while glumes of *E. smithii* are widest at the base with 3-5 nerves.

Elymus spicatus (Pursh) Gould, Bluebunch Wheatgrass [*Agropyron spicatum* (Pursh) Scribn. & Smith, *Pseudoroegneria spicata* (Pursh) Löve, including *A. inerme* (Scribn. & Smith) Rydb.]. Stems 30-100 cm high, forming tussocks; leaves 1-3 mm wide, often rolled; inflorescence 8-15 cm long; spikelets with 6-8 flowers, 1 per node, little overlapping on the spike; glumes 5-10 mm long, pointed; lemmas with a bent awn at maturity or (rarely) awnless. June-Aug. Fig. 331.

Abundant in montane grasslands, common on sparsely vegetated subalpine slopes; East, West. AK to Ont. south to CA, AZ, TX. The unawned form occurs in the St. Mary Valley and is reported for Granite Park (Standley 1921).

Elymus trachycaulus (Link.) Shinn., Bearded Wheatgrass, Slender Wheatgrass [*Agropyron subsecundum* (Link) Hitchc., *A. caninum* (L.) Beauv. var *andinum* (Scribn. & Smith) Hitchc., *A. trachycaulum* (Link) Malte var. *unilaterale* (Cassidy) Malte]. Stems 20-80 cm tall, erect, forming tussocks; leaves flat, 2-8 mm wide; spike with overlapping spikelets, 12-17 mm long, 1 per node; awn 0-20 mm long. July-Aug.

Ssp. *latiglumis* (Scribn. & Smith) Barkw. & Dewey, with awns < 5 mm long and an inflorescence 3-8 cm long, is abundant in moist, alpine and subalpine meadows, rocky slopes and open forest; ssp. *trachycaulus*, with awns < 5 mm long and a spike 10-25 cm long, is uncommon in montane and lower subalpine grasslands, meadows and open forest; ssp. *subsecundus* (Link) Gould, with awns > 5 mm long and a 1-sided inflorescence 7-15 cm long, is common in montane and subalpine grasslands along the east edge of the Park. East, West. AK to Lab. south to CA, NM, Mex.

Festuca L., Fescue

Mostly bunch- or tussock-forming perennials; ligules short, membranous; leaf sheaths open; spikelets several-flowered, stalked, in a branched inflorescence; glumes unawned, shorter than the lemmas, usually unequal; lemmas rounded on back, awned from the tip or unawned; palea as long as lemma. Reference: Aiken & Darbyshire (1990).

1. Spikelets containing small bulb-like plants .... *F. vivipara*
1. Spikelets not containing bulblets 2

2. Lemmas ≥ 7 mm long, unawned or minutely awned................... *F. scabrella*
2. Lemmas < 7 mm long, awned 3

3. Leaves flat, 4-10 mm wide; moist forests ....*F. subulata*
3. Leaves 1-2 mm wide with inrolled margins; more open habitats 4

4. Base of stems reddish, arched outward; lower leaf sheaths becoming broken and stringy with age .... *F. rubra*
4. Stems erect; lower leaf sheaths mostly not reddish, remaining hard and intact 5

5. Awns 4-12 mm long, usually longer than the lemma .... *F. occidentalis*
5. Awns 1-4 mm long, mostly shorter than the lemma 6

6. Inflorescence usually 10-20 cm long; anthers 2-4 mm long ...... *F. idahoensis*
6. Inflorescence 1-10 cm long; anthers < 2 mm long 7

7. Lemmas green; inflorescence usually 2-10 cm long; anthers ≥ 1 mm long .....
.... *F. saximontana*
7. Lemmas purplish; inflorescence 1-3 cm long; anthers < 1 mm long 8

8. Stems densely but inconspicuously short-hairy just below the inflorescence ..
....*F. baffinensis*
8. Stems glabrous or sparsely hairy .... *F. brachyphylla*

Festuca baffinensis Polunin. Similar to *F. brachyphylla* but stems 5-15 cm high, short-hairy. Aug.

Uncommon in dry alpine turf and fellfields; East. AK to Greenl. south to CO.

Festuca brachyphylla Schultes [*F. ovina* L. var. *brevifolia* (R. Br.) Wats.]. Stems 5-15 cm tall; leaves ≤ 1 mm wide, inrolled; spikelets purplish, short-stalked in a narrow inflorescence; glumes 2-3 mm long; lemmas 2-3 mm long, the awns 1-3 mm long; anthers < 1 mm long. July-Aug.

Common in stony soil of exposed ridges and slopes, rock outcrops, and talus near or above treeline; East, West. Circumpolar south to CA, NM, ME. *F. baffinensis* has short-hairy upper stems, and *F. saximontana* is a taller plant.

Festuca idahoensis Elmer, Idaho Fescue. Stems 25-80 cm high; leaves to 1 mm wide, inrolled; spikelets 5- to 7-flowered in a narrow inflorescence 6-12 cm long with ascending branches; glumes 3-7 mm long; lemmas 5-8 mm long with awns 2-4 mm long; anthers 2-4 mm long. June-Aug.

Abundant in montane and subalpine grasslands, common on open slopes to treeline; East, West. B.C., Alta. south to CA, CO. A dominant plant of fescue grasslands, often with sagebrush in the North Fork Flathead drainage.

Festuca occidentalis Hook. Stems slender, 30-80 cm high; leaves < 1 mm wide, inrolled; foliage light green; spikelets mostly 4- to 5-flowered, well separated on ascending branches of a diffuse inflorescence; glumes 2-5 mm long; lemmas 4-6 mm long, the fine awns 4-12 mm long. June-Aug.

Uncommon in open, montane, and lower subalpine coniferous forest, often in compacted soil of trails and old roads; East, West. B.C. to Ont. south to CA, ID, WY, MI. The tall stems compared to the short tufts of light green basal leaves help distinguish this species.

Festuca rubra L., Red Fescue. Stems 20-70 cm tall, loosely clumped from short rhizomes; leaves 1-3 mm wide, inrolled or folded, the lower sheaths reddish; inflorescence narrow with short ascending branches; glumes 2-8 mm long; lemmas 5-8 mm long, awns 1-3 mm long. Aug.

Rare around buildings and campgrounds, uncommon in montane to subalpine meadows and open forest; East, West. Circumboreal south through most of U. S. Introduced and native forms both occur in Montana; plants of low-elevation, disturbed sites in the Park are probably exotic genotypes.

Festuca saximontana Rydb. [*F. ovina* L. var. *rydbergii* St-Yves]. Stems 10-40 cm high; leaves to 1 mm wide, inrolled; spikelets light green, short-stalked in a narrow inflorescence; glumes 2-3 mm long; lemmas 2-4 mm long, awns 1-3 mm long; anthers 1-2 mm long. June-Aug.

Uncommon in stony soil of subalpine and lower alpine meadows; East, West. AK to Newf. south to OR, UT, CO, NE. *F. brachyphylla* usually has purplish lemmas, and *F. idahoensis* is usually larger with larger lemmas and anthers.

Festuca scabrella Torr., Rough Fescue [*F. campestris* Rydb., *F. altaica* Trin. ssp. *scabrella* (Torr.) Hult.]. Stems 30-90 cm high, forming large tussocks; leaves 2-4 mm wide, usually inrolled, rough-surfaced; inflorescence open with ascending and spreading branches; glumes 5-7 mm long; lemmas 7-9 mm long, awns minute or absent. June-July. Fig. 332.

Abundant in montane and subalpine grasslands; East, West. B.C. to Man. south to OR, CO, ND. The dominant plant of montane grasslands.

Festuca subulata Trin. Stems 50-100 cm high; leaves flat, 4-10 mm wide; spikelets drooping, 3- to 5-flowered in a diffuse inflorescence; glumes 3-7 mm long; lemmas ca. 7 mm long with fine awns 5-20 mm long. July-Aug.

Common in moist, montane forest including aspen; East, West. AK to CA, UT, WY. Often growing with *Elymus glaucus* and *Bromus vulgaris*.

Festuca vivipara (L.) Small [*F. ovina* L. var. *vivipara* L.], Stem up to 20 cm high; leaves to 1 mm wide, inrolled; foliage glabrous; inflorescence narrow with erect branches; spikelets of 2 glumes, ca. 3 mm long, and 3-5 sterile lemmas each enfolding a small bulb-like grass plant. July-Aug.

Uncommon in moist to wet, alpine turf, often on cool slopes; East. Boreal N. America south to B.C. and MT. The taxonomic status of these asexual plants is uncertain (Aiken and Derbyshire 1990).

Glyceria R. Br., Manna Grass

Rhizomatous perennials; ligules membranous; leaves with closed sheaths; spikelets several-flowered; inflorescence with whorled branches; glumes unequal, shorter than the lemmas, 1-veined, unawned; lemmas rounded on back with 7 parallel nerves, unawned; palea about as long as the lemma.

Puccinellia pauciflora strongly resembles a *Glyceria*, but the larger glume has 3 veins.

1. Spikelets cylindrical, ≥ 7 mm long; branches of inflorescence erect ....
.... **G. borealis**
1. Spikelets ovate ≤ 6 mm long; inflorescence open, pyramidal 2
2. Leaves 2-5 mm wide .... **G. striata**
2. Some leaves > 5 mm wide 3
3. Spikelets 3-5 mm long; ligule bumpy upon magnification .... **G. elata**
3. Spikelets 5-6 mm long; ligule smooth .... **G. grandis**

Glyceria borealis (Nash) Batchelder [*Panicularia borealis* Nash]. Stems 50-100 cm tall; ligule 5-10 mm long, pointed; leaves 3-6 mm wide; spikelets 7-14 mm long, cylindrical; inflorescence narrow with mostly erect branches; lemmas 3-5 mm long. July-Aug.

Common in standing water of montane marshes and shallow lake margins; East, more common West. AK to Newf. south to CA, NM, SD, PA. The linear spikelets and narrow inflorescence are diagnostic.

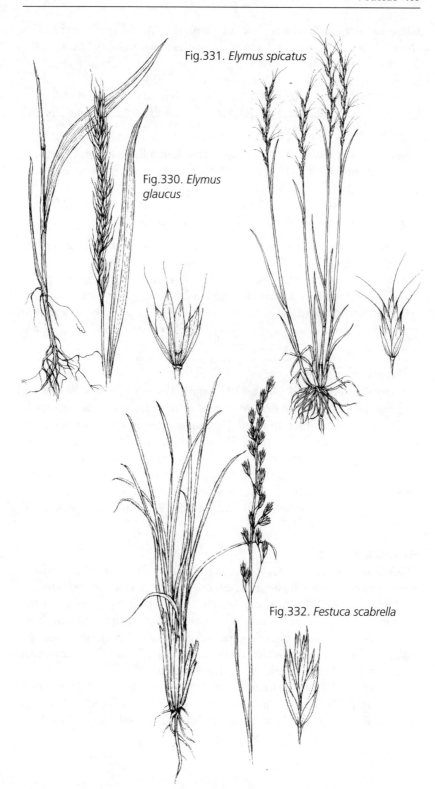

Fig.331. *Elymus spicatus*

Fig.330. *Elymus glaucus*

Fig.332. *Festuca scabrella*

Glyceria elata (Nash) Jones [*Panicularia nervata* (Willd.) Kuntze var. *elata* (Nash) Piper]. Stems 60-120 cm high; ligules 3-6 mm long, blunt; leaves flat, 6-10 mm wide; spikelets 3-5 mm long in a diffuse, pyramidal inflorescence; lemmas ca. 2 mm long. Aug. Fig. 333.

Common in wet soil of montane stream banks and around wetlands; East, West. B.C., Alta. south to CA, NM. *G. grandis* has larger spikelets and *G. striata* has shorter ligules.

Glyceria grandis Wats. ex Gray. Stems 90-150 cm high; ligules blunt, 4-9 mm long; leaves flat, 6-15 mm wide; spikelets 5-6 mm long in an open, pyramidal inflorescence; lemmas ca. 2.5 mm long.

Montane marshes and slow streams; reported for St. Mary Lake (Maguire 1934). AK to Newf. south to CA, NM, IA, VA. See *G. eleta.*

Glyceria striata (Lam.) Hitchc., Fowl Mannagrass [*Panicularia nervata* (Willd.) Kuntze]. Stems 30-70 cm high, often forming tussocks; ligules 1-3 mm long; leaves 2-5 mm wide; spikelets 3-4 mm long, often purplish; inflorescence open, pyramidal; lemmas to 2 mm long. July-Aug.

Common in montane marshes, stream banks, and wet meadows, especially around ponds; East, West. AK to Newf. south to Mex. *G. elata* has longer ligules and wider leaves.

Helictotrichon Besser

Helictotrichon hookeri (Scribn.) Henrard. Tussock-forming perennial; stems 10-50 cm high; leaves mostly basal with open sheaths and blades 2-4 mm wide; ligules membranous 1-3 mm long; spikelets shiny with 3-6 flowers, short-stalked in a spike-like inflorescence; glumes nearly equal, 10-14 mm long, pointed; lemmas 10-12 mm long, barely shorter than the glumes, with a bent awn 10-15 mm long from the back and hairs at the base; palea shorter than the lemma. June-Aug. Fig. 334.

Uncommon in montane to alpine grasslands and meadows; East. Alta. to Man. south to NM, ND, MN.

Hierochloe R. Br.

Hierochloe odorata (L.) Beauv., Sweetgrass [*Torresia odorata* (L.) Hitchc.]. Rhizomatous perennial; stems 25-60 cm high; ligules membranous, 3-5 mm long; leaves 3-5 mm wide, vegetative basal ones long, those on the stem short; inflorescence pyramidal with spreading branches; spikelets shiny bronze, 3-5 mm long with 3 flowers, only 1 of which bears a seed; glumes equal, slightly longer than the lemmas; lemmas 3-5 mm long, unawned; palea nearly as long as the lemma. May-Aug. Fig. 335.

Uncommon in moist montane to subalpine meadows and depressions in grasslands; East, West. Circumboreal south to CA, NM, SD, PA. Plants produce flowering stems early, then produce long, basal leaves. These later

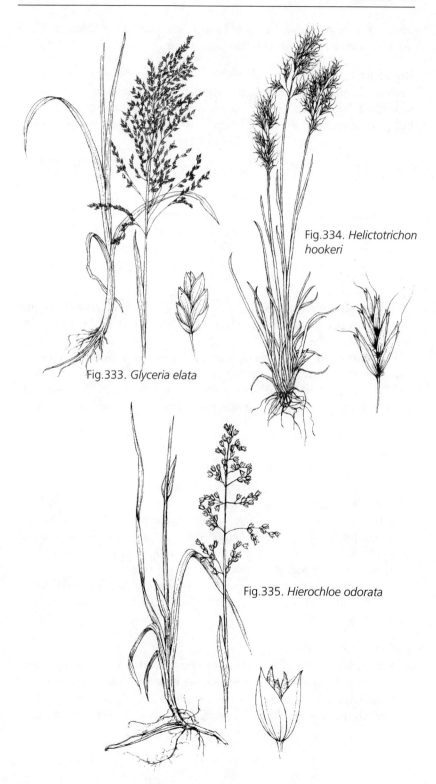

Fig.334. *Helictotrichon hookeri*

Fig.333. *Glyceria elata*

Fig.335. *Hierochloe odorata*

leaves smell of vanilla and are dried and braided by Native Americans. It is used as incense in ceremonies and as perfume.

Hordeum L., Barley
Tussock-forming perennials; ligules short, membranous; leaf sheaths open; spikelets 1-flowered, sessile, 3 per node in a spike inflorescence 5-10 cm long, only the middle one fertile; glumes awl-shaped, awned; lemmas awned; palea shorter than lemma. See *Elymus elymoides.*

1. Awn of central lemma ≤ 15 mm long .... *H. brachyantherum*
1. Awn of central lemma ≥ 18 mm long .... *H. jubatum*

Hordeum brachyantherum Nevski [*H. nodosum* L., *H. jubatum* L. ssp. *breviaristatum* Bowden]. Stems 20-80 cm tall; leaves 2-6 mm wide; glumes 5-18 mm long; awn of central lemma 4-14 mm long. July-Aug.

    Uncommon in vernally wet soil around montane ponds and depressions; east margin of the Park and Marias Pass area. AK to Newf. south to CA, NM.

Hordeum jubatum L., Foxtail Barley. Stems 20-60 cm tall; leaves 2-4 mm wide; glumes 35-90 mm long; awns of the central lemma 20-50 mm long. July-Aug. Fig. 336.

    Uncommon in vernally wet soil around ponds and lakes along the east margin of the Park; also along roads, East, West. AK to Newf. south to Mex. Buffalo wallows and disturbed pond margins were probably this plant's native habitat; now it is just as common along roads and in other disturbed areas.

Koeleria Pers.
Koeleria macrantha (Ledeb.) Schultes, Junegrass [*K. cristata* Pers., *K. pyramidata* (Lam.) Beauv.]. Tussock-forming perennial; stems 20-50 cm high; ligules membranous, short; leaves 1-2 mm wide; spikelets with 2-4 flowers, short-stalked in a narrow, branched inflorescence 2-8 cm long, the branches spreading in flower but erect before and after; glumes nearly equal, pointed; lemmas 3-5 mm long, rounded on back, ca. as long as the glumes, sometimes awn tipped; palea shorter than the lemma. June-July. Fig. 337.

    Grasslands; common montane, uncommon higher; East, West. Circumboreal south to Mex. The inflorescence is much wider in flower than before or after.

Lolium L., Ryegrass
Lolium multiflorum Lam., Italian Ryegrass. Short-lived perennial; stems 20-80 cm high, forming small tufts; ligules membranous, short; leaves 4-8 mm wide; spikelets with 5-15 flowers, sessile, born edgewise to the axis of the spike inflorescence; glume 7-12 mm long, solitary in all but the terminal

spikelet; lemmas shorter than the glume, short-awned; palea as long as the lemma. July.

Rare along roads and around residences; collected once near West Glacier. Introduced from Europe as a lawn grass.

Melica L., Oniongrass
Perennial with clustered stems; ligules membranous; leaf sheaths closed; spikelets with 2-several flowers, the upper sterile and folded together; inflorescence branched; glumes unequal, pointed; lemmas longer than the glumes; palea shorter than or equal to the lemma.

1. Lemmas with awns; stems not greatly swollen at the base .... *M. smithii*
1. Lemmas unawned; stems bulbous-based 2
2. Lemmas long-pointed, long-hairy on the veins .... *M. subulata*
2. Lemmas rounded at the tip, glabrous 3
3. Longest glume <9 mm long, distinctly shorter than lowest lemma ....
.... *M. spectabilis*
3. Longest glume 9-13 mm long, ca. as long as lowest lemma .... *M. bulbosa*

Melica bulbosa Geyer ex Porter & Coult. [*M. bella* Piper]. Stems 30-80 cm tall, bulbous-based, densely clustered on a short rhizome; ligule 3-4 mm long, ragged; leaves 2-4 mm wide; inflorescence narrow with short, erect branches; spikelets 10-15 mm long; longest glume 9-13 mm long; the lemmas barely longer than the glumes, unawned. July.

Rare in montane and lower subalpine grasslands; East and the Marias Pass area. B.C. to CA, CO.

Melica smithii (Porter) Vasey. Stems 30-80 cm high; ligules 3-9 mm long, blunt; leaves lax, 4-12 mm wide; inflorescence sparse, diffuse with spreading branches; spikelets 15-20 mm long; glumes often purplish, lemmas ca. 10 mm long with an awn 3-6 mm long. June-July.

Uncommon in moist to wet, montane and lower subalpine open forest; East, West. B.C. to Ont. south to OR, WY, MI.

Melica spectabilis Scribn. Stems 30-60 cm high, bulbous-based, well-spaced and attached to the rhizome by a short stalk; ligules 1-3 mm long; leaves 2-3 mm wide; inflorescence narrow with short, ascending, wavy branches; spikelets 9-12 mm long; longest glume 6-7 mm long; the lemmas slightly longer, purple tipped, unawned. June-July. Fig. 338.

Common in moist, montane and lower subalpine grassland and open forest; East, West. B.C., Alta. south to CA, CO.

Melica subulata (Griseb.) Scribn. Stems 30-80 cm high, mostly bulbous-based; ligules 1-5 mm long, ragged; leaves flat 2-5 mm wide; inflorescence narrow with few, long, ascending branches; spikelets 15-20 mm long; lemmas 9-13 mm long, long-tapered, unawned, sparsely long-hairy. June-July.

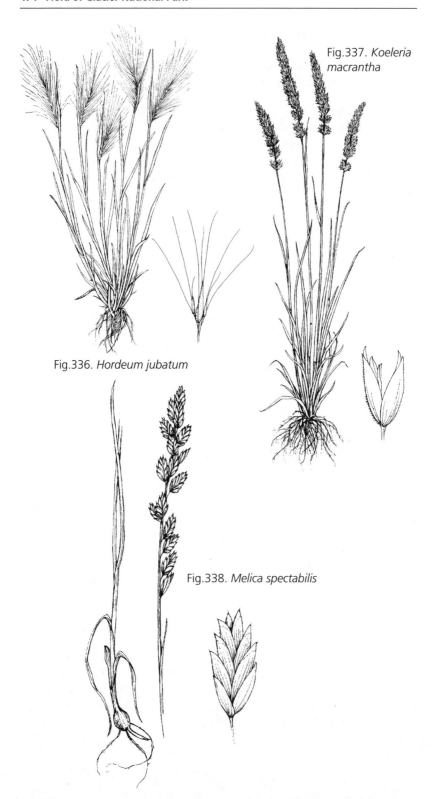

Fig.337. *Koeleria macrantha*

Fig.336. *Hordeum jubatum*

Fig.338. *Melica spectabilis*

Common in moist, open, montane coniferous forest; East, West. AK to CA, ID, WY.

Muhlenbergia Schreb., Muhly

Annuals or rhizomatous perennials; ligules membranous; leaf sheaths open; inflorescence branched; spikelets 1-flowered, stalked; glumes (exclusive of awns) shorter or equal to the lemma; lemma rounded on back, pointed but unawned; palea nearly as long as lemma.

1. Inflorescence dense, cylindrical; spikelets 5-6 mm long .... *M. glomerata*
1. Inflorescence narrow but more diffuse; spikelets 2-3 mm long 2
2. Glumes < 1 mm long; stems rooting at the nodes; plants not rhizomatous ... . *M. filiformis*
2. Glumes > 1 mm long; plants rhizomatous .... *M. richardsonis*

Muhlenbergia filiformis (Thurb.) Rydb. Annual to short-lived perennial; stems 3-10 cm high, rooting at the nodes, mat-forming; ligules 1-3 mm long; leaves 1-3 mm wide; inflorescence narrow; spikelets ca. 2 mm long. July.

Rare in vernally wet, shallow soil, montane; known from 1 location near Grinnell Lake. B.C. to CA, NM, SD. *M. richardsonis* is similar but has larger spikelets.

Muhlenbergia glomerata (Willd.) Trin. Perennial; stems 30-60 cm high, often branched at the base; ligules ca. 1 mm long; leaves flat 2-4 mm wide; inforescence dense, cylindrical; spikelets 4-6 mm long. July-Aug.

Rare on hummocks in montane, calcareous fens; known only from north of Apgar. Yuk. to Newf. south to NV, CO, NE, IA, WV. The spike-like inflorescence is reminiscent of *Alopecurus*.

Muhlenbergia richardsonis (Trin.) Rydb. [*M. squarrosa* (Trin.) Rydb.]. Mat forming, rhizomatous perennial; stems 10-40 cm high, erect or ascending; ligules 1-2 mm long; leaves ca. 1 mm wide, inrolled; inflorescence narrow; spikelets 2-3 mm long. July-Aug. Fig. 339.

Rare in montane grasslands, sometimes around ponds or on hummocks in fens; East, West. B.C. to N. B. south to CA, NM, NE, MI. See *M. filiformis*.

Oryzopsis Michx., Rice Grass

Tuft- or tussock-forming perennials; ligules membranous, short; leaf sheaths open; spikelets 1-flowered, stalked, in a narrow inflorescence with erect branches; glumes equal; lemma as long as glumes, awned; palea as long as the lemma and enclosed by it.

1. Leaves flat, 3-7 mm wide; glumes 6-8 mm long .... *O. asperifolia*
1. Leaves with rolled margins; glumes ca. 4 mm long .... *O. exigua*

Oryzopsis asperifolia Michx. [*Achnatherum asperifolium* (Michx.) Barkw.]. Stems 30-70 cm tall; ligules short; leaves flat, 3-7 mm wide; inflorescence

5-8 cm long; glumes 6-8 mm long; awn of lemma 5-10 mm long. June-July. Fig. 340.

Common in moist montane forest; West. B.C. to Newf. south to WA, ID, NM, SD. The clumps of broad, deep green leaves are distinctive.

Oryzopsis exigua Thurb. [*Achnatherum exigua* (Thurb.) Barkw.]. Stems 10-30 cm high with old stems at the base; ligules 2-4 mm long; leaves 1 mm wide, inrolled; inflorescence 3-7 cm long; glumes 4 mm long; awn of lemma 4-5 mm long. June-Aug.

Common in stony, sparsely vegetated soil of grasslands, open forest, rocky slopes, and ridges, montane to lower alpine; East, West. B.C. to Alta. south to OR, UT, CO.

Panicum L.

Panicum capillare L., Witch Grass. Annual; stems erect to spreading, 10-50 cm high; ligules of long hairs; foliage long-hairy; leaf blades 5-12 mm wide; spikelets 2-flowered but appearing 1-flowered, in a diffuse inflorescence with spreading, whorled branches; glumes unequal, one very small, the other as long as the sterile lemma, 2-3 mm long; fertile lemma hard with a pointed tip, shorter than the long glume, unawned; palea enfolded in the lemma. Aug.

Rare in disturbed soil around roads and buildings; collected once at West Glacier. B.C. to Que., south to most of U. S. This plant's native habitat is pond margins and riverine gravel bars.

Phalaris L., Canary Grass

Phalaris arundinacea L., Reed Canarygrass. Rhizomatous perennial with stems to 200 cm high; ligules membranous, 2-5 mm long; leaves 6-15 mm wide; spikelets stalked, with 1 fertile and 2 sterile flowers in tightly branched, spike-like clusters in a narrow inflorescence 6-18 cm long; glumes equal, ca. 5 mm long; sterile lemmas minute; fertile lemma 3-4 mm long, rounded at the tip. July-Aug. Fig. 341.

Common in montane wet meadows, marshes, stream banks, and roadside ditches; East, West. Circumboreal south to most of U. S. Standley (1921) observed it only around East Glacier, but it is now much more common. Although the plant is native to Canada and the northern U. S., many of our plants are probably derived from European cultivars introduced for forage production (Merigliano and Lesica 1998).

Phleum L., Timothy

Tussock-forming perennials; ligules membranous, 1-3 mm long; leaves flat, 4-8 mm wide, the sheaths open; spikelets 1-flowered, short-stalked, congested in a branched, but spike-like, cylindrical inflorescence; glumes equal, V-shaped in cross section with a line of thick, long hairs on the

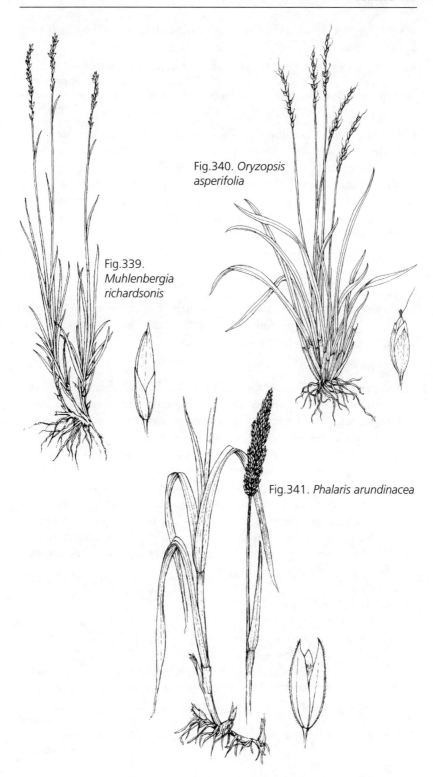

Fig.340. *Oryzopsis asperifolia*

Fig.339. *Muhlenbergia richardsonis*

Fig.341. *Phalaris arundinacea*

margins, awned; lemma smaller than glumes, blunt-tipped; palea as long as the lemma.

1. Inflorescence > 5 cm long, < 1 cm wide; stems swollen at the base ....
.... *P. pratense*
2. Stems not bulbous-based; inflorescence < 5 cm long, but often > 1 cm wide
.... *P. alpinum*

Phleum alpinum L., Alpine Timothy [*P. commutatum* Gaudin]. Stems 10-40 cm high; inflorescence 15-30 mm long; glumes 3-5 mm long with awns 2-3 mm long; lemma 2-3 mm long. July-Aug. Fig. 342.

Moist meadows, grasslands, and stony slopes, often in disturbed or sparsely vegetated soil; abundant subalpine and alpine, uncommon lower; East, West. AK to Newf. south to CA, NM, Europe. An early successional species useful in high-elevation revegetation.

Phleum pratense L., Timothy. Stems 40-90 cm high with a swollen base; inflorescence 5-12 cm long; glumes 3-4 mm long with awns ca. 1 mm long; lemma 2 mm long. July-Aug.

Montane grasslands and roadsides; abundant East, common West. Native to Europe. Apparently introduced in parts of the Park by horse packers attempting to increase forage for their stock.

Poa L., Bluegrass

Mainly perennial; ligules membranous; leaf sheaths closed below; leaf blades with prow-like tips; spikelets several-flowered, stalked in a branched inflorescence; flowers sometimes unisexual; glumes unawned; lemmas longer than glumes, unawned, usually hairy; palea slightly shorter than lemma. References: Arnow (1981), Kellogg (1985), Soreng (1985).

This group consists of many poorly differentiated species, the result of a predominance of asexual reproduction in many of the sections as well as frequent hybridization. A perfectly reliable and easy-to-use key is probably impossible. Kellogg (1985) reduced many formerly recognized taxa to a single species (*P. secunda*); however, many very similar species separated by difficult characters are still recognized in other sections of the genus.

1. Spikelets converted to bulb-like plantlets .... *P. bulbosa*
1. Spikelets with flowers not bulblets.. 2

2. Spikelets ≤ 4 mm long 3
2. Spikelets > 4 mm long 8

3. Plants montane to lower subalpine 4
3. Plants near or above treeline 6

4. Plants of grasslands or open forest, forming tufts or tussocks .... *P. interior*
4. Plants rhizomatous or stoloniferous, sometimes in wet habitats 5

5. Stems arising from rhizomes; plants of meadows, grasslands, roadsides, lawns... *P. pratensis*
5. Plants stoloniferous; plants of marshes wet meadows, and streamsides ....
.... *P. palustris*

6. Lemmas with long, tangled cobweb-like hairs at the base............. *P. interior*
6. Lemmas sometimes hairy but without long tangled hairs at the base 7

7. Lemmas glabrous .... *P. lettermanii*
7. Lemmas short-hairy .... *P. glauca*

8. Stems arising from rhizomes, sometimes in small tufts 9
8. Plants forming bunches or tussocks; rhizomes lacking............................. 12

9. Lemmas glabrous or nearly so; anthers aborted .... *P. wheeleri*
9. Lemmas hairy, usually with long tangled hairs at the base 10

10. Plants near or above treeline... *P. arctica*
10. Plants montane to lower subalpine 11

11. Stems flattened .... *P. compressa*
11. Stems round in cross section .... *P. pratensis*

12. Spikelets > 6 mm long 13
12. Spikelets < 6 mm long 15

13. Spikelets not compressed, nearly cylindrical before and after flowering, > 2
times as long as wide; lemma rounded on back .... *P. secunda*
13. Spikelets somewhat compressed, mostly < 2 times as long as wide; lemma
with a ridge down the middle of the back... 14

14. Lemmas hairy on the lower half... *P. stenantha*
14. Lemmas not hairy .... *P. cusickii*

15. Lemmas 2-3 mm long... *P. glauca*
15. Lemmas > 3 mm long.. 16

16. Anthers > 1 mm long 17
16. Anthers ≤ 1 mm long 19

17. Lemmas with long, tangled cobweb-like hairs at the base............. *P. arctica*
17. Lemmas sometimes hairy but without long tangled hairs at the base 18

18. Spikelets not compressed, nearly cylindrical before and after flowering, > 2
times as long as wide; lemma rounded on back .... *P. secunda*
18. Spikelets compressed, < 2 times as long as wide; lemma with a ridge down
the middle of the back .... *P. alpina*

19. Plants montane to lower subalpine, often along trails or in disturbed areas .
.... *P. annua*
19. Plants near or above treeline... 20

20. Long tangled hairs at base of lemma; branches of inflorescence spreading
or reflexed .... *P. leptocoma*
20. Hairs at base of lemma short; inflorescence branches erect or ascending ....
.... *P. laxa*

Poa alpina L., Alpine Bluegrass. Stems 10-30 cm high, forming small tussocks; ligule 1-4 mm long; leaves short, 2-5 mm wide; inflorescence congested, pyramidal with spreading to ascending branches; spikelets flattened, purplish, 3-6 mm long; glumes unequal; lemmas 3-4 mm long, hairy on lower half. June Aug. Fig. 343.

Abundant in moist to dry turf, meadows, and grassland as well as moraine, and rocky slopes, mostly near or above treeline, occasionally lower;

East, West. Circumpolar south to OR, UT, CO, MI. An early successional species that persists as the turf becomes more dense; useful in revegetation because of its great ecological flexibility.

Poa annua L., Annual Bluegrass. Annual or short-lived perennial; stems 3-15 cm long, erect to prostrate, forming small bunches; ligules 1 mm long; leaves 1-3 mm wide; inflorescence small, open with spreading lower branches; spikelets 3-6 mm long; glumes unequal; lemmas 2-3 mm long, long-hairy lower on the veins. June-Aug.

Locally common in shaded, sparsely vegetated soil along trails and roads and around buildings and campgrounds; East, West. Introduced from Europe. Some plants persist through winter and flower a second time.

Poa arctica R. Br. [*P. grayana* Vasey]. Stems 10-30 cm high, forming small tufts on short rhizomes; ligules 1-3 mm long; leaves 1-3 mm wide, mostly basal; inflorescence narrowly pyramidal with ascending or spreading branches; spikelets 4-8 mm long, purplish; glumes unequal; lemmas 3-5 mm long, flattened, hairy on the lower half, usually with cobwebby hairs at the base. July-Aug.

Common in moist turf, wet cliffs or fens, usually on cool slopes, alpine or rarely lower; East. Our plants are ssp. *grayana* (Vasey) Löve et al. Circumpolar south to OR, UT, NM.

Poa bulbosa L. Stems 20-40 cm high, forming small tussocks; ligules 1-4 mm long; leaves 1-3 mm wide; inflorescence spike-like; flowers converted to bulb-like plantlets to 15 mm long; glumes 2-3 mm long.

Rare along roads and in other disturbed areas; collected once near Apgar. Introduced from Europe.

Poa compressa L., Canada Bluegrass. Stems flattened, 20-50 cm high from prolific rhizomes; ligules ca. 1 mm long; leaves 2-4 mm wide; inflorescence narrow at first but becoming narrowly pyramidal with short, spreading branches at flowering; spikelets 4-6 mm long; glumes nearly equal; lemmas 2-3 mm long, hairy on margins and midvein. July-Aug.

Uncommon in disturbed, sparsely vegetated, often compacted soil along trails and roads and around buildings and campgrounds; East, West. Introduced from Europe. The flattened stems are distinctive.

Poa cusickii Vasey [*P. epilis* Scribn.]. Stems 15-40 cm high, forming tussocks; ligules short; leaves 1-3 mm wide; inflorescence congested with ascending branches; spikelets 5-8 mm long: flowers mainly female; glumes nearly equal; lemmas 4-6 mm long, sometimes purplish with membranous margins, densely to sparsely hairy on the margins and midvein. June-Aug.

Common in montane grassland to alpine meadows and turf; East, West. B.C. to Man. south to CA, UT, CO.

Poa glauca Vahl. [*P. rupicola* Nash]. Stems 5-25 cm high, forming small tussocks; ligules ca. 1 mm long, blunt; leaves 1-3 mm wide; inflorescence narrowly open with ascending branches; spikelets purplish, 3-5 mm long; glumes slightly unequal; lemmas ca. 3 mm long, hairy on the margins. July-Aug.

Common in drier alpine turf and fellfields; East, less common West. Our plants are ssp. *rupicola* (Nash) Boivin. Circumpolar south to CA, NM, SD. See *P. laxa*.

Poa interior Rydb. [*P. nemoralis* L. var. *interior* Butters & Abbe, *P. glauca* Vahl in part]. Stems 20-60 cm high, forming small tussocks; ligules < 1 mm long; leaves 1-2 mm wide; inflorescence narrowly open with ascending branches; spikelets 3-4 mm long; glumes slightly unequal; lemmas 2-3 mm long, hairy on the margins with long tangled hair at the base. July-Aug.

Common in montane and lower subalpine grasslands and dry, open forest; East, West. B.C. to Man. south to WA, AZ, TX.

Poa laxa Haenke. Stems 8-25 cm high, forming small tussocks; ligules 2-4 mm long; leaves 1-2 mm wide; inflorescence somewhat narrow with nearly glabrous, ascending branches; spikelets 4-6 mm long; glumes unequal; lemmas 3-5 mm long, long-hairy on the midvein and margins. Aug.

Rare in alpine turf; known only from north of Logan Pass. Our plants are ssp. *banffiana* Soreng. Circumpolar south to MT, VT. This recently described taxon (Soreng 1991) is closely allied to *P. leptocoma* and *P. glauca*; however, the inflorescence branches of the former are spreading or reflexed, while those of the latter have minute prickles.

Poa leptocoma Trin. [*P. paucispicula* Scribn. & Merr.]. Stems 15-50 cm high, forming small, loose tufts; ligules 1-2 mm long; leaves 2-3 mm wide; inflorescence with spreading or ascending branches; spikelets flattened, 5-7 mm long, purplish; glumes unequal; lemmas 3-5 mm long, hairy on lower margins and midvein with cobwebby hairs at the base. Aug.

Uncommon in moist meadows and turf near or above treeline; East, West. Our plants are var. *paucispicula* (Scribn. & Merr.) Hitchc. AK to CA, NM; Asia. Reports of *P. reflexa* Vasey & Scribn. are referable here.

Poa lettermanii Vasey. Stems 2-10 cm high, forming small tussocks; ligules 1-3 mm long; leaves ca. 1 mm wide; inflorescence narrow with erect or ascending branches; spikelets purplish 4-5 mm long; glumes nearly equal 2-4 mm long; lemmas glabrous 2-3 mm long. Aug.

Rare in moist alpine turf; collected once near Logan Pass. B.C. to Alta. south to CA, UT, CO.

Poa palustris L. Stems loosely clustered, 40-100 cm high, sometimes prostrate and rooting at the nodes; ligules 3-5 mm long; leaves 1-3 mm wide; inflorescence long, open with spreading, whorled branches; spikelets

3-4 mm long; glumes barely unequal; lemmas 2-3 mm long, hairy on the lower margins and midvein with cobwebby hairs at the base. July-Aug.

Common in moist or wet meadows and along slow streams, sometimes in open montane forest; East, West. Circumboreal south to CA, NM, MO, VA. A. S. Hitchcock (1950) considers this plant native, while C. L. Hitchcock (Hitchcock et al. 1969) believes it is introduced.

Poa pratensis L., Kentucky Bluegrass. Stems 20-80 cm high, forming tufts from spreading rhizomes; ligules 1-3 mm long, blunt; leaves 1-4 mm wide, usually folded; inflorescence narrow to pyramidal with whorled branches; spikelets 3-6 mm long; glumes unequal; lemmas 3-4 mm long, hairy on margins and midvein with cobwebby hairs at the base. June-Aug.

Common in moist to wet meadows, grasslands and open forest, abundant in disturbed areas along roads, trails, campgrounds and buildings, montane to lower subalpine; East, West. Circumboreal south through most of U. S. Generally thought to be introduced from Europe, Boivin and Löve (1960) consider one form (*P. agassizensis* Boivin & Löve) to be native in our area. It seems likely that our plants are a mixture of indistinguishable native and introduced forms.

Poa secunda Presl, Sandberg Bluegrass [*P. ampla* Merr., *P. gracillima* Vasey, *P. juncifolia* Scribn., *P. lucida* Vasey, *P. nevadensis* Vasey, *P. sandbergii* Vasey, *P. scabrella* (Thurb.) Benth.]. Stems 15-90 cm high, forming small tussocks; ligules 1-6 mm long; leaves 1-4 mm wide, sometimes inrolled; inflorescence narrow to open with erect to spreading branches; spikelets 4-7 mm long, nearly cylindrical; glumes slightly unequal; lemmas 3-5 mm long rounded on back, glabrous to hairy on the lower part. June-Aug. Fig. 344.

Ssp. *juncifolia* (Scribn.) Soreng [=*P. ampla, P. juncifolia*], 30-90 cm tall with spikelets 5-7 mm long, is common in montane grasslands; East. Ssp. *secunda* has two forms in the Park: one, < 30 cm tall with a narrow inflorescence and spikelets 4-7 mm long [=*P. sandbergii, P. nevadensis, P. scabrella*], is common in montane to alpine grasslands and rocky slopes and ridges; East, West. The other [=*P. gracillima*] is taller with an open inflorescence and spikelets 5-7 mm long; it occurs in grasslands, meadows, and turf and is abundant in the alpine and subalpine and common lower; East. Yuk. to Newf. south to Mex., Chile. Kellogg (1985) united the numerous taxa of the "*P. sandbergii* complex" under *P. secunda*. Many of these entities seem very distinct when viewed at a local scale but merge together when viewed over the entire range of the complex.

Poa stenantha Trin. Stems 30-50 cm high, forming tussocks; ligules 1-4 mm long; leaves 1-2 mm wide; inflorescence somewhat open with spreading, whorled branches; spikelets 6-8 mm long; glumes unequal; lemmas 4-6 mm long, long-hairy on the lower half. July-Aug.

Uncommon in stony soil of montane to subalpine grasslands and meadows; East, West. AK to Alta. south to OR, CO.

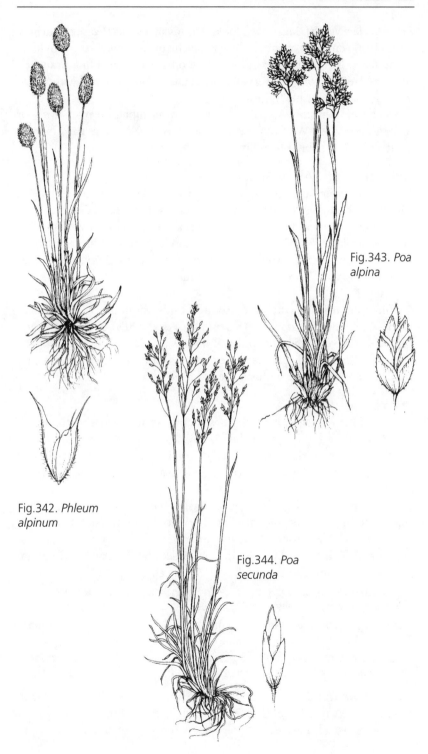

Fig.343. *Poa alpina*

Fig.342. *Phleum alpinum*

Fig.344. *Poa secunda*

Poa wheeleri Vasey [*P. nervosa* (Hook.) Vasey var. *wheeleri* (Vasey) Hitchc.]. Stems 20-60 cm high, in small clusters on rhizomes; ligules to 1 mm long, blunt; leaves 1-3 mm wide; inflorescence open with spreading branches; spikelets 3-6 mm long; flowers all female; glumes nearly equal; lemmas 3-5 mm long, sparsely hairy. June-Aug.

Uncommon in stony soil of montane to lower subalpine open forest or rocky slopes; East, West. B.C., Alta south to CA, NM. The rhizomatous habit and forest habitat help distinguish this bluegrass.

Puccinellia Parl., Alkali Grass

Puccinellia pauciflora (Presl) Munz. Rhizomatous perennial with stems 15-100 cm high; ligules membranous, 5-6 mm long, ragged; leaves 3-12 mm wide; inflorescence open with spreading or ascending, whorled branches; spikelets 3- to 7-flowered, 4-5 mm long, purplish; glumes unequal, shorter than the lemmas, rounded at the tip, the longer with 3 veins; lemmas 2-3 mm long with 5 prominent nerves, blunt with ragged edges at the tip. July-Aug. Fig. 345.

Uncommon in marshes and around ponds, montane; collected near Camas Creek and Marias Pass. AK south to CA, NM, SD. Similar to *Glyceria* spp. which have 1-veined glumes. *P. distans* (L.) Parl. and *P. nuttalliana* (Schult.) Hitchc., with shorter ligules and less prominent lemma veins, are common around saline ponds just east of the Park.

Stipa L., Needle Grass

Tussock-forming perennials; ligules membranous; leaf sheaths open; spikelets 1-flowered in a branched inflorescence; glumes narrow, nearly equal; lemma cylindric, awned from the tip, hardened with a sharp-pointed base (callus); palea enclosed in the larger lemma. Reference: Barkworth et al. (1979).

The awns twist when dry and untwist when moist, so just the daily changes in humidity enable the seed to drill itself into the soil. Barkworth and Everett (1987) propose placing our species of *Stipa* into two genera, *Stipa* and *Achnatherum*; however, their evidence that these are more natural groups is not compelling enough to make the change at this time.

1. Inflorescence diffuse with long spreading branches .... **S. richardsonii**
1. Inflorescence narrow with erect branches 2
2. Awns 10-15 cm long; glumes > 15 mm long .... **S. comata**
2. Awns < 4 cm long 3
3. Leaf sheaths long-hairy at the top .... **S. viridula**
3. Leaf sheaths not long-hairy at the top .... **S. nelsonii**

Stipa comata Trin. & Rupr., Needle-and-Thread [*Hesperostipa comata* (Trin. & Rupr.) Barkworth]. Stems 30-70 cm high; ligules 2-4 mm long; leaves 1-2 mm wide, usually inrolled; inflorescence with erect to ascending branches;

glumes 15-20 mm long; lemmas 8-12 mm long, the awn 10-15 cm long. July.

Rare in grasslands along the east margin of the Park. B.C. to Ont. south to CA, TX, NE.

Stipa nelsonii Scribn. [*S. occidentalis* Thurb. var. *nelsonii* (Scribn.) Hitchc., *S. o.* var. *minor* (Vasey) Hitchc., *S. columbiana* Macoun, *Achnatherum nelsonii* (Scribn.) Barkworth]. Stems 30-80 cm tall; ligules 1-2 mm long; leaves 1-3 mm wide, usually inrolled; inflorescence narrow with short, erect branches; glumes 7-12 mm long; lemmas 4-6 mm long, the awn 2-3 cm long. June Aug. Fig. 346.

Common in montane grassland and open forest; East, West. Our plants are ssp. *dorei* Barkworth & Maze. Yuk. to Sask. south to CA, NV, UT, CO. *S. viridula* has a ring of long hairs on the tops of the leaf sheaths that is lacking in *S. nelsonii*.

Stipa richardsonii Link [*Achnatherum richardsonii* (Link) Barkworth]. Stems 40-80 cm tall; ligules < 1 mm long; leaves inrolled, 1-2 mm wide; inflorescence diffuse, spikelets on the tips of long, spreading branches; glumes 7-10 mm long; lemmas 5-6 mm long, the awn 2-3 cm long. July-Aug.

Uncommon in montane grasslands; East, West. Yuk. to Man. south to WA, ID, CO, SD. The diffuse inflorescence with long-awned spikelets is diagnostic.

Stipa viridula Trin., Green Needlegrass [Achnatherum viridula (Trin.) Barkworth]. Stems 40-80 cm high; ligules ≤ 1 mm long; leaves 1-3 mm wide, sometimes inrolled; inflorescence narrow with short, erect branches; glumes 7-10 mm long; lemmas 5-6 mm long, the awn 2-3 cm long.

Montane grasslands along the east edge of the Park. B.C. to Man. south to AZ, NM, KS, MN. Standley (1921) reports that this species is frequent, but no collections have been made during the last 80 years.

Trisetum Pers.

Tussock-forming perennials; ligules membranous, ragged; leaf sheaths open; spikelets mostly 2- or 3-flowered, stalked in a branched inflorescence; glumes unequal; lemmas mostly awned from the back, longer or equal to the glumes, forked at the tip; palea slightly shorter than the lemma.

1. Lemmas without awns .... ***T. wolfii***
1. Lemmas with awns 2

2. Awns of the lemmas 5-6 mm long 3
2. Awns of the lemmas > 8 mm long 4

3. Stems to 50 cm high; inflorescence dense, spike-like .... ***T. spicatum***
3. Stems mostly > 50 cm high; inflorescence narrow but not spike-like ....
.... ***T. orthochaetum***

4. Glumes 3-5 mm long, shorter than the lemmas .... *T. cernuum*
4. Glumes 5-7 mm long, as long as the lowest lemma .... *T. canescens*

Trisetum canescens Buckl. Stems 50-80 cm high; ligules 1-4 mm long; leaves 3-7 mm wide; inflorescence narrow with ascending branches; spikelets 2- or 3 flowered; glumes 5-7 mm long; lemmas 5-7 mm long, the bent awn 10-14 mm long. July-Aug.

Uncommon in montane meadows, fens, and forest openings; East, West. B.C. to Alta. south to CA, ID, MT.

Trisetum cernuum Trin. Stems 50-90 cm high; ligules 1-3 mm long; leaves 5-10 mm wide; inflorescence diffuse with wavy branches; spikelets usually 3-flowered; glumes 3-5 mm long; lemmas 5-6 mm long with hairs at the base ca. 2 mm long, the awn ca. 10 mm long. July-Aug.

Common in moist montane and lower subalpine coniferous forest; East, West. AK to Alta. south to CA, ID, MT. Often with other tall forest grasses, such as *Bromus vulgaris* and *Elymus glaucus.*

Trisetum orthochaetum Hitchc. Stems 50-100 cm high; ligules 3-5 mm long; leaves 4-7 mm wide; inflorescence open with ascending branches; spikelets mostly 3-flowered; lemmas 5-6 mm long, as long as the glumes, the awn 5-6 mm long. July.

Rare in openings of wet lower submontane forest; collected once near Swiftcurrent Lake. Otherwise known only from the Bitterroot Mtns. southwest of Missoula, this grass is thought to be a sterile hybrid between *T. wolfii* and *T. canescens.*

Trisetum spicatum (L.) Richt. Stems 10-60 cm high; foliage often short-hairy; ligules 1-2 mm long; leaves 1-4 mm wide; inflorescence spike-like; spikelets 2-flowered; lemmas 3-5 mm long, shorter than the longer glume, the bent awn 5-6 mm long. July-Aug. Fig. 347.

Abundant in stony soil of grasslands, meadows, drier turf, and talus, upper montane to alpine; East, West. Circumpolar south to CA, Mex. The inflorescence is reminiscent of Junegrass.

Trisetum wolfii Vasey. Stems 50-100 cm high; ligules 2-4 mm long; leaves 2-6 mm wide; inflorescence narrow with erect or ascending branches; spikelets 2-flowered; lemmas 4-6 mm long, shorter than the glumes, unawned. July-Aug.

Uncommon in moist montane to subalpine meadows and open forest; East West. WA to Alta. south to CA, UT. Our only awnless *Trisetum.*

Triticum L., Wheat
Triticum aestivum L. Annual; stems 30-60 cm high, branched at the base; ligules membranous, short; leaves with open sheaths and blades 1-2 cm wide; spikelets 2- to 5-flowered, sessile, 1 per node in a spike inflorescence

5-12 cm long; glumes 6-8 mm long, pointed or awned; lemmas longer than the glumes with an awn up to 8 cm long; palea as long as the lemma. July-Sept.

Rare, introduced along roads and railroad tracks, does not persist for more than 1 or 2 years.

POTAMOGETONACEAE: PONDWEED FAMILY

Potamogeton L., Pondweed

Submergent or floating-leaved aquatic perennials; stems flaccid; leaves mostly with entire or wavy margins, alternate or opposite above with membranous, stem-sheathing stipules at the base of the leaf; submerged leaves and floating leaves often different; flowers small, bisexual, sessile in stalked, cylindric, interrupted or solid spikes from leaf axils; sepals 4, minute; stamens 4; ovaries 4, each with a sessile stigma; fruit a seed surrounded by hard flesh (achene), more-or-less ovoid with a short beak on top.

Plants often produce free-floating turions (large buds) that sink and overwinter in mud until forming new plants the following summer. The turions, root tubers, and fruits are important foods for water fowl. All of our species occur in the montane zone. *P. foliosus* Raf., similar to *P. pusillus* but without glands at the leaf bases, occurs just north and west of the Park.

1. Leaves clasping the stem at the base 2
1. Leaves not clasping the stem 3

2. Stems zig-zag; some leaves > 10 cm long .... *P. praelongus*
2. Stems mostly straight; leaves usually < 10 cm long .... *P. richardsonii*

3. At least the floating leaves ≥ 5 mm wide 4
3. All leaves < 5 mm wide 8

4. Submersed leaves thread-like, 1-2 mm wide but > 10 cm long .... *P. natans*
4. Submersed leaves > 2 mm wide 5

5. Submersed leaves sickle-shaped, > 2 cm wide .... *P. amplifolius*
5. Submersed leaves usually < 2 cm wide, not sickle-shaped 6

6. Floating leaf blades long-tapered to the base, little different than the
   submerged leaves ....*P. alpinus*
6. Floating leaves rounded at the base, markedly more elliptic than the
   submerged ones 7

7. Submerged leaves > 10 cm long .... *P. epihydrus*
7. Submerged leaves mostly < 10 cm long .... *P. gramineus*

8. Stems flattened and thin-edged 9
8. Stems round in cross section 10

9. Leaves with ca. 15 veins .... *P. zosteriformis*
9. Leaves with < 10 veins ....*P. friesii*

10. Leaves attached to the stipule that sheaths the stem (pulling the leaf pulls
    the stipule away from the stem) 11
10. Leaves attached directly to the stem with a pair of globose glands at the
    point of attachment 13

11. Sheathing stipule inflated, > 2 times the width of the stem .... *P. vaginatus*
11. Stipule closely sheathing the stem, not much wider 12
12. Fruit with a tiny (0.5 mm) beak; leaves long-tapered to the tip...................
.... *P. pectinatus*
12. Fruit not beaked; leaves rounded to the tip................................ *P. filiformis*
13. Stipules veiny and fibrous, shredding into persistent fibers .... *P. friesii*
13. Stipules membranous 14
14. Mature fruit < 2.5 mm long, < 1.6 mm wide .... *P. pusillus*
14. Mature fruit > 2.5 mm long, > 1.6 mm wide .... *P. obtusifolius*

Potamogeton alpinus Balbis [*P. lucens, P. tenuifolius* Raf.]. Stems to 1 m long; submerged leaves strap-shaped, 5-12 cm long, 5-12 mm wide; floating leaves (often lacking) similar, narrowly elliptic, reddish, 4-7 cm long; stipules wide at the base, 15-25 mm long; spikes 15-30 mm long; fruits 3-4 mm long with a curved beak.

Common in shallow water of ponds and lakes; East, West. Circumboreal south to CA, UT, CO.

Potamogeton amplifolius Tuckerman. Stems to 1 m long; submerged leaves narrowly elliptic, 8-20 cm long, 2-5 cm wide, folded up along the midvein, sickle-shaped in outline; floating leaves long-petioled, the blades narrowly elliptic, 5-10 cm long; stipules to 10 cm long; spikes to 6 cm long; fruit 4-5 mm long, 3-ridged with a stout beak.

Uncommon in shallow water of fen ponds; East, West. B.C. to Newf. south to CA, MT, SD, VA. The large sickle-shaped leaves just below the surface are distinctive.

Potamogeton epihydrus Raf. Stems flattened, to 1.5 m long; submersed leaves strap-shaped, 10-20 cm long, 3-10 mm wide; floating leaves long-petiolate, the blades elliptic, 4-8 cm long; stipules 1-3 cm long; spikes 2-4 cm long; fruit 3-4 mm long with a ridge on back.

Rare in shallow water of ponds; reported from near Lake McDonald by Maguire (1934). AK to Newf. south to CA, ID, CO, eastern U. S.

Potamogeton filiformis Pers. [*P. interior* Rydb.]. Stems slender, to 50 cm long; leaves all submerged, thread-like, to 10 cm long, joined to the sheathing stipule (not the stem); spikes 1-2 cm long, composed of separate clusters; fruit 2-3 mm long.

Uncommon in shallow water of lakes, ponds, and slow streams; East, West. Circumboreal south to CA, AZ, CO, SD.

Potamogeton friesii Rupr. Stems flattened, to 1 m long; leaves all submerged, to 8 cm long, 1-3 mm wide, strap-shaped with a pair of globose glands at the base; stipules 7-15 mm long, often shredded; spikes 1-2 cm long; fruits 2-3 mm long.

Locally common in shallow water of lakes and ponds; known from Lake Sherburne and Camas Creek areas. Circumboreal south to WA, MT, SD, IN, VA.

Potamogeton gramineus L. [*P. heterophyllus* Schreb.]. Stems 30-60 cm long; submerged leaves strap-shaped, 3-9 cm long, 3-10 mm wide; floating leaves petiolate, narrowly elliptic, 2-6 cm long; stipules to 4 cm long; spikes 15-25 mm long; fruits 2-3 mm long with a ridge on back. Fig. 348.

Common in shallow water of ponds and lakes; East, West. Circumboreal south to most of U. S. Our only species occurring in habitats that dry up by the end of summer.

Potamogeton natans L. Stems to 2 m long; submerged leaves linear, tubular, 10-20 cm long, 1-2 mm wide; floating leaves petiolate, the blade elliptic, 5-10 cm long; stipules 4-10 cm long; spikes to 5 cm long; fruit 3-5 mm long with a beak to 1 mm long.

Rare in shallow to deep water of ponds and lakes; known only from near Polebridge. Circumboreal south to most of U. S.

Potamogeton obtusifolius Mert. & Koch. Stems to 1 m long; leaves all submerged, often reddish, 3-9 cm long, 1-4 mm wide with a pair of globose glands at the base; stipules 5-15 mm long; spikes 8-15 mm long; fruits 3-5 mm long.

Rare in shallow water of ponds and small lakes; known only from near Marias Pass. Circumboreal south to WA, MT.

Potamogeton pectinatus L. Stems highly branched, 30-40 cm long; leaves all submerged, thread-like, to 12 cm long and 1 mm wide, attached to the sheathing stipule (rather than the stem); spikes 1-3 cm long, of widely separated flower clusters; fruits 3-4 mm long.

Rare in shallow, often saline water of ponds; reported for near East Glacier (Standley 1921). Circumboreal south to Mex. This pondweed becomes much more common in warmer climates just outside the Park boundaries. See *P. vaginatus*.

Potamogeton praelongus Wulfen. Stems 2-3 m long, somewhat zigzag instead of straight; leaves all submerged, clasping the stem, narrowly lance-shaped, 10-25 cm long, 2-3 cm wide; stipules 4-10 cm long; spikes 25-50 mm long; fruits 4-5 mm long.

Uncommon in deep water of larger lakes; East, West. AK to Newf. south to CA, UT, CO, IN, NY. The zigzag stems are distinctive. See *P. richardsonii*.

Potamogeton pusillus L. [*P. p.* var. *mucronatus* (Fieb.) Graeb., *P. berchtoldii* Fieb., *P. panormitanus* Biv.]. Stems to 1 m long; leaves all submerged, 1-5 cm long, ca. 1 mm wide, strap-shaped with a pair of globose glands at the base; stipules 3-10 mm long, easily detached; spikes to 1 cm long; fruit 2-3 mm long.

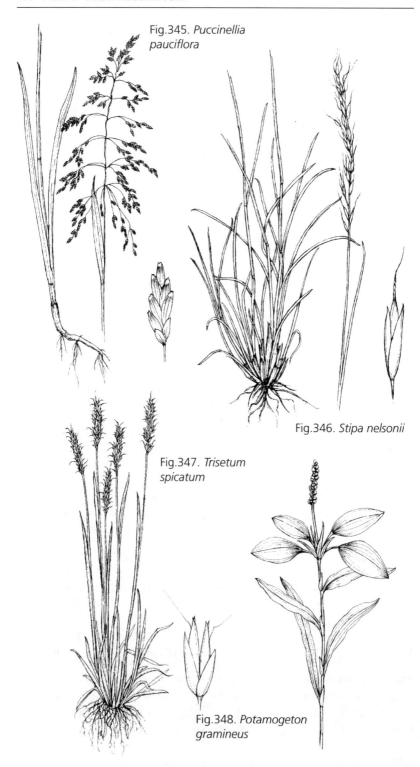

Fig.345. *Puccinellia pauciflora*

Fig.346. *Stipa nelsonii*

Fig.347. *Trisetum spicatum*

Fig.348. *Potamogeton gramineus*

Common in shallow water of ponds, lakes, and slow streams; East, West. Circumboreal south to most of U. S.

Potamogeton richardsonii (benn.) Rydb. Stems 30-60 cm long; leaves all submerged, narrowly lance-shaped with wavy margins, clasping the stem, to 10 cm long, 2 cm wide; stipules 1-2 cm long, often shredded; spikes 15-30 mm long; fruits 2-4 mm long.

Common in shallow water of ponds and slow streams; East, West. AK to Lab., south to CA, CO, IA, NY. *C. praelongus* has zigzag stems.

Potamogeton vaginatus Turcz. Stems highly branched, to 70 cm long; leaves all submerged, linear, 2-10 cm long, 1-2 mm wide, attached to the stipule (rather than the stem); stipule swollen, sheathing the stem; spikes 3-8 cm long, of 5-9 separate clusters; fruits ca. 3 mm long, ridged on back.

Uncommon in shallow to deep water of larger lakes; reported for Two Medicine and St. Mary lakes (Maguire 1934). AK to Newf. south to OR, WY, NY; Eurasia. The sheaths of *P. pectinatus* are not swollen, but clasp the stem tightly.

Potamogeton zosteriformis Fern. [*P. compressus* L.]. Stems to 60 cm long, flattened with ridged edges; leaves all submerged, linear, to 20 cm long, 2-5 mm wide; stipules 1-3 cm long, whitish; spikes 10-25 mm long; achenes 4-5 mm long with a curved beak.

Rare in water of small montane lakes; known only from near Apgar (Standley 1921). AK to Que. south to CA, MT, eastern U. S. The wing-edged, flattened stems are distinctive.

SCHEUCHZERIACEAE FAMILY

Scheuchzeria L.

Scheuchzeria palustris L. Perennial, rhizomatous herb; stem ascending, 20-40 cm tall; leaves alternate, grass-like, tubular, erect with broad, sheathing bases; flowers 3-12, bisexual, stalked in a narrow inflorescence; sepals and petals 3, alike, greenish-white, 3 mm long, spreading; stamens 6; ovaries 3; fruits 3, spreading ovoid capsules, 5-8 mm long with 1-2 seeds each. July-Aug. Fig. 349.

Rare in montane sphagnum fens in the Lake McDonald area. Our plants are var. *americana* Fern. Circumboreal south to CA, ID, MT, IA, NJ.

TYPHACEAE: CATTAIL FAMILY

Rhizomatous emergent perennials; leaves grass-like; flowers unisexual, subtended by a small bract, densely clustered in globose or cylindrical heads, male above female; sepals and petals absent or undifferentiated; male flowers of 1-7 stamens; female flowers with 1 ovary and style (ours); fruit a hard coated, beaked seed (achene).

Sparganium is often placed in its own family, the Sparganiaceae.

1. Flowers in cylindrical heads > 10 cm long .... **Typha**
1. Flowers in globose heads < 3 cm long .... **Sparganium**

Sparganium L., Burr-reed

Leaves with enlarged, sheathing bases; flowers densely clustered in globose heads well-separated in a narrow or branched inflorescence; stamens 2-5; fruit beaked, narrowed at both ends, clustered in a mace-like head.

S. eurycarpum Engelm., with flat leaves and > 5 male heads, is reported for just west of the Park.

1. Leaves V-shaped in cross section, 6-10 mm wide .... ***S. emersum***
1. Leaves flat, 2-6 mm wide 2

2. Male heads 2-5; female heads ≥ 1 cm wide .... ***S. angustifolium***
2. Male heads 1, sometime 2; female heads ca. 5 mm wide .... ***S. natans***

Sparganium angustifolium Michx. Stems 20-100 cm long, often floating; leaves as long or longer; male heads 2-5, sessile; female heads 2-4, sessile or stalked, 10-15 mm wide at maturity; achenes 3-6 mm long, the beak 1-4 mm long. July-Aug.

Uncommon in montane lakes and ponds, often associated with fens; East, West. Circumboreal south to CA, NM, PA. See *S. emersum*.

Sparganium emersum Rehmann [*S. multipedunculatum* (Morong) Rydb., *S. e.* var. *multipedunculatum* (Morong) Reveal]. Stems 20-40 cm high, usually emergent; leaves V-shaped in cross section, 20-50 cm long; male heads 3-8, sessile; female heads 3-5, sessile or stalked, 2-3 cm wide at maturity; achenes 4-5 mm long, slightly contracted in the middle, the beak 3-4 mm long. Aug. Fig. 350.

Uncommon in shallow water of montane lakes, ponds, and fens; East, West. Our plants are ssp. *emersum*. Circumboreal south to CA, NM, SD, PA. May not be truly distinct from *S. angustifolium*.

Sparganium natans L. [*S. minimum* Fries]. Stems slender, 10-80 cm long; leaves 3-5 mm wide; male heads usually 1, sessile; female heads 1-4, sessile or stalked, to 10 mm wide at maturity; achenes 2-3 mm long, the beak ca. 1 mm long. July-Sept.

Uncommon in shallow to deep water of montane lakes and ponds in the North Fork Flathead valley. Circumboreal south to CA, UT, MT, ND, TN, NJ.

Typha L., Cattail

Typha latifolia L. Stems 1-2 m tall; leaves 1-2 cm wide; leaves and stems spongy; flower heads cylindrical; female head dark brown, > 10 cm long, to 25 mm wide at maturity; male head light brown, somewhat shorter, above and contiguous with female; achene minute with numerous long, slender hairs at the top. July-Aug. Fig. 351.

Fig.349. *Scheuchzeria palustris*

Fig.350. *Sparganium emersum*

Fig.351. *Typha latifolia*

Uncommon in shallow to moderately deep water of montane marshes, ponds, lakes, and ditches; East, West. Circumboreal south to most of N. America. Roots were eaten by Native Americans, and the fluffy seeds were used to dress wounds and soften cradleboards. The long hairs allow the achenes to float and attach to waterfowl.

Index

Abies, 90-91
  bifolia, 90
  grandis, 90
  lasiocarpa, 14, 29, 90-91, **93**
Academy of Natural Sciences, 23
Aceraceae, 57, 98
Acer, 98
  douglasii, 98
  glabrum, 13, 29, 98, **102**
Achillea, 116-17
  lanulosa, 117
  millefolium, 14, 116-17, **119**
  nobilis, 116-17
Achnatherum
  asperifolium, 475
  exigua, 476
  nelsonii, 485
  richardsonii, 485
  viridula, 485
Actaea, 297, 299
  rubra, 297, 299, **302**
Adder's-tongue family, 65, 78-82
Adenocaulon, 115, 117
  bicolor, 117, **119**
Adiantum, 83-84
  aleuticum, 83-84, **86**
  pedatum, 83
Agastache, 258, 259
  urticifolia, 259
Agoseris, 114, 117-18
  aspera, 118
  aurantiaca, 117-18
  elata, 117
  glauca, 117-18, **119**
  graciliens, 117
  graminifolia, 117
  heterophylla, 117
  pumila, 118
  scorzoneraefolia, 118
  villosa, 118
Agropyron, 448, 450
  andinum, 466
  caninum, 466
  cristatum, 450
  dasystachyum, 465
  desertorum, 450
  inerme, 465
  latiglumis, 466
  repens, 465
  smithii, 465
  spicatum, 465
  subsecundum, 466
  subsecundus, 466
  trachycaulum, 466
  unilaterale, 466
Agrostis, 449, 450-52
  alba, 451
  exarata, 450-51, **453**
  hiemalis, 451
  humilis, 451

  palustris, 451
  scabra, 450-51
  stolonifera, 13, 450-52
  thurberiana, 451-52
  variabilis, 451-52
Aira
  atropurpurea, 461
  cespitosa, 461
  elongata, 461
Alaska, 28
Alberta, 28
Alder, 13, 15, 18, 31, 168, 170
Alexanders, heart-leaved, 110
Alfalfa, 240
Alisma, 382
  americanum, 382
  brevipes, 382
  gramineum, 382
  plantaga-aquatica, 382, **397**
  trivale, 382
Alismataceae, 56, 382
Allium, 426-28
  cernuum, 426, **427**
  fibrosum, 428
  fibrillum, 426
  geyeri, 426, 428
  nuttallii, 428
  schoenoprasum, 426, 428, **430**
  sibiricum, 428
  textile, 426, 428
Allocarya
  californica, 177
  orthocarpa, 177
Alnus, 168, 170
  crispa, 170
  incana, 13, 168, 170
  sinuata, 170
  tenuifolia, 168
  viridis, 18, 31, 168, 170, **173**
Alopecurus, 448, 452
  aequalis, 452, **453**
  alpinus, 452
  aristulatus, 452
  occidentalis, 452
Alpine Bistort, 288-89
Alpine Bluegrass, 479-80
Alpine Dryad, 17, 21, 315
Alpine Fireweed, 269
Alpine Hawkweed, 152
Alpine Timothy, 478
alpine zone, 20-21
Alsike Clover, 243-44
Altyn formations, 11
Alum-root, 346-47
Alyssum, 179, 180
  alyssoides, 180
  desertorum, 180
Amaranthaceae, 60, 98-99
Amaranthus, 98-99
  blitoides, 98
  graecizans, 98-99

  retroflexus, 98-99
Ambrosia, 115, 118
  artemisifolia, 118
  elatior, 118
  psilostachya, 118
Amelanchier, 313, 314
  alnifolia, 14, 31, 314, **316**
American Bistort, 286
American Brooklime, 372
Amerorchis, 438, 439
  rotundifolia, 439
Amsinckia, 171-72
  barbata, 172
  menziesii, 171-72
Anacardiaceae, 58
Anaphalis, 114, 118, 120
  margaritacea, 118, 120, **126**
Androsace, 295
  chamaejasme, 295
  lehmanniana, 295
  occidentalis, 295
  puberulenta, 295
  septentrionalis, 295, **298**
  subumbellata, 295
Anemone, 297, 299-300
  drummondii, 299
  globosa, 299
  lithophila, 299, **302**
  multifida, 299-300
  nuttalliana, 300
  occidentalis, 299-300, **302**
  parviflora, 21, 299-300
  patens, 299-300
  piperi, 299
  tetonensis, 299-300
Angelica, 100-101
  arguta, 14, 18, 100-101
  dawsonii, 100-101, **102**
  lyallii, 100
Angelica, 100-101
Annual Bluegrass, 480
Anogra nuttallii, 277
Antennaria, 114, 120-23
  alpina, 120-21
  anaphaloides, 120-21
  aprica, 122
  aromatica, 30, 120-21
  chlorantha, 121
  corymbosa, 120-21
  flavescens, 123
  howellii, 120-21
  luzuloides, 120-21
  media, 120-22
  microphylla, 122, 123
  monocephala, 120, 122
  neglecta, 121
  parvifolia, 120, 122
  pulcherrima, 120, 122
  pulvinata, 121
  racemosa, 120, 122, **126**
  rosea, 120, 122-23, **126**
  rosea sensu stricto, 123
  umbrinella, 120, 123

Apiaceae, 59, 61, 99-110
Apocynaceae, 62, 110-11
Apocynum, 110-11
  ambigens, 111
  androsaemifolium, 110-11,
   113
  cannabinum, 110-11
  pumilum, 111
Appekunny formations, 11
Apple, Rocky Mountain, 15, 29
aquatic plants, 55
Aquilegia, 297, 299, 300-301
  flavescens, 17, 29, 300, **302**
  jonesii, 30, 300
Arabis, 179, 180-82
  bourgovii, 180
  confinis, 180
  divaricarpa, 180-81
  drummondii, 180-81, **185**
  glabra, 180-81
  hirsuta, 180-81
  holboellii, 180-81
  lemmonnii, 180, 182
  lignipes, 181
  lyallii, 180, 182
  lyrata, 180, 182
  nuttallii, 180, 182, **185**
  retrofracta, 181
  sparsiflora, 180, 182
Araceae, 382-83
Araliaceae, 58, 61, 111-12
Aralia, 111-12
  nudicaulis, 16, 111-12, **113**
Arborvitae, 89
Arceuthobium, 380
  americanum, 380, **381**
  douglasii, 380
arctic air masses, 10
Arctic-Alpine Flora, 31-34
Arctic Bellflower, 196-97
Arctium, 115, 123
  minus, 123
Arctostaphylos, 221
  adenotricha, 221
  uva-ursi, 14, 31, 221, **222**
Arenaria, 201-2
  capillaris, 202, **204**
  congesta, 202
  formosa, 202
  laricifolia, 205
  lateriflora, 205
  nuttallii, 205
  obtusiloba, 205
  propinqua, 205
  rossii, 205
  rubella, 205
  sajanensis, 205
  serpyllifolia, 202
argillites, 11
Arnica, Heart-leaf, 125
Arnica, 14, 17-18, 115, 123-27
  alpina, 30, 124
  amplexicaulis, 124
  angustifolia, 124

chamissonis, 124
cordifolia, 15, 18, 124-25,
  126
diversifolia, 123, 125
foliosa, 124
fulgens, 124-25
gracilis, 125
latifolia, 18, 21, 124-25
longifolia, 124-**26**
mollis, 21, 123, 127
parryi, 123, 127
rydbergii, 124, 127
sororia, 29, 124, 127, **132**
tomentosa, 124
Arrow-grass family, 424
Arrowhead, 382
Arrowleaf, 135
Artemisia, 115, 127-29
  absinthium, 128
  biennis, 128
  campestris, 128, **132**
  discolor, 129
  diversifolia, 129
  dracunculoides, 128
  dracunculus, 128-29
  flocosa, 129
  forwoodii, 128
  frigida, 127-29
  gnaphaloides, 129
  ludoviciana, 127-29
  michauxiana, 128-29, **132**
  spithamaea, 128
  tridentata, 14, 29, 127-29
Arum family, 382-83
Asparagus, 428
  officinalis, 428
Ash, mountain, 18, 29, 330, 332
Aspen, 14-15, 30, 336
Asphodel, False, 434
Aspidotis, 84
  densa, 84
Aspleniaceae, 54
Asplenium, 65
  trichomanes-ramosum, 65
  viride, 65
Aster, 13, 116, 129-35, 150
  alpigenus, 130-31
  apricus, 133
  ascendens, 130-31
  campestris, 130-31
  chilensis, 131
  conspicuus, 15, 29, 130-31,
   132
  crassulus, 133
  eatonii, 130-31, **134**
  engelmannii, 29, 130-31,
   133, **134**
  falcatus, 130, 133
  foliaceus, 130, 133
  fremontii, 135
  frondeus, 133
  hesperius, 130
  junciformis, 130
  laevis, 130, 133

meritus, 135
modestus, 130, 133
nelsonii, 131
occidentalis, 130, 135
sayianus, 133
sibiricus, 130, 135
Aster family, 58-60, 62, 112-68
Asteraceae, 58-60, 62, 112-68
Astragalus, 232-35
  aboriginum, 234
  adsurgens, 233
  agrestis, 233
  alpinus, 13, 233-34, **237**
  americanus, 233-34
  australis, 233-34
  bourgovii, 233-34, **237**
  canadensis, 233-34
  dasyglottis, 233
  drummondii, 233, 235
  flexuosus, 31, 233, 235
  forwoodii, 234
  goniatus, 233
  macounii, 235
  rogginsii, 233, 235
  striatus, 233
  tenellus, 233, 235
  vexilliflexus, 233, 235
Athyrium, 66, 68, 70
  alpestre, 66
  distentifolium, 66
  filix-femina, 31, 66, 68, **69**,
   70
Atriplex, 212
  hastata, 212
  patula, 212
Autumn Willow, 343
avalanche chutes, 18
Avena, 449, 454
  sativa, 454
Avens, 317-18
  Mountain, 314-15

Baldhip Rose, 30, 327-28
Balsamorhiza, 115, 135
  sagittata, 29, 135, **138**
Balsam Poplar, 336
Balsamroot, 135
Baltic Rush, 418-19
Bamberg, Sam, 23
Baneberry, 297, 299
Barbarea, 179, 182-83
  orthoceras, 183
  vulgaris, 183
Barberry family, 168
Barley, 472
Barnyard Grass, 461, 463
Basil, Wild, 262
Basin Wildrye, 464
Bastard Toad-flax, 345
Batchelder, 468
Batrachium flaccidum, 306
Bead Lily, 429
Beaked Sedge, 409
Bearberry, 14, 17, 198, 221

Bearded Wheatgrass, 464
Beardtongue, 368-70
Beargrass, 17, 436
Beavers, 15
Bebb Willow, 338-39
Beckmannia, 448, 454
   erucaeformis, 454
   syzigachne, 454, **458**
Bedstraw, 335
Bee Balm, 260
Bee Plant, Rocky Mountain, 197
Beech Fern, 87
Bee Plant, 197
beetles, 20
Beggar's-ticks, 135-36
Bellflower family, 196-97
Bellis, 116, 135
   perennis, 135
Belt Series, 11
Bentgrass, 450-52
Berberidaceae, 58
Berberis, 168
   aquifolium, 168
   repens, 14, 168, **173**
Bergamot, Wild, 260
Besseya, 356, 358
   wyomingensis, 358, **363**
Betony, 372-374
Betula, 170-71
   fontinalis, 170
   gladulosa, 170
   occidentalis, 13, 170
   papyrifera, 15, 30, 170-71, **173**
Betulaceae, 58, 168-71
Bidens, 116, 135-36
   cernua, 135-36
Big Sagebrush, 129
Bilberry, Dwarf, 229
Bindweed, 216
   Black, 286
Biosphere Reserve, 9
Birch family, 13, 15-16, 30, 168-71
Biscuitroot, 103-5
Bistort, 286, 288-89
Bitter Cherry, 326
Bittercress, 184
Bitter Dock, 290
Bitterroot, 292, 294
Bittersweet family, 212
Blackberry, 328-30
Black Bindweed, 286
Black Birch, 170
Black Cottonwood, 13, 30, 336
Black Elderberry, 199
Blackfeet Indian Reservation, 9
Black Hawthorn, 314
Black Medick, 239-40
Black Snakeroot, 108, **109**, 110
Blacktail Hills, 13
Black Twinberry, 198
Bladder Campion, 207
Bladder Fern, 66-67

Bladderpod, 191
Bladderwort family, 262, 264
Blankenship, Joseph, 22
Blanketflower, 147
Blazing Star, 154-55
Blazing-star family, 265
blister rust, 20, 92, 94
Bluebells, 175-76
Blueberry, 229
Bluebunch Wheatgrass, 13, 464
Blue Elderberry, 199
Blue-eyed Grass, 417
Blue-eyed Mary, 361
Blue Flax, 265
Bluegrass, 478-84
Bluejoint, 457
Blue Lettuce, 154
Blue Scorpion-grass, 177
Blueweed, 174
Bob Marshall Wilderness complex, 9
Bog Birch, 170
Bog Orchid, 444-47
Boisduvalia glabella, 273
Borage family, 64, 171-77
Boraginaceae, 64, 171-77
Boreal Flora, 30-34
Botrychium, 79-82
   boreale, 82
   hesperium, 79-80
   lanceolatum, 79-80
   lunaria, 79-80, **81**
   minganense, 79-80
   montanum, 79-80
   multifidum, 79-80, 82
   paradoxum, 79, 82
   pinnatum, 79, 82
   silaifolium, 80
   simplex, 79, 82
   virginianum, 79, **81**, 82
Bracken, 65
bracts, 99
Bramble, 328-30
Brassica, 179, 183
   arvensis, 183
   campestris, 183
   juncea, 183
   kaber, 183
Brassicaceae, 61, 177-94
Brickellia, 115, 136
   grandiflora, 136, **138**
British Columbia, 28
Brome Grass, 454-56
Bromus, 14, 449, 454-56
   anomalus, 455
   carinatus, 29, 454-55
   ciliatus, 455, **458**
   commutatus, 454-55
   inermis, 35, 454-55
   marginatus, 455
   polyanthus, 455
   mollis, 454, 456
   pumpellianus, 454, 456
   richardsonii, 455

   secalinus, 454, 456
   tectorum, 454, 456
   vulgaris, 455-56
Brook Grass, 459
Brooklime, American, 372
Broomrape family, 277
Buck-bean family, 267
Buckbrush, 199, 312
Buckler Fern, 68
Buckthorn family, 31, 312
Buckwheat family, 56, 58, 60, 282-91
Buffaloberry, 14, 17, 31, 219-20
Bugleweed, 259-60
Bugloss,, Viper's, 174
Bull Thistle, 139-40
Bulrush, 414
Bunchberry, 216
Bupleurum, 99
   americanum, 101, **102**
Burdock, 123
Bur-marigold, 135-36
Burr-reed, 492
Bursa bursa-pastoris, 184
Butter-and-Eggs, 362
Buttercup family, 55, 57, 59-61, 63, 297-311
Butterfly-weed, 274
Butterweed, 158, 160-63
Butterwort, 264

Cabbage, Skunk, 382-83
Calamagrostis, 449, 456-59
   canadensis, 15, 457, **458**
   inexpansa, 457, 459
   neglecta, 457, 459
   purpurascens, 17, 457
   rubescens, 15, 457
   stricta, 457, 459
   vaseyi, 457
calcium, 11
Callitrichaceae, 56, 194, 196
Callitriche, 194, 196
   autumnalis, 194
   hermaphroditica, 194, 196, **200**
   palustris, 196
   verna, 194, 196
Calochortus, 426, 428-29
   apiculatus, 429, **430**
   macrocarpus, 429
Calypso, 438, 439
   bulbosa, 439, **442**
Camas, 429
   Death, 436-38
Camas Road, 15
Camassia, 426, 429
   quamash, 429, **430**
Camelina, 179, 183-84
   microcarpa, 183-84
Camissonia, 268
   breviflora, 268
Campanula, 196-97
   glomerata, 196

rotundifolia, 196, **200**
uniflora, 196-97
Campanulaceae, 62-63, 196-97
Campe orthoceras, 183
Campion, 206-7
Canada, 9, 28
  Boreal flora and, 30-31
Canada Bluegrass, 480
Canada Buffaloberry, 14, 17, 31,
  219-20
Canada Fleabane, 140
Canada Thistle, 35, 139
Canada Wildrye, 464
Canarygrass, 35, 476
Caper family, 197
Capnoides
  aureum, 246
  sempervirens, 246
Capparaceae, 61, 197
Caprifoliaceae, 57, 63, 197-201
Capsella, 179, 184
  bursa-pastoris, 184
Caraway, 101
Cardamine, 179, 184
  breweri, 184
  oligosperma, 184
  pensylvanica, 184, **185**
Carduus, 115, 136
  nutans, 136
Carex, 383-409
  aenea, 396
  albonigra, 387, 389-90
  aperta, 386, 390
  aquatilis, 386, 390
  arcta, 388, 390
  atherodes, 386, 390
  athrostachya, 388, 391
  atrata, 391, 395
  atrosquama, 387, 390-91
  aurea, 386, 391, **397**
  backii, 385, 387, 391
  bebbii, 388, 391-92
  bipartita, 398
  brevior, 388, 392
  brunnescens, 389, 392
  buxbaumii, 15, 387, 392
  canescens, 389, 392
  capillaris, 21, 386, 393
  chordorrhiza, 31, 387, 393
  concinna, 385, 393
  concinnoides, 385, 393
  crawfordii, 389, 393-94
  curta, 392
  cusickii, 387, 394
  danaensis, 400
  deweyana, 389, 394
  diandra, 387, 394
  digyna, 396, 398
  dioica, 396
  disperma, 387, 394
  douglasii, 384, 388, 395
  eleocharis, 408
  eleusinoides, 399
  elynoides, 30, 385, 395

epapillosa, 387, 395
erecta, 395
exsiccata, 409
festivella, 401
filifolia, 385, 395
flava, 386, 395-96
foenea, 389, 396
geyeri, 384, 396
goodenovii, 399
group classification of,
  384-89
gynocrates, 384-85, 396
halleri, 400
haydeniana, 388, 396, **397**
heliphila, 31, 385, 396, 398
hepburnii, 401
hoodii, 387, 398
incurviformis, 400
inops, 396
interior, 389, 398
kellogii, 399
lachenalii, 388, 398
laeviculmis, 388-89, 398
lanuginosa, 402
lasiocarpa, 15, 385, 399
lenticularis, 386, 389, 399,
  407
leptalea, 385, 399
leptopoda, 394
limosa, 31, 386, 399-400
livida, 31, 386, 400
magellanica, 402
maritima, 387, 400
media, 387, 400
mertensii, 387, 400
microptera, 388, 401
nardina, 20, 385, 401, **407**
nebrascensis, 386, 401
nigricans, 19, 21, 31, 385,
  401-2
norvegica, 400
nova, 390
nubicola, 396
obtusata, 385, 402
oederi, 409
pachystachya, 388-89, 402
paupercula, 31, 386, 402
paysonis, 17, 21, 386, 402,
  407
pellita, 385, 402
pensylvanica, 396, 398
petasata, 388-89, 403
petricosa, 385, 403
phaeocephala, 21, 388, 403
physocarpa, 406
platylepis, 388-389, 403
plectocarpa, 399
podocarpa, 21, 386, 403-4
polygama, 392
praegracilis, 384, 388, 404, 406
praticola, 389, 404
preslii, 388-89, 404
pyrenaica, 385, 404
raynoldsii, 386, 405

rossii, 385, 405
rostrata, 31, 386, 405
rupestris, 20-21, 385, 405
sartwellii, 387, 405-6
saxatilis, 386, 406
scirpoidea, 21, 339, 384-85,
  406, **407**
siccata, 387, 406
simulata, 384, 387, 406
spectabilis, 386, 406, 408
stenophylla, 384, 387, 408
stipata, 387, 408
substricta, 390
tahoensis, 388-89, 408
tenuiflora, 31, 388, 408-9
tolmiei, 402
utriculata, 15, 386, 409,
  412
vesicaria, 386, 409
viridula, 386, 409
xerantica, 408
Carum, 100
  carvi, 101
  gairdneri, 108
Caryophyllaceae, 59, 62-63, 201-
  12
Cascade Mountains, 28
Cassiope, 220, 221
  tetragona, 21, 221, **224**
Castilleja, 17, 358-60
  cusickii, 359
  hispida, 359
  lanceifolia, 360
  lauta, 360
  lutea, 359
  lutescens, 359
  miniata, 359-60
  occidentalis, 359-60
  rhexifolia, 359-60, **363**
  sulphurea, 359-60
  vreelandii, 360
Catabrosa, 450, 459
  aquatica, 459
Catchfly, 206-9
Cattail family, 491-94
Ceanothus, 312
  sanguineus, 312
  velutinus, **311**, 312
Cedar family, 13, 16, 30, 88-89
Celastraceae, 57, 212
Centaurea, 115, 116, 136-37
  jacea, 136-37
  maculosa, 35, **138**
  montana, 136-37
Cerastium, 201, 202-3
  alpinum, 203
  arvense, 13, 21, 202-3, **204**
  beeringianum, 202-3
  fontanum, 202-3
  strictum, 202
  vulgatum, 203
Chamomile, Wild, 155
Chamomilla suaveolens, 155
Chaenorrhinum, 358, 361
  minus, 361

Chamerion, 268-69
  angustifolium, 18, 268-69, 270
  latifolium, 21, 268-69
Cheatgrass, 456
Cheeses, 267
Cheilanthes, 84
  gracillima, 84
  siliquosa, 84
Cheirinia
  cheiranthoides, 190
  inconspicua, 191
Chenopodiaceae, 59-60, 212, 214-16
Chenopodium, 212, 214-15
  album, 214
  atrovirens, 215
  botrys, 214
  capitatum, 214
  fremontii, 214-15, **218**
  gigantospermum, 214
  glaucum, 214-15
  humile, 215
  leptophyllum, 214-15
  rubrum, 214-15
  salinum, 215
  simplex, 214
Cherry, 326
Chickweed, 202-3, 209-12
Chimaphila, 221, 223
  umbellata, 16, 18, 31, 221, 223, **224**
Chinese Mustard, 183
Chives, 428
Chokecherry, 13, 326-27
Chrysanthemum, 116, 137
  leucanthemum, 137, **138**
Chrysopsis villosa, 150
Cicely, Sweet, 107-8
Cicuta, 100, 101, 103
  bulbifera, 101, 103
  douglasii, 101, 103, **106**
  maculata, 103
  occidentalis, 103
Cinna, 449, 459
  latifolia, 459, **462**
Cinquefoil, 14, 17, 35, 318, 320-26
Circaea, 268, 269
  alpina, 269, **275**
Cirsium, 115, 137, 139-40
  arvense, 35, 137, 139
  hookerianum, 17, 137, 139, **145**
  lanceolatum, 139
  undulatum, 139
  vulgare, 139-40
Clarkia, 268, 269, 271
  pulchella, 269, 271, **275**
Claytonia, 291-292
  lanceolata, 17, 291, **293**
  linearis, 294
  megarhiza, 21, 291-92
  parvifolia, 294

  perfoliata, 291-92
  sibirica, 291-92
Clematis, 297, 301, 303
  columbiana, 301
  ligusticifolia, 301
  occidentalis, 301, 303, **307**
Cleome, 197
  serrulata, 197
climate, 9-11
  alpine zone and, 20-21
  montane zone and, 13-16
  subalpine zone and, 17-20
Clintonia, 425, 429
  uniflora, 15-16, 18, 30, 429, **430**
Clover, 243-44
  Owl, 365-66
  Sweet, 240
Club-moss family, 75-78
Clubrush, 414
Cockerell, 152
Coeloglossum, 439
  viride, 439, **442**
Cogswellia
  macrocarpa, 105
  sandbergii, 105
  triternata, 105
Coleosanthus grandiflorus, 136
Collinsia, 358, 361
  parviflora, 361
Collomia, 280
  linearis, 280, **283**
Colorado, 28
Coltsfoot, 156
Columbia Falls, 22
Columbine, 300-301
Comandra, 345
  livida, 345
  pallida, 345
  umbellata, 345, **348**
Common Chickweed, 210
Common Dandelion, 167
Common Flax, 265
Common Juniper, 88
Common Sunflower, 148
Common Tansy, 35, 166
Coneflower, 158
Conifers, 15-16, 88-97
Conimitella, 345, 346
  williamsii, 346
Continental Divide, 19
  climate and, 9-11
  fire and, 16
  grasslands and, 13-14
  montane zone and, 13
Convolvulaceae, 62, 64, 216
Convolvulus, 216
  arvensis, 216, **218**
Conyza, 114, 140
  canadensis, 140
Corallorhiza, 438-40
  innata, 440
  maculata, 440, **442**
  mertensiana, 440

  multiflora, 440
  striata, 440
  trifida, 440
Coral-root, 439-40
Cordilleran Flora, 28-30, 33-34
Cornaceae, 57, 61, 216-17
Corn Lily, 434, 436
Cornus, 216-17
  canadensis, 16, 216, **218**
  sericea, 216-17, **218**
  stolonifera, 13, 16, 216
Corydalis, 244, 246
  aurea, 246
  sempervirens, 246, **249**
Cotton Grass, 411-13
Cottonwood, 13, 16, 30, 336
couplets, 25-26
Cow Parsnip, 103
Cranberry, Low-bush, 201
Crane's-bill, 250
Crassulaceae, 62-63, 217
Crataegus, 313, 314
  douglasii, 13, 314, **316**
Crazyweed, 240, 242-43
Creeping Buttercup, 309
Creeping Charlie, 259
Creeping Juniper, 88-89
Creeping Wintergreen, 223
Crepis, 114, 140-41
  atrabarba, 140
  elegans, 140, **145**
  intermedia, 140-41
  nana, 21, 140-41
  runcinata, 140-41
  tectorum, 140-41
Crested Wheatgrass, 450
Cretaceous age, 11
Crocus, Wild, 300
Cronartium ribicola, 20
Crowfoot, 304-10
Cruciferae, 59
Cryptantha, 171, 172, 174
  affinis, 172
  celosioides, 172
  nubigena, 172
  sobolifera, 172, **173**
  spiculifera, 172, 174
  torreyana, 172, 174
Cryptogramma, 84-85
  acrostichoides, 84-85, **86**
  crispa, 85
  densa, 84
  stelleri, 84-85
Cudweed, 147-48
Cupressaceae, 54-55, 88-89
Curlycup Gumweed, 148
Curly Dock, 290
Currant, 92, 94, 251-52
Cut-leaf Daisy, 143
Cynoglossum, 17, 174
  officinale, 174
Cyperaceae, 56, 383-415
Cypripedium, 438, 441
  montanum, 441, **442**
  passerinum, 441

Cystopteris, 66, 67
  fragilis, 67, **69**, 71
  montana, 67
Cythera bulbosa, 439

Dactylis, 449, 459
  glomerata, 459
Daisy, 135, 137, 141-46
Daisy Fleabane, 146
Dalmation Toadflax, 35
Damm, Christian, 23
Dandelion, 117-18, 166-67
Danthonia, 449, 459-61
  americanum, 460
  californica, 460
  intermedia, 13, 460, **462**
  parryi, 460
  pinetorum, 460
  spicata, 460
  unispicata, 460-61
Death Camas, 436-38
DeBolt, Ann, 23
Deciduous Forest, 30
Delphinium, 297, 303-4
  bicolor, 303, **307**
  depauperatum, 303
  glaucum, 303-4
  nuttallianum, 303
Dendroctonus ponderosae, 20
Dennstaedtiaceae, 54, 65
DeSanto, Jerry, 23
Deschampsia, 449, 461
  atropurpurea, 21, 461, **462**
  caespitosa, 461
  elongata, 461
Descurainia, 179, 186
  incisa, 186, **192**
  intermedia, 186
  pinnata, 186
  richardsonii, 186
Desert Parsley, 103-5
Devil's Club, 112
Dewberry, 330
Dianthus, 201, 203
  armeria, 203
Dicentra, 244, 246
  uniflora, 246, **249**
dichotomous keys, 25-26
Dicots, 98-381
Digitalis, 356, 361
  purpurea, 361
Diphasiastrum, 75,, 76
  alpinum, 76
  complanatum, 76, **81**
  sitchense, 76
Diplotaxis, 179, 186
  erucoides, 186
  muralis, 186
Disporum
  hookeri, 431
  reganum, 431
  trachycarpum, 431
discoid flowers, 112
Dock, 289-91

Dodecatheon, 295, 296
  acuminatum, 296
  conjugens, 296
  pauciflorum, 296
  pulchellum, 296, **298**
Dogbane family, 110-11
Dog Mustard, 190
Dogwood family, 216-17
Doll's-eyes, 297, 299
dolomite, 11
Dondia depressa, 216
Dooryard Knotweed, 286
Dotted Gay-feather, 154-55
Douglas fir, 15-16, 18, 29, 95-96
Draba, 179, 186-90
  albertina, 187-88
  andina, 189
  aurea, 187-88, **192**
  breweri, 187-88
  cana, 188
  crassifolia, 187-88
  densifolia, 187-88
  glacialis, 188
  incerta, 187-88
  lanceolata, 188
  lonchocarpa, 187-89
  macounii, 187, 189
  mccallae, 188
  nemorosa, 187, 189
  nivalis, 187, 189
  oligosperma, 187, 189
  paysonii, 187, 189, **192**
  praealta, 187, 190
  verna, 187, 190
Dracocephalum, 258, 259
  parviflorum, 259, **261**
Dragonhead, 259
Drosera, 219
  anglica, 219
  rotundifolia, 219, **222**
Drosseraceae, 31, 62, 219
Dryad, 314-15
Dryas, 314-15
  drummondii, 315
  octopetala, 17, 315, **316**,
    339
Drymocalis
  arguta, 321
  glandulosa, 322
Dryopteridaceae, 54, 65-71
Dryopteris, 31, 66, 67-68, 70
  austriaca, 67-68
  carthusiana, 67-68, **69**
  cristata, 65, 67-68
  dilatata, 67
  expansa, 67-68
  filix-mas, 66, 67-68
  linnaeana, 70
  spinulosa, 67-68
Duckweed family, 425
Dulichium, 31, 383, 410
  arundinaceum, 31, 410
Dutch Clover, 244
Dwarf Bilberry, 229

Dwarf Huckleberry, 229
Dwarf Mistletoe, 380-**81**
Dwarf Willow, 21, 31

East Glacier, 10
Echinochloa, 448, 461, 463
  crus-galli, 461, 463
Echinopanax horridum, 112
Echium, 171
  vulgare, 174
Elaeagnaceae, 57-58, 219-20
Elaeagnus, 219
  commutata, 13, 219, **222**
Elder, Marsh, 152, 154
Elderberry, 18, 31, 198-99
Eleocharis, 55, 383, 410-11
  acicularis, 410
  palustris, 410, **412**
  quinqueflora, 411
  pauciflora, 410-11
  tenuis, 410-11
Elephant's-head, 366
elevation, 12
Elk Horns, 269, 271
Elk Sedge, 396
Elk Thistle, 139
Elodea, 415
  canadensis, 415, **422**
Elrod, Morton, 22
Elymus, 448, 463-66
  canadensis, 463-64
  cinereus, 463-64
  condensatus, 464
  elymoides, 463-64
  glaucus, 14, 18, 463-64,
    469
  innovatus, 463-65
  lanceolatus, 464-65
  piperi, 464
  repens, 463, 465
  smithii, 464-65
  spicatus, 13, 17, 29, **469**
  trachycaulus, 17
Enchanter's Nightshade, 269
endemics, 33-34
Endlicher, 96
Engelmann Spruce, 18-19, 29, 92
English Daisy, 135
English Plaintain, 278
Epilobium, 268, 271-74
  alpinum, 17, 271, 272, 273
  anagallidifolium, 271-72,
    275
  angustifolium, 269
  brachycarpum, 271-72
  ciliatum, 271-72
  clavatum, 271-72
  glaberrimum, 271-72
  glandulosum, 272, 273
  halleanum, 271-73
  hornemannii, 271, 273
  lactiflorum, 271, 273
  latifolium, 269
  leptocarpum, 271, 273

leptophyllum, 273
palustre, 271, 273
paniculatum, 272
platyphyllum, 272
pygmaeum, 271, 273-74
saximontanum, 271, 274
suffruticosum, 274
watsonii, 272, 274
Epipactis, 438, 441
gigantea, 441
Equisetaceae, 54, 71-75
Equisetum, 13, 31
arvense, 16, 72, **73**, 74
ferrisii, 71
fluviatile, 15, 72, 74
hyemale, 71-72
kansanum, 72
laevigatum, 72, 74
litorale, 72, 74
palustre, 72, 74
praealtum, 72
pratense, 72, 74
scirpoides, 16, 71, 74
sylvaticum, 72, 74
trachyodon, 71
variegatum, 71, 74-75, **81**
Ericaceae, 19, 21, 57-63, 220-30
Erigeron, 116, 141-46
acris, 142
asper, 143
caespitosus, 142-43
canadensis, 140
compositus, 141, 143, **145**
conspicuus, 146
glabellus, 142-43
humilis, 142-43
jucundus, 142
lackschewitzii, 142-43
lanatus, 21, 30, 142-44
lonchophyllus, 142, 144
macranthus, 146
ochroleucus, 142, 144
peregrinus, 17, 19, 142, 144, **145**
philadelphicus, 142, 144-**145**
ramosus, 146
simplex, 142, 146
speciosus, 142, 146
strigosus, 142, 146
subtrinervis, 142, 146
unalaschkensis, 143
uniflorus, 146
Eriogonum, 14, 282, 284
androsaceum, 21, 30, 284
depressum, 284
flavum, 17, 284, **287**
ovalifolium, 21, 282, 284
piperi, 284
subalpinum, 284
umbellatum, 282, 284
Eriophorum, 31, 383, 411-13
alpinum, 415
angustifolium, 411, 413

chamissonis, 411, **412**
gracile, 411
polystachion, 411, 413
viridicarinatum, 411, 413
Erodium, 248
cicutarium, 250
erosion, 12
Erucastrum, 179, 190
gallicum, 190
Erysimum, 179, 190-91
cheiranthoides, 190
inconspicuum, 190-91, **192**
Erythonium, 425, 429, 431
glandiflora, 14, 17-19, 21, 429, 431, **433**
Euphorbia, 230
esula, 35, 230, **237**
Euphorbiaceae, 59, 230
Euphrasia, 358, 361
arctica, 361
European Mountain Ash, 332
Evening-primrose family, 60-61, 268-77
Everlasting, 118, 120
Explorer's Gentian, 247
Eyebright, 361

Fabaceae, 60, 63, 230-44
Fairy-bell, 431-32
Fairy Slipper, 439
False Asphodel, 434
False Flax, 183-84
False Hellebore, 434, 436
False Solomon's seal, 432
Fanweed, 194
fascicles, 92
fellfields, 20
fens, 15
Ferns, 65-87
Desert Parsley, 104
House, 205
Water-milfoil, 252-53
Fescue, 13, 466-68
Festuca, 449-50, 466-68
altaica, 468
baffinensis, 466-67
brachyphylla, 21, 31, 466-67
breviflora, 467
campestris, 468
idahoensis, 13, 17, 29, 466-67
occidentalis, 466-67
ovina, 467
rubra, 466-67
rydbergii, 467
saximontana, 466-68
scabrella, 13, 17, 466, 468, **469**
subulata, 466, 468
vivipara, 447, 466, 468
Fiddleheads, 65
Fiddleneck, 171-72
Field Bindweed, 216

Field Chickweed, 202-3
Field Mint, 260
Figwort family, 55, 57, 59, 63-64, 356-74
Filago, 114, 146-47
arvensis, 146-47
Filix fragilis, 67
Fir, 14, 19, 90
Douglas, 15-16, 18, 29, 95-96
fire and, 20
Grand, 16
subalpine, 14, 18-19, 29
fire, 14-16
subalpine forests and, 19-20
Fireweed, 268-69
Flag, 415, 417
Flathead River, 9
coniferous forests and, 15-16
montane zone and, 13
Flax, False, 183-84
Flax family, 265
Fleabane, 141-146
Flora of Glacier National Park, Montana
(Standley), 22
floristic analysis
introduced plants and, 34-35
plant geography and, 27-34
by region, 27
synopsis of, 26
flower head types, 112
Foamflower, 356
Fool's Huckleberry, 18, 29, 225
forests
aspen, 14-15
coniferous, 15-16
krummholz, 19
riparian, 13
subalpine zone and, 18-20
Forget-me-not, 176-77
Fowl Mannagrass, 470
Foxglove, 361
Foxtail, 452-53
Foxtail Barley, 472
Fragaria, 313, 315, 317
bracteata, 315
glauca, 315
platypetala, 315
vesca, 315
virginiana, 315, 317, **319**
Fragile Fern, 67
Fringecup, 347
Fringed Gentian, 248
Fringed Loosestrife, 296
Fringed Sagewort, 129
Fritillaria, 426, 431
pudica, 431, **433**
Fritillary, 431
Fumariaceae, 59-60, 63, 244, 246
Fumitory family, 244, 246

Gaillardia, 116
  aristata, 147, **149**
Galium, 334-35
  aparine, 334
  bifolium, 334-35
  boreale, 334-35, **344**
  mexicanum, 334
  tinctorum, 335
  triflorum, 13, 334-35
  verum, 334-35
gametophyte, 65
Garden Wall, 17
Gaultheria, 221, 223
  humifusa, 223, **224**
  ovatifolia, 223
Gaura, 268, 274
  coccinea, 274
Gay-feather, dotted, 154-55
Gayophytum, 268, 274, 276
  diffusum, 274, **275**, 276
  humile, 274, 276
  intermedium, 274
  nuttallii, 274
  racemosum, 274, 276
Gentian family, 246-48, **249**
Gentiana, 246
  acuta, 248
  affinis, 247
  amarella, 248
  calycosa, 19, 247, **249**
  glauca, 31, 247
  propinqua, 248
  prostrata, 247-48
Gentianaceae, 64, 246-48
Gentianella, 248
  amarella, 248, **249**
  crinita, 248
  procera, 248
  propinqua, 248
Gentianopsis, 248
  macounii, 248
Geocaulon, **345**
  lividum, 345
geographic patterns
  Arctic-Alpine, 31-40
  Boreal, 30-31
  Cordilleran, 27-30
  Great Plains, 32
  historical, 33-34
  introduced plants, 34-35
  local, 32-33
geology, 11-12
Geraniaceae, 61-62, 248, 250
Geranium family, 61-62, 248, 250
Geranium, 248, 250
  bicknellii, 250
  rechardsonii, 14, 250
  viscosissimum, 14, 19, 250,
    254
Geum, 314, 317-18
  aleppicum, 317
  macrophyllum, 317, **319**
  rivale, 317-18
  strictum, 317
  triflorum, 14, 317-18, **319**

Giant Helleborine, 441
Giant Hyssop, 259
Gill-over-the-Ground, 259
Ginseng family, 111-12
Glacier Lily, 14, 429, 431
Glacier National Park
  alpine zone and, 20-21
  Arctic-Alpine Flora, 31-32
  author's methods and, 24-
    27
  Boreal Flora and, 30-31
  botanical exploration of,
    22-23
  climate of, 9-11
  Cordilleran Flora and, 28-
    30
  geographic patterns in,
    27-40
  geology of, 11-12
  introduced plants and, 34-
    35
  location of, 9
  meadows and, 17-20
  subalpine zone and, 17-20
glaciers, 11-12
Glecoma, 258
  hederacea, 259
Glyceria, 450, 468, 470
  borealis, 468
  elata, 468, 470, **471**
  grandis, 468, 470
  striata, 468, 470
Glycosoma occidentalis, 107
Glycyrrhiza, 232, 236
  lepidota, 236, **237**
Gnaphalium, 114, 147-48
  canescens, 147-48
  macounii, 147, **149**
  microcephalum, 147
  palustre, 147-48
  viscosum, 147
Goatsbeard, 167-68
Going-to-the-Sun crossing, 9, 18
Golden Aster, 150
Golden Pea, 243
Goldenrod, 163, 165
Golden-smoke, 246
Goldenweed, 158
Goodyera, 438, 441, 443
  oblongifolia, 16, 441, 443,
    446
  repens, 441, 443
Gooseberry family, 92, 94, 251-52
Goosefoot family, 212, 214-16
Goose Grass, 334
Gramineae. *See* Poaceae
Grand Fir, 16
Grape, Oregon, 14-15, 168
Grapefern, 65, 82
Grass family, 56, 447-87
  Blue-eyed, 417
  Brook, 459
  Cotton, 411-13
  Goose, 334

group classification of,
    447-50
grasslands, 21, 29
  Arctic-alpine flora and,
    31-32
  montane zone and, 13-14
  subalpine zone and, 17
Grass-of-Parnassus, 350
Gratiola, 358, 361-62
  ebracteata, 361-62
Great Falls, 22
Great Plains flora, 32
Green Alder, 18, 31, 170
Green Needlegrass, 485
Grindelia, 116, 148
  perennis, 148
  squarrosa, 148, **149**
Grinnell formations, 11
Grossularia inermis, 251
Grossulariaceae, 58, 251-52
Groundsel, 158, 160-63
Groundsmoke, 274, 276
Grouse Whortleberry, 230
Gumbo Lily, 276
Gumweed, 148
Gymnocarpium, 65, 67, 68, 70
  disjunctum, 31, 68, 70, **73**
  dryopteris, 70
Gymnosperms, 88-97

Habenaria
  dilatata, 445
  elegans, 444
  hyperborea, 445
  obtusata, 445
  orbiculata, 445
  saccata, 445, 447
  stricta, 445, 447
  unalascensis, 444
  viridis, 439
Hackelia, 171, 174-75
  floribunda, 174-75
  jessicae, 175
  micrantha, 174-75, **178**
Hair Grass, 461
Hairy Golden Aster, 150
Halenia deflexa, 246
Halerpestes cymbalaria, 306
Haloragaceae, 55, 252-53
Haplopappus
  carthamoides, 158
  lanceolatus, 158
  lyallii, 112
Harebell, 196
Harvey, LeRoy H., 23
Hawksbeard, 140-41
Hawkweed, 150-52
Hawthorn, 13, 314
Heart-leaf Arnica, 125
Heart-leaved Alexanders, 110
Heath family, 19, 21, 220-30
Heather, 221, 226
Hedge-hyssop, 361-62
Hedge Nettle, 262

Hedysarum, 232, 236
 alpinum, 236
 americanum, 236
 boreale, 236
 cinerascens, 236
 sulphurescens, 17, 20, 236,
  241
Helena formations, 11
Helianthus, 116, 148, 150
 annuus, 148
 fascicularis, 150
 laetiflorus, 150
 nuttallii, 148, 150, **153**
 petiolaris, 148, 150
 rigidus, 148, 150
 subrhomboideus, 150
Helictotrichon, 449, 470
 hookeri, 470, **471**
Hellebore, False, 434, 436
Helleborine, 441
Hemieva ranunculifolia, 355
Hemlock, 16, 30, 96, 101, 103
Hemp, Indian, 110-11
Hemp Nettle, 259
Heracleum, 100, 103
 lanatum, 103
 spondylium, 13-14, 18,
  103, **106**
Hermann, Frederick J., 23
Hesperostipa comata, 484
Heterotheca, 115, 150
 villosa, 150, **153**
Heuchera, 345, 346-47
 cylindrica, 346, **348**
 flabellifolia, 346
 glabella, 346
 parvifolia, 346-47
Hieracium, 114, 150-52
 albiflorum, 18, 151, **153**
 aurantiacum, 151
 caespitosum, 35, 151
 canadense, 152
 cynoglossoides, 151
 gracile, 152
 pratense, 151
 scabriusculum, 152
 scouleri, 151-52
 triste, 21, 151-52, **153**
 umbellatum, 151-52
Hierochloe, 449, 470-72
 odorata, 470-72, **471**
Hippuridaceae, 56, 59, 253
Hippuris, 253
 vulgaris, 253, **257**
Hitchcock, A. S., 22-23
Hitchcock, C. Leo, 23
Holly Fern, 70
Hollyhock, Wild, 267
Holodiscus, 313, 318
 discolor, 30, 318, **319**
Honeysuckle family, 18, 29, 197-
 201
Hop Clover, 244

Hordeum, 448, 472
 brachyantherum, 472
 breviaristatum, 472
 jubatum, 472, **474**
 nodosum, 472
Horehound, Water, 259-60
Horsemint, 259-60
Horsetail family, 71-75
Horseweed, 140
Hound's Tongue, 174
House Fern, 205
Huckleberry, 15-16, 18, 29, 229
 Fool's, 18, 29, 225
Hudsonian zones, 12
Hungarian Brome, 455
Huperzia, 75
 haleakalae, 77
 miyoshiana, 77
 occidentalis, 77
Hydrangea family, 253
Hydrangeaceae, 57, 253
Hydrocharitaceae, 56, 415
Hydrophyllaceae, 64, 253, 255-56
Hydrophyllum, 255
 capitatum, 255, **257**
Hymenoxys, 116, 152
 richardsonii, 152
Hypericaceae, 60, 62, 256, 258
Hypericum, 256, 258
 formosum, 21, 256, 258,
  261
 perforatum, 35, 256, 258
 scouleri, 258
Hypopitys
 latisquama, 225
 monotropa, 225
Hyssop, 259

Ibidium romanzoffianum, 447
ice age, 11-12
Idaho fescue, 13, 467
igneous formations, 11
Iliamna, 265, 267
 rivularis, 267, **270**
Indian Hemp, 110-11
Indian Pipe, 225-26
indusium, 66
introduced plants, 34-35
involucels, 99
Iridaceae, 56, 415, 417
Iris family, 56, 415, 417
Iris missouriensis, 415, 417, **422**
Isoetaceae, 54, 55
Isoetes, 75
 bolanderi, 75
 echinospora, 75
Italian Ryegrass, 472-73
Iva, 115, 152, 154
 xanthifolia, 152, 154, **157**
Ivy, Poison, 99

Jacob's Ladder, 282
Jerusalem Oak, 214
jet stream, 10

Jim Hill Mustard, 194
Jones, Marcus, 22
Juncaceae, 56, 417-24
Juncaginaceae, 56, 424
Juncoides
 campestre, 424
 glabratum, 423
 parviflorum, 424
 piperi, 424
 spicatum, 424
Juncus, 417-23
 albescens, 423
 alpinoarticulatus, 418
 alpinus, 418
 arcticus, 418
 balticus, 418-19
 biglumis, 417, 419
 bufonius, 417, 419
 castaneus, 418-19
 confusus, 418-19, 421
 drummondii, 17, 19, 21,
  418, 420
 ensifolius, 418, 420
 filiformis, 418, 420
 longistylis, 418, 420
 mertensianus, 418, 421,
  422
 nevadensis, 418, 421
 nodosus, 418, 421
 parryi, 418, 421
 regelii, 418, 421
 saximontanus, 420
 tenuis, 418, 421, 423
 tracyi, 420
 triglumis, 417, 423
Juneberry, 314
Junegrass, 472
Juniper, 19, 30-31, 88-89
Juniperus, 88-89
 communis, 19, 31, 88, **93**
 horizontalis, 88-89
 scopulorum, 88-89
 sibirica, 88

Kalmia, 220, 223
 microphylla, 21, 223
 occidentalis, 223
 polifolia, 223, **224**
Kentucky bluegrass, 14, 35, 482
King's Crown, 217
Kinnikinnick, 221
Kirkwood, Joseph E., 22
Knapweed, 35, 136-37
Knotweed, 285-89
Kobresia, 383, 413
 bellardi, 413
 myosuroides, 413
 simpliciuscula, 21, 413
Koeleria, 449-50, 472
 cristata, 472
 macrantha, 472, **474**
 pyramidata, 472
krummholz forests, 19

Lace Fern, 84
Lactuca, 114, 154
  biennis, 154, **157**
  pulchella, 154
  serriola, 154
  spicata, 154
  virosa, 154
Ladies' Tresses, 447
Lady Fern, 66
Lady's-slipper, 441
Lake McDonald, 15-16
Lake Sherburne, 10
Lamb's Quarters, 214
Lamiaceae, 63-64, 258-62
Lappula, 171, 175
  echinata, 175
  floribunda, 174
  occidentalis, 175
  redowskii, 175
  squarrosa, 175
Larch, 16, 18-19, 30, 91
Larix, 90, 91
  lyallii, 18-19, 30, 91
  occidentalis, 15-16, 30, 91, **93**
Larkspur, 303-4
Lathyrus, 238
  bijugatus, 238
  lanszwertii, 238
  latifolius, 238
  ochroleucus, 238, **241**
Laurel, 223
Lavauxia flava, 276
Leafy Spurge, 35, 230
Least Moonwort, 82
Leather-leaf, 347
Ledum, 220, 223, 225
  glandulosum, 223, 225,
  227
Lemna, 425
  minor, 425, **427**
  triscula, 425
  turionifera, 425
Lemnaceae, 55, 425
Lentibulariaceae, 55, 262, 264
Leontodon
  ceratophorum, 166
  laevigatum, 166
  lyratum, 167
  taraxacum, 167
Lepargyrea canadensis, 219
Lepidium, 179, 191
  densiflorum, 191
Leptarrhena, 346, 347
  pyrolifolia, 347, **348**
Leptotaenia multifida, 104
Lesquerella, 179, 191
  alpina, 191
  spathulata, 191
Lettuce, 154, 156, 158, 294
Leucanthemum vulgare, 137
Lewisia, 291, 292, 294
  pygmaea, 292, **293**
  rediviva, 292, 294
  triphylla, 292, 294

Lewis Range, 9
Leymus cinereus, 464
Liatris, 115, 154-55
  punctata, 31, 154-55
Lichens, 23
Licorice, 83, 236
ligulate flowers, 112
Ligusticum canbyi, 99
Lilac, Wild, 312
Liliaceae, 56, 425-38
Lily family, 14, 56, 425-38
Lily, Gumbo, 276
Limber Pine, 92, 94-95
Limestone, 11
Limosella, 356, 362
  aquatica, 362
Linaceae, 62, 265, **266**
Linanthus, 280
  septentrionalis, 280, **283**
Linaria, 358, 362
  dalmatica, 35, 362
  vulgaris, 362, **363**
Linnaea, 197-98
  borealis, 15-16, 18, 35, 197-98,
  200
Linum, 265
  lewisii, 265, **266**
  perenne, 265
  usitatissumum, 265
Lip Fern, 84
Listera, 438, 443-44
  borealis, 443
  caurina, 443
  convallarioides, 443
  cordata, 443-44, **446**
Lithophragma, 346, 347
  bulbifera, 347
  glabrum, 347
  parviflorum, 347, **348**
Lithospermum, 171, 175
  ruderale, 175, **178**
Livingston Range, 9
Loasaceae, 265
Lobelia, 197
  kalmii, 197
Locoweed, 232-35
Lodgepole Pine, 14-18, 29, 94
Logan Pass, 17, 19
Logging Creek, 16
Lolium, 448, 472-73
  multiflorum, 472-73
Lomatium, 100, 103-5
  ambiguum, 103-4
  cous, 104
  dissectum, 104
  foeniculaceum, 103-5
  macrocarpum, 104-5
  sandbergii, 104-5, **106**
  triternatum, 103, 105, **106**
Lonicera, 197, 198
  ciliosa, 198
  involucrata, 198, **200**
  utahensis, 18, 29, 198
Loosestrife, 296-97

Lousewort, 366-68
Low-bush Cranberry, 201
Lucerne, 240
Lupine, 238-39
Lupinus, 232, 238-39
  argenteus, 238, 239, **241**
  flexuosus, 239
  laxiflorus, 238
  lepidus, 238
  leucophyllus, 238
  leucopsis, 239
  polyphyllus, 239
  sericeus, 14, 238-39
Luzula, 423-24
  arcuata, 423
  campestris, 424
  comosa, 424
  glabrata, 423
  hitchcockii, 18-19, 423-24,
  427
  multiflora, 423-24
  parviflora, 423-24
  piperi, 19, 21, 423-24
  spicata, 21, 423-24
  wahlenbergii, 424
Lychnis
  alba, 207
  apetalata, 209
Lycopodiaceae, 54, 75-78
Lycopodiella inundata, 75
Lycopodium, 31, 76
  alpinum, 76
  annotinum, 78, **81**
  clavatum, 77-78
  complanatum, 76
  dendroideum, 78
  lagopus, 77-78
  obscurum, 78
  selago, 77
  sitchense, 76
Lycopus, 258-60
  americanus, 259-60
  uniflorus, 259-260, **261**
Lysichiton, 382-83
  americanus, 382-83, **397**
  kamtschatcensis, 382
Lysimachia, 295-97
  ciliata, 296
  thyrsiflora, 296-97, **298**

McCune, Bruce, 23
McLaughlin, William T., 23
McMullen, J. L., 23
Madder family, 334-35
Madia, 114, 116, 155
  glomerata, 155, **157**
Mad Wolf Mountain, 17
Maguire, Bassett, 22-23
Mahonia repens, 168
Maianthemum
  racemosum, 432
  stellatum, 432
Maidenhair Fern family, 83-85
Malacosoma, 15

Male Fern, 68
Mallow family, 265, 267
Malva, 265, 267
   neglecta, 267
Malvaceae, 60, 63, 265, 267
Mannagrass, 468, 470
Maple family, 13, 15, 98
Mare's-tail, 253
Marias Pass, 13
Marigold, 135-36
Mariposa, 428-29
Marsh Elder, 152, 154
marshes, 15
Marsh Fern family, 87
Matricaria, 115-16, 155
   discoidea, 155
   maritima, 155
   matricarioides, 155
Meadow Aster, 131
Meadow Hawkweed, 35, 151
meadows, 17
   avalanche chutes and, 18
   forests and, 18-20
Meadowsweet, 332-34
Medicago, 232, 239-40
   lupulina, 239
   sativa, 239-40
Medick, 239-40
Melampyrum, 358, 362
   lineare, 362, **367**
Melandrium apetalum, 209
Melica, 449-50, 473-75
   bella, 472
   bulbosa, 473
   smithii, 473
   spectabilis, 473, **474**
   subulata, 473
Melilotus, 232, 240
   alba, 240
   officinalis, 240, **241**
Mentha, 258, 260
   arvensis, 260, **261**
   canadensis, 260
Mentzelia, 265
   dispersa, 265, **266**
Menyanthaceae, 63, 267
Menyanthes, 267
   trifoliata, 267, **270**
Menziesia, 220, 225
   ferruginea, 18, 29, 225, **227**
   glabella, 225
Merriam zones, 12
Mertensia, 171, 175, 176
   longiflora, 176
   oblongifolia, 176, **178**
   viridis, 176
Microseris, 114, 155-56
   cuspidata, 156
   nutans, 155-56, **157**
Microsteris, 280-81
   gracilis, 280-81, **283**
Mimulus, 358, 364-65
   breviflorus, 364
   caespitosus, 364

floribundus, 364
guttatus, 364-65
hallii, 364
lewisii, 17, 364-65, **367**
moschatus, 364-65
tilingii, 364-65
Miner's Candle, 174
Miner's Lettuce, 294
Mint family, 258-62
Minuartia, 201, 203, 205
   austromontana, 205
   elegans, 205
   nuttallii, 203, 205
   obtusiloba, 20, 31, 203,
      205, **208**
   rossii, 203, 205
   rubella, 203, 205
Missouri Goldenrod, 165
Mistletoe family, 380-**81**
Mist-maiden, 256
Mitella, 345-47, 349-50
   breweri, 349, **353**
   nuda, 16, 349
   pentandra, 349
   stauropetala, 349
   trifida, 349-50
   violacea, 349
Mitrewort, 347, 349-50
Moehringia, 201, 205-6
   lateriflora, 205-6, **208**
Moldavica parviflora, 259
molecular genetics, 27
Monarda, 258, 260
   fistulosa, 260, **263**
   menthaefolia, 260
Moneses, 221, 225
   uniflora, 225, **227**
Monkey-flower, 364-65
Monocots, 382-494
Monolepis, 212, 215
   nuttalliana, 215
Monotropa, 220, 225-26
   hypopitys, 225, **227**
   uniflora, 225-26
Montana State University, 22
montane zone, 32, 35
   aspen forests of, 14-15
   Bracken family, 65
   coniferous forests of, 15-
      16
   grasslands of, 13-14
   riparian forests of, 13
   wetlands of, 15
Montia, 291, 294
   linearis, 294
   parvifolia, 294, **298**
   perfoliata, 292
   sibirica, 292
Montiastrum lineare, 294
Moonwort, 65, 82
Mooseberry, 201
Morning-glory family, 216
Moss Campion, 207
Moss Gentian, 247-48

Mountain Ash, 18, 29, 330, 332
Mountain Avens, 314-15
Mountain Big Sagebrush, 14
Mountain Box, 212
Mountain Brome, 455
Mountain Dandelion, 117-18
Mountain Heather, 226
Mountain Lover, 18, 212
Mountain Maple, 13, 15
Mountain Pine Beetles, 20
Mountain Sorrel, 284-85
Mountain Tarweed, 155
Mouse-ear Chickweed, 202-3
Mousetail, 304
Mudstones, 11
Mudwort, 362
Mugwort, 127-29
Muhlenbergia, 448-49, 475
   filiformis, 475
   glomerata, 475
   richardsonis, 475, **477**
   squarrosa, 475
Musineon, 100, 105
   divaricatum, 105
Mustard family, 61, 177-94
Myatt, Mona, 23
Myosotis, 171, 176-77
   alpestris, 176-77, **178**
   laxa, 176-77
   micrantha, 176-77
   scropioides, 176-77
   sylvatica, 171, 176
Myosurus, 297, 304
   lepturus, 304
   minimus, 304, **307**
Myriophyllum, 252-53
   exalbescens, 252-53, **254**
   spicatum, 252

Navarretia, 280, 281
   intertexta, 281
Nebraska Sedge, 401
Needle-and-Thread, 484-85
Needlegrass, 14, 484-85
Nemophila, 253, 255
   breviflora, 255
Nepeta, 259, 260
   cataria, 260
Nettle family, 375
   Hedge, 262
   Hemp, 259
New York Botanical Garden, 22
Nightshade, Enchanter's, 269
Ninebark, 318, 320
Nodding Onion, 426
Northern Bedstraw, 335
Northern Divide, 9
Northern Great Plains, 9
North Fork Flathead Inside Road,
   14
Northwestern University, 23
Nortia altissima, 194
Nothocalais, 114, 156
   cuspidata, 156

Nuphar, 267
  luteum, 267, **270**
  polysepalum, 267
  variegatum, 267
Nymphaea polysepala, 267
Nymphaeaceae, 55, 267

Oak Fern, 68, 70
Oak, Jerusalem, 214
Oatgrass, 13, 459-61
Oats, 454
Oceanspray, 30, 318
Oenothera, 268, 276-77
  biennis, 277
  breviflora, 268
  caespitosa, 276
  flava, 276
  nuttallii, 276-77
  strigosa, 277
  villosa, 276-77, **279**
Old Man in the Spring, 160, 163
Old Man's Whiskers, 318
Oleaster family, 219-20
Onagraceae, 60-61, 268-77
One-flowered Wintergreen, 225
Onion, 426-28
Oniongrass, 473-75
Ophioglossaceae, 54, 65, 78-82
Ophioglossum, 82
  pusillum, 82, **86**
  vulgatum, 82
Ophrys
  caurina, 443
  convallarioides, 443
  cordata, 444
Oplopanax, 111, 112
  horridus, 112, **113**
Opulaster malvaceus, 318
Orange Hawkweed, 151
Orchard Grass, 459
Orchid family, 56, 438-47
Orchidaceae, 56, 438-47
Orchis rotundifolia, 439
Oregon Grape, 14-15, 168
Oreobroma pygmaea, 292
Oreocarya glomerata, 172
Orobanchaceae, 62, 277
Orobanche, 277
  fasciculata, 277
  sedi, 277
  uniflora, 277, **279**
Orpine, Red, 217
Orthilia, 221, 226
  secunda, 31, 226, **227**
Orthocarpus, 358, 365-66
  luteus, 365
  tenuifolius, 365-66, **367**
Oryzopsis, 449, 475-76
  asperifolia, 475-76, **477**
  exigua, 475-76
Osmorhiza, 100, 107-8
  brevipes, 107
  chilensis, 13, 107
  depauperata, 107-8

divaricata, 107
occidentalis, 14, 29, 107,
  109
purpurea, 107-8
Owl Clover, 365-66
Oxeye Daisy, 137
Oxyria, 282, 284-85
  digyna, 31, 284-85, **287**
Oxytropis, 232, 240, 242-43
  alpicola, 242
  borealis, 240, 242
  campestris, 240, 242, **245**
  cusickii, 242
  deflexa, 240, 242
  gracilis, 242
  monticola, 242
  parryi, 240
  podocarpa, 240, 242
  sericea, 240, 242
  spicata, 242
  splendens, 240, 243
  viscida, 242
Oyster-root, 167-68

Pachylophus caespitosus, 276
Pacific storms, 9-10
Pacific Yew, 16, 96
Paintbrush, 358-60
Palouse prairie, 13
Panicularia
  borealis, 468
  elata, 470
  nervata, 470
Panicum, 449, 476
  capillare, 476
Papaveraceae, 59, 278
Papaver, 278
  pygmaeum, 30, 33, 278,
  279
  radicatum, 278
Paper Birch, 15-16, 30, 170-71
Parnassia, 345, 350
  fimbriata, 350, **353**
  kotzebuei, 350
  montanensis, 350
  palustris, 350
  parviflora, 350
Parrot's-beak, 366
Parsley family, 99-110
Parsley Fern, 84-85
Parsnip, 103, 108, 110
Pascopyrum smithii, 465
Pasqueflower, 300
Pastinaca, 100, 108
  sativa, 108
Pathfinder, 117
Paxistima myrsinites, 18, 212, **213**
Pea family, 60, 63, 230-44
Pearlwort, 206
Pedicularis, 358, 366-68
  bracteosa, 366, **367**
  contorta, 366
  groenlandica, 366
  racemosa, 366, 368

Pellaea densa, 84
Pemble, Richard, 23
Pennycress, 194
Penstemon, 358, 368-70
  albertinus, 368, **371**
  attenuatus, 368-69
  confertus, 368-69
  ellipticus, 17, 368-69, **371**
  eriantherus, 368-69
  linearifolius, 369
  lyallii, 368-69
  nitidus, 368-70
  procerus, 368, 370
  virens, 368
Pentaphylloides, 313, 318
  floribunda, 318
  fruticosa, 14, 318, **324**
Peppergrass, 191
Peramium decipiens, 443
Perideridia, 100, 108
  gairdneri, 108, **109**
Periwinkle, 110-11
Persoon, 414-15
Petasites, 115-16, 156
  frigidus, 156
  nivalis, 156
  sagittatus, 156, **159**
Phacelia, 255-56
  hastata, 17, 255-56
  leptosepala, 255
  leucophylla, 255
  linearis, 255-56
  lyallii, 255-56, **257**
  sericea, 255-56
Phalaris, 449, 476
  arundinacea, 35, 476, **477**
Phegopteris, 87
  connectilis, 87
  polypodioides, 87
Philadelphus, 253
  lewisii, 253, **257**
Phleum, 448, 476, 478
  alpinum, 17, 478, **483**
  commutatum, 478
  pratense, 14, 35, 478
Phlox family, 280-82
Phlox, 281
  alyssifolia, 281
  caespitosa, 281
  hoodii, 281, **283**
Phyllodoce, 18-19, 220, 226
  aleutica, 226
  empetriformis, 21, 226
  glanduliflora, 21, 226, **231**
Physaria, 179, 191, 193
  didymocarpa, 191, 193,
  195
  saximontana, 191, 193
Physocarpus, 313, 318, 320
  malvaceus, 318, 320, **324**
Picea, 14, 90, 92
  canadensis, 92
  engelmannii, 15, 29, 92, **93**
  glauca, 13, 16, 30, 92

Pigweed family, 98-99
Pinaceae, 54, 89-96
Pine family, 13, 14-20, 29, 30, 89-97
Pine
    Prince's, 221, 223
    Running, 76
Pineapple Weed, 155
Pinedrops, 226, 228
Pinegrass, 457
Pinesap, 225
Pinguicula, 262, 264
    vulgaris, 264, **266**
Pink family, 59, 62-63, 201-12
Pink Wintergreen, 228
pinnules, 65
Pinus, 90, 92, 94-95
    albicaulis, 18, 29, 94
    contorta, 14, 29, 94, **97**
    flexillis, 94-95
    monticola, 16, 30, 94-95
    ponderosa, 13, 29, 94-95
Piperia, 438-39, 444
    elegans, 444
    unalascensis, 444, **446**
Pipsissewa, 221, 223
Plagiobothrys, 171
    scopulorum, 177
    scouleri, 177, **185**
Plaintain family, 278, 280, 382
Plantaginaceae, 64, 278, 280
Plantago, 278, 280
    canescens, 278, **279**
    lanceolata, 278
    major, 278, 280
    septata, 278
Platanthera, 438-39, 444-47
    dilatata, 444-45, **446**
    elegans, 444
    hyperborea, 445
    obtusata, 444-45
    orbiculata, 444-45
    saccata, 445, 447
    stricta, 445, 447
Pleistocene age, 33-34
Plum, 326
Pneomonanthe
    affinis, 247
    calycosa, 247
Poa, 450, 478-84
    alpina, 20-21, 31, 479-80, **483**
    ampla, 482
    annua, 479-80
    arctica, 479-80
    bulbosa, 447, 478, 480
    compressa, 479-80
    cusickii, 479-80
    epilis, 480
    glauca, 479, 481
    gracillima, 482
    grayana, 480
    interior, 478, 481
    juncifolia, 482

laxa, 479, 481
leptocoma, 479, 481
lettermanii, 479, 481
lucida, 482
nemoralis, 481
nervosa, 484
nevadensis, 482
palustris, 478, 481-82
paucispicula, 481
pratensis, 13-14, 35, 478-79, 482
rupicola, 481
sandbergii, 482
scabrella, 482
secunda, 479, 482, **483**
stenantha, 479, 482
wheeleri, 479, 484
Poaceae, 56, 447-87
    group classification of, 447-50
Podfern, 84
Poison Ivy, 99
Polebridge, 10, 15
    grasslands and, 13-14
Polemoniaceae, 64, 280-82
Polemonium, 280-81, 282
    parvifolium, 282
    pulcherrimum, 282
    viscosum, 21, 282, **283**
Polygonaceae, 56, 58, 60, 282-91
Polygonum, 282, 285-89
    achoreum, 285
    amphibium, 285-86
    austiniae, 286
    aviculare, 285-86
    bistortoides, 17, 285-86, **287**
    coccineum, 286
    convolvulus, 285-86
    douglasii, 285-86, **287**
    engelmannii, 286
    erectum, 285
    kelloggii, 288
    lapathifolium, 285, 288
    minimum, 285, 288
    polygaloides, 285, 288
    ramosissimum, 285, 288
    viviparum, 20-21, 31, 285, 288-89, **293**
    watsonii, 288
Polypodiaceae, 54, 65, 83
Polypodium, 83
    hepserium, 83
    vulgare, 83
Polypody, 83
Polystichum, 66, 70
    andersonii, 70
    lonchitis, 31, 70, **73**
    munitum, 70
Ponderosa Pine, 13, 16, 29, 94-95
Pondweed family, 487-91
Popcorn Flower, 177
Poplar, 336
Poppy, 278

Populus, 336
    balsamifera, 13, 30, 336
    tremuloides, 14, 30, 336, **344**
Portulacaceae, 61, 63-64, 291-94
Potamogeton, 487-91
    alpinus, 487-88
    amplifolius, 487-88
    baginatus, 488, 491
    berchtoldii, 489
    compressus, 491
    epihydrus, 487-88
    filiformis, 488
    friesii, 487-89
    gramineus, 487, 489, **490**
    heterophyllus, 489
    interior, 488
    lucens, 488
    mucronatus, 489
    natans, 487, 489
    obtusifolius, 488-89
    panormitanus, 489
    pectinatus, 488-89
    praelongus, 487, 489
    pusillus, 488-89
    richardsonii, 487, 491
    tenuifolius, 488
    zosteriformis, 487, 491
Potamogetonaceae, 31, 56, 487-91
Potato family, 374
Potentilla, 314, 320-26
    anserina, 320-21
    argentea, 321
    arguta, 320-22
    blaschkeana, 322
    concinna, 320, 322
    dichroa, 322
    diversifolia, 20-21, 321-22, **324**
    filipes, 322
    fruticosa, 318
    glandulosa, 17, 21, 320, 322
    glaucophylla, 322
    gracilis, 321-23
    hippiana, 320, 323
    ledebouriana, 326
    monspeliensis, 323
    nivea, 321, 323, **324**, 325
    norvegica, 321, 323
    nuttallii, 322
    ovina, 321, 323, 325
    palustris, 320, 325
    pensylvanica, 320, 325
    pltyloba, 325
    pulcherrima, 322
    quinquefolia, 321, 325
    recta, 35, 321, 325-26
    uniflora, 321, 326
    viridescens, 322
Poverty Oatgrass, 460
Povertyweed, 152, 154, 215
Prairie Coneflower, 158

Prairie Gentian, 247
Prairie Rose, 327-28
Prairie Smoke, 318
Prairie Star, 347
Precambrian age, 11
precipitation, 10
Prenanthes, 114, 156, 158
    sagittata, 156, 158, **159**
Prickly Lettuce, 154
Prickly Rose, 327
Primrose family, 64, 294-97
Primulaceae, 64, 294-97
Prince's Pine, 221, 223
Prosartes, 425, 431-32
    hookeri, 431
    trachycarpa, 29, 431-32,
    433
Prunella, 259, 260, 262
    vulgaris, 260, 262, **263**
Prunus, 313, 326-27
    corymbulosa, 326
    emarginata, 326
    melanocarpa, 326
    pensylvanica, 326
    virginiana, 13, 326-27, **331**
Pseudoroegneria spicata, 465
Pseudotsuga, 90, 95-96
    menziesii, 15, 29, 95-96, **97**
    mucronata, 95
Pteridaceae, 54, 83-85
Pteridium, 65
    aquilinum, 65, **69**
Pteridophytes, 65-87
Pterospora, 220, 226, 228
    andromedea, 226, 228, **231**
Ptilocalais nutans, 155
Puccinellia, 450, 484
    pauciflora, 484, **490**
Pulsatilla
    ludoviciana, 300
    occidentalis, 300
Purple Mountain Saxifrage, 355
Purslane family, 291-94
Pussy-toes, 120-23
Pyrola, 221, 228-29
    asarifolia, 228, **231**
    bracteata, 228
    chlorantha, 228
    minor, 228
    picta, 228-29
    secunda, 226
    uniflora, 225
    virens, 228
Pyrrocoma, 115, 158
    carthamoides, 158, **159**
    lanceolatus, 158

Quackgrass, 464
Quaking Aspen, 336
Quamasia quamash, 429
Quillwort, 75

radiate flowers, 112
Radicula
    curvisiliqua, 193
    lyrata, 193
    palustris, 193
Rafinesque, 103-5, 107-8, 117-18
Ragged Robin, 269, 271
Ragweed, 118
Ragwort, 158, 160-63
rain, 10
Ranunculaceae, 55, 57, 59-61, 63,
    297-311
Ranunculus, 297, 304-10
    abortivus, 305
    acris, 35, 305
    aquatilis, 304, 306
    bongardi, 309
    cardiophyllus, 305-6
    cymbalaria, 305-306
    eschscholtzii, 17, 21, 31,
        305-6, **307**
    flammula, 305-6, 308
    glaberrimus, 305, 308
    gmelinii, 304, 308
    helleri, 306
    inamoenus, 305, 308
    macounii, 305, 308
    oreganus, 308
    orthorhynchus, 305, 308-9
    pedatifidus, 305, 309
    purshii, 308
    pygmaeus, 305, 309
    repens, 304, 309
    reptans, 306
    saxicola, 306
    suksdorfii, 306
    testiculatus, 304, 309
    uncinatus, 305, 309, **311**
    verecundus, 305, 310
Raspberry, 328-30
Ratibida, 116, 158
    columnifera, 158, **159**
Rattlesnake Orchid, 441, 443
Razoumofskyia americana, 380
Red Bench, 15
Red Cedar, 16
Red Clover, 244
Red Fescue, 467
Red Orpine, 217
Red Osier Dogwood, 13, 216-17
Red Raspberry, 329
Redtop, 451
Reed, Wood, 459
Reed Canarygrass, 35, 476
Reedgrass, 456-59
Rein Orchid, 444
Rhamnaceae, 57-59, 312
Rhamnus, 312
    alnifolia, 31, 312, **316**
Rhinanthus, 358, 370
    borealis, 370
    crista-galli, 370, **371**
    minor, 370
Rhodiola integrifolia, 217

Rhus, 99
    radicans, 99
Ribes, 251-52
    hendersonii, 252
    hudsonianum, 251
    inerme, 251
    irriguum, 252
    lacustre, 31, 251-52, **254**
    oxyacanthoides, 251-52
    setosum, 252
    viscosissimum, 251-52, **254**
Rice Grass, 475-76
riparian forests, 13
River Birch, 13
river terraces, 16
Rock Brake, 84-85
Rockcress, 180-82
Rocket, Yellow, 183
Rock Harlequin, 246
Rock-jasmine, 295
Rock Rose, 276
Rocky Mountain Apple, 15, 29
Rocky Mountain Bee Plant, 197
Rocky Mountain Juniper, 89
Rocky Mountain Maple, 98
Rocky Mountains, 9, 28, 33-34
Romanzoffia, 253
    sitchensis, 256
Rorippa, 179, 193
    curvisiliqua, 193
    islandica, 193
    palustris, 193, **195**
    tenerrima, 193
Rosa, 313, 327-28
    acicularis, 14, 327, **331**
    arkansana, 31, 327-28
    bourgeauiana, 327
    gymnocarpa, 16, 30, 327-
        28
    nutkana, 327-28
    say, 327
    ultramontana, 328
    woodsii, 327-28
Rosaceae, 14, 16, 30, 58-59, 61-
    62, 312-34
Rose family, 14, 16, 30, 58-59, 61-
    62, 312-34
Rose, Rock, 276
Rosentretter, Roger, 23
Roseroot, 217
Rosy Pussy-toes, 122-23
Rough Fescue, 468
Rowan Tree, 332
Rubiaceae, 5, 59, 63, 334-35
Rubus, 313, 328-30
    acaulis, 329
    arcticus, 329
    idaeus, 329, **331**
    leucodermis, 329
    parviflorus, 15, 329-30
    pedatus, 329-30
    pubescens, 16, 329-30, **331**
    strigosus, 329

Rumex, 282, 289-91
  acetosa, 289
  acetosella, 289-90
  crispus, 289-90
  maritimus, 289-90
  mexicanus, 290
  obtusifolius, 289-90
  occidentalis, 289-90
  paucifolius, 289-90
  salicifolius, 289-91, **293**
  triangulivalvis, 290
Running Pine, 76
Rush family, 56, 417-24
Rush
  Scouring, 71-75
  Spike, 410
Russian Thistle, 215-16
Rydberg, P. A., 22
Ryegrass, 472-73

Sagebrush, 14, 29, 127-29
Sagewort, 127-29
Sagina, 201, 206
  nivalis, 206
  saginoides, 206, **208**
Sagittaria, 55, 382
  cuneata, 382
St. Johnswort, 35, 256, 258
St. Mary, 10-11, 15
Salicaceae, 58, 335-43
Salix, 336-43
  anglorum, 338
  arctica, 19-21, 31, **344**
  barrattiana, 338-39
  bebbiana, 15, 338-39
  boothii, 337, 339
  brachycarpa, 337-40
  candida, 337, 340
  caudata, 341
  commutata, 337-38, 340
  drummondiana, 13, 15,
    337, 340, **344**
  eriocephala, 13, 338, 340
  exigua, 13, 337, 341, 342
  farriae, 338, 341
  geyeriana, 15, 338, 341
  glauca, 338, 341
  interior, 341
  lasiandra, 337, 341-42
  lucida, 341
  mackenzieana, 340
  melanopsis, 337, 342
  monticola, 342
  myrtillifolia, 339
  nivalis, 342
  petrophylla, 338
  phylicifolia, 342
  planifolia, 338, 342
  pseudomonticola, 338, 342
  pseudomyrsinites, 339
  reticulata, 20-21, 31, 337,
    342
  rigida, 340
  scouleriana, 18, 338, 343

  serissima, 337, 343
  sitchensis, 337, 343
  subcoerulea, 340
  vestita, 337, 343, **344**
Salsify, 167-68
Salsola, 212-16
  australis, 215
  iberica, 215
  kali, 215
  pestifer, 215-16
Sambucus, 197-99
  cerulea, 199
  melanocarpa, 199
  racemosa, 18, 31, 199, **200**
Sammons, Irene, 23
Sandalwood family, 345
Sandbar Willow, 341
Sandberg Bluegrass, 482
Sand Spurry, 209
Sandwort, 201-3, 205-6
Sanguisorba, 314, 330
  annua, 330
  occidentalis, 330
Sanicula, 99, 108, 110
  graveolens, 108
  marilandica, 108, **109**, 110
Santalaceae, 60, 345
Sarsaparilla, 111
Saskatoon, 314
Satureja, 258, 262
  vulgaris, 262
Saussurea, 112
  americana, 112
  densa, 112
Savory, 262
Saxifraga, 346, 351-55
  adsendens, 352
  arguta, 354
  bronchialis, 21, 351-52,
    353
  caespitosa, 352
  cernua, 351-52
  debilis, 355
  ferruginea, 351-52
  hyperborea, 355
  lyallii, 351, 354, **357**
  mertensiana, 351, 354
  nivalis, 351, 354
  occidentalis, 351, 354
  odontoloma, 351, 354
  oppositifolia, 351, 355
  oregana, 355
  rivularis, 351, 355
  subapetala, 30, 351, 355
Saxifragaceae, 60-62, 345-56
Saxifrage family, 60-62, 345-56
Scenic Point, 19
Scheuchzeriaceae, 491
Scheuchzeria, 56, 491
  palustris, 31, 491, **493**
Schoenoplectus, 383, 413-14
  acutus, 414, **416**
  subterminalis, 414
Schuyler, Alfred, 23

Scirpus, 383, 414
  acutus, 414
  cespitosus, 415
  hudsonianus, 415
  microcarpus, 414, **416**
  occidentalis, 414
  subterminalis, 414
Scorpion-grass, 176-77
Scouler Willow, 18
Scouring Rush, 71-75
Scrophulariaceae, 55, 57, 59, 63-
  64, 356-74
Scutellaria, 258, 262
  epilobifolia, 262
  galericulata, 262, **263**
Sea Blite, 216
Sedge family, 15, 21, 23, 56, 383-
  415
sedimentary formations, 11
Sedum, 217
  douglasii, 217
  integrifolium, 217
  lanceolatum, 217
  roseum, 217
  stenopetalum, 217, **218**
Sego Lily, 428-29
Selaginellaceae, 54, 85-87
Selaginella, 85-87
  densa, 85, 87
  scopulorum, 85-87, **86**
  standleyi, 85, 87
  wallacei, 85, 87
Self-heal, 260, 262
Senecio, 115, 158, 160-63
  burkei, 161
  canus, 160-61, **164**
  conterminus, 161
  cymbalaria, 160-61
  cymbalarioides, 21, 160-61
  foetidus, 161
  fremontii, 160-61, **164**
  hydrophiloides, 160-61
  indecorus, 160-62
  integerrimus, 160, 162
  megacephalus, 160, 162
  ovinus, 161
  pauperculus, 160, 162
  pseudaureus, 160, 162
  resedifolius, 161
  streptanthifolius, 160, 162
  subnudus, 161
  triangularis, 19, 29, 160,
    162-63, **164**
  vulgaris, 160, 163
Serapias gigantea, 441
Sericotheca discolor, 318
serotiny, 92
Serviceberry, 14, 16, 18, 31, 314
Shea, Dave, 23
Shear, 456
Shepherdia, 219-20
  canadensis, 14, 31, 219-20, **222**
Shepherd's Purse, 184
Shinleaf, 228-29

Shooting Star, 295-96
Shrubby Cinquefoil, 14, 17, 318
Sibbaldia, 314, 330, 333
    procumbens, 21, 330, 333
Sickletop, 366, 368
Sidebells Wintergreen, 226
Sieversia ciliata, 318
Silene, 201, 206-9
    acaulis, 20-21, 31, 206-7, **208**
    cserei, 206-7
    cucubalus, 206-7
    latifolia, 206-7
    menziesii, 206-7
    parryi, 206-7, 209, **213**
    pratensis, 207
    uralensis, 206, 209
Silky Lupine, 239
Silverberry, 13, 219
Silverweed, 321
Sisymbrium, 179, 194
    altissimum, 194
Sisyrinchium, 415, 417
    angustifolium, 417
    montanum, 417, **422**
    mucronatum, 417
    sarmentosum, 417
Sitanion hystrix, 464
Sitka Alder, 170
Sium, 100
    cicutaefolium, 110
    suave, 110, **113**
Siyeh formations, 11
Skullcap, 262
Skunk Cabbage, 382-83
Sky Pilot, 282
Slough Grass, 454
Smartweed, 285-89
Smelowskia, 179, 194
    americana, 194
    calycina, 20, 194, **195**
    lobata, 194
Smilacina, 425, 432
    racemosa, 432
    stellata, 13, 432, **433**
Smithsonian Institution, 22-23
Smooth Brome, 35, 455
Snakeroot, 108-10
Snapdragon family, 55, 57, 59, 63-64, 356-74
Snowberry, 14-15, 31, 199
snow fields, 21
Snow Willow, 342
Soapberry, 219-20
soil properties, 11, 19, 34
    montane zone and, 13-16
    turf and, 20-21
Solanaceae, 57, 64, 374
Solanum, 374
    dulcamara, 374, **377**
    sarrachoides, 374
Solidago, 115, 163-65
    canadensis, 163, **164**
    ciliosa, 165

concinna, 165
decumbens, 165
dilatata, 165
elongata, 163
gigantea, 163, 165
missouriensis, 163, 165
multiradiata, 21, 163, 165, **169**
purshii, 165
scopulorum, 165
serotina, 163
simplex, 163, 165
spathulata, 1265
Solomon's Plume, 432
Solomon's-seal, 432
Sonchus, 114, 165-66
    arvensis, 165-66
    asper, 165-66
Sophia
    hartwegiana, 186
    intermedia, 186
Sorbus, 313, 330, 332-3
    aucuparia, 332
    sambucifolia, 332
    scopulina, 18, 29, 332-33
    sitchensis, 332
sori, 65-66
Sorrel, 284-85, 289-91
Sow Thistle, 165-66
Sparganium, 31, 55, 492
    angustifolium, 492
    emersum, 492, **493**
    minimum, 492
    multipedunculatum, 492
    natans, 492
Speedwell, 372-74
Spergularia, 201, 209
    rubra, 209
Sperry Glacier, 22
Sphaeralcea rivularis, 267
Spike-moss family, 85-87
Spike Rush, 410
Spiraea, 16, 29, 313, 332-34
    avalanche chutes and, 18
    betulifolia, 15, 29, 332
    densiflora, 332-34
    douglasii, 332, 334
    lucida, 332
    forests and, 18
Spiranthes, 438, 447
    romanzoffiana, 447, **453**
Spleenwort, 65
sporophores, 78
Spot Mountain, 17
Spotted Coral-root, 440
Spotted Knapweed, 35, 137
Spring Beauty, 291
Spruce, 13, 14-16, 18-19, 29, 30, 92
Spurge, 230
spur shoots, 92
Squirreltail, 464

Stachys, 259, 262
    palustris, 262, **263**
    scopulorum, 262
Standley, Paul, 22-23, 32
Steer's-head, 246
Steironema ciliatum, 296
Stellaria, 201, 209-12
    alpestris, 210
    americana, 21, 30, 209-10, **213**
    borealis, 209-10
    calycantha, 209-10
    crassifolia, 209-10
    crispa, 209, 211
    laeta, 211
    longifolia, 209, 211
    longipes, 209, 211, **213**
    media, 209, 211
    monantha, 211
    obtusa, 209, 211
    umbellata, 209, 212
Stenanthium, 426, 432, 434
    occidentale, 432, **435**
Stickseed, 174-75
Sticky Currant, 252
Sticky Geranium, 250
Stinking Currant, 251
Stipa, 14, 449, 484-85
    columbiana, 485
    comata, 484-85
    minor, 485
    nelsonii, 29, 484-85, **490**
    occidentalis, 485
    richardsonii, 484-85
    viridula, 484-85
stolons, 376
Stonecrop family, 217
Stoneseed, 175
Strawberry, 315, 317
Strawberry Blite, 214
Streptopus, 425, 434
    amplexifolius, 434, **435**
Striped Coral-root, 440
Suaeda, 212, 216
    calceoliformis, 216
    depressa, 216
Subalpine Fir, 14, 18-19, 29
Subalpine Larch, 18-19, 30, 91
subalpine zone
    avalanche chutes and, 18
    forests and, 18-20
    meadows and, 17-18
Suksdorfia, 345, 355
    ranunculifolia, 355, **357**
    violacea, 355
Sulphur Cinquefoil, 35, 325-26
Sumac family, 99
Sun Cups, 268
Sundew family, 219
Sunflower family, 58-60, 112-68
sunshine, 10-11
Swamp Currant, 31, 252
swamps, 15
Sweet Cicely, 107-8

Sweet Clover, 240
Sweetgrass, 470-**71**, 472
Sweet Pea, 238
Sweet-scented Bedstraw, 335
Sweetvetch, 236
Sword Fern, 70
Symphoricarpos, 14, 197, 199
    albus, 15, 31, 199, **204**
    occidentalis, 199
Synthyris wyomingensis, 358

Tall Buttercup, 35, 305
Tall Lettuce, 154
Tall Pussy-toes, 121
Talus, 21
Tanacetum, 115, 166
    vulgare, 35, 166
Tansy, 166
Tansymustard, 186
Taraxacum, 114, 166-67
    ceratophorum, 166, **169**
    laevigatum, 166
    lyratum, 166-67
    officinale, 166-67
Taraxia breviflora, 268
Tarragon, 128-29
Tarweed, 155, 171-72
Taxaceae, 55, 96
taxonomic philosophy, 24-25
Taxus, 96
    brevifolia, 16, 96, **97**
Tellima, 346, 356
    grandiflora, 356, **357**
temperature, 9-12
    alpine zone and, 20-21
    montane zone and, 13-16
    subalpine zone and, 17-20
tent caterpillars, 15
Thalictrum, 297, 310
    occidentale, 15, 18-19, 29,
        310, **311**
    venulosum, 310
Thelypteridaceae, 54, 87
Thelypteris phegopteris, 87
Thermopsis, 232, 243
    rhombifolia, 243, **245**
Thick-spike Wheatgrass, 464
Thimbleberry, 15-16, 329-30
    avalanche chutes and, 18
    forests and, 18
Thin-leaved Alder, 168, 170
Thistle, 35, 136-37, 139-40, 165-
    66, 215-16
Thlaspi, 179, 194
    arvense, 194, **195**
Thoroughwax, 101
Thoroughwort, 136
Thuja, 89
    plicata, 13, 30, 89, **93**
Tiarella, 346, 356
    trifoliata, 16, 18, 30, 356, **363**
    unifoliata, 356
Tickle Grass, 451

Timber Oatgrass, 460
Timothy, 14, 35, 476, 478
Toadflax, 362
    Bastard, 345
    Dalmation, 35
Tofieldia, 426, 434
    glutinosa, 434
    intermedia, 434
    occidentalis, 434, **435**
    pusilla, 434
Tolmachevia integrifolia, 217
Tonestus lyalli, 112
Torresia odorata, 470, 472
Tower Mustard, 181
Townsendia, 116, 167
    condensata, 167
    parryi, 167, **169**
Toxicodendron, 99
    rydbergii, 99, **102**
Tragopogon, 114, 167-68
    dubius, 167-68, **169**
Trail Plant, 117
Trapper's Tea, 223, 225
Treacle Mustard, 190
Trembling Aspen, 336
Triantha
    montana, 434
    occidentalis, 434
Trichophorum, 383, 414-15
    alpinum, 415
    caespitosum, 415, **416**
Trifolium, 232, 243-44
    agrarium, 243, **245**
    aureum, 243
    hybridum, 243-44
    pratense, 243-44
    procumbens, 243-44
    repens, 243-44
Triglochin, 424
    palustris, 424, **427**
Trillium, 425, 434
    ovatum, 434, **437**
Trimorpha acris, 142
Triple Divide Peak, 9
Trisetum, 449-50, 485-86
    canescens, 486
    cernuum, 18, 486
    orthochaetum, 34, 485-86
    spicatum, 17, 485-86, **490**
    wolfii, 485-86
Triticum, 448, 486-87
    aestivum, 486-87
Trollius, 297, 310
    albiflorus, 310, **311**
    laxus, 310
trophophores, 78-79
Tsuga, 90, 96
    heterophylla, 16, 30, 96, **97**
Tufted Hairgrass, 461
Tumble Mustard, 194
Tumbleweed, 215-16
turf, 20-21, 29
    Arctic-alpine flora and,
        31-32

Twayblade, 443
Twinberry, 198
Twinflower, 197-98
Twinpod, 191, 193
Twisted-Stalk, 434
Two Medicine valley, 15, 17, 22
Typha, 492, 494
    latifolia, 492, **493**, 494
Typhaceae, 56, 491, 491-94

Umbach, Levi M., 22
umbels, 99
United Nations Education,
    Scientific and Cultural
    Organization (UNESCO), 9
University of Montana, 22-23
Urtica, 375
    dioica, 18, 375, **377**
    lyallii, 375
Urticaceae, 59, 375
U.S. Congress, 9
U.S. Forest Service, 9, 23
Utah Honeysuckle, 18, 29
Utricularia, 31, 262, 264
    intermedia, 264
    macrorhiza, 264
    minor, 264
    vulgaris, 264, **266**

Vaccinium, 220, 229-30
    caespitosum, 229
    globulare, 229
    membranaceum, 15, 29,
        229, **231**
    myrtilloides, 18, 229-30
    myrtillus, 29, 229-30, **231**
    scoparium, 18, 29, 229-30
Vagnera
    amplexicaulis, 432
    racemosa, 432
    sessiliflora, 432
    stellata, 432
Vahlodea atropurpurea, 461
Valerian family, 59, 63, 375-76
Valeriana, 375-76
    dioica, 375
    edulis, 30, 375-76
    occidentalis, 375-76
    septentrionalis, 375
    sitchensis, 17, 19, 375-76,
        377
Valerianaceae, 59, 63, 375-76
Variegated Wintergreen, 228-29
vegetation
    alpine zone and, 20-21
    Arctic-alpine flora and, 31
    aspen forests, 14-15
    Boreal flora and, 30-31
    coniferous forests, 15-16
    Cordilleran flora and, 28-
        30
    elevation and, 12
    fellfields, 20
    geological influence and,
        11

grasslands, 13-14
Great Plains flora and, 32
meadows, 17-20
montane zone and, 13-16
riparian forests, 13
snow fields, 21
subalpine zone and, 17-20
talus, 21
turf, 20-21
wetlands, 15
Veratrum, 425, 434, 436
  eschscholtzii, 434, 436
  viride, 18, 30, 434, 436,
    437
Verbascum, 356, 370
  blattaria, 370
  thapsus, 370
Verbena, 376
  bracteata, 376
Verbenaceae, 63-64, 376
Veronica, 358, 372-74
  alpina, 374
  americana, 372
  arvensis, 372
  catenata, 372-73
  chamaedrys, 372-73
  officinalis, 35, 372-73
  peregrina, 372-73
  scutellata, 372-73
  serpyllifolia, 372-74
  verna, 372, 374
  wormskjoldii, **371**, 372,
    374
Vervain family, 376
Viburnum, 197, 201
  edule, 201, **204**
  pauciflorum, 201
Vicia, 232, 244
  americana, 244, **245**
Vinca, 110-11
  major, 111
Viola, 378-80
  adunca, 378
  canadensis, 378-79
  glabella, 378-79, **381**
  linguaefolia, 379
  macloskeyi, 378-79
  montanensis, 378
  nephrophylla, 378-79
  nuttallii, 378-79
  orbiculata, 16, 30, 378-79,
    381
  palustris, 378-80
  purpurea, 378, 380
  renifolia, 378, 380
  sororia, 379
Violaceae, 60, 376, 378-80
Violet, 376, 378-80
Viper's Bugloss, 174
Virginia Grapefern, 82
Virgin's-bower, 301, 303
Vreeland, Frederick K., 22

Wake-Robin, 434
Wallflower, 190-91
Wandering Daisy, 144
Water Birch, 170
Water Crowfoot, 306
Water Hemlock, 101, 103
Water Horehound, 259-60
Waterleaf, 253, 255-56
Water-lily family, 267
Water-milfoil family, 252-53
Water-parsnip, 110
Water-plantain family, 382
Water Smartweed, 286
Water-starwort family, 194, 196
Waterton Lakes National Park, 9
Waterweed family, 415
Wavy-leaved Thistle, 139
Western Aster, 135
Western Hemlock, 16, 30, 96
Western Larch, 15-16, 18, 30, 91
Western Red Cedar, 13, 16, 30, 89
Western Wheatgrass, 464
Western White Pine, 30, 95
West Glacier, 10, 16
wetlands, 15
Wheat, 486-87
Wheatgrass, 450, 463-66
White Angelica, 100-101
Whitebark Pine, 18-20, 29, 92, 94
White Campion, 207
White Clover, 244
White-flowered Hawkweed, 151
White Geranium, 250
White Lady's-slipper, 441
White Lettuce, 156, 158
White Mountain Heather, 221
White Pine, 16, 18, 92, 94
White Pine blister rust, 20
White Spruce, 13, 30, 92
White Sweet Clover, 240
White Thistle, 139
Whitlow-wort, 186-190
Whortleberry, 18, 29, 229, 230
Wild Basil, 262
Wild Bergamot, 260
Wild Chamomile, 155
Wild Crocus, 300
Wild Gooseberry, 252
Wild Hollyhock, 267
Wild Licorice, 236
Wild Lilac, 312
Wild Parsnip, 108
Wild Sarsaparilla, 111
Wild Strawberry, 315, 317
Wildrye, 463-66
Williams, R. S., 22
Willow family, 13, 15, 18, 20-21,
  31, 58, 335-43, **344**
Willow Herb, 271-74
Willow
  Wolf, 219
wind, 10-12
Windflower, 299-300

Wintercress, 182-83
Wintergreen, 223, 225-26, 228-29
Wiregrass, 418-19
Witch Grass, 476
Wolf Willow, 219
Wood Fern family, 65-71
Woodland Strawberry, 315
Woodnymph, 225
Wood Reed, 459
Wood Rush, 423-24
Woodsia, 66, 71
  oregana, 71
  scopulina, 71, **73**
Wood's Rose, 327-28
woody plants, 57-59
Wooly Weed, 151-52
World Heritage Sites, 9
Wormwood, 127-29

Xerophyllum, 426, 436
  tenax, 17-19, 30, 436, **437**

Yarrow, 116-17
Yellow Angelica, 100-101
Yellow Bedstraw, 335
Yellow Clover, 243
Yellow Coral-root, 440
Yellowcress, 193
Yellow Mountain Avens, 315
Yellow Rattle, 370
Yellow Rocket, 183
Yellow Sweet Clover, 240
Yellow Water Lily, 267
Yew family, 16, 96

Zigadenus, 426, 436-38
  elegans, 436, **437**
  venenosus, 436, 438
Zizia, 99, 110
  cordata, 110
  aptera, 110